Chemicals as Intentional and Accidental Global Environmental Threats

NATO Security through Science Series

This Series presents the results of scientific meetings supported under the NATO Programme for Security through Science (STS).

Meetings supported by the NATO STS Programme are in security-related priority areas of Defence Against Terrorism or Countering Other Threats to Security. The types of meeting supported are generally "Advanced Study Institutes" and "Advanced Research Workshops". The NATO STS Series collects together the results of these meetings. The meetings are co-organized by scientists from NATO countries and scientists from NATO's "Partner" or "Mediterranean Dialogue" countries. The observations and recommendations made at the meetings, as well as the contents of the volumes in the Series, reflect those of participants and contributors only; they should not necessarily be regarded as reflecting NATO views or policy.

Advanced Study Institutes (ASI) are high-level tutorial courses to convey the latest developments in a subject to an advanced-level audience

Advanced Research Workshops (ARW) are expert meetings where an intense but informal exchange of views at the frontiers of a subject aims at identifying directions for future action

Following a transformation of the programme in 2004 the Series has been re-named and re-organised. Recent volumes on topics not related to security, which result from meetings supported under the programme earlier, may be found in the NATO Science Series.

The Series is published by IOS Press, Amsterdam, and Springer, Dordrecht, in conjunction with the NATO Public Diplomacy Division.

Sub-Series

A. Chemistry and Biology	Springer
B. Physics and Biophysics	Springer
C. Environmental Security	Springer
D. Information and Communication Security	IOS Press
E. Human and Societal Dynamics	IOS Press

http://www.nato.int/science
http://www.springer.com
http://www.iospress.nl

Series C: Environmental Security

Chemicals as Intentional and Accidental Global Environmental Threats

edited by

Lubomir Simeonov
Bulgarian Academy of Sciences, Sofia,
Bulgaria

and

Elisabeta Chirila
Ovidius State University, Constanta,
Romania

 Springer

Published in cooperation with NATO Public Diplomacy Division

Proceedings of the NATO Advanced Study Institute on
Chemicals as Intentional and Accidental Global Environmental Threats
Borovetz, Bulgaria
16–27 November 2005

A C.I.P. Catalogue record for this book is available from the Library of Congress.

ISBN-10 1-4020-5097-6 (PB)
ISBN-13 978-1-4020-5097-8 (PB)
ISBN-10 1-4020-5096-8 (HB)
ISBN-13 978-1-4020-5096-1 (HB)
ISBN-10 1-4020-5098-4 (e-book)
ISBN-13 978-1-4020-5098-5 (e-book)

Published by Springer,
P.O. Box 17, 3300 AA Dordrecht, The Netherlands.

www.springer.com

Printed on acid-free paper

All Rights Reserved
© 2006 Springer
No part of this work may be reproduced, stored in a retrieval system, or transmitted in any form or by any means, electronic, mechanical, photocopying, microfilming, recording or otherwise, without written permission from the Publisher, with the exception of any material supplied specifically for the purpose of being entered and executed on a computer system, for exclusive use by the purchaser of the work.

Contents

Preface	xi
Acknowledgements	xiii
List of Contributors	xv
Environmental Risk Assessment *Liviu-Daniel Galatchi*	1
Sampling and Sample Pretreatment for Environmental Analysis *Elisabeta S. Chirila, Camelia Draghici, Simona Dobrinas*	7
Lessons Learned from Industrial Chemical Accidents: Italian and International Initiatives *Sara Visentin*	29
Selection of Effective Technologies for Management of Contaminated Lands *Thomas McHugh, John Connor*	45
Europe as a Source of Pollution – The Main Factor for the Eutrophication of the Danube Delta and Black Sea *Liviu-Daniel Galatchi, Marian Tudor*	57
Effective Approaches for Modeling Ground Water Load of Surface Applied Chemicals *Marnik Vanclooster*	65
Fate of Chemicals in the Aquatic Environment *Bogdana Koumanova*	93
Vulnerability, Resilience and Adaptation: Response Mechanisms in an Environmental Emergency – The Asian Tsunami in Thailand and Hurricane Katrina in the United States *Heike Schroeder, Dayna Yocum*	105

One Approach for Cognitive Assessment on Group of School
Children Exposed to Lead in the Community of Veles
Mihail Kochubovski — 127

Ecological and Professional Chemical Dangers
for Reproductive Health
Olga Sivochalova, Eduard Denisov — 137

Technological Transfer. Miniature Laser Mass Spectrometer
for Express Analysis of Environmental Samples
Lubomir Simeonov, Georgy Managadze — 149

Incidental and Accidental Pollution in Italy
Michele Arienzo — 163

Pesticides as Global Environmental Pollutants
Kosta Vassilev, Veska Kambourova — 173

Regulatory use of Pesticide Concentrations in the Environment
in the EU
Jan Linders B.H.J. — 193

Threats on the Seaside Lakes Water and Their Acute
and Long-Term Consequences
Liviu-Daniel Galatchi, Simona Dobrinas, Elisabeta Chirila — 201

Evaluating Health Risks and Prioritising Response Actions
for Contaminated Lands
Thomas McHugh, John Connor — 219

SAFEFOODNET: An Ongoing Project on Chemical Food Safety:
Network for Enlarging Europe
Marianna Elia, Angelo Moretto, Sara Visentin — 237

Degradation of *N*-Nitrosodimethylamine (NDMA)
in Landscape Soils
*Michele Arienzo, Jay Gan, Frederick Ernst, Stephen Qin,
Svetlana Bondarenko, David Sedlak* — 247

Pollutants Effects on Human body – Toxicological Approach
*Gheorghe Coman, Camelia Draghici, Elisabeta Chirila,
Mihaela Sica* — 255

The Use of Models for the Evaluation of Chemical Attenuation in the Environment
Charles Newell, Thomas McHugh   267

Mechanism of the remediation (detoxification) of chemicals (pesticide, heavy metals, other toxic chemical compounds)
Youseff Zeroual, Mohamed Blaghen   271

Environmental Protection and Public Health Projects. Educational and Training Programs
Camelia Draghici   281

Planning and Execution of a Pilot Phytoremediation Experiment
Biana Simeonova, Lubomir Simeonov   297

Determination of protein complexed paralytic shellfish poisoning. Mechanism of the remediation (detoxification) of chemicals (pesticide, heavy metals, other toxic chemical compounds)
Nadia Takati, Driss Mountassif, Hamid Taleb, Mohamed Blaghen   303

European Institutions for Controlling Chemical Air Pollution: An Analysis of CLRTAP-European Union Interplay
Heike Schroeder, Dayna Yocum   321

Development of a Pesticides Biosensor Using Carbon-Based Electrode Systems
Adina Arvinte, Lucian Rotariu, Camelia Bala   337

Study of the Content of Chemicals in Landfill Leachate
Andreana Maximova, Bogdana Koumanova   345

Aspects of Eutrophication as a Chemical Pollution with Implications on Marine Biota at the Romanian Black Sea Shore
Alice Sburlea, Laura Boicenco, Adriana Cociasu   357

Investigation of the Constanta Surface Water's Pollution Sources
Alina Coman, Elisabeta Chirila, Ionela Popovici Carazeanu   361

The Potential Use of *Festuca* Cultivars and Lignite for phytostabilization of Heavy Metal Polluted Soils
Jacek Krzyżak, Tyler Lane, Anna Czerwińska   367

Development of Analytical Method for Determination of PCBs in Soil by Gas Chromatography with Electron-Capture detector
Anna Dimitrova, Teodor Stoichev, Tomislav Rizov, Anastasiya Kolarska, Nikolay Rizov, Alexandar Spasov — 375

Method for the determination of Some Selected Steroid Hormones and Bisphenol A in Water at Low ng/l Level by On-line Solid-Phase Extraction Combined with Liquid Chromatography-Tandem Mass Spectrometry
Borislav Lazarov, J.A. van Leerdam — 379

Evaluation of Biocide – Free Antifouling Systems
Lucica Barbes — 393

Removal of Dissolved Organic Compounds by Granular Activated Carbon
Bernd Schreiber, Diana Wald, Viktor Schmalz, Eckhard Worch — 397

External and Internal Exposure to Carbon Disulfide at the Working Place
Christina Kopcheva, Teodor Panev, Tzveta Georgieva, Vidka Nikolova, Todor Popov — 401

Accumulation of Copper, Cadmium, Iron, Magnesium and Zink in Three Development Stages of Red Pepper
Simona Dobrinas, Semaghiul Birghila, Marius Belc — 409

Determination of Organochlorine and Polycyclic Aromatic Hydrocarbons Pesticides in Honey From Different Regions in Romania
Simona Dobrinas, Semaghiul Birghila, Valentina Coatu — 413

Evaluation of the Women Fertility Health and the Environment. Nickel Exposure in Industry
Elena Makarova-Zemlyanskaya, Ludmila Dueva, Marina Fesenko — 417

Study on the Influence of Vegetation on the Quality of Aquatic Systems
Silviya Lavrova, Bogdana Koumanova — 421

Heavy Metals Concentrations in Aquatic Environment and Living
Organisms in the Danube Delta, Romania
*Mihaela-Iuliana Tudor, Marian Tudor, Cristina David, Liliana
Teodorof, Dana Tudor, Orhan Ibram* 435

Determination of Cyanides from Distilled Alcoholic Drinks
Semaghiul Birghila, Naliana Lupascu 443

The Quality of the Drinking Water – the Main Factor
into the Socio-Hygienic Monitoring
Grigore Friptuleac, Sergiu Cebanu, Vladimir Bernic 447

Detection of Heavy Metals and Organic Pollutants
from Black Sea Marine Organisms
Gabriela Stanciu, Simona Lupşor 451

Methylmercury in Surface Sediments in Coastal Environments.
Comparison Between a Dynamic Macrotidal Estuary
and a Microtidal Lagoon
*Teodor Stoichev, David Amouroux, Olivier F.X. Donard,
Christo Daiev* 455

A Quick and Easy Retrieval of Information on Pesticide Toxicity
to Humans and Environment: PESTIDOC
*Francesca Vellere, Teresa Mammone, Romilde Basla
and Claudio Colosio* 459

Study of Biofilm Formation on Different Pipe Materials in a
Model of Drinking Water Distribution System and its Impact
on Microbiological Water Quality
Zvezdimira Tsvetanova 463

Modeling the transport and fate of Contaminants
in the environment: soil, water and air
Maria De Lurdes Dinis, António Fiúza 469

Equilibrium studies of Pb(II), Zn(II) and Cd(II)
ions onto granular activated carbon and natural zeolite
Mirko Marinkovski, Liljana Markovska, Vera Meshko 477

Nitrate Pollution of the Lesnovska River Caused by Filtration
of Chemicals by Agricultural Areas
Svetlana Bozhinova, Grigor Velkovski 487

Pesticide Applications and Sustainable Agricultural
Development in Armenia
Vardan Sargsyan, Arman Sargsyan   493

Evaluation of biofilms occuring in drinking water distribution
system of Balatonfüred
Zsuzsa Ludmány, Matyas Borsányi, Marta Vargha   501

Subject Index   509

PREFACE

The book contains the contributions at the NATO Advanced Study Institute on Chemicals as Intentional and Accidental Global Environmental Threats, which took place in Borovetz, Bulgaria, November 16-27, 2005.

A diverse group of scientists, representing the fields of ecology, chemistry, medicine, epidemiology, public health, toxicology, risk assessment, environmental protection and management, modelling, environmental remediation technologies came together to discuss the chemicals threats for the global environment.

The intentional or accidental release of chemicals into environment poses a global threat to public health and security. There is a forecast that terrorists could begin to use more advanced technologies including chemicals to threaten or attack civil and political institutions. Terrorism is expected to increase the threat of chemicals in environment through direct release of chemical agents or sabotage of existing chemical facilities. The intentional release of chemicals into environment can cause widespread panic, injury and mortality due to the potentials of exposure of large population. In addition, the management of accidental chemical releases and historical chemical contamination represents a significant economic burden to society.

The ASI was aimed at performing a critical assessment of the existing knowledge of chemical threats to environmental security with special reference to the prevention of chemical releases, rapid detection, risk assessment and effective management of emergency situations and long-term consequences of chemical releases. The technologies evaluated at the ASI concern prevention and management of both intentional and accident releases of chemicals to the environment. The participation of lecturers with different scientific fields and young scientists allowed for a multi-media focus covering threats to food, air, water, and soil. The exchange of experience of the lecturers and the young scientist's ideas in this field contributed to the finding of answers to critical questions and identification of the future research needs and new approaches in management and structuring of the preventive systems and methodologies.

Lubomir SIMEONOV	Elisabeta CHIRILA
Central Solar-Terrestrial Laboratory	Faculty of Physics, Chemistry and Petroleum Technology
Bulgarian Academy of Sciences	Ovidius State University
Sofia, Bulgaria	Constanta, Romania

ACKNOWLEDGEMENTS

The organizers wish to extend their gratitude to the NATO Science Committee for the approval of the ASI and the Financial Award provided, which made the event possible.

Special thanks go to Dr. Fausto Pedrazzini, Programme Director of Panel Chemistry / Biology / Physics of the NATO Public Diplomacy Division for his continious help and advices for the organization of the event

The editors are grateful to the Publishing editor of the book Mrs. Annelies Kersbergen and to the personal of Springer / NATO Publishing Unit for the operative collaboration and help for the publication of the results from the NATO Advanced Study Institute on Chemicals as Intentional and Accidental Global Environmental Threats, held in Borovetz, Bulgaria, November 16-27, 2005.

LIST OF CONTRIBUTORS

Arman Sargsyan
Production Management and Environmental Issues
Department Information Technology and Systems
Yerevan State University
6 Orbeli Str., Yerevan 375028
Armenia

Marnik Vanclooster
Department of Environmental Sciences
Faculty of Bioengineering, Agronomy and Environment
Universite Catholique de Loivain
Croix du Sud 2, BP 2, B-1348 Louvain –la Neuve
Belgium

Adreana Maximova
Department of Chemical Engineering
University of Chemical Technology and Metallurgy
8 Kliment Ochridski Str., Sofia 1756
Bulgaria

Anna Dimitrova
Chemical Analyses in the Environment
Laboratory of Soil and Waste
National Center of Public Health Protection
Ac. Gechov Bul. 15, Sofia 1431
Bulgaria

Bogdana Koumanova
Department of Chemical Engineering
University of Chemical Technology and Metallurgy
8 Kliment Ochridski Str., Sofia 1756
Bulgaria

Borislav Lazarov
Chemicals in the Environment
Environmental Hygiene
National Center of Public Health Ptotection
Ac. Gechov Bul. 15, Sofia 1431
Bulgaria

Hristina Kopcheva
Chemical Analyses, Spectrometry and Chromatography
Air Pollution Laboratory
National Center of Public Health Protection
Ac. Gechov Blvd. 15, Sofia 1431
Bulgaria

Kosta Vasilev
National Centre for Public Health Protection
Dimirar Nestirov Bul. 15, Sofia 1431
Bulgaria

Lubomir Simeonov
Solar Terrestrial Influences Laboratory
Bulgarian Academy of Sciences
G.Bonchev Str. Block 29, Sofia 1113
Bulgaria

Sylvia Lavrova
Dep. Chemical Engineering
University of Chemical Technology and Metallurgy
8 Kliment Ochridski Str., Sofia 1756
Bulgaria

Svetlana Bozhinova,
Institute of Water Problems
Bulgarian Academy of Sciences
G. Bonchev Str, Block 1, Sofia 1113
Bulgaria,

Teodor Stojchev
Department Food Composition
National Center of Public Health Protection
Ac. Gechov Bul 15, Sofia 1431
Bulgaria

Zvezdimira Tsvetanova
Institute of Water Problems
Bulgarian Academy of Sciences
G. Bonchev Str, Block 1, Sofia 1113
Bulgaria

Bernd Schreiber
Institute of Water Chemistry
Technical University of Dresden
40 Zellescher Weg; Dresden 01062
Germany

Zsuzsa Ludmany
Department of Water Hygiene,
Fodor Jozsef National Public Health Centre
National Institute of Environmental Health
Budapest, Gyali ut. 2-6, H-1097
Hungary

Franceska Vellere
International Center for Pesticide and Health Risk Prevention
Via Magenta 25; 20020 Busto Garolfo-Milan
Italy

Marianna Elia
Dipartimento di Medicina Ambientale e Sanità Pubblica
Medicina del Lavoro,
Università degli Studi di Padova
Via Giustiniani 2, 35128 Padova
Italy

Michele Arienzo
Dipartmento di Scienze del Suolo della Pianta e dell' Ambiente
Via Universita 100, 80055 Portici
Italy

Sara Visentin
International Center for Pesticide and Health Risk Prevention
Via Magenta 25; 20020 Busto Garolfo-Milan
Italy

Sergiu Cebanu
Hygiene Department
State Medical and Pharmaceutical University
1 Mihai Viteazul Str., Chisinau 2019
Moldova

Mohamed Blaghen
Laboratory of Microbilogy, Biotechnolgy and Environment,

Faculty of Science,
University Hassan II AOn-Chock;
Km 8, Route d'Eljadida. Maarif B.P 566 Casablanka 20100
Morocco

Jan Linders B.H. J.
RIVM-SEC,
P.O.Box 1HL-3720 BA Bilthoven
Netherlands

Anna Czerwińska
Environmental Biotechnology Department
Institute for Ecology of Industrial Areas
6 Kossutha Str., Katowice 40844
Poland

Maria de Lurdes Dinis
CIGAR-Research Center in Environment and Resources
Engineering Faculty, Oporto Yniversity
Rua Aquilino Ribeiro 117T, S. Mamede de Infesta 4465024, Matosinhos
Portugal

Mihail Kochubovski
Department of Water and Communal Hygiene,
Republic Institute for Health Protection,
50 Divizija Str. 6; Skopje 1000
Republic of Macedonia

Mirko Marinkovski
Department of Chemical and Control Enineering
Faculty of Technology and Metallurgy
16 Ruger Boskovic Str., Skopje 1000
Republic of Macedonia

Adina Arvinte
Department of Analytical Chemistry
Faculty of Chemistry, University of Bucharest
Panduri Str. No.90, 030018 Bucharest-5
Romania

Alina Coman
Chemistry Department
Ovidius University

124 Mamaia Blvd, 900527 Constanta
Romania

Camelia Draghici
Analytical and Environmental Chemistry
Transilvania University
50 Iuliu Maniu Blvd, Brasov
Romania

Elisabeta Chirila
Chemistry Department,
Ovidius University,
124 Mamaia Blvd, 900527 RO, Constantza
Romania,

Lavinia Mihai
Chemistry Department
Ovidius University
124 Mamaia Blvd, 900527 Constanta
Romania

Liviu-Daniel Galatchi
Department of Ecology and Environmental Protection
Ovidius University
124 Mamaia Bul., Constanta 900024
Romania

Lucica Barbes
Chemistry Department
Ovidius University
124 Mamaia Blvd, 900527 Constanta
Romania

Marian Tudor
Danube Delta Research National Institute
165 Babadag Street, Tulcea 820112, Tulcea
Romania

Naliana Lupascu,
Chemistry Department
Ovidius University
124 Mamaia Blvd, 900527 Constanta,
Romania

Simona Dobrinas,
Chemistry Department
Ovidius University
124 Mamaia Blvd, 900527 Constanta
Romania

Simona Lupsor
Chemistry Department
Ovidius University
124 Mamaia Blvd, 900527 Constanta
Romania

Elena Makarova-Zemlyanskaya
Research Institute of Occupational Health
Russian Academy of Medical Sciences
31 Prospect Budennogo, Moscow 105275
Russian Federation

Olga Sivochalova
Institute of occupational Health,
31 Pospect Budenogo, Moskow 105275
Russian Federation.

Thomas McHugh
Groundwater Services , Inc.
2211 Norfolk
Suite 1000, HoustonTX 77098
USA

Heike Schroeder
Institutional Dimensions of Global Environmental Change (IDGEC)
Donald Bren School of Environmental Science and Management
University of California,
Santa Barbara, CA 93106-5131
USA

ENVIRONMENTAL RISK ASSESSMENT

LIVIU-DANIEL GALATCHI
Department of Ecology and Environmental Protection
Ovidius University of Constanta
124 Mamaia Boulevard, 900527 Constanta, Romania
E-mail: galatchi@univ-ovidius.ro

Abstract. Environmental risk assessment evaluates the quantitative and qualitative characteristics of the environment, in order to highlight the risk on environment and human health due to the potential presence or use of specific pollutants. Risk is a probability of an adverse direct or indirect effect, on the environment or human health. It is a combination of the probability of occurrence of an event and the possible extent of that event's adverse effects and consequences, in terms of adverse effects on the ecosystem and human injury. Risk is defined as the probability of an event to occur, related to the seriousness and extent of its consequences. An environmental risk assessment should be conducted, by adopting a systematic approach, when it is determined that a management action may have consequences to either the state of the environment or human health or well-being.

Keywords: risk assessment; hazard identification; dose-response assessment; exposure assessment; ecosystem concentrations; system boundary

1. Introduction

Ecological risk assessment and the human health risk assessment are the two related disciplines of which the environmental risk assessment is comprised. Both the ecological and the human health risk assessment, involves four steps:
- determination if a chemical, physical, or biological entity could induce adverse effects on individuals, populations, communities, or ecosystems, being linked to a particular effect on the environment (*hazard identification*);

- determination of the relationship between the magnitude of the exposure and the probability of occurrence of the effects (*dose-response assessment*);
- determination of the extent of exposure, of the condition under which an organism comes into actual contact with a stressor (*exposure assessment*);
- description of the nature and the magnitude of risk (*risk characterization*).

2. The Interactive Nature of Environmental and Human Health Risk Assessment

There are several successive interactive stages by which an environmental risk assessment proceeds. Problem identification is the most important activity, describing the resources, which could be affected, and the possible consequences of the action.

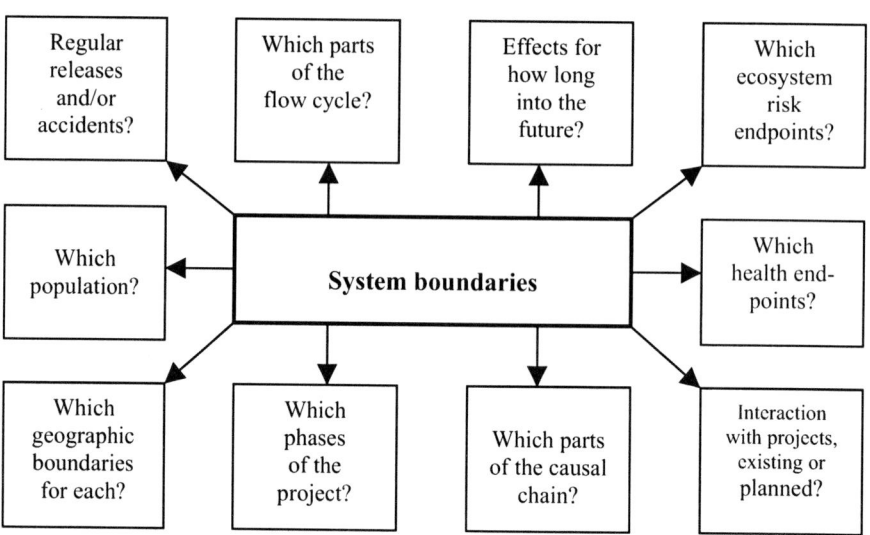

FIGURE 1. Questions related to system boundaries.

The first tier is the scoping activity, which identifies concerns and issues that have potential risks. Further detailed studies are done if there are identified unacceptable risks. Related to the levels of analysis, system boundaries (Fig. 1), and risk expressions (Fig. 2), through which the scoping activity proceeds, there some questions to be asked. The second tier is screening, in which the data describing the dose-response relationships of exposure scenarios is analysed. This is when we identify the risk issues. During the third tier, there are performed modeling of exposure and direct measure of effects (Vadineanu, 2000).

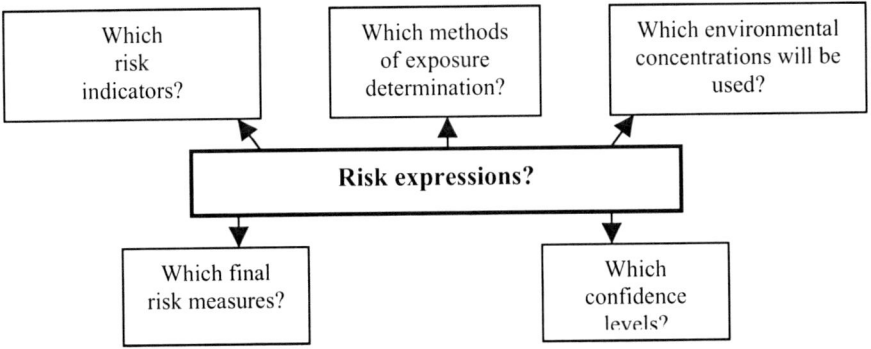

FIGURE 2. Questions related to risk expressions.

3. Risk Characterization and Comparison

Risk characterization serves to organize the information gained through the two main actions performed in environmental risk assessment: identifying and reducing the risks (Fig. 3).

The risk characterization should also contain a description and estimate of the uncertainty of the assessment.

It is recommended to include different points of views when performing the environmental risk assessment (Fig. 4).

Useful comparison could be related to risks supposed by alternative technologies, sites or projects, which might provide an equivalent economic growth benefits. Also, important reference points to interpret new risks, if they are comparable, are the familiar risks. When the development projects are enough large to offer benefits, the risk may be acceptable.

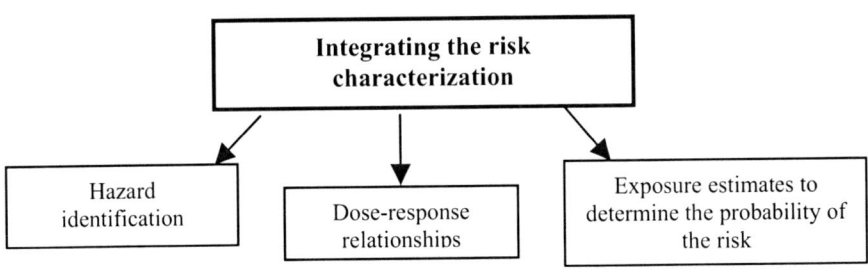

FIGURE 3. Information integrated by the risk characterization.

One of the issues to be considered for comparison of the risks is costs/benefits of the risk reduction.

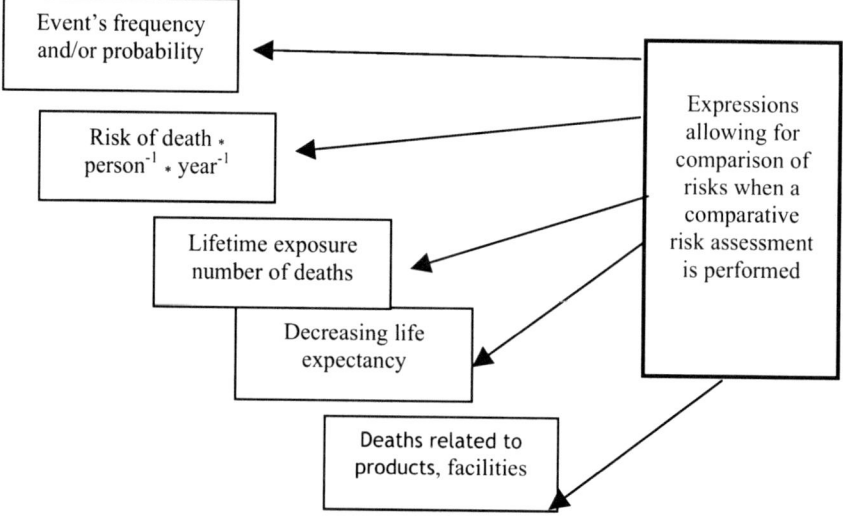

FIGURE 4. Expressions allowing for comparison of risks.

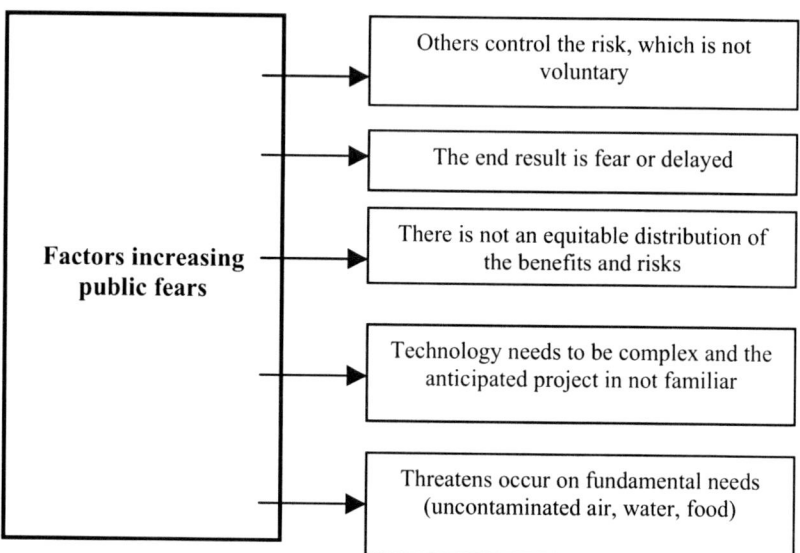

FIGURE 5. Factors that can lead to strong opposition to projects. This figure refers to false public perceptions and the need to engage in effective risk communication.

4. Communication and Management in Risk Assessment

The reaction of the public to a risk that has been recognized, analyzed, and categorized depends on how they take perceive and understand the information (Fig. 5).

This is why the decision analysis should be the form through which the risk information is communicated, stating the available options, risks, costs, benefits, and their distribution within the society. Risk management involves an evaluation of cost/effectiveness of risk reduction measures so that priorities can be made. Evaluating the measures in order to reduce the risk and implementing the cost-effective ones, is to manage the risk.

There are three important phases of managing the risks (Fig. 6).

The above phases show the necessity of identifying the hazard to people and environment, the quantification of the possibility to occur such hazards, the magnitude of the events, and then it is necessary to establish the limits of risk acceptability.

The last phase consists of the design and implementation of measures of risk control.

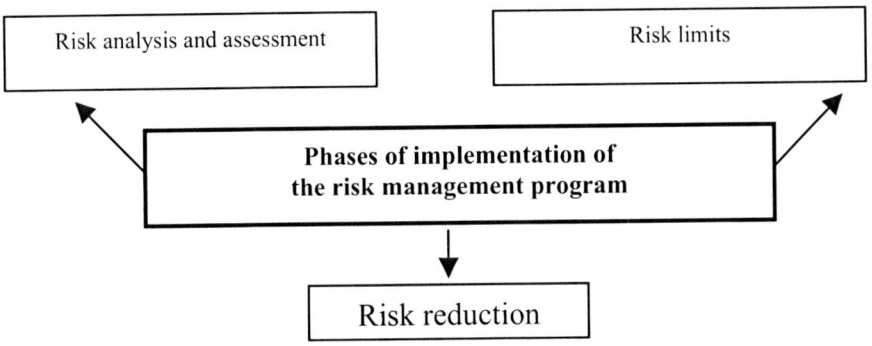

FIGURE 6. The three phases of the risk management.

5. Assessing the Environmental Risks

The risk assessment comprises of three steps: hazard identification, consequence analysis, quantitative analysis and frequency assessment (Fig. 7).

An emergency plan is supposed to be comprised of two components, on-site measures and off-site measures.

The on-site emergency plan includes:
- the required mechanisms for alarm and communications;

- appointment of qualified personnel;
- details of the emergency control centers.
- The off-site emergency plan should be based on the below features: organization, communications, specialized emergency equipment and knowledge, voluntary organizations, chemical and meteorological information, humanitarian arrangements, public information, and assessment (Guvernul Romaniei, 2001).

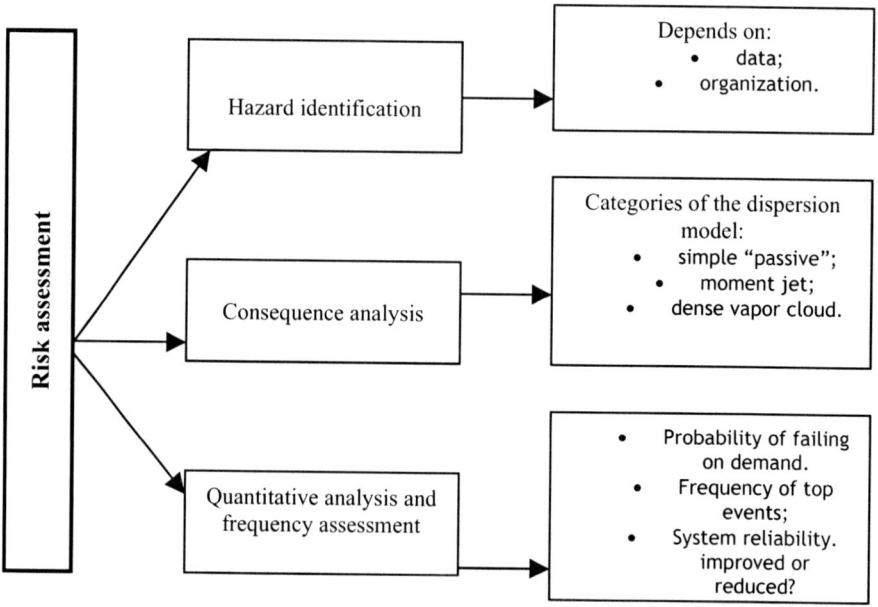

FIGURE 7. Steps of the risk assessment. This introduces a management issue concerning a chemical release to the environment (an industrial accident).

ACKNOWLEDGEMENTS

We thank Professor Stoica-Preda Godeanu for helpful suggestions during our research.

References

Guvernul Romaniei, 2001, "Strategia Nationala pentru Dezvoltare Durabila", Bucharest, pp. 1–120.
Vadineranu, A., 2000, "Dezvoltarea durabila", University of Bucharest Press, Bucharest, pp. 1–224.

SAMPLING AND SAMPLE PRETREATMENT FOR ENVIRONMENTAL ANALYSIS

ELISABETA S. CHIRILA[1]*, CAMELIA DRAGHICI[2] AND SIMONA DOBRINAS[1]
[1]*Ovidius University, Chemistry Department, 124 Mamaia Blvd, RO-900527 Constanta - 3, Romania;*
[2]*Transilvania University, Chemistry Department, 50 Iuliu Maniu Str., 500091 Brasov, Romania*

*To whom correspondence should be addressed: echirila@univ-ovidius.ro

Abstract. The sample pretreatment methods are constantly developing aiming to respond to the large variety of samples source and composition. For example, in order to preconcentrate toxic metals from liquid samples the most used methods are the extraction, based on extraction agents like chelates and ion-pairing are, used after the samples' digestion. Organic pollutants have different ways for extraction and preconcentration depending on the structure, concentration of the interest analyte and on the sample's provenience, like: liquid-liquid extraction (LLE), solid phase extraction (SPE), solid phase microextraction (SPME), microwave-assisted extraction (MAE), supercritical fluid extraction (SFE) etc. The available methods and techniques used for sampling and sample preparation in order to determine organic and inorganic pollutants in air, water, soil and biota samples are presented. To emphasize the sampling and sample preparation process, some original approaches concerning the accumulation factors of inorganic and organic pollutants in the Romanian Black seacoast ecosystem are presented.

Keywords: sampling air; water; soil; biota; VOCs; OCPs; PAHs; toxic metals; extraction; preconcentration

1. Introduction

Environment analysis, like any other analytical process, must follow three major steps:
- sampling and sample preparation for measurements (simply called "sampling");
- measuring;
- data processing.

The samples are portions of environment, from both biotope (air, water, soil) and biogenesis (plants, animals) that are representative for the place and the moment of sampling (Kramer, 1995).

The sampling strategies must insure the chemical integrity of the analyzed components; therefore they are sometimes rather difficult. This is the reason why at the beginning of each procedure, the objectives must be clearly stated. Sampling, preservation and sample preparation depend on the nature of the analyzed components (organic or inorganic), on their concentration (major components, minor, or traces) and on the analytical technique (Chirila and Draghici, 2003; Chirila, 2004). Moreover, knowing the sample history might help the optimal choice of all the steps of the process (Baiulescu et al., 1999; Crompton, 2001).

2. Sampling

When a new study designs a sampling program, addressed to an environmental problem, the environment units (i.e. the target population and the sampling population) must be defined.

The unit can be differently defined, considering the objectives, the measurements, the regulations, the costs or other criteria. For example, the vegetation (dried, calcined and prepared for analysis) from a 2m x 2m area can be a unit. A representative unit is the one selected for measuring the target population, so that combined with other representative units it should offer a faithful image of the investigated phenomenon.

The target population is a set of N population units among which interferences are occurring, while **the sampling population** is a set of population units available for measurements.

Therefore, when designing the sampling plan, specialists must be consulted, sometimes interdisciplinary groups, gathering chemists, hydrologists, biologists and environment specialists.

Each sample (air, soil, water, organisms) must be numbered and labeled, containing the following information:
- collection data and hour;
- place/site;

- name, surname of the person who did the sampling.
- A special file must accompany the sample, also containing specific information:
- objective of sampling;
- observations on the external pollution indices and, briefly, the type of analysis that will follow;
- meteorological data;
- nature and dimension of the pollution source (if any).

2.1. AIR SAMPLING

Atmospheric samples contain various species, inorganic or organic compounds, low molecular or high molecular substances, ranging from gaseous to nonvolatile substances, which are adsorbed on solid particles.

Air pollutants drawing depends on the nature of the pollutant and its aggregation state. For organic pollutants in air, one of the following techniques can be used (Niessner, 1993; Lewis and Gordon, 1996):
- passing through columns or filters;
- reaction with treated substrates;
- condensation in cooling traps.

Often atmospheric sampling is performed using canister methods. The canister is filled either by vacuum release, or by pumping, using stainless steel or Teflon diaphragm pumps. A similar method is the one using collapsible Teflon or Tedlar sample bags. The bags are filled by pumping and than transported for analysis to a laboratory.

A second method of sample acquisition, based on use of a solid-phase adsorbent as an analyte trap, is widely used for low volatility species that are less suitable for sampling using canister methods because of the increased capacity of analyte condensation on the container walls. The adsorbent used in the trap is chosen in such a way to introduce an element of selectivity to the trapping mechanism, although in practice a trap-all approach is commonly used.

Inorganic adsorbents are also commonly used, notably potassium carbonate and magnesium perchlorate. Adsorbents like these, however, have limited absorbtion capacity and often require frequent regeneration or replacement. A combination of initial condensation and a second stage of adsorbent scrubber often provide sufficient capacity to dry a sample stream of air for many hours or days. Continuous drying may be achieved using permeation membranes such as Nafion.

Particles are commonly retained on membrane filters or impactors. Gaseous components and volatile organic compounds (VOCs) are collected by trapping, either cryogenically (for the most volatile substances) or on sorbents of increasing retentivity (for the less volatile ones).

The constituents of particulates are separated into approximately organic and inorganic substances by Soxhlet extraction or ultrasonication with an appropriate organic solvent (methanol, dichloromethane, cyclohexane, etc.) or by supercritical fluid extraction (SFE) with CO_2. SFE is becoming increasingly accepted for the extraction of analytes because it is more rapid, reliable and efficient for a great variety of matrices than the older techniques and it is environmentally friendly by reducing the need for organic solvents.

2.1.1. *Absorbent Columns*

For the quantitative evaluation of vaporous molecular compounds in a gaseous sample, a very efficient and fast technique is used, consisting in a single use of a small sealed column (4 x 400mm), presented in Fig. 1. The column contains molecular sieves or different absorbents such as active carbon with graphite structure or organic functional polymers. A flow pump insures the aspiration of a fixed gas volume, with a flow between 0.1-1 L/min. The absorbed compounds are recuperated via solvent extraction (often with CS_2), or via thermal desorption, which has the advantage of also diluting the compounds and avoiding impurities. This last technique can be further adapted for gas chromatography. The column is introduced in special oven that may increase temperature, for a few seconds, up to 350°C. The resulted compounds are directly transferred to the chromatograph injector. Alternatively, extraction tubes, with a low enough diameter to be directly introduced in the injector, are used.

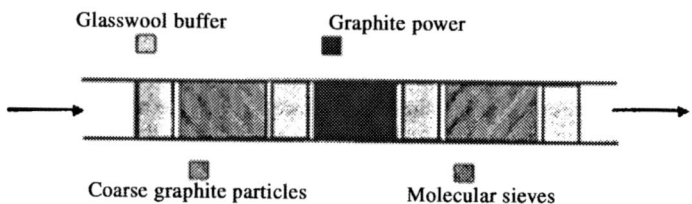

FIGURE 1. Collecting column.

2.1.2. *Adsorbents*

The active carbon and the silica gel can adsorb many vapor phase organic compounds. In certain conditions the quantity introduced in a tube that corresponds to a volume of aspired air allows the calculation of the average concentration of the analyzed compound. Desorption is done using a solvent or a solvents mixture and compounds are afterwards separated and quantitatively determined by gas chromatography, for example.

2.1.3. Filters

For collecting air samples containing pollutants as smoke, powders or dust there is recommended to pass the sample through a filter made of different materials, with different pore sizes. The pollutant quantity is evaluated gravimetrically and is then extracted from the filter with appropriate solvents.

2.1.4. Absorption in Liquids

Absorbing components is done in a slow process, by bubbling air in a liquid, in an absorber. The shapes and dimensions of the absorbers are different, some of them being similar with the laboratory bubbling vessels. The absorbents are mixtures of solvents able to solve almost all pollutants existing in air. These liquids must have low volatility and good water, fog and aerosols retention, for example glycerin, ethylene glycol, dimethylsulfoxide or isopropanol.

2.2. WATER SAMPLING

Considering their further analysis, water samples can be of three types:
- for physical-chemical analysis;
- for biological analysis;
- for bacterial analysis.

The samples differ according to their collection, quantity, the used devices and the preservation method.

2.2.1. Sampling for Physical-Chemical Analysis

For the physical-chemical analysis volumes between 100mL and 3L are collected using manual or automatic devices. The water samples are collected in colorless glasses, chemically resistant (pyrex glass with fixed composition) but lately, because of their fragility polyethylene vessels with the following advantages replaced these:
- mechanical resistance;
- avoiding the ion exchange between water and the recipient.

Some authors mention that the polymer vessels are adequate for silica and metal ions analysis (Na^+, K^+, Ca^{+2}, Mg^{+2}, Ag^+, Cd^{+2}, Cu^{+2}) but are unsuitable for dissolved permeable gases and for phosphates, that can be adsorbed.

2.2.2. Sampling for Biological and Bacterial Analysis

The water for biological analysis is collected in 1L to 10L vessels. The collection techniques differ according the known lifetime – biotope:
- for zoobentos (the bottom of the waters) – dredgers are used;
- for phytoplankton (in the water) the plankton net made of silk sieve sacks is used.

Water for bacterial analysis is collected in glass sterilized vessels with 100-300 mL volume.

2.3. SOIL SAMPLING

If there is deep and wide spread soil contamination, samples are collected from a 20 cm^2 surface with a showel, to a dept of 1-5 cm. For an approximate density of contamination in a region, diagonal sampling is recommended from different areas. There are available iron plate patterns with 20 or 40 cm^2 holes.

After the split spoon is opened and a fresh surface is exposed to the atmosphere, the sample collection process should be completed in a minimal amount of time. Visual inspection and an appropriate screening method may be selected to determine the interval of the soil core to be sampled. Removing a sample from a material should be done with the least amount of disruption (disaggregation) as possible. Additionally, rough trimming of the sampling location surface layers should be considered if the material may have already VOCs lost (been exposed for more than a few minutes) or if other waste, different soil strata, or vegetation may contaminate it.

Removal of surface layers can be accomplished by scraping the surface using a clean spatula, scoop or knife. When inserting a clean coring tool into a fresh surface for sample collection, air should not be trapped behind the sample. Undisturbed sample are obtained by pushing the barrel of the coring tool into a freshly exposed surface and removing the corer once filled. Then the exterior of the barrel should be quickly wiped with a clean disposable towel to ensure a tight seal and the cap snapped on the open end. The sampler should be labeled, inserted into the sealable pouch and immediately cooled to $4 \pm 2°C$.

2.4. PLANTS AND ANIMALS SAMPLING

The organisms used in environment monitoring must fulfill the following criteria: easy to identify, with adequate dimensions, long living, abundant, less mobile etc.

The results of many monitoring programs depend on the collecting method. Therefore, the sampling procedures must be discussed between the environment specialist, the statisticians and the analysts. Different species can accumulate different elements in different extent therefore sampling must be done by people that can identify the species. The samples are stored in inert material recipients and can be preserved through freezing, chemical substances, sterilization, etc. Before storage, if heavy metals are monitored, the samples must be washed with deionized water.

3. Sample Preparation

This step aims to bring the samples in a suitable form, ready for analysis but in some cases it means also the purification or concentration of the analyte. The preparation techniques are very different according to the nature of the pollutant and concentration. Some techniques, used in preparing the samples for the analysis of the inorganic pollutants (especially heavy metals) and organic pollutants are presented.

3.1. SAMPLE PREPARATION FOR INORGANIC POLLUTANTS ANALYSIS

Heavy metals represent the majority of the inorganic pollutants from air, water, soil or biota. In Fig. 2 the flow diagram of the sample preparation steps for metal analysis is presented.

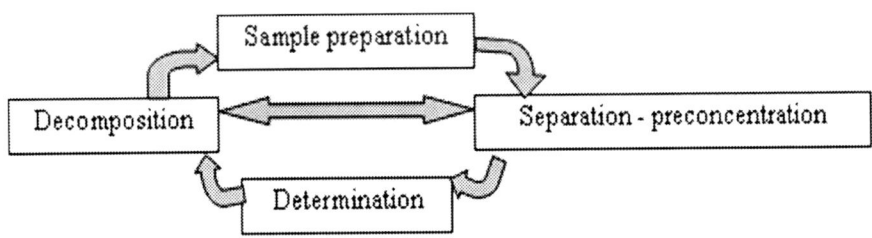

FIGURE 2. Inorganic preparation flow diagram.

Most of the quantitative techniques for traces analysis are using liquid samples. Therefore, the complex solid samples (which can be particulate matter from air, soil, and biota) or the organic liquids are subject of a disaggregating process (called digestion or mineralization). The digestion/extraction of the elemental species from the solid samples can be the most difficult aspect of the analysis. Adequate techniques are necessary to convert the solid matrix into solution. The major error source (and the only one) is the digestion process (Anderson, 1999).

3.1.1. Decomposition

Solid sample digestion can be done using:

- wet digestion with HCl, HF (if silicates are present), HNO_3, royal water, $HClO_4$ or H_2SO_4 with H_2O_2;
- melting with fondants such as NaOH or $K_2S_2O_8$;
- calcinations followed by solving the ash in an acid;
- solving at high pressure, in digestion bombs;
- microwave digestion, a new and modern technique, combining pressure and temperature.

From the solution resulted after digestion, a known sample quantity is analyzed using different methods, like spectrometry or electrometric techniques.

3.1.2. Separation-Preconcentration

Trace elemental analysis is under constant development and the challenging analyses of today become the routine of tomorrow. In order to analyze the trace components, liquid samples must pre-concentrated by: column retention (Akatsuka et al., 1998, Santelli et al., 1994), ion exchange (Anezaki et al., 1998, Ferreira et al., 2000), precipitation (Atanassova et al., 1998, Chen et al., 1997) or chelation followed by solvent extraction.

The experimental process of inorganic extraction of a neutral complex, regardless of the type of complex, is essentially the same. The neutral complex interaction with the aqueous phase, including but not limited to the solubility, depends on the charge and polarity of the overall complex. The first step is to generate a neutral complex with the analyte of interest using chelates or ion pair agents. The most used compounds for inorganic extractions are presented below.

As chelating agents: 8-hydroxyquinoline (and derivatives), dimethylglyoxime, diphenyldithiocarbazone, sodium diethyldithiocarbamate, sodium N, N' phenyl acetyldithiocarbamate, acetylacetone, thenoyltrifluoroacetone, ammonium N-Nitroso-N-phenylthydroxylamine(cupferron), di- n-butyl phos-phoric acid, di(2-ethylhexyl)phosphoric acid, 1-nitroso-2-naphthol.

As ion-pair agents:

- chelated ion-pairs (ethylenediaminetetraaceticacid (EDTA)/halide or 1,10-phenoanthroline/perchlorate);
- non-chelated ion-pairs (tetraalkylammonium salts, tetraphenylarsonium salts);
- halide ion-pairs (HCl, HF, HI).

A small volume of organic solvent is added to the sample mixture. For example, 1L aqueous sample may be extracted into 40 mL of organic solvent. Extraction can be performed in a separatory funnel or by using a mechanical shaker table. The pH of the mixture may need to be manipulated, depending on the exact extraction scheme used. In addition, masking agents may be used to obtain specificity. After mixing, the two phases are separated and the procedure is generally repeated several times. The organic phase is combined from each extraction. The concentration of the elements in the sample is increased by 1-3 orders of magnitude in the organic phase. The extract can then be further pre-concentrated if needed (back-extraction, evaporation, etc.) or analyzed directly, for example by flame AAS.

Extraction using chelates. Inorganic extraction, utilizing chelates, for preconcentration and/or analytical separation, has been exploited for many instrumental systems. For flame AAS analysis, the inorganic solvent extraction is often designed to increase the concentration of elements of interest and, most importantly, to reduce the concentration of alkali and alkaline earth elements (i.e. leave most of them in the aqueous phase). This separation is especially necessary for many natural water samples such as seawater, brines etc. Trace element analyses of clinical samples such as blood, urine etc., also benefit of inorganic extraction, for flame AAS analysis as well as other determination techniques (ET-AAS, ICP-AES, etc.). Radiochemistry separations for NAA often use inorganic extraction techniques that use chelating schemes. Extraction schemes have also been developed, which leave the analyte of interest in the aqueous phase and remove interfering compounds through the organic phase. This technique has limited applicability due to the limited solubility of the chelate in the organic solvents.

Extraction using ion association (ion-pair). Neutral complexes can be formed through ion association (ion-pair) and extracted from the aqueous solution into an organic solvent. Ion associations with inorganic extracts encompass a wide range of extraction schemes. General sub-groups include chelated ion-pairs, non-chelated ion-pairs and halide-cation ion-pairs. The halide-cation pairs are typically extracted into oxygen-containing solvents, such as methylisobutyl ketone, diethylether and alcohols.

The nature of the solvent is of special importance for inorganic extractions. There are several criteria that should be evaluated when choosing a solvent for an inorganic extraction. The solvent should have the following characteristics:

- immiscible with aqueous solution (i.e. low solubility in water); for convenience, density' water if the sample is drawn off;
- extracts the desired metal chelates;
- does not form emulsions;
- compatible with the analytical determination technique;
- environmentally safe and nontoxic;
- available in an acceptably uncontaminated state.

3.2. SAMPLE PREPARATION FOR THE ANALYSIS OF ORGANIC POLLUTANTS

The preservation techniques aim to minimize the sample degradation occurring between collection and analysis, which is caused by physical processes, like volatilization, adsorption, or diffusion are, or chemical processes, like photochemical or biochemical degradation (Parr et al., 1988).

In order to apply an analytical method it is compulsory to avoid interferences between the analyte and the components of the matrix. Therefore, most of the time, the analysis begins with the separation of the components from the sample matrix. The gaseous pollutants, trapped in solid or liquid matrix will be extracted using the methods described in this chapter. The results of the analysis depend on the preparation step. The influence oft his step is comparable with the precision of the analytical devices. So, designing the preparation method received all the attention and uses all the recent developments in chemistry and automatics.

Preparing the sample for organic pollutants analysis can be done following three routes:

- extraction;
- chromatography;
- derivatisation techniques.

3.2.1. *Extraction from Liquid Samples*

Economical, as well as technical features should be taken into consideration when choosing between liquid-liquid or solid-phase extraction for a particular sample. In this sense, liquid-liquid and solid-phase extraction techniques should be considered as complementary approaches, and although the general trend is towards the replacement of liquid-liquid extraction methods by solid-phase extraction, this is never likely to be a complete replacement.

Liquid – liquid extraction. The usual extraction technique requires a solvent or a solvent mixture, able to solve all the water pollutants. The easiest method is to extract the water pollutants by mixing the sample with the organic solvent in a separation funnel. According to the extraction conditions, the extract may contain:

- low polarity pollutants;
- low volatile pollutants (universal extraction for neutral semivolatiles);
- acid or alkaline compounds (selective extraction).

This technique is long lasting and uses toxic solvents. Sometimes emulsions are formed making the water/organic phase separation very difficult. Therefore in the past years the separation funnels have been replaced by extraction columns (e.g. the Extrelute column, Merck). The extraction columns are filled with a special material with a large pores volume. The Kiesselgur matrix is chemically inert and can be used on a broad pH domain (1...13).

The column extraction is based on the following principle: the water sample is distributed as a film at the contact with the adsorbent. Afterwards, the elution with immiscible water solvents is performed (e.g. diethylether, ethylacetate or chlorinated solvents). All the lyphophyle pollutants are extracted from the aqueous into the organic phase. During the process the aqueous phase remains on the stationary phase and emulsion formation is no longer possible. The extract can be vaporized and the pollutants will be analyzed.

Solid-phase extraction. This technique replaces the classic liquid-liquid extraction and is based on the observation that the organic analytes can be adsorbed on a specific substrate, in a micro column. On this route, the interfering compounds from the matrix are avoided, and a selective concentration from 100 to 5000 times is achieved, the method being able to isolate even traces. The micro columns contain adsorbents with different particle dimensions that allow the use of low pressures, forcing the sample and the eluting agent to pass through it. Solid-phase extraction for liquid samples became a widely used laboratory technique with the introduction of disposable sorbent cartridges containing porous siloxane-bonded silica particles, sized to allow sample processing by gentle suction (Fig. 3).

A typical solid-phase extraction cartridge consists of a short column (generally an open syringe barrel) containing a sorbent with a nominal particle size of 50-60μm, packed between porous metal or plastic frits. A large number of sorbents are used today, corresponding to the desire for general purpose, class-specific and even compound-specific extractions. Slow sample processing rates for large sample volumes, low tolerance to blockage by particles and sorbed matrix components, and problems arising from the low and variable packing density of cartridge devices spawned the development of alternative sampling formats based on disc technology.

The particle-loaded membranes consist of a web of polytetra-fluoroethylene (PTFE) microfibrils, suspended in which are sorbent particles of about 8-10μm diameter. The membranes are flexible with a homogeneous structure containing 80% (w/w) or more of sorbent particles formed into circular disks of 0.5mm thick with diameters from 4 to 96 mm. For general use they are supported on a sintered glass disc (or other support) in a standard filtration apparatus using suction to generate the desired flow through the membrane.

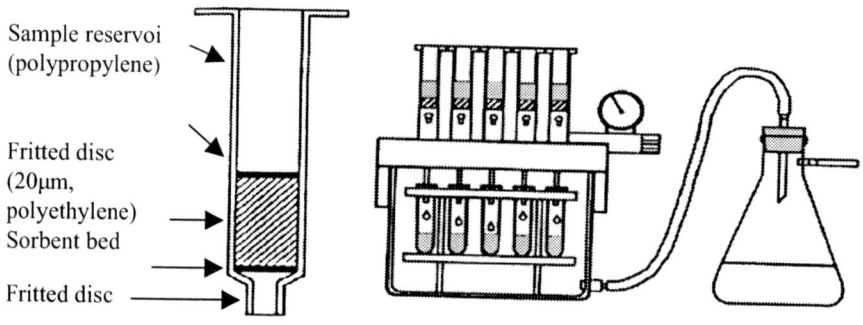

FIGURE 3. Schematic diagram showing the typical construction of a solid-phase extraction cartridge and a vacuum manifold for parallel sample processing.

The solid phase extraction follows four steps:
- *adsorbent conditioning,* preparing the cartridge for further interaction with the sample;
- *sample injection*, with the analytes adsorption;
- *washing,* for the matrix component removal;
- *elution*, with the selective desorption of the monitored compounds and the eluent collection.

Solid phase microextraction (SPME) represents a modern alternative for sample preparation; eliminating most of the disadvantages related to water sample preparation, SPME is a new sample enrichment technique that can easily transfer the analytes to the GC inlet. Since the invention of the technique in 1989 by J. Pawliszyn, its applications have dramatically increased. It has been used mainly for environmental water analysis (Pawliszyn, 1997). The basic equipment of SPME is simple. As shown in Fig. 4, a fused-silica rod is connected to a stainless steel tube that can be withdrawn inside a syringe needle, after sampling, for protection and transfer to GC inlets.

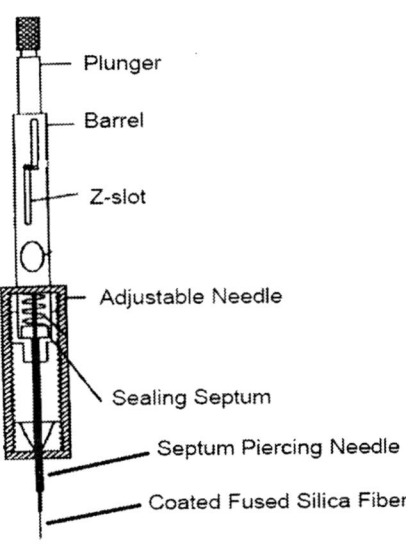

FIGURE 4. Schematic diagram of SPME assembly.

The SPME devices contain a melted silica fiber, covered with polydimethylsiloxane, fixed on a stainless steel plunger and a bar, like in a microseringe. The fiber is introduced in a protecting cap, able to perforate the septum, obturating the sample flask. The organic analytes are adsorbed on the molten silica phase; magnetic stirring increases adsorption. About 15 minutes are necessary for the adsorption equilibrium. After adsorption, the silica fiber is introduced back in the cap, extracted from the flask and then is coupled to a GC injector where the analytes are thermally desorbed, cryogenic focused at the entrance of the column and separated. The SPME can use *split-splitless* or *on column* injectors.

Many factors may influence the method precision: the place of the fiber in the injector (for each desorption it must be in the warmest region), the temperature of the cryogenic focusing (stabile and quite low) the time between extraction and desorption (the volatile analytes may get lost in the fiber). Table 1 presents the SPME fibers and their recommended use.

TABLE 1. SPME fibers and their recommended use.

Fiber coating material	Abbreviation	Recommended use
Polydimethylsiloxane	PDMS	Nonpolar compounds
Carboxen/polydimethylsiloxane	Carboxen/PDMS	Very volatile compounds
Polyacrylate	PA	General
Polydimethylsiloxane/divinylbenzene	PDMS/DVB	General
Carbowax/divinylbenzene	CW/DVB	Polar compounds
Carbowax/templated resin	CW/TPR	Polar compounds

Headspace extraction is used for the pollutants trapped in a matrix that cannot be introduced, as such, in a chromatograph. There are two techniques:

- *static headspace analysis*, which is probably the simplest solvent-free sample preparation technique, has been used for decades to analyze volatile organic compounds. The sample (liquid or solid) is placed in a vial and the vial is sealed. The vial is then heated and the volatile compounds are driven into the headspace. Equilibrium between the headspace and the sample matrix is reached. A portion of the vapor from the headspace is injected into a GC.
- *the dynamic technique* uses a carrier gas (helium) for eluting the volatile parts to a collector where they are adsorbed and concentrated. A thermal desorption follows in the collector, allowing the gas components to enter the GC. The sample can be recovered by stripping.

Purge and trap extraction is used for organic non-polar volatile compounds extraction for GC analysis. An inert gas is bubbled in the water sample moving the organic volatiles into the vapor phase. These are trapped in active carbon and/or condensed. The trap containing the adsorbent is passed into a heated desorption chamber that allows desorption of the retained compounds. This is not always a fast process (as needed for GC) but cryogenic focusing may be used. It is very important to use highly pure purge gas. Purging water media may raise difficult problems because usually low water quantities are allowed in the column.

3.2.2. *Extraction from Solid Samples*

It is well known that common extraction techniques for solid matrices include (Poole et al., 1990; Poole and Poole, 1996):

- solvent extraction,
- Soxhlet extraction,
- microwave-assisted extraction (MAE),
- accelerated-solvent extraction (ASE).
- supercritical fluid extraction (SFE),
- ultrasound extraction (sonication),

Solvent extraction. The solid samples must firstly be grinded and milled for enhancing the specific surface. A sample is then covered with the solvent, mixed for 15-20 min, left for decantation and filtered. This operation is several times repeated. This is a long-lasting technique that requires large solvent quantities. In the field, the fine grinded sample is introduced in a column having a glass wool rod and filter paper at the bottom (as filtering cartridge). The extraction mixture is introduced by the upper part, and the force passing through the filter cartridge is applied using a syringe that perforates the rubber rod of the sampling flask.

Soxhlet extraction allows use of large amount of sample (e.g. 10-30g), no filtration is required after the extraction, the technique is not matrix dependent, and many Soxhlet extractors can be set up to perform in unattended operation. The most significant drawbacks of Soxhlet extraction are: long extraction times (e.g. up to 24-48 h), large amount of solvent usage (300-500 mL per sample), and the need for evaporation after sample extraction. The compounds that can be determined using Soxhlet extraction as pretreatment step are organochlorine pesticides (OCPs), polycyclic aromatic hydrocarbons (PAHs), polychlorinated biphenyls (PCBs), polychlorinated dibenzo-dioxins (PCDDs), polychlorinated dibenzo-furans (PCDFs), phenols, benzidines, nitrosamines, phthalate esters etc. from solid samples (soil, sediment, sludge, biological tissue, etc.) that must be previously homogenized (mix, ground and sieve) and also from particulates, vapor, smoke and liquid samples previously filtered on glass or quartz fiber or using polyurethane foam plugs (PUFs), XAD resins, Tenax GC etc.

The improvements in Soxhlet extractions are: high-pressure Soxhlet extraction (HPS), automated Soxhlet extraction (Soxhlet HT and Buchi B811) and focused microwave-assisted Soxhlet extraction (Soxwave and FMASE) as shown by Luque de Castro MD and Garcia-Ayuso in 1998. The solvents used for extraction are polar species (dimethyl formamide, tetrahydrofuran, acetonitrile, acetone, methanol, ethanol), non-polar species (hexane, iso-octane, ethers, light petroleum), aromatics (benzene, toluene) or others of general use (dichloromethane, ethyl acetate, isopropyl alcohol, chloroform) and the extraction time varies from 8 to 48 h depending on the matrix, analyte and solvent.

Microwave assisted extraction (MAE) is a well-known technique, firstly used for inorganic pollutants, lately extended to organic compounds. It is one of the "green" preparation methods, fast, accurate and with minimum solvent use. The microwave process gave higher yields than the traditional steam distillation process. In 1993, Onuska and Terry published the first data on the use of MAE for pollutants from environmental samples. They successfully extracted organochlorine pesticide residues from soils and sediments using a 1:1 mixture of isooctane-acetonitrile using sealed vials. The samples were irradiated for five 30 s intervals.

In 1994 Lopez-Avila et al., published their work to expand the use of MAE to 187 volatile and semivolatile organic compounds from soils. The compounds included polyaromatic compounds, phenols, organochlorine pesticides and organophosphorus pesticides. MAE uses microwaves that can easily penetrate into the sample pores causing the solvent trapped in the pores to heat evenly and rapidly. In contrast to conventional heating MAE is a promising technique because:

- it is fast (e.g. 20-30 min per batch of as many as 12 samples);
- uses small amounts of solvents, compared to Soxhlet and sonication extraction (30 mL in MAE versus 300-500 mL in Soxhlet extraction);
- allows full control of extraction parameters (time, power, temperature);
- stirring of the sample is possible;
- allows high temperature extraction;
- no drying agents are needed, since water absorbs microwaves very fast and thus can be used to heat up the matrix.

MAE has also several drawbacks that contributed to its slow acceptance such as:

- extracts must be filtered after extraction, which slows down the operation;
- polar solvents are needed;
- cleanup of extracts is needed, because MAE is very efficient (e.g. 'everything' gets extracted);
- the equipment is moderately expensive.

Accelerated solvent extraction (ASE) is a fairly new extraction method that was approved recently by the U.S. Environmental Protection Agency (EPA) as Method 3545. The extraction is done in a closed vessel at elevated temperatures ($50°$ to $200°C$) and pressures (1500-2000 psi).

This technique is attractive because it is fast (e.g. extraction time is approximately 15 min per sample), uses minimal solvent (15-40 mL), no filtration is required after the extraction, and the instrumentation allows extraction in unattended operation. At least 24 samples can be processed sequentially and different sample sizes can be accommodated (e.g. 11, 22, and 33-mL vessels are available).

Supercritical fluid extraction (SFE) has certain advantages due to the properties, interimate between liquid and gas. Particularly, the viscosity of a supercritical phase is in the gas magnitude order while solubility properties (repartition coefficients, K) correspond to the liquid phase. E.g. at 160 bar and $60°C$ carbon dioxide has the density 0.7 g/mL and the polarity close to toluene, and is called therefore "dense gas".

As an extraction fluid, supercritical carbon dioxide is mostly used because its critical parameters can be rather easy obtained ($T_c = 31°C$, $P_c = 74$ bar) and is non-polluting. Other supercritical fluids are nitrogen protoxide ($T_c = 36°C$, $P_c = 71$ bar), ammonia ($T_c = 132°C$, $P_c = 115$ bar) and water ($T_c = 374°C$, $P_c = 217$ bar). The supercritical fluids can be removed at low temperatures, without any toxic wastes but the necessary high pressure can be dangerous in industrial applications. The extractors are tubular devices, pressure resistant where the (semi) solid sample is placed.

There are two experimental techniques:
- *off-line*, by depressing the supercritical fluid, which, coming into the gas phase releases the concentrated analytes; it follows the selective extraction step, using the classic solvents
- *on-line*, by the direct analysis of the extract, under pressure, sent to GC or HPLC.

The newer extraction techniques such as SFE, MAE, and ASE are very attractive because they are a lot faster, use much smaller amounts of solvents, and are environmentally friendly techniques. For example, SFE uses carbon dioxide or modified carbon dioxide (e.g., carbon dioxide containing a small amount of an organic solvent known as modifier) for extraction. Carbon dioxide is a nontoxic, nonflammable, and environmentally friendly solvent. Furthermore, the extraction selectivity can be controlled by varying the pressure and temperature of the supercritical fluid and by the addition of modifiers.

Ultrasound extraction (sonication) is based on the conversion of AC current at 50/60 Hz into electrical energy at 20 kHz and its transformation in mechanical vibrations. Due to the cavity, microscopical vapor bubbles are formed and, after implosion, they produce strong shockwaves into the sample. For isolating the (semi)volatile organic compounds, the liquid-liquid ultrasound technique is applied to samples such as soils, sediments, coal, etc. The process is also useful for the biological materials destruction (Loconto, 2001). Sonication extraction is faster than Soxhlet extraction (30-60 min per sample) and allows extraction of a large amount of sample with a relatively low cost, but it still uses about as much solvent as Soxhlet extraction, is labor intensive, and filtration is required after extraction.

3.2.3. Derivative Techniques

These techniques contain a step where the chemical structure of the components is affected, for a better identification and increased separation efficiency. A good derivative agent must have:

- selective reactions, with certain functional groups;
- high reaction rate;
- good reaction efficiency;
- stabile reaction products, easily detectable.

The ideal agents are selective and non-toxic, have very small retention times, high efficiency and no interferences. E.g. the derivative process of chlorophenols into their pentaflorinebenzoles gives a higher sensitivity and selectivity for GC on capillary columns with electrons capture detectors (Patnaik, 1996).

4. Sampling and Sample Pretreatment for Inorganic and Organic Pollutants Determination in the Romanian Black Seacoast Ecosystem

In order to establish the accumulation factors of inorganic and organic pollutants in the Romanian Black seacoast ecosystem, water, sediment and biota samples were collected along the coast and pretreated in order to measure pollutants concentrations as follows.

4.1. SAMPLING FOR INORGANIC POLLUTANTS ANALYSIS

4.1.1. *Water Samples*

Water samples were filtered on quantitative filter paper and stored at 4^0C in plastic bottles. Typically, an aliquot of 500mL of a sample was adjusted to pH 5 by proper additions of diluted CH_3COOH or NaOH. After adjusting the pH, 5mL of ammonium pyrrolidine dithiocarbamate (APDC) 1% was added to the sample, which then was placed in a separating funnel and manually stirred for a few minutes. The addition of 10mL of methyl isobutyl ketone (MIBK) was used for the separation (extraction) process. The organic extracts were made up to 25mL in a volumetric flask and finally analysed for metals concentrations using an atomic absorption spectrometer (Guguta et al., 2004).

4.1.2. *Solid Samples (Sediment and Biota)*

The sediment and biota samples were carefully washed with distilled water, dried and then mortared. Solid samples (0.1–0.9 grams) were digested with Digesdahl device Hach by adding 8 mL of 65% nitric acid suprapure (provided by Merck, Germany) and 10 mL hydrogen peroxide (provided by Euromedica, Romania) at 170^0C. The obtained solutions have been filtered into 50 mL glass bottles and stored in plastic bottles. The metal concentrations were determined using atomic absorption spectrometry (Chirila et al., 2004).

4.2. SAMPLING FOR ORGANIC POLLUTANTS ANALYSIS

4.2.1. *Water Samples*

All of the surface seawater samples were collected in 1L glass bottles capped with glass caps, from different sites on the Black Sea Coast: The collection of the seawater samples was made from the surface, approximately to 10 m from the coast. Samples were filtered to remove particulate material and kept refrigerated at 4°C away from light prior to extraction, which was done within 24h. The extracts were analyzed before two weeks of collection.

All of the water samples were handled by liquid-liquid extraction (LLE) with hexane in the separation funnel. All glassware was thoroughly cleaned and rinsed with acetone and then hexane before use. The extracts were concentrated by rotary evaporator and then the sulfur-containing compounds were removed with elemental copper using ultrasonication bath.

In order to separate the aliphatic and aromatic fractions, a liquot of the extract was applied to 5 g of activated aluminum oxide and silica-gel column, respectively 5 g of activated Florisil column for pesticides, topped with 1 cm of anhydrous sodium sulfate, which was pre-washed with n-hexane as described in EPA method 8270C and 3600C. The columns were eluted with n-hexane-dichlormethane (3:1). Each fraction was concentrate to 1 mL using the Kuderna–Danish concentrator. The concentrated aliquot was blown down with nitrogen, the internal standards (9, 10 dihidroanthracene and trichlorobiphenyl) were added, and the final volume was injected in a HP5890 gas chromatograph equipped with an electron capture detector (ECD) for oregano chlorine pesticides (OCPs) analysis (1µL) or interfaced to a HP5972 mass selective detector for PAH analysis (0.5µL), as showed by Dobrinas et al. in 2004a.

4.2.2. Solid Samples (Biota)

Different species of mussels and fishes were carefully collected, wrapped in polyethylene bags and frozen at 4^0C.

For organic pollutants extraction, a 5-10 g amount of fish and mussel liver was homogenized with anhydrous sodium sulfate and extracted with n-hexane in a Soxhlet device. The extracts were filtered and concentrated by rotary evaporator (Dobrinas et al., 2004b).

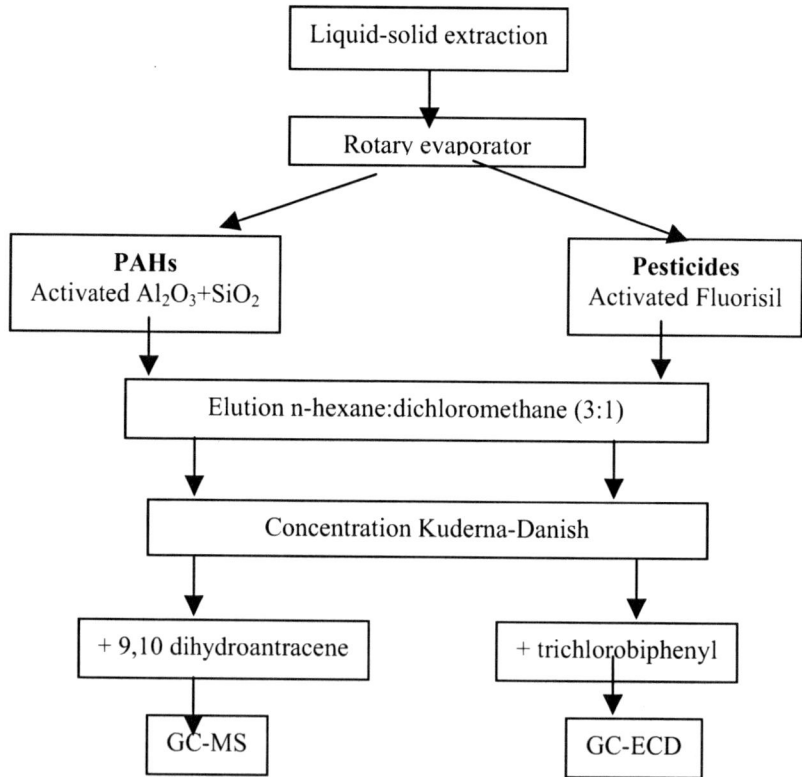

FIGURE 5. Biota samples preparation.

Then the obtained extracts were prepared for PAHs and OCPs determination as above. Figure 5 presents the preparation of samples in order to measure organic pollutants from biota.

References

Akatsuka, K., Yoshida, Y., Nobuyama, N., Hoshi, S., Nakamura, S. and Kato, T., 1998, Preconcentration of Trace Cadmium from Seawater Using a Dynamically Coated Column of Quaternary Ammonium Salt on C_{18}-bonded Silica Gel and Determination by Graphite-furnace Atomic Absorption Spectrometry", *Analytical Sciences*, 14, 529–533.

Anderson, K.A., 1999, *Analytical Techniques for Inorganic Contaminants*, AOAC International.

Anezaki, K., Chen, X., Ogasawara, T., Nukatsuka, I. and Ohzeki, K., 1998, Determination of Cadmium and Lead in Tap Water by Graphite-furnace Atomic Absorption Spectrometry After Preconcentration on a Finely Divided Ion-exchange Resin as the Pyrrolidinedithiocarbamate Complexes, *Analytical Sciences*, 14, 523–527.

Atanassova, D., Stefanova, V. and Russeva, E., 1998, Co-precipitative Pre-concentration with Sodium diethyldithiocarbamate and ICP-AES Determination of Se, Cu, Pb, Zn, Fe, Co, Ni, Mn, Cr and Cd in Water, *Talanta*, 47, 1237–1234.

Baiulescu, G.E., Dumitrescu, P. and Zugrăvescu, Gh., 1990, *Sampling*, Ellis Horwood London.

Belardi, R. and Pawliszyn J., 1989, The Application of Chemically Modified Fused Silica Fibres in Extraction of Organics from Water Matrix Samples, and their Rapid Transfer to Capillary Column, *Water Pollution Research Journal of Canada*, 24, 179–191.

Chen, H., Jin, J. and Wang, Y., 1997, Flow- injection on-line Coprecipitation-preconcentration System Using Copper (II) diethyldithiocarbamate as Carrier for Flame Atomic Absorption Spectrometric Determination of Cadmium, Lead and Nickel in Environmental Samples, *Analytica Chimica Acta*, 353, 181–188.

Chirila E. and Draghici, C., 2003, Pollutants analysis. I. Water Quality Control, Transilvania University press, Brasov.

Chirila, E., Bavaru, A., Carazeanu, I., 2004, About Accumulation Factors of Cd, Cr, Cu and Pb in the Black Sea Coast Biota, *Rapp. Comm. Int. Mer .Medit*, 37, 179.

Chirila, E., 2004, Sampling, in Colbeck, I, Drăghici, C. and Perniu, D. (Eds), *Polution and Enviromental Monitoring*, The Publishing House of the Romanian Academy, Bucharest, 109–128.

Crompton, T.R., 2001, *Determination of Metals and Anions in Soils, Sediments and Sludges*, Spon Press, Taylor & Francis Group.

Dobrinas S., Birghila, S., Coatu, V., Stoica, A.I. and Baiulescu, G.E., 2004a, Contamination of the Black Sea Coast by Polycyclic Aromatic Hydrocarbons and Organochlorine Pesticides, *Revista de Chimie (Bucharest)*, 55, 483–487.

Dobrinas, S., Coatu, V. and Chirila, E., 2004b, Organochlorine Pesticides and PAHs Compounds Occurrence in Mussel and Fish Liver from Black Sea., *Rapp. Comm. Int. Mer. Medit.*, 37, 188.

Ferreira, S.L.C., Ferreira, J.R., Danlas, A.F., Lemos, V.A., Araujo, N.M.L. and Spinola Costa, A.C., 2000, Copper Determination in Natural Water Samples by Using FAAS

After Preconcentration onto Amberlite XAD-2 Loaded with Calmangite, *Talanta*, 50, 1253–1259.

Guguta, C., Draghici, C., Chirila, E. and Carazeanu, I., 2004, Occurrence of Some Metals in Different Marine Water by FAAS, *Ovidius University Annals of Chemistry*, 15, 39–42.

Kramer, K.J.M., 1995, Quality Assurance of Sampling and Sample Handling for Trace Metal Analysis in Aquatic Biota, în *Quality Assurance in Environmental Monitoring*, Quevauviller, Ph., (Ed), VCH.

Lewis, R.G. and Gordon, S.M., 1996, Sampling for Organic Chemicals in Air, in Keith, L., (Ed.) *Principles of Environmental Sampling*, 2nd Edition, ACS.

Loconto, P.R., 2001, *Trace Environmental Quantitative Analysis*, Marcel Dekker Inc. Ed.

Lopez-Avila, V., Young, R. and Beckert, W., 1994, Microwave-assisted Extraction of Organic Compounds from Standard Reference Soils and Sediments, *Analytical Chemistry* 66(7): 1097–1106.

Luque de Castro, M.D. and Garcia-Ayuso, L.E., 1998, Soxhlet Extraction of Solid Materials: an Outdated Technique with a Promising Innovative Future, *Analytica Chimica Acta* 369, 1–10.

Niessner, R., 1993, Sampling Techniques for Air Pollutants, in Barcelo, D. (Ed), *Environmental Analysis. Techniques, Applications and Quality Assurance*, pp. 3–22, Elsevier.

Onuska, F. and Terry, K., 1993, Extraction of Pesticides from Sediments Using Microwave Technique, *Chromatographia*, 36, 191–194.

Parr, J., Bollinger, M., Callaway, O. and Carlberg, K., 1988, Preservation Techniques for Organic and Inorganic Compounds in Water Samples, in Keith, L., (Ed.), *Principles of Environmental Sampling*, 2nd Edition, ACS.

Patnaik, P., 1996, *Handbook of Environmental Analysis*, CRC, Boca Raton.

Pawliszyn, J., 1997, *Solid Phase Microextraction: Theory and Practice*. New York: Wiley-VCH.

Poole, C.F. and Poole, S.K., 1996, Trends in Extraction of Semivolatile Compounds from Solids for Environmental Analysis, *Analytical Communications*, 33, 11H-14H.

Poole, S.K., Dean, T.A., Oudsema, J.W. and Poole, C.F., 1990, Sample Preparation for Chromatographic Separations: an Overview, *Analytica Chimica Acta*, 236, 3–42.

Santelli, E.R., Gallego, M., and Valcarcel, M., 1994, Preconcentration and Atomic Adetermination of Copper Traces in Waters by On-line Adsorbtion-elution on an Activated Carbon Minicolumn, *Talanta*, 41, 817–823.

LESSONS LEARNED FROM INDUSTRIAL CHEMICAL ACCIDENTS: ITALIAN AND INTERNATIONAL INITIATIVES

SARA VISENTIN
U.O. International Centre for Pesticides and Health Risk
Prevention, Ospedale e Polo Universitario Luigi Sacco
Via G. B. Grassi 74, 20157 Milano
E-mail: sara.visentin@icps.it

Abstract. Chemicals dangerous for the environment and human health can be released in the environment due to several reasons: accidental spills (storage or transport conditions) or uncontrolled processes at high temperature and pressure, which often lead to or originate from fires or explosions. The focus of this paper is mainly on chemical accidents of industrial origin. Some of the current activities and initiatives undertaken at the international, European and Italian mainly concerning databases on chemicals, accidents and industries which have the potential to pose concern to the environmental and human safety are summarized. It is concluded that sharing of information, dissemination of lessoned learned and implemented; networking among authorities and experts as well as the harmonization of instruments (enforcement measures and educational activities) represent important factors to improve prevention strategies.

Keywords: industrial accidents; hazardous chemicals; database; international initiatives; environmental pollution; lessons learned

1. Introduction

In general terms a chemical accident is the result of an uncontrolled release into the external environment of one or more chemical compounds that are hazardous to health of harmful to property or the environment. The risks from a chemical accident largely depend on the properties of the compound itself, the quantities being handled, the process used, and the nature of the accident or the release, as well as the surroundings (including meteorological conditions), population density, and the emergency response measures taken to minimize the consequences. Chemical accidents may result from a fire, an explosion or a leak that leads to a sudden release of hazardous substances. They may occur during manufacture, storage, transport (by rail, road, air, water and pipeline), use and disposal. The causes of chemical accidents are quite diverse. They may be related to a process, but may also arise from machinery failure or human error. For example, a reactor vessel may rupture owing to wrong temperature and/or pressure, leading to the unintentional production and release of hazardous compounds. Gas releases may result from the accidental mixing of chemicals. Fires may lead to the formation of toxic compounds that are released into the environment as a result of the fire or the efforts to put it out (UNEP, 1992).

The World Health Organization (WHO, 2001) defines chemical incidents as accidental or intentional events that threaten to expose or do expose responders and/or members of the public to a chemical hazard. Incidents can be sudden and acute, when hazardous chemicals are 'overtly' released into the environment. Incidents can also have an apparently slow onset, when there is a 'silent' release. The result could be substantial pollution of the immediate neighborhood, which might have health effect similar to the long-term consequences of short-term, major accidents.

External factors affect the probability of an accident. In fixed installations (such as chemical plants, storage rooms, warehouses and reprocessing sites) natural events such as floods or extreme weather as well as, for example, power cuts may lead to an uncontrolled release of chemicals. The risk of chemical accidents during road transport depends on factors such as traffic density, road conditions and speed limits; the effects depend on where an accident occurs (WHO, 1995).

In principle, accidental releases of hazardous chemicals may affect all environmental media (air, water and soil) and consequently the food-chain, independent of the type of release and the environmental medium initially affected. Explosions or leaks of inflammable materials can release quantities of toxic chemicals into the atmosphere, the distribution of which depends on the height of the plume and the meteorological conditions. The chemicals released may subsequently contaminate land, water, food crops or livestock. While most chemical accidents occurring in fixed installations tend to be contained within the industrial compound itself, the effects generated by major chemical accidents may cross national borders.

There is a wide range of issues associated with chemical accident prevention, preparedness, and response, including reporting and follow-up activities. Since acquiring knowledge is the essential prerequisite for planning actions for adequate strategies, the present paper will summarize some of the current activities and initiatives undertaken at the international, European and Italian level with emphasis on the available European legislation and existing databases on chemicals, accidents and industries which have the potential to pose concern to the environmental and human safety. Lessons can be learned from the knowledge gained from investigation, study or other activities in regard to the technical, behavioral, cultural, management, or other factors that led, could have led, or contributed to the occurrence of an accident. In order to translate lessons learned into effective practice, attention should be also paid to their effective dissemination. Sharing of information, dissemination of lessons learned and implemented; networking among governmental authorities and experts as well as the harmonization of instruments, such as enforcement measures and educational activities, represents important factors to improve strategies in this field (OECD, 2005).

2. Historical Accidents Leading to Current Practices for Managing Process Safety

Table 1 summarizes the catastrophic accidents that gave rise to the lessons learned that prompted different countries to adopt their present regulatory and industry measures for managing process safety (OECD, 2005).

The Bhopal tragedy reinforced by the 1989 Phillips accident were needed before the USA adopted regulatory measures such as the OSHA Process Management Standard (PSM) and EPA Risk Management Program (RMP) and industry trade associations adopted the Canadian Responsible Care program. The terrible accident at Flixborough led to major changes in the UK regulations. In the following paragraph the dramatic ICMESA accident (the "Seveso accident") is described since it prompted the attention on the need of the prevention of such disasters in Europe and led to the Seveso Directive 82/501/CEE (CEC, 1982). After recent major catastrophic events (Baia Mare, Enschede, and Toulouse), further improvements were incorporated into subsequent Seveso Directives (CEC, 1996, 2003) and various country regulations, which form the foundation of the European policy on the subject.

3. The Seveso Accident

3.1. THE INSTALLATIONS IN QUESTION

The Seveso accident is summarized from the dossiers "Seveso 20 years after" (FLA, 1998) and from the report "Seveso, twenty years after" (La Roche, Ltd,1997) which include several details about the event and

TABLE 1. Some Major Chemical Accident Events in the last Quarter Century (adapted from OECD, 2005).

Location of Accident	Date	Type of Event	Some Resulting Consequences	Regulatory Response
Flixborough, UK	1974	Explosion and fire	28 killed, over 100 injured	COMAH 1984
Seveso, Italy	1976	Runaway reaction	Large Dioxin environment contamination massive evacuations, Large animal kill	Initial Seveso Directive
Bhopal, India	1984	Runaway MIC reaction	≈ 2500 people killed and 100,000 injured, high litigation costs	USA Emerg. Planning & Community Right to know Act- CMA CAER Program
Basel, Switzerland	1986	Warehouse Fire	Massive contamination of Rhine and very large fish kill	Changes in Seveso Directive
Pasadena, USA	1989	Explosion and fire	23 deaths, ≈ 100 injured Over $1 billion in losses	Triggered 1990 USA CAAct & RMP & PSM process Stds
Longford, Victoria, Australia	1998	Explosions and fires	Two deaths, gas supply to Melbourne cut for 19 days. Losses over $1.3 Billion	Process Regulatory initiatives Victoria
Enschede, The Netherlands	2000	Explosion and fire	22 deaths, ≈ 1000 injured, 350 houses and factories destroyed	Changes in Seveso Directive
Toulouse, France (Oppau & Texas City)	2001	Explosion and fire	30 deaths, ≈ 2000 injured, 600 homes destroyed, 2 schools demolished	Changes in Seveso Directive

response measures as well as follow-up programmes undertaken for the investigation, recovery and monitoring of the area.

The ICMESA was a chemical plant (Industrie Chimiche Meda Società), of 170 workers, property of the firm Givaudan S.A. in Geneva, located 15 km North of Milan. It produced intermediate compounds for the cosmetic and pharmaceutical industry among which 2,4,5 trichlorophenol (TCP), a toxic inflammable compound used for the chemical synthesis of herbicides. At the plant, TCP was generally produced at 150-160°C by means of an exothermic thermostat reaction.

3.1.1. *The Accident*

On Friday 9th of July
16:00   The TCP reaction vessel is filled with the various starting materials.

On the 10th of July 1976,
2:30   According to the temperature diagram the reaction is completed.
04:45   The foreman in charge gives the order to interrupt a distillation which is not completed. The heating is turned off and the vessel contents mixed for a further 15 minutes. The last measured temperature is 158°C. (water cooling is not turned on, because the reaction is believed to be finished).
06:00   The night shift is over. The workers leave the factory, and only the cleaning and maintenance crew remains behind.
12:37   The rupture disk in the safety valve burst a result of excessive pressure, caused by an exothermic reaction in the TCP vessel. A chemical mixture in the form of an aerosol cloud escapes into the air in a south easterly direction.
13:00   A foreman present at the plant calls the deputy head of production, who arrives at the plant then minutes later.
13:45   A foreman turns on the cooling system, thereby stopping the escape of the mixture.

3.1.2. *Consequences*

A mixture of several different pollutants (about 3000 kg) including dioxin, was released into the atmosphere. As for the dioxin content (confirmed on July, 17) in the toxic cloud, technical literature reports different evaluations, ranging from 300 g to 130 kg. Quite uncommonly for that time of year, a 5 m/sec wind was blowing; the cloud contaminated the ground according to a linear path for about 6 km, south-eastward from ICMESA.

July, 10. ICMESA warned the inhabitants not to eat any local vegetables or fruits. On July 13 a few small animals died and on the 14 the first signs of skin irritation occurred in children.

Evacuation of 736 persons leaving in the most polluted zone. Day time evacuation of pregnant women and children of less contaminated part (26 July-August 2).

By 1978 some 77.000 animals were slaughtered.

Abortion: In view of the risk of giving birth to a deformed child, some women decode to have an abortion (August).

Long term health effects: follow-up studies showed the increase of cancer incidence attributable to dioxin exposure.

3.1.3. Action Taken

Institutional measures

In collaboration with Regione Lombardia, the Italian Ministry of Health formed the Technical-Scientific Central Committee with the task to plan immediate measures and establish clean-up procedures. Four Commissions were created to monitor the exposed population and animals, perform analyses, plan clean-up strategies. The final programmes included four areas of intervention:

1. the assessment of the extent of the pollution in soil, water, and vegetation to plan measures for decontamination and recovery of both soil and buildings, which could also prevent further dispersion of contaminants;
2. health care measures for animals and population exposed;
3. social assistance, including the provision of shelters for evacuated people;
4. the recovery of public buildings, facilities and agricultural soils to the conditions prior to the accident;

In order to perform these activities in close coordination, the Seveso Special Office, with 70 employees was established in 1977.

Assessment and management of the environmental damage

The whole polluted area was classified in three zones (A, B and R) by the detected level of dioxin (Table 2).

Boundary values were established following preexisting natural or artificial borders, in accordance with the contamination pattern. Since zone A had a wide range of concentration measures, smaller zones (A1-A8) were identified with a more homogeneous dioxin values. Experimental research showed TCDD was highly persistent in soil and air and it was not revealed to undergo microbial degradation. The monitoring of dioxin distribution in the upper soil layer over 17 months proved that the substance decreased to 50% in the first 5 months, due to photodecomposition, and then it tended to balance. This information drove all of the subsequent recovery activities. In zone A, the entire 40 cm top soil layer was removed and stored in two 300,000 m³ special controlled burrows located nearby the accident site. Zone A was completely recovered in 1977. Part of the former A area (sub-zones A1-A5) was subsequently by a park, the Bosco delle Querce.

TABLE 2. Surface, number of inhabitants and levels of dioxin in zone A, B and R.

Zone	Surface (ha)	Inhabitants	TCDD concentration ($\mu g/m^2$)	Average boundary value ($\mu g/m^2$)
A	83.7	706	580.4 – 15.5	
A/B				50
B	269.4	4613	4.3 – 1.7	
B/R				5
R	1430	30774	1.4 – 0.9	

The A6 and A7 hosted about 67% of the evacuated population. The total extension was 32 hectares; the minimum distance from ICMESA was 1200 m. TCDD levels in the soil were about 270 $\mu g/m^2$. In this case, the removal of 25 cm of soil decreased of about 90% the level of TCDD.

Rehabilitation of zones B and R started in 1977. Ploughing and mixing 7 cm soil layer increased photodecomposition of TCDD, whose levels dropped down considerably.

Assessment of exposure and risk for residents

Following the indications of the regional Medical-Epidemiological Commission, all inhabitants of highly contaminated zone A were evacuated. The trigger dioxin concentration in soil was than 50 $\mu g/m^2$. In zone B, a typical agricultural-industrial setting, the contamination levels were lower than 5 $\mu g/m^2$. In zone R, they were lower than 2 $\mu g/m^2$. Both B and R zones were given strict sanitary rules to follow, including forbearance to eat local agricultural products and poultry or other animals and the daytime evacuation of children and pregnant women (Table 3).

In 1984, a new risk assessment was carried out for zone B inhabitants. Compared with that of the emergency period, the new assessment made use of better, more reliable, analytical information. The following exposure routes were considered:

- casual ingestion of soil
- absorption from dermal contact with soil
- inhalation of contaminated dust

- ingestion of drinkable water
- ingestion of vegetables grown in home gardens
- consumption of animal products (primarily chickens and rabbits) from the area.

TABLE 3. Measures designed to limit human exposure in zones B and R.

ZONE B RESIDENTS
Intensification of personal hygiene
No animal breeding or vegetable planting
Daily relocation of children up to 12 years and pregnant women
Abstention from procreation
Minimization of air dust level (vehicle speed limit, 30 km/hr)
Careful emptying of vacuum cleaners
Hunting prohibited for approximately 8 years
ZONE R RESIDENTS
Intensification of personal hygiene
No animal breeding or vegetable planting
Pets to be fed with food from areas other than zones A, B and R
Hunting prohibited for approximately 8 years

Moreover, in order to achieve more accurate results on risk assessment, further parameters were used, such as TCDD transmigration index from soil to vegetation and vegetables, residents' nutritional habits, including the consumption of eggs and milk, which were not considered in previous analyses.

Sanitary investigations

The programmes for sanitary investigations were developed according to three lines:

- Systematic longitudinal sanitary control of people exposed or at-risk, in order to guarantee secondary prevention and possible therapy; several departments specialized in pediatrics, internal medicine, gynecology, neonatology, dermatology, occupational medicine, ophthalmology, neurology, immunology, genetics and laboratories were involved.
- epidemiological surveillance of inhabitants, to identify incidence and prevalence of different pathologies and of rare events.
- laboratory research on TCDD effects, to contribute to a proper planning of the surveillance program. Since TCDD exposure was only indirectly assessable (according to residence area) at the time of the accident, that is a direct method of assessing TCDD in the serum was not available,

tests with altered metabolism values would have been of extreme importance. More than 20 laboratory tests were performed involving analysis on the hepatic, lung, bone and immune system functions, on lipid metabolism, on the nerve systems, as well as studies on the ratio of spontaneous abortions and on the presence of congenital malformation events. Tests were performed on a sample of about 17,000 people in 1976-1984.

Following these investigations, an International Steering Committee was created. The need of additional investigations on possible long-time effects was stressed.

In 1987, researchers from the Center for Disease Control in Atlanta (USA), updated a methodology to assess dioxin concentration in the serum. Serum sample frozen at $-20°C$ in 1976 were analyzed and results confirmed the high exposure of inhabitants, especially in the A zone, with sometimes extremely high TCDD concentrations.

Epidemiological investigations showed no difference in congenital malformation rates between exposed and unexposed populations, though there had been a possible increase in malignancy after 10 year from the accident (Bertazzi P.A. et al., 1989). A 20 year follow-up showed a clear and consistent excess of lymphohemopoietic neoplasms in both genders. Moreover, an overall increase in diabetes was reported notably among women. Chronic and respiratory diseases were moderately increased suggesting a link with accident-related stressors and chemical exposure (Bertazzi P.A. et al., 2001)

The disposal of toxic waste

After first emergency interventions, the main concern was the disposal of toxic waste from the ICMESA exploded reactor. In the spring of 1982, the vessel was emptied under strict safety precautions. The 41 barrels containing residues were transferred to destination. On the other hand, materials from recovery interventions, building demolition and soil removal were stored in two burrows placed in the towns of Seveso (tank A) and Meda (tank B). Tank A hosted material from the Seveso area and also ruins from ICMESA plant, for a total of about 200,000 m^3. Tank B, located near the Certesa stream, hosted materials from the contaminated area placed north of ICMESA and TCDD contaminated mud originating from the Seveso purification plant. The total volume of material was about 80,000 m^3.

3.1.4. *Lessons Learned*

Poor involvement of high executives in risk management is often mentioned as a factor contributing to higher levels of risk. More involvement of high ranked officials in routine risk management should allow for a stronger control of intermediate management decisions. This was one of the main criticisms of the Seveso accident, where the directors of the corporation ultimately responsible were unfamiliar with the hazards.

Owing to the lack of knowledge about the impact on man and animals of the barely perceptible substance in the polluted cloud, it was some time before a definitive assessment of the damage could be made, thus naturally exacerbating the anxiety and concern in the population affected.

Zoning regulation is certainly one of the oldest and most obvious forms of risk reduction. By splitting the territory and isolating hazardous plants, public exposure to risk is diminished. One of the main problems at Seveso was that several municipalities extended up to the plant's boundary.

4. The European Legislation

The Seveso event enlightened the fact that contemporary legislation did not protect sufficiently workers and the general population from chemical risk of industrial origin. The European Community to formulate Directive 82/501/CEE of June 24, 1982 on industrial activities at relevant risk of accident, known as the "Seveso Directive". The aim was to reduce the source and incidence of technological risks by analyzing the causes, inspecting the establishments and creating an accident prevention system, which could assure internal and external safety of industrial sites. The directive defines:

- "industrial activity", an establishment in which dangerous substances are handled or stored, including their transport inside the plant;
- "producer", anyone who carries on an industrial activity, thus introducing a wide definition to prevent elusion of responsibility on the basis of functions delegations;
- "major accident" an occurrence such as a major emission, fire, or explosion resulting from uncontrolled developments in the course of the operation of any establishment and leading to serious danger to human health and/or the environment, immediate or delayed, inside or outside the establishment, and involving one or more dangerous substances;
- "dangerous substance" means a substance, mixture or preparation classified according to the kind of plant in which it is treated or stored and to the dangerousness of single industrial activities (annexes II, III and IV).

The directive reported the duties of producers and administrative bodies. Producers were required to attain every safety measure to prevent accidents, reporting the types and amounts of chemicals stored or produced and other factors such as information about the plant's location, kind of process, plant type, personnel exposed to risk, safety devices and existing emergency measures). Public administration would first evaluate the authorization of new plants, receive notifications, process information, and verify the suitability of emergency and safety plans.

The subsequent new directive no. 96/82/CE, strengthened the system of control measures widening the extent of its application and encouraging the exchange of information among Member States. Every plant having enough quantities of hazardous substances to be considered at risk of major accident. The distinction between storage and production of hazardous substances was removed. Notably, attention was paid to the management of risk and accident, since it was recognized that several accidents were attributable to management mistakes. Definitions are significantly modified: there was no reference to industrial activity, while definitions for "establishment" and "plant" were given.

In the light of recent industrial accidents (Toulouse, Baia Mare and Enschede) and studies on carcinogens and substances dangerous for the environment, the Seveso II Directive was extended by the Directive 2003/105/EC. The most important extensions of the scope of that Directive were to cover risks arising from storage and processing activities in mining, from pyrotechnic and explosive substances and from the storage of ammonium nitrate and ammonium nitrate-based fertilizers.

The Seveso Directive is administered by the European Union through the Major Accident Hazards Bureau services (MAHB) located within the Joint Research Centre (JRC) in Ispra, Italy. The Bureau also oversees the European Community's Documentation Centre Industrial Risk (CDCIR) and manages the Major Accident Reporting System (MARS) with the aim to create a repository of information and facilitate the exchange between the members of the European Community. MARS follows the requirements of the Seveso II directive and collects information about major chemical incidents as well as the response and results. Member States are required to report the events by using standardized forms.

The network known as IMPEL, which stands for the "European Union Network for the Implementation and Enforcement of Environmental Law", was created in 1992 in order to encourage the exchange of information and the comparison of personal experience, and to facilitate a more coherent approach as regards the implementation, application and monitoring of environmental law. Thirty countries - all Member States of the European Union, the two acceding countries Bulgaria and Romania, the two candidate countries Croatia and Turkey as well as Norway - and the European Commission now participate in the network. One of the IMPEL multi-annual projects led by France regards the organization of workshops/seminars to exchange and share experiences on lessons learned from accidents. Reports since 1999 are available at URL http://europe.eu.int/comm/environment/impel.

5. International Initiatives

The Inter-Organization Programme for the Sound Management of Chemicals (IOMC) was established in 1995 following recommendations made by the 1992 UN Conference on Environment and Development to strengthen co-operation and increase international co-ordination in the field of chemical safety. Purpose of IOMC is to promote co-ordination of the policies and activities pursued by the Participating Organizations, jointly or separately, to achieve the sound management of chemicals in relation to human health and the environment. Participating organizations are: the Food and Agriculture Organization of the United Nations (FAO), the International Labour Organization (ILO), the Organization for Economic Co-operation and Development (OECD), the United Nations Environment Programme (UNEP); the United Nations Industrial Development Organization (UNIDO); the United Nations Institute for Training and Research (UNITAR); the World Health Organization (WHO); observer organizations are the United Nations Development Programme (UNDP and World Bank. Purpose of IOMC is to promote co-ordination of the policies and activities pursued by the Participating Organizations, jointly or separately, to achieve the sound management of chemicals in relation to human health and the environment.

A detailed inventory of IOMC projects by lead organization is available at the U.R.L. http://www.who.int/iomc/en/. Among them, the International Programme on Chemical Safety (WHO/IPCS) leads the activity "Alert & Response Mechanisms for Chemical Accidents" in cooperation WHO Member States, WHO Regional Offices IPCS Participating Institutes, European Commission, and UK Health Protection Agency. The purpose is to strengthen systems for surveillance, emergency preparedness & response and to consider the inclusion of biological, chemical or radiological events of 'international concern' in a revision of the International Health Regulations (2005). Specific outputs include the set up of a System for Global Chemical Incident Alert & Response and of a database of global chemical incidents of public health significance. IPCS activities in this respect started in 2002 with the creation of a joint operation center with the existing WHO Global Alert and Response (GAR) team for infectious diseases. The outbreak verification team screens information about disease outbreaks of potential international concern received from a wide range of sources, including the Global Outbreak Alert and Response Network (GOARN), ChemiNet, the Global Public Health Information Network

(GPHIN), WHO regional and country representatives, official government sources, WHO Collaborating Centres, non-governmental organizations, inter-governmental organizations, news media, eyewitnesses, and others. These outbreaks may be of chemical, biological, radiological or unknown origin. The team carries out a risk assessment to determine whether there is a need to alert the government concerned and whether assistance should be offered in response to the outbreak. In the same year, IPCS (WHO, 2003) started to compile a database of global chemical incidents database to identify sentinel events and provide alerts, describing the public health consequences resulting from acute incidents and providing a mechanism for capacity strengthening. This database is compiled from various sources and includes details of: the date the incident occurred; the location and type of incident; the chemical(s) released; the public health impact of the incident; the public health action taken; and whether the event met the revised International Health Regulations criteria for an event of potential international public health concern. During the first phase of this work, from 1 August 2002 to 30 April 2003, approximately 25,000 events were scrutinized: of these, 364 (1.5%) were identified as being eligible for inclusion in the global database. Of the 364 events, 27 (7.4%) met the criteria for chemical incidents of potential international concern.

Among other initiatives, IPCS provides information on chemicals, diagnosis and treatment through the IPCS INCHEM database available on the web free of charge at http://www.inchem.org/. IPCS INCHEM is a means of rapid access to internationally peer reviewed information on chemicals commonly used throughout the world, which may also occur as contaminants in the environment and food. Another essential source of toxicological information on chemicals is the IPCS INTOX database (available free of charge at http://www.intox.org/databank/index.htm). It includes the following internationally peer-reviewed documents: Poisons Information Monographs, Treatment Guides Antidote Monographs, Environmental Health Criteria Monographs, International Chemical Safety Cards, WHO/FAO Pesticide Data Sheets and a worldwide list of poison centers.

A valuable initiative has been undertaken by the Center for Research on the Epidemiology of Disasters (CRED, a World health Organization Collaborating Center since 1980, supporting the WHO Global Programme for Emergency Preparedness and Response) and the Office of U.S. Foreign Disaster Assistance (OFDA). The initiative responds to the need for complete and verified data on disasters and their human and economic impact, by making available a specialized, validated database (EM-DAT) with the purpose to facilitate preparedness, reducing vulnerability and improving management. EM-DAT is searchable free of charge at http://www.em-dat.net/ and contains essential core data on the occurrence and effects of over 12,800 mass disasters in the world from 1900 to present. The database is compiled from various sources, including UN agencies, non-governmental organizations, insurance companies, research institutes and press agencies. Disasters covered by EMDAT are of natural as well as technological origin. A disaster is entered in the database if matching at least one of the following criteria: 10 or more people reported killed; 100 people reported affected; declaration of a state of emergency; call for international assistance.

The programme "Awareness and Preparedness for Emergencies at Local Level (APELL)" was instituted since 1987 within the United Nations Environment Programme (UNEP). Its purpose is to enable governments, in co-operation with industry, to work with local leaders to identify the potential hazards in their communities and to prepare measures to respond and control emergencies which might threaten public health, safety and the environment. The programme aims to provide the community with information about APELL by giving all stakeholders (community, industry and local government) handbooks and technical reports and promoting seminars and workshops for comprehensive understanding of needs, priorities and possible actions which should be pursued for the implementation of APELL. The APELL database of disasters was originally initiated by OECD (time span 1970-1997). UNEP, with OECD's permission, extended the list to cover the years 1990-1997. Several institutions and organizations provided the information (OECD, MHIDAS, TNO, SEI, UBA-Handbuch Stoerfaelle, SIGMA, Press Reports, UNEP, BARPI). Agreed inclusion criteria are: 25 death or more; or 125 injured or more; 10000 evacuated or more; or 10 thousand people or more deprived of water. The list includes natural and technological disasters.

Among the availability of different international database on chemical incidents it is worth mentioning the ARIA database maintained by the Bureau d'Analyses des Risques et des Pollutions Industrielles (BARPI) of the French Ministry of the Environment (searchable free of charge at http://aria.ecologie.gouv.fr/index2.html) covering the period 1992-2003 and the UK Major Hazard Incident Data Service (MHIDAS) covering the last 20 years and earliest most significant incidents. Its use is on a fee basis (http://www.hse.gov.uk/infoserv/mhidas.htm).

6. Italy: State of the Art and Perspectives

The "Seveso Directive", along with its later modifications, has been adopted in Italy, even if with about 4 years delay, through DPR 175/1988, which has established rules for the control of industrial activities under relevant accident risk. Seveso II entered in force in 1996 with the DL

334/99. The Seveso chapter has been approved in September 2006 and its publication in the Italian Official journal will be the next step before its application.

The main responsible Italian authorities are: the Ministry of the Environment (prevention); Ministry of Defence for Civil Protection and Disaster Relief (response); Ministry of Health (urgent medical care) and Ministry of the Interior (for transport issues).

The national bodies involved in the assessment of the risk from chemical accidents are: the National Institute for Public Health (ISS); the National Institute for workers' safety (ISPESL) and the National Agency for the Protection of the Environment. The last agency prepares periodic scientific reports on the risk from chemical accidents in Italy and maintains the database on industrial accidents on behalf of the Ministry of the environment that holds the inventory of the plants at risk of chemical accidents.

Another source of data on chemical accidents is the National Fire Brigades, whose database has inside the information on several chemical industrial accidents.

A project of the Ministry of the Environment, the agency for the protection of the environment and the Fire Brigades in under development to provide a common database on accidents compatible with MARS. Results of a pilot phase show the importance of this initiative, since it aims at reducing duplicate efforts and is toward the international harmonization of adequate tools for planning preventing activities.

7. Conclusion

The paper reports some of the initiatives undertaken in the field of chemical accidents of industrial origin. The accent was on the necessity to share available information between countries about the occurrence of the events and lessons learned. It is recognized that there is still the need to improve the harmonization of reporting tools towards the homogenization of data collection. The issue of possible transboundary effects of chemical accidents has not been covered in this report. However, a number of reporting and alert systems have been established in Conventions signed by representatives of several countries in Europe and world-wide.

Although not deeply discussed in this paper, preventive strategies may comprise the mapping of the plants at risks following legislative requirements and the training on chemical emergencies to different audiences: general population, medical officers, and plant managers. Training may include: information on chemical hazard and risk assessment for humans and the environment. The difference between the concepts of hazard and risk should be communicated and clarified to the general population and to managers in order to gain adequate knowledge as prerequisite of preventing and response strategies.

References

APAT, (2002), *Mappatura del rischio industriale in Italia*, Agenzia per la protezione dell'ambiente e per i servizi tecnici, Roma.

Bertazzi P.A., Consonni D., Bachetti S, Rubagotti M, Baccarelli A, Zocchetti C, Pesatori A, (2001), *Health effects of Dioxin Exposure: a 20-year mortality study*, Am J of Epidemiol, 153 (11): 1031–1044.

Bertazzi P.A., Zocchetti C., Pesatori A.C., et al. (1989) *Ten-year mortality study of the population involved in the Seveso incident in 1976*. Am J Epidemiol, 129:1187–200.

CEC (1997) *Council Directive 96/82/EC on the control of major-accident hazards* OJ No L 10 of 14 January 1997.

CEC (2003) *Directive 2003/105/EC of the European Parlament and of the council of 16 December 2003 amending council directive 96/82/EC*, Official Journal, 31 December 2003.

CEC, (1982) *Council Directive 82/501/EC on the major-accident hazards of certain industrial activities* OJ No L 230 of 5 August 1982.

FLA (1998), *Seveso 20 years after. From dioxin to the Oak Wood*, Fondazione Lombardia per l'Ambiente, Dossier N. 33, Milano 1998.

La Roche 1997), Seveso-Twenty Years After, La Roche Ltd, Switzerland, 2nd edition (available at URL www.roche.com/com_hist_1965)

OECD 1991, *The State of the Environment*, Organisation for Economic Cooperation & Development, Paris.

OECD, 2003 *Guiding Principles for Chemical Accident Prevention, Preparedness and Response, Guidance for Industry (including Management and Labour), Public Authorities, Communities, and other Stakeholders*, OECD Environment, Health and Safety Publications Series on Chemical Accidents No. 10.

OECD, 2005 *Report of the OECD Workshop on Lessons Learned from Chemical Accidents and Incidents, 21–23 September 2004, Karlskoga, Sweden*, OECD Environment, Health and Safety Publications in the series on Chemical Accidents, No. 14, ENV/JM/MONO(2005)6.

UNEP (1992) *Hazard identification and evaluation in a local community*, UNEP technical report N. 12, United Nations Environment Programme, Paris.

WHO (2003), Weekly Epidemiological Record, N° 38, 19 September 2003.

WHO, 2001 *Guidance on the Public Health Response to Biological and Chemical Weapons*, WHO, Geneva.

WHO European Centre for Environment and Health (1995), *Concern for Europe's Tomorrow. Health and the Environment in the WHO European Region*, Wissenschaftliche Verlagsgesellschaft mbH, Stuttgart, Germany.

WHO, 2005 *The Fifty-eight World Health Assembly resolution containing the Revised International Health Regulation* (document A58/55), 23 May 2005.

SELECTION OF EFFECTIVE TECHNOLOGIES FOR MANAGEMENT OF CONTAMINATED LANDS

THOMAS MCHUGH* AND JOHN CONNOR
Groundwater Services, Inc.
2211 Norfolk, Suite 1000, Houston, Texas, 77098, USA

*To whom correspondence should be addressed. tcmchugh@gsi-net.com

Abstract. In the past, many remedies implemented at contaminated sites have failed to achieve site remedial action objectives either because not all of the objectives were defined prior to selecting the remedy, or a remedy was selected that was not capable of achieving the site objectives. In order to ensure the selection of effective remedies, this report outline a process for i) the identification of risk-based and non-risk remedial action objectives and ii) the evaluation of potential technologies to identify an acceptable remedy that will achieve all of the site remedial action objectives. By providing for the separation of risk-based and non-risk remedial action objectives, the remedy selection process allows the user to both i) identify a range of remedies that are capable of protecting human health and ecological resources and ii) understand the additional constraints imposed by the non-risk objectives.

Keywords: contaminated land; risk assessment; remedy selection

1. Introduction

Over the next 30 years, it is estimated that $209 billion will be spent on the remediation of 294,000 contaminated sites in the United States (USEPA, 2004). In the past, a large percentage of remediation systems have failed to achieve all of the remedial action objectives either because the people managing the remediation did not understand these objectives or because the selected remediation technology was not able to achieve these objectives. In some cases, a remedial action objective (such as restoration of groundwater to drinking water quality) cannot be achieved by any available technology. In order to ensure the efficient and effective remediation of contaminated sites, a formalized remedy selection proves is needed to i) identify all remedial action objectives for a site and ii) identify technologies capable of achieving these objectives. This report describes a process for the selection of an appropriate remedy for a contaminated site that will address all identified remedial action objectives.

1.1. REQUIRED SITE INFORMATION

Prior to remedy selection, a site assessment should be completed resulting in: i) the establishment of remedial action objectives, ii) a determination that a remedial action is required to achieve these objectives, iii) an identification of site areas requiring a remedial action, and iv) a conceptual site model that reflects the results of the site assessment. The remedial action objectives are assumed to have been established using Risk-Based Corrective Action (RBCA) or another risk-based assessment method that results in the identification of appropriate remedial action objectives based on an evaluation of sources of contamination, human health and ecological exposure pathways, and potential receptors. Remedial action objectives may be established using ASTM (2002), Standard Guide for Risk-Based Corrective Action Applied at Petroleum Release Sites; ASTM (2000), Standard Guide for Risk-Based Corrective Action; or a similar process to characterize site risks based on the potential for unsafe exposure to contaminants. The remedial action objectives identified in this manner are typically based on protection of human health and ecological resources but may also include non-risk objective such as resource protection standards and the prevention of aesthetic or nuisance impacts.

Each risk-based remedial action objective for an exposure pathway will typically include numeric remedial action levels for each contaminant. Remedial action levels may also be developed for non-risk remedial action objectives such as resource protection standards. The non-risk remedial action levels may include thickness or mobility criteria for non-aqueous phase liquids (NAPL).

1.2. OVERVIEW OF REMEDY SELECTION

To ensure that the selected remedy will achieve all of the remedial action objectives, a step-wise remedy selection process can be used as follows:
- *Development of risk-based remedial action objectives* that includes identification of complete exposure pathways and numeric remedial action levels.

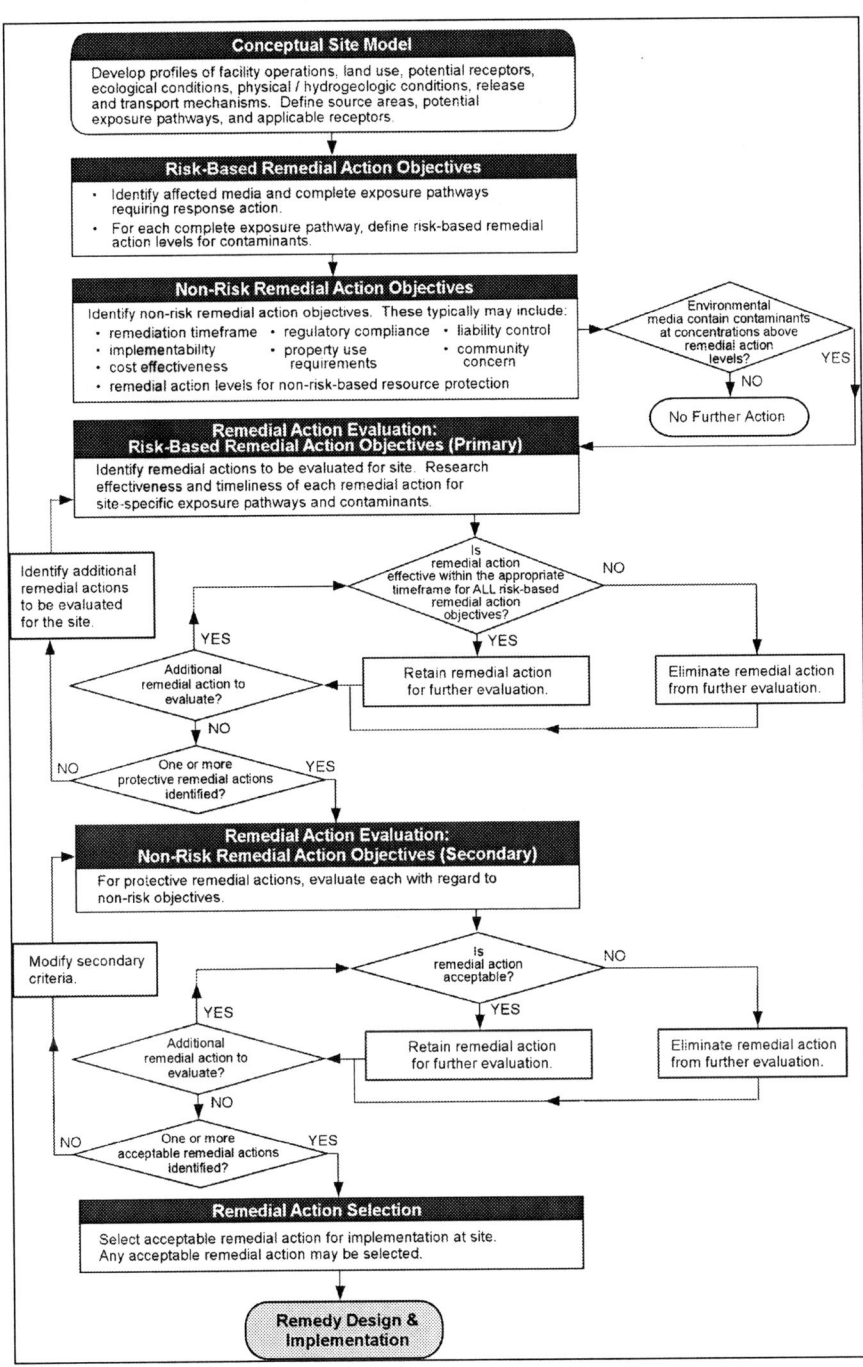

FIGURE 1. Remedy Selection Process.

- *Development of non-risk remedial action objectives* based on resource protection and other non-risk considerations. Resource protection objectives typically include numeric remedial action levels while other non-risk criteria are typically non-numeric and may include: remediation timeframe, implementability, cost effectiveness, regulatory compliance, property use requirements, liability control, and community concern.
- *Evaluation of protectiveness* to identify protective remedial actions that will be effective and timely for each risk-based remedial action objective for the site.
- *Evaluation of the retained remedies using the non-risk remedial action objectives* to identify acceptable remedial actions that satisfy the minimum level for each non-risk criterion.
- *Remedial action selection* to choose the acceptable remediation technology to be implemented at the site.
- *Remedy design and implementation* to ensure that the selected remedy is effectively implemented at the site and satisfies the remedial action objectives and the secondary criteria.

This process is intended for use in the selection of final remedial actions, but may also be used in the selection of emergency response measures and interim measures provided that risk-based remedial action objectives and non-risk remedial action objectives are available for the evaluation of such measures.

2. Identification of Remedial Action Objectives

The first step in the remedy selection process is the identification of remedial action objectives that reflect the full range of current and future site management goals including:

- *Risk-based Remedial Action Objectives* – A set of objectives based on protection of human health and ecological resources developed for the site that identifies the contaminants, affected environmental media, complete exposure pathways, and risk-based remedial action levels.
- *Non-risk Remedial Action Objectives* – A set of objectives based on non-risk considerations for current and future site management. These objectives may include resource protection standards or nuisance and odour control standards with numeric remedial action levels. In addition, non-risk objectives may include performance standards not based on contaminant concentrations such as: remediation timeframe,

implementability, cost effectiveness, regulatory compliance, property use requirements, liability control, and community concern.

2.1. USE OF THE CONCEPTUAL SITE MODEL

An effective conceptual site model is essential for the identification of risk-based and non-risk remedial action objectives. The conceptual model should identify source areas; complete, potentially complete, and incomplete exposure pathways; and receptors. In addition, the conceptual model should identify contaminants, affected environmental media, and specific areas within the affected environmental media to be addressed by the selected remedial action. Although a conceptual model should be developed prior to initiation of the risk-based remedy selection process, the conceptual model should be updated as needed during the remedy selection process to reflect any changes in the understanding of the site. ASTM (2003), Standard Guide for Developing Conceptual Site Models for Contaminated Sites, provides guidance for the development of an effective conceptual model.

2.2. IDENTIFICATION OF RISK-BASED REMEDIAL OBJECTIVE

Risk-based remedial action objectives are used to identify remedial actions that will be protective of human health and the environment. These objectives should be developed using a risk-based framework that includes: i) identification of contaminants, ii) affected environmental media, iii) complete exposure pathways and resource protection requirements, and iv) remedial action levels. Remedial action levels reflect the concentrations of contaminants in the source area below which no remedial action is required in order to protect human health and ecological resources. These remedial action levels may be established using ASTM (2002), Standard Guide for Risk Based Corrective Action Applied at Petroleum Release Sites; ASTM (2000), Standard Guide for Risk-Based Corrective Action; or any similar risk-based framework.

2.3. IDENTIFICATION OF NON-RISK REMEDIAL ACTION OBJECTIVES

Non-risk remedial action objectives are used to identify remedial actions that will satisfy the current and future non-risk requirements for the site. While risk-based remedial action objectives ensure long-term protection of human health and the environment, non-risk objectives address all other site remedial action requirements and constraints. The non-risk remedial action objectives should cover all non-risk site constraints that will define an acceptable remedy. These non-risk objectives may include remedial action levels (i.e., numeric criteria) that are not directly tied to human or ecological exposure such as:

- Application of drinking water standards to non-drinking water resources: remedial actions levels for drinking water are applied to water resources that will not be used as drinking water in the foreseeable future.
- Aesthetic impact standards: remedial action levels for the prevention of tastes or odours, unacceptable visual impacts, or other nuisance impacts.
- NAPL removal requirements: NAPL thickness or mobility criteria for groundwater resources where human exposure will not occur in the foreseeable future.

However, more commonly, non-risk remedial action objectives are not directly tied to site contaminant concentrations and therefore do not include remedial action levels. Examples of non-numeric remedial objective include:

- Timeliness: remedial action will be completed within a timeframe that meets the site-specific requirements.
- Implementability: remedial action can be implemented and will protect human health and the environment during implementation.
- Surety: the level of confidence that the remedial action will achieve the remedial action objectives at the site.
- Cost: remedy cost is acceptable.
- Regulatory compliance: remedy satisfies regulatory requirements.
- Property use compatibility: remedy allows for acceptable current and future property use.
- Liability control: remedy controls current and future liability associated with site.
- Community acceptance: remedy is acceptable to third party stakeholders.

2.4. REQUIREMENTS FOR REMEDIAL ACTIONS

A remedial action is required if environmental media contain contaminants at concentrations above the risk-based or non-risk remedial action levels. If all contaminant concentrations are below the remedial action levels, then no further action is required.

3. Evaluation of Remedial Actions

For sites where a remedy is required to achieve the remedial action objectives, one or more potential remedies must be identified for implementation at the site. These potential remedies are then evaluated against the remedial action objectives in order to identify a remedy that will achieve all of these goals.

In risk-based remedial action screening, potential remedial actions are screened to first identify protective remedial actions which can achieve all of the risk-based remedial action objectives and then to identify acceptable remedial actions that can also achieve all non-risk objectives. The separate evaluation of potential remedial actions using risk-based objectives and then non-risk objectives allows for a clear understanding of the full range of options available to protect human health and ecological resources, and how those options are constrained by non-risk considerations such as property use, liability control, or non-risk regulatory requirements. In addition, the non-risk objectives for a site may be balanced against cost or other considerations in a way that is unacceptable for risk objectives based on health protection.

3.1. TYPES OF REMEDIAL ACTION TECHNOLOGIES

Remedial action technologies can be categorized into three groups based on the mechanism used to achieve the remedial action objectives:

<u>Removal</u>: A remedial action technology to take environmental media away from the site to another location for storage, processing, or disposal in accordance with all applicable requirements.

<u>Decontamination</u>: A remedial action technology based on permanent and irreversible treatment processes to an environmental medium so that the threat of release of contaminants at concentrations above the remedial action levels is eliminated.

<u>Control</u>: A remedial action technology to apply physical and/or institutional controls to prevent exposure to contaminants present in environmental media at concentrations above the remedial action levels. Control measures must be combined with appropriate maintenance, monitoring, and any necessary further remedial action to satisfy the remedial action objectives and be protective of human health and the environment.

Within a risk-based remedial action framework, any of these three types of technologies may be considered for selection provided that they are effective at protecting human health and ecological resources. For a complex site, a remedial action may consist of multiple technologies (e.g., clay soil cap or monitored natural attenuation) representing one or more classes of remedial action (i.e., removal, decontamination, or control) that are combined in order to achieve all of the remedial action objectives. The user must research the effectiveness of each potential remedial action in order to determine whether the remedial action is capable of achieving all of the risk-based remedial action objectives.

3.2. REMEDIAL ACTION EFFECTIVENESS

For each potential remedy identified for evaluation, the effectiveness must be evaluated in order to determine whether the remedy is capable of achieving all of the remedial action objectives. The evaluation of effectiveness for each remedial action should be made within the context of the conceptual site model (i.e., the contaminants, physical and hydrogeologic conditions, and other site-specific factors affecting technology effectiveness). A number of resources are available to assist with the evaluation of technical effectiveness of potential remedial actions:

- Federal Remediation Technologies Roundtable (FRTR): http://www.frtr.gov/
- USEPA Technology Innovation Program: http://clu-in.org.
- Ground-Water Remediation Technologies Analysis Center (GWRTAC): http://www.gwrtac.org/
- USEPA Remediation and Characterization Innovative Technologies (REACH IT): http://www.epareachit.org/
- NAVFAC Environmental Restoration and BRAC Website: http://enviro.nfesc.navy.mil

Each potential remedial action must be evaluated with respect to its ability to achieve each risk-based and non-risk remedial action objective and to prevent unsafe exposure at all times. The timeframe available to achieve the remedial objectives is likely to vary from site to site. For example, a remedy for contaminated drinking water must prevent short-term exposure for users of the existing water supply. However, for a remedial action objective applying drinking water standards to groundwater not currently used for drinking water, the remedy must be within a timeframe based on the potential future use of this groundwater.

3.3. REMEDIAL ACTION SCREENING: RISK-BASED OBJECTIVES

A protective remedial action can achieve the risk-based remedial action objectives through removal, decontamination, and/or control of environmental media containing contaminant concentrations above the risk-based remedial action levels. To be retained as a protective remedial action, the remedial technologies must be capable of achieving the remedial action objectives. Potential remedial actions that are determined to be protective with an acceptable level of confidence are retained for further evaluation using the non-risk remedial action objectives. Potential remedial actions which are not protective are eliminated from further evaluation. One or more protective remedial technologies must be identified as part of the risk-based remedial action screening. If all potential remedial technologies are eliminated during the risk-based screening, then additional potential remedial actions must be identified and evaluated through the screening process.

3.4. REMEDIAL ACTION SCREENING: NON-RISK OBJECTIVES

All remedial actions identified as protective based on the evaluation using risk-based remedial action objectives should be included in the non-risk remedial action screening. An acceptable remedial action achieves the risk-based remedial action objectives and satisfies the minimum acceptable standard for each non-risk remedial action objective. Remedial actions that are determined to be acceptable are retained for potential selection. Remedial actions which are not acceptable are eliminated from further evaluation.

3.5. TECHNICAL IMPRACTICABILITY

Technical impracticability exists when no potential remedial action will satisfy both the risk-based remedial action objectives and the non-risk remedial action objectives. If all potential remedial actions are eliminated through remedial action screening, then a determination of technical impracticability may be appropriate. Following a technical impracticability determination, the <u>non-risk</u> remedial action objectives must be modified in order to allow for the selection of a remedial action that will satisfy the modified objectives and will be protective of human health and the environment.

3.6. REMEDY SELECTION

Any acceptable remedial action (i.e., any remedial action technology that satisfies the risk-based and non-risk remedial action objectives) may be selected for implementation at the site. If more than one acceptable remedial action is identified, the remedial action which does the best job of satisfying the non-risk objectives will typically be selected for implementation (i.e., the remedial action with the best combination of cost, timeliness, surety, and other secondary considerations).

4. Remedial Action Design and Implementation

In order to ensure that the selected remedial action achieves all of the remedial action objectives, the user must continue to consider all of these objective during the remedy design and implementation.

4.1. REMEDIAL ACTION DESIGN

To verify that the selected remedy will achieve all risk-based and non-risk objectives, bench scale and/or field pilot testing activities may be performed prior to, or during the remedial action design. In addition to engineering design (which is not covered in this report), remedial action design will include remedy monitoring methods to verify remedy effectiveness and/or remedy completion. This monitoring may include:

- Verification sampling (removal or decontamination technologies)
- Point of compliance monitoring (control technologies)
- Integrity monitoring (control technologies)

The type of monitoring method selected depends on the type of remedial action being implemented. The monitoring methods selected should be capable of verifying remedy completion and monitoring remedial action effectiveness during remedy implementation (if needed).

Remedial action monitoring criteria provide the basis for determining remedy effectiveness and/or completion:

- <u>Remedy completion</u>: Has remedial action achieved the remedial action objectives?
- <u>Remedy effectiveness</u>: Is remedial action progressing towards achieving the remedial action objectives?

Remedy completion is typically demonstrated by comparison of source media or exposure media concentrations to the remedial action limits. In contrast, remedy effectiveness criteria are used to determine whether the selected remedial action needs to be modified or replaced in order to achieve the remedial action levels while continuing to satisfy the other objectives (e.g., cost, timeliness, etc.). For example, a pump and treat remedy may require modification or replacement if contaminant concentrations in groundwater stabilize prior to achievement of the remedial action levels. In addition, remedy effectiveness monitoring criteria may be used at the time of system start-up in order to optimize system operation.

4.2. REMEDIAL ACTION COMPLETION

The remedy is complete when the remedy monitoring has demonstrated that the remedial action has achieved the remedial action objectives. Control remedies may require post-response care and monitoring to ensure the continued effectiveness of the remedial action following completion. When a remedy does not achieve all of the remedial action objectives, a determination of technical impracticability may be made based on remedy effectiveness monitoring (i.e., the selected remedial action is not effective and no alternative acceptable remedial action can be identified). Following a technical impracticability determination, the non-risk remedial action objectives must be modified in order to allow for the selection of a remedial action that will satisfy the modified objectives and will be protective of human health and the environment.

Post response care involves continued operation and/or maintenance of control technologies to ensure continued effectiveness. Monitoring methods and criteria will typically be the same as those used to demonstrate remedy completion. However, lower intensity monitoring will typically be sufficient to provide assurance of continued remedy effectiveness. Monitoring may be terminated when the long-term effectiveness of the remedy has been demonstrated.

References

ASTM, 2000, E 2081 - 00, Standard Guide for Risk-Based Corrective Action. http://www.astm.org.
ASTM, 2002, E 1739 – 95 (2002), Standard Guide for Risk Based Corrective Action Applied at Petroleum Release Sites. http://www.astm.org.
ASTM, 2003, E 1689-95 (2003)e1, Standard Guide for Developing Conceptual Site Models for Contaminated Sites. http://www.astm.org.
ASTM, 2004, E 1943 – 98 (2004), Standard Guide for Remediation of Groundwater by Natural Attenuation at Petroleum Release Sites. http://www.astm.org.

USEPA, 1990, The Feasibility Study: Detailed Analysis of Remedial Action Alternatives, Directive 9355.3-01FS4, March 1990, Office of Solid Waste and Emergency Response, U.S. Environmental Protection Agency

USEPA, 1993, Guidance for Evaluating Technical Impracticability of Ground-Water Restoration, Directive 9234.2-25, September 1993, Office of Solid Waste and Emergency Response, U.S. Environmental Protection Agency

USEPA, 2004, Cleaning Up the Nation's Waste Sites: Markets and Technology Trends, Office of Solid Waste and Emergency Response, EPA 542-R-04-015, September, 2004.

EUROPE AS A SOURCE OF POLLUTION – THE MAIN FACTOR FOR THE EUTROPHICATION OF THE DANUBE DELTA AND BLACK SEA

LIVIU-DANIEL GALATCHI[1] * AND MARIAN TUDOR[2]
[1]*Department of Biology, Ovidius University of Constanta, 124, Mamaia Blvd., 900527 Constanta, Romania*
[2]*Department of Ecology and Environmental Protection, Danube Delta National Institute for Research and Development 165 Babadag Str, 820112 Tulcea, Romania*

*To whom correspondence should be addressed. galatchi@univ-ovidius.ro

Abstract. The Danube - Black Sea Region contains the single most important non-oceanic body of water in Europe. Every year, about 350 cubic kilometers of river water pour from the Danube into the Black Sea from an area of 2 million square kilometer basin, covering about one third of the area of continental Europe. Over 160 million people live in this basin.

Keywords: Danube River; Danube Delta; Black Sea; pollution; eutrophication.

The Danube River is by far the single most important contributor to the nutrient pollution of the Black Sea.

The Danube is the most international river basin in the world, which makes co-ordinated action even more important and challenging. The strategic importance of the region is increasing in the context of an enlarged Europe.

Until now the Danube has been an important link in Central Europe as well as the border between EU and the Balkans and Black Sea Region. With the EU enlargement, a number of the Danube countries will be a member of the European Union and the Danube will become a central axis of Europe while the Black Sea will become a coastal area of the Union.

In environmental and health terms, the Danube - Black Sea region suffers from very acute from very acute problems. The Danube is subject to increasing pressure affecting the supply of drinking water, irrigation, industry, fishing, tourism, power generation and navigation. All too often it is also the final destination of wastewater disposal. These intensive uses have created severe problems of water quality and quantity, and reduction of biodiversity in the basin.

The Danube River has its source in a spring in the castle of Donaueschingen. This was done by agreement, because there are claims that the source of the Danube River is the source of the river Breg. Another opinion puts the beginning of the Danube River at the confluence of the Brigach and the Breg rivers. Its length from the confluence of the Breg and Brigach rivers - close to Donaueschingen - to the zero station at Sulina is 2,880 km.

The Danube receives its increasing flow from the rivers Lech; Isar; Inn; Enns; Mura-Mur/ Drava-Drau, coming from the Alps, the northern Morava, the rivers coming from the Carpathian mountains, like Vah; Nitra; Hron, the Tisza - with all its tributaries, those from the Dinarian ridge (the Sava and the southern Morava with all their tributaries), and from the rivers coming from the outer Carpathian mountains in Romania (Jiu; Olt; Arges) and the Balkan mountains (Iskar; Yantra). The last tributaries are in Romania, the Siret and the Prut rivers.

The waters of the Danube River and its tributaries combine to make up river-related ecosystems of high economic, social and environmental value. The River Basin includes numerous important natural areas such as wetlands (including floodplains), with a high number of endangered endemic plant and animal species. The river network supports drinking water supply, agriculture, industry, fishing, tourism and recreation, power generation and navigation but it also receives the waste waters for a region with a population of about 85 million in eleven different countries, which will become thirteen with the inclusion of Bosnia-Herzegovina and Yugoslavia. During the period of centralized planning the main emphasis of Central and Eastern European Countries was on production and policies took little or no account, of the degradation of the environment. The economic transition now going on, including industrial restructuring and agricultural reform, created an opportunity to change this situation and to prevent, reduce and control pollution and waste generation substantially to the benefit of the environment and of peoples' quality of life. The breathing space provided by the transition can be used to ensure that environmental concerns are properly integrated into industrial, agricultural and other sector policies in the future. The development of public awareness and of environmental policies that will contribute to the sustainable development of countries concerned is a challenge to countries in transition.

The basin of the Danube River with about 200 km^3 contributes per year to the receiving Black Sea; its flow is as big as the one of the Volga River.

The catchments area of the Danube River presently covers the territories of Albania, Austria, Bosnia-Herzegovina, Bulgaria, Croatia, the Czech Republic, the Federal Republic of Germany, Hungary, Italy, Macedonia, Moldova, Poland, Romania, the Slovak Republic, Slovenia, Switzerland, Ukraine and the Federal Republic of Yugoslavia. 13 of these 18 riparian States hold in the Danube Basin territories bigger than 2,000 km^2.

The Danube River discharges into the Black Sea through a delta which is the second largest natural wetland area in Europe. The waters of the Danube River basin and its tributaries combine to make up an aquatic ecosystem of high economic, social and environmental value. It includes numerous important natural areas including wetlands and floodplain forests. It supports the drinking water supply, agriculture, industry, fishing, tourism and recreation, power generation, navigation and the end disposal of waste waters for a densely populated region of Europe.

A large number of dams, dikes, navigation locks and other hydraulic structures have been built to serve some of these important human activities, including over forty major reservoirs on the Danube River itself. But such structures have caused changes in flow pattern and damage to the functions and biodiversity of the river system. Furthermore, the intensity of agricultural, industrial and urban uses has created problems of water quality and quantity, and reduced biodiversity in the basin. These changes have caused significant environmental damage, such as reduced sediment transport, increased erosion and reduced self-purification capacity, including public health aspects in connection with drinking water supply of the population, recreation and bathing.

The most important problems (not in order of importance) affecting the health of the Danube River ecosystems and the water users in the basin are the high nutrient loads (nitrogen and phosphorus), changes in river flow patterns and sediment transport regimes, contamination with hazardous substances including oils, competition for available water, microbiological contamination, and contamination with substances causing heterotrophic growth and oxygen depletion.

The aquatic habitats of the basin are part of a single system, so that harmful activities in one section affect other sections. For example, the degradation of the delta and the north-west shelf region of the Black Sea are caused by eutrophication (to a great extent) from the cumulative inflow of nutrients from the Danube River. Nutrient and pollution loads coming from the river must be reduced if the health of the whole system, including that of the delta and Black Sea, are to be restored. This can be helped by natural buffering systems such as wetlands and floodplains, which also contribute to biological diversity.

Wastes from cities and industries, chemical fertilizers, and manure from intensive and large-scale livestock operations have poured into the river system and the groundwater, raising nutrient levels and causing eutrophication. Other highly polluting activities in the Danube River basin include petrochemicals processing, iron and metal processing, timber, paper and pulp, and municipal solid waste disposal. Microbiological contamination is a problem throughout the river basin. It is generally caused by the discharge of urban waste and storm waters, as well as by livestock and agricultural run-off. Inadequate waste treatment and disposal mean that urban and industrial discharges contribute significant quantities of substances causing heterotrophic growth and oxygen depletion.

During the period of centralized planning systems, the central and eastern European countries did not develop full environmental protection policies which responded to the degradation of the river environment. Legal standards for environmental quality were often uninforced or unenforceable.

Significant progress towards reducing pollution has been made in the economically most developed countries. The middle and lower basin countries have invested in municipal water supply and waste water treatment facilities, although in many cases their performance has been below design levels because of lack of maintenance. The economic transition these countries are now undergoing has caused industrial production to decline and has thus reduced pollution loads. Comparable changes in agriculture have reduced the amounts of chemicals used which end up in surface and groundwater. There is now an opportunity to take advantage of the breathing space provided by the transition and ensure that environmental concerns are integrated into industrial and agricultural policies before economic activity picks up.

Apart from Germany and Austria, all other Danube countries are undergoing fundamental transformation of their political, legal, administrative, economic and social systems. This transformation can bring about a better consideration of environmental needs. Industrial restructuring and agricultural reforms should reduce pollution and waste substantially by reversing the former regimes' emphasis on production. Policy and administrative reforms should also create opportunities for improvement of water quality and supply.

The Danube River Basin is populated by slightly more than 80 million inhabitants. The data are all based on information provided by the States that co-operate under the umbrella of the DRPC.

The Danube River Basin covers an area of about 802,890 km². The Danube is the largest river discharging into the Black Sea not only from the point of view of average annual flow of 6800 m³/s, but also in respect to the sediment load. According to GEF (1996) the Danube is the biggest polluter of the Black Sea especially in relation to the load of suspended solids and total nitrogen. The nutrient discharge increased manifold from the early 1960s up to the late 1980s and are considered responsible for the eutrophication of rivers and lakes and, especially for the Western Black Sea and its coastal areas which are strongly influenced by the Danube. Effective measures to reduce nutrient inputs need the knowledge of their quantities, their sources and their regional distribution within the sub-basins. The nutrient state of a river system depends on natural characteristics, the level and structure of the nutrient emitted into the river system, caused by geogenic background and anthropogenic activities. The analysis of the present state of input and load situation within different scales of a river basin is therefore one important pre-requisite for deriving quality criteria and management plans.

One of the main environmental problems related to the water bodies in the Danube – Black Sea region is due to the high content of nutrients that flows into the Black Sea, both via the rivers and directly from land-based sources. This leads to eutrophication of the rivers and the sea, which is recognised as one of the principal causes of their degradation. The eutrophication has consequences for biodiversity in the water bodies and in the surrounding wetlands and forests, and also for human health in the region. Eutrophication is the main problem threatening the biodiversity and the economic potential of the riverine ecosystems, the Danube delta and the Black Sea. It is caused by the high load of nutrients (nitrogen and phosphorus). The riverine ecosystems and the unique Danube delta have experienced rapid eutrophication in recent decades. The nutrient load from the Danube into the delta increased several-fold for nitrogen and phosphorus from 1960 to 1990 (Oosterberg, Menting, Hanganu, Gridin, Tudor, 1998). Shallow lakes and slow-flowing channels have experienced a shift in vegetation from submerged higher plants, water lilies etc. to planktonic algae. The decomposition of organic material following extreme seasonal algae growth has caused oxygen depletion in the bottom layers and reduced the diversity of the flora and fauna. The composition of the fish population has shifted from commercially valuable species towards economically less valuable algae eating species. The total number of fish species has been reduced from about 100 in 1960 to 75 in 1990.

Coastal eutrophication caused by high nutrient loads reduces the productivity of wetlands and fisheries. It is estimated that the Danube contributes about half of the total river load of nutrients to the Black Sea. The shallow north-western shelf of the sea into which the Danube discharges is where the damage is most evident. Fisheries have been drastically reduced; biodiversity has suffered; and the quality of the beaches has declined because of excessive growth of algae and jellyfish (Suciu, Constantinescu, David, 2002).

Eutrophication of the wetlands and shallow water bodies along the main stream of the Danube River and some tributaries shows that these ecosystems are filtering the nutrients transported to the Black Sea.

However, the increased load of nutrients to the river system and the radically reduced area of functioning wetlands limit the absorptive capacity of the wetlands (Stiuca, Staras, Tudor, 2002).

The total load of nutrients to the Black Sea from the Danube can be estimated from water monitoring data from the Bucharest Convention network. The flow-corrected figures for 1991 are about 540000 tones of inorganic nitrogen and about 45000-50000 tones of phosphorus (Chirila, Godeanu, Godeanu, Galatchi, Capota, 2002). The actual load in a year is highly influenced by the total water volume discharged. More nutrients are transported in years with high precipitation than in dry years. There is considerable uncertainty about the emissions of nutrients to the Danube River system from different sources and different countries. Data from country reports give about 700000 tones/year of total nitrogen and about 90000 tones/year of total phosphorus. These data may represent only part of the total discharges, since several sources of surface and groundwater run-off have not been included in some of the country studies. The contribution from Bosnia-Herzegovina and the F.R. Yugoslavia is also not included. There is a major discrepancy between these figures and monitoring data in the Danube delta. Evidently, considerable amounts of nutrients are transformed (e.g. by denitrification) and/or retained in the reservoirs and floodplains/wetlands along the river. Much more detailed studies are needed to develop a reasonably good understanding of the retention and denitrification function of the river ecosystem and the relative contribution of the regions in the basin to the delta and Black Sea load.

The sources of nutrients are the same as those for the river ecosystems. The share of each main sector has been estimated at:

- Agriculture: about 50% of the total load;
- Cities, rural towns and villages: about 25% of the total load;
- Industry: about 20-30% of the total load, including atmospheric deposition and background load (Strauss et al., 2003).

Diffuse sources of nutrients, primarily from agriculture and background loads, are dominant in the upstream countries of Germany, Austria, the Czech Republic and Slovakia. Further downstream, the point sources become more important, except in Moldova. This is probably a temporary situation, as the economic transition in the downstream countries has caused an abrupt drop in the use of fertiliser.

The retention capacity of the riverine wetlands for nitrogen and phosphorus is not known. However, the discrepancy between the estimates of total load to the Danube river systems and the calculated transport into the Black Sea (based on measurements in the delta) indicate that even today's significantly reduced river wetlands have an important function in reducing the total amount of nutrients transported (Galatchi, L.D., 2004). Restoring wetlands and managing them for the purpose of maximizing nutrient removal may be one of the most promising measures to reduce eutrophication. A nutrient balance of the Danube delta for 1991 showed a nutrient retention of 65000 tones of nitrogen and 3300 tones of phosphorus.

ACKNOWLEDGMENTS

We would like to express our gratitude for the support received from our colleagues working at the "Danube Delta" National Institute for Research and Development in Tulcea, and at the Ovidius University of Constanta, Romania.

References

Chirila E., Godeanu S., Godeanu M., Galatchi L.D., Capota P., 2002, Analytical characterization of the Black Seacoast lakes, *Environmental Engineering and Management Journal*, 2, 205–212.

Galatchi L.D., 2004, Current Situation and New Strategies for Sustainable Development of the Danube Delta in Romania, Ukraine and Moldova, online http://levis.sggw.waw.pl/wethydro/contents/sic/pdf/liviu_galatchi.pdf, Center of Excellence in wetland Hydrology WetHYDRO, Warsaw University, Poland, 12 pages.

Oosterberg W., Menting G., Hanganu J., Gridin M., Tudor M., 1998, Filtering capacity of Mustaca reedbed, *Raport RIZA*, 1998.

Stiuca, R., Staras, M., Tudor, M., 2002, The ecological restoration in the Danube Delta. An alternative for sustainable management of degraded wetlands, *Proceedings of the 34th IAD Conference*, p. 707–720.

Strauss P. et al., 2003, Evaluated model on estimating nutrient flows due to erosion/runoff in the case study areas selected, *Deliverable D2.1 of the project "Nutrient Management in the Danube Basin and its Impact on the Black Sea" (daNUbs) supported under contract EVK1-CT-2000-00051 by the Energy, Environment and Sustainable Development (EESD) Programme of the 5th EU Framework Programme*, http://danubs.tuwien.ac.at/.

Suciu R., Constantinescu A., David C., 2002, The Danube Delta: Filter or bypass for the nutrient input into the Black Sea?, *Large Rivers Vol. 13*, No. 1–2, p 165–173.

EFFECTIVE APPROACHES FOR MODELING GROUND WATER LOAD OF SURFACE APPLIED CHEMICALS

MARNIK VANCLOOSTER
Department of Environmental Sciences and Land Use Planning,
Université Catholique de Louvain
Croix du Sud 2, BP2, B-1348 Louvain-la-Neuve, Belgium
E-mail: vanclooster@geru.ucl.ac.be

Abstract. Land applied chemicals from different origin may have a negative impact on the quality of surface and groundwater. Migration of dangerous chemicals through the vadose zone is therefore a possible pollution pathway for vulnerable drinking water resources. Effective approaches for protecting groundwater resources builds on a thorough understanding of the flow and transport of chemicals in the vadose zone. Unsaturated flow and transport models may therefore be used as tools to predict flow and transport supporting risk management and operational decision making. In this paper, we show how models for pollutant dispersion modelling in soils have been developed. The developments are based on the use of flow and transport models for assessing risks in the agricultural sector, in particular to evaluate risks associated with the use of agro-chemicals. We believe, however, that the concepts and approaches presented in these models are sufficiently generic for modelling chemical load of vulnerable groundwater bodies also from non-agricultural diffuse source. Particular emphasis in this report is put on (i) the limitations of the current modelling approaches for management applications, (ii) the problems associated with the estimation of the required modelling data and parameters, and (iii) the transfer of scientific know-how into operational decision-making tools.

Keywords: soil solute transport; diffuse pollution modelling; groundwater protection

1. Introduction

Soil and groundwater are important natural resources that should be protected from human induced pollution e.g. caused by a chemical terror attack. Efficient protection requires a thorough understanding of the fate and transport mechanisms of solutes in the vadose zone. Solutes in this context is referred to dissolved substances entering the soil and subsoil from diffuse and point pollution sources together with the soil water, while the vadose zone is referred to the unsaturated soil body covering eventually a subsurface saturated groundwater body. Given the unsaturated nature of the vadose zone (i.e. soil pores are partially filled with soil gas), at least three distinct phases characterize the vadose zone: a solid, a gaseous and a liquid phase.

Solutes are carried through the soil zone, where they are subjected to a range of phase exchange (e.g. sorption) and transformation processes. Along with the boundary conditions at the soil surface, these transport, phase exchange and transformation processes will determine the ultimate concentrations and fluxes of substances in the vadose zone and groundwater.

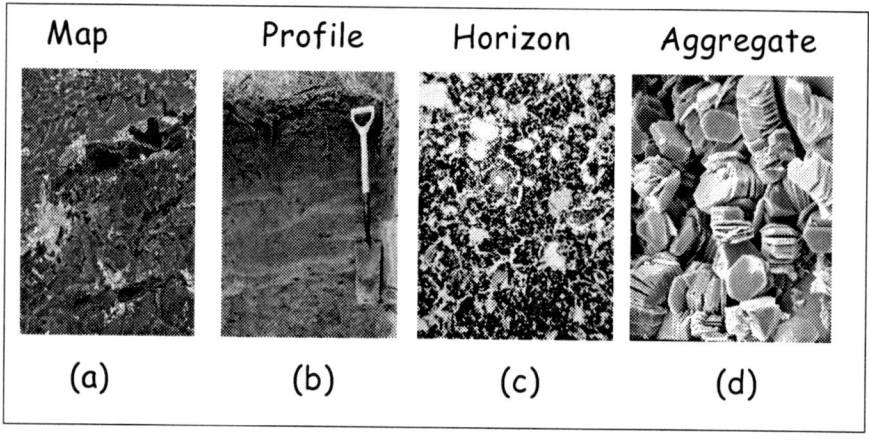

FIGURE 1. Illustrations of soil variability at different spatial scales. At the regional scale, variability is expressed by the difference of different soil mapping units within a soil map sheet (a). At the scale of the soil profile, variability is observed by the presence of different diagnostic horizons (b). Within the soil horizons, variability is presented by the heterogeneous appearance of soil aggregates (c). Within aggregates, microscopic variability is observable in the porous structure (d).

The basic thermodynamic principles for describing flow and contaminant transport in the vadose zone are well established, but the complexity of the processes results in different conceptualisations of flow and contaminant transport in current models. The modelling and characterisation of solute fate and transport in soils is complicated by the space-time variability of the underlying processes in particular in the vadose zone (Fig. 1). In addition, any experimental technique is operational at a certain scale, which is not necessarily the scale at which the process can reasonably be described, neither the scale at which a prediction is needed. The expressions of the solute fate and transport are therefore often considered as scale-dependent, and scaling is needed to model and characterise the transport processes at the larger spatial and temporal scales.

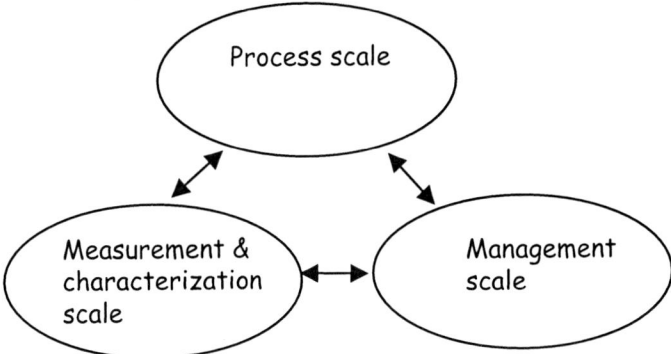

FIGURE 2. The scale paradigm. Process occur at different scale as compated to the scale at which measurement techniques are operational. The management scale for which models should be developed differ often from the measurement and process scale.

This is referred to as the scaling paradigm (Fig. 2). The scaling is of utmost importance given the fact that most environmental problems occur at the larger spatio-temporal scales, while most characterization techniques and well-validated modelling concepts are only operational at the lower scale.

2. The Continuum Approach: From Core to Field and From Field to Region

For dealing with variability of soil properties at the larger scale, a continuum approach is implemented. Thereby a representative elementary volume (REV) is considered to exist and material properties related to flow and transport are defined at the centre of this REV. Thermodynamic principles related to conservation of mass and momentum are further applied on the REV to obtain governing flow and transport equations. The spatial and time integration of these equations may than allow to describe fate and transport at the larger continuum scale.

Two remarks should be formulated here. First, in most operational models dealing with chemical transport in soil, the REV pertains to the macroscopic scale. The macroscopic scale, often referred to as the local scale, corresponds to the scale where in the plot of the expected value of an essential property determining flow and transport (e.g. hydraulic conductivity) versus scale, a first nearly constant level along some appreciable length scale is reached. This can be illustrated with a simple theoretical experiment where the porosity in a porous medium is determined by sampling the porous medium with sampling windows of different sizes (Fig. 3).

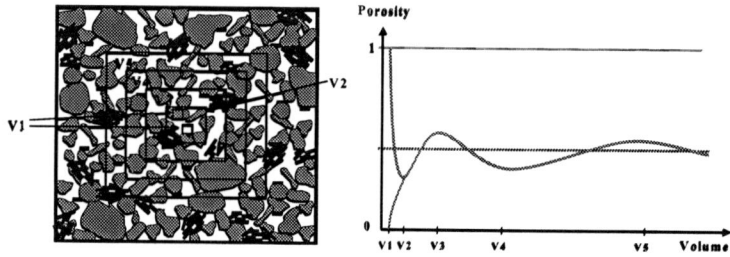

FIGURE 3. Sampling the porous medium with different sizes of the sampling window for determining the porosity.

In practice, this local scale is considered to correspond to the size of the characterization techniques of local soil properties, let's say a small laboratory column. As such the microscopic pore scale variability is no longer explicitly modelled but encoded through effective flow and transport properties at the macroscopic level. The effective macroscopic properties contain of course the signature of the lower level microscopic variability. As such macroscopic effective moisture retention function, hydraulic conductivity or hydrodynamic dispersivity is determined by microscropic pore size distribution, connectivity and tortuosity within the macroscopic sample.

Second, the space-time integration can only be performed rigorously when the space-time course of the effective macroscopic properties within the continuum is known. Integration of local models at larger scale have recently been made possible by introducing spatially distributed modelling technology (Fig. 4). With spatial distributed modelling, a spatially heterogeneous three-dimensional soil-groundwater body is disaggregated in spatial homogeneous units for which a local homogeneous model is implemented.

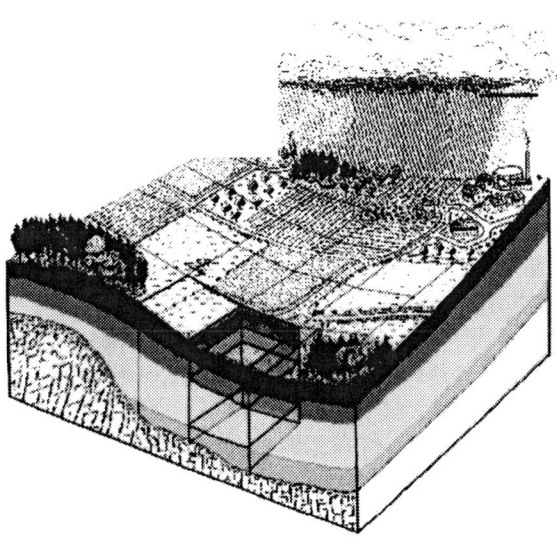

FIGURE 4. The spatially distributed model for large scale diffusion modelling. A homogeneous soil column is isolated in the heterogeneous soil-subsoil system. For the local homogeneous soil column, a local scale model is implemented.

Characterizing the flow and transport properties for each calculation unit has definitely been facilitated by the development of GIS technology. Yet, in practice, sizes of elementary units in regional applicable spatially distributed models are large compared to characterization scale of most soil properties. Hence, the theoretical basis for assigning local parameters to the effective properties of the calculation unit within a spatially distributed model is often weak and subjected to a lot of uncertainty and imprecision. Although quite some studies recently addressed this issue more rigorously by implementing a stochastic continuum approach for modelling flow and transport (see further).

3. Classical Approaches for Local Scale Modelling of Flow and Transport in Soils

The physical laws of mass, energy, and momentum conservation are the building blocks for describing flow and transport in soils. Fundamental thermodynamic laws are combined with appropriate flux formalisms such as Darcy's law or Fick's law, to yield coupled equations for flow and transport in soils.

The Richards equation is one of the most popular formalism used to describe water flow in chemical transport models:

$$C(h)\frac{\partial h}{\partial t} = -\text{div}[-k(h).\nabla H] - Sw \qquad (1)$$

where H, the total hydraulic head (L); h, the matric head (L); Sw, the sink-source term for water (T-1); k(h), the hydraulic conductivity relationship (L T-1); C(h), the differential moisture capacity (L-1): and t, the time (T). The hydraulic head is related to the soil moisture content of the soil through the soil specific soil moisture retention characteristic. The slope of this moisture retention is the differential moisture capacity, or $C(h) = \partial\theta/\partial h$. For non hysteretic rigid soils, different closed form expressions of the moisture retention and hydraulic conductivity relationships are available. Reviews of approaches for characterizing the basic soil hydraulic properties are given in Leij, F.J. and van Genuchten, M.T. (1999).

Chemical substances can be present in different phases of the soil. The resident chemical concentration [M L^{-3}] is defined as the mass storage of the chemical substance per unit volume. The volume averaged total resident concentration C_t^r is the volume-weighed sum of the resident concentrations in the separate phases:

$$C_t^r = \theta C_l^r + C_a^r + \alpha C_g^r + \eta C_n^r \qquad (2)$$

where θ [L^3 L^{-3}] is the volume average water content, η is the nonaqueous liquid volume fraction [L^3 L^{-3}], α is the volumetric air content, the subscripts l, a, g and n define respectively the dissolved, adsorbed, gaseous and nonaqueous phases. Usually, C_a^r is defined as the product of the soil dry bulk density by the mass density of adsorbed chemicals (mass of chemicals by mass of soil). For inert non-volatile substances in a non-saturated soil, composed of only one single liquid phase (i.e. the soil water), the resident fluid concentration of the chemical substance will therefore be $C_l^r = C_t^r/\theta$.

In this report, we will focus on 'inert' solutes that do not partition between different phases but remain in the water phase. Performing risk analysis for inert solutes can be considered as a worst case of all potential contaminants.

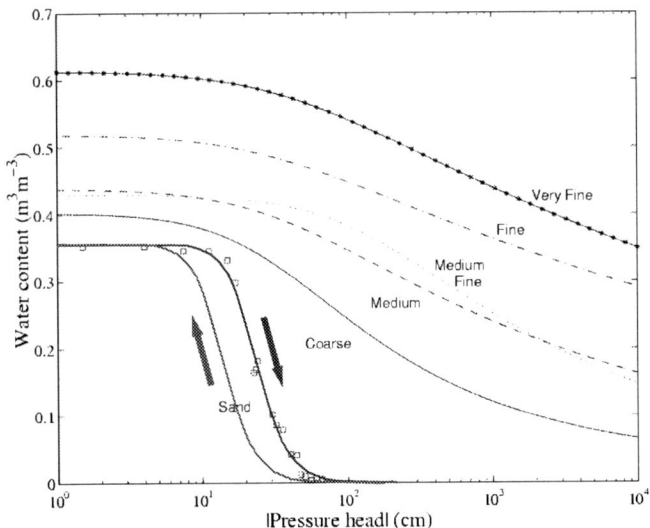

FIGURE 5. Examples of soil moisture retention curves for different textured soils. Heavily textured soils such as clay soils will, in contrast to light textured sandy soils, have a larger total porosity and saturated moisture content, but also large residual moisture contents.

For mobile solutes, flux averaged concentrations are defined as the ratio of solute flux to soil liquid flux through a given area, i.e.

$$C_l^f = J_s / J_w \qquad (3)$$

where J_s, is the area averaged solute flux vector [M L^{-2} T^{-1}], and J_w, the area averaged soil water flux vector [L T^{-1}]. When water flow in the unsaturated zone is predominantly in the vertical direction and when solutes are applied uniformily over a wide area, the horizontal components of the water flow andsolute flux vectors can be neglected.

Since this often applies to leaching experiments and to practical applications dealing with diffuse pollution, transport in soils is often treated as a one-dimensional problem. Using a 1-D mass balance equations, both concentration definitions are related through:

$$\frac{\partial C_t^r}{\partial t} + \frac{\partial J_w C_l^f}{\partial z} = 0 \qquad (4)$$

Convection or advection is the phenomenon where the solutes move with the water. The convective solute flux, $J_{s,c}$ [M L^{-2} T^{-1}] is given by:

$$J_{s,c} = J_w C_l^r \tag{5}$$

where J_w is the area averaged Darcian water flux vector.

The advance of a solute plume through a soil is characterized by an average velocity, which, in case of inert solute, equals the averaged pore water velocity $\overline{v}$:

$$\overline{v} = J_w / \theta_{eff} \tag{6}$$

where θ_{eff} [L^3 L^{-3}] the effective transport volume. When all the water is accessible for the solutes, θ_{eff} equals the soil volumetric moisture content θ. However, due to electric repulsion of ions by an oppositely charged solid phase, very slow diffusive mixing between mobile and stagnant pore water, and exclusion of larger molecules from small pores, θ_{eff} may be much smaller than θ.

When solute plume migrates in a soil, it will disperse essentially due to two mechanisms: molecular diffusion and hydrodynamic dispersion. Molecular diffusion is a microscopic process, induced by thermal agitation and molecular collisions, referred to as Brownian motion. Brownian particle displacement of a solute added to a stagnant liquid phase leads to a Gaussian distribution of particle locations that results from a large number of independent and zero mean particle displacements. Because of the independence of the particle displacements, the variance of the particle location distribution is proportional to the time: $\sigma_x^2 = 2.D_0.t$, where t [T] is the time and D_0 [L^2 T^{-1}] the Brownian diffusion coefficient. For sufficiently low concentrations, Brownian motion in a stagnant liquid leads to a diffusive flux J_{dif} [M L^{-2} T^{-1}] that dissipates concentration gradients and can be modelled by Fick's law:

$$J_{dif} = -D_0 \cdot \frac{\partial C_l^r}{\partial z} \tag{7}$$

A similar process will occur in a static fluid in a porous medium. The diffusion process in this case will be hindered by the presence of the solid phase. The cross sectional area across which diffusion can take place is reduced by a factor that is equal to the volumetric water content, θ. The tortuosity of the pores increases the microscopic distance across which diffusive transport must take place to dissipate macroscopic concentration gradients and reduces the diffusive flux by a tortuosity factor ξ:

$$J_{dif} = -D_0 \cdot \xi \cdot \theta \cdot \frac{\partial C_1^r}{\partial z} \qquad (8)$$

Because the tortuosity of the water filled pore space increases with decreasing soil water content, ξ is a decreasing function of θ.

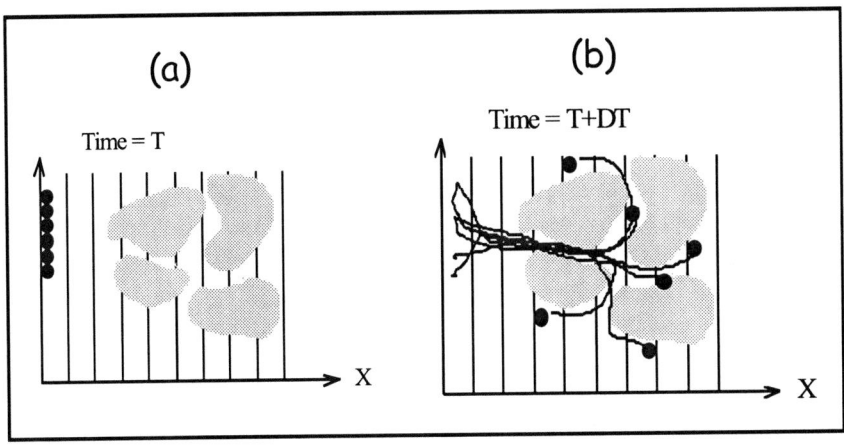

FIGURE 6. Illustration of the hydrodynamic dispersivity induced by the variability of velocities within pores of different sizes. (a) Solute particles are released in variable flow field at time T. (b) Solute particles are observed at different location at time T + DT and generates a dilution of concentrations.

When water flows through the porous medium, the deviation of the local scale advection velocities around the mean advection velocity $\overline{v}$ induces hydrodynamic dispersion. At the pore scale, streamline density increases from the pore wall to the pore center. For real and irregularly shaped soil pores however, this within pore variability will be more pronounced, as streamline densities change between pore bodies and pore necks. The heterogeneous pore size distribution with larger flow velocities in larger pores leads to an important additional variability of velocities (Fig. 6).

The advection velocity variability generates additional 'advective particle displacements', relative to the mean particle displacement $\overline{v}$. Analogous to Brownian motion, the effect of a large number of independent 'advective particle displacements' on solute transport can be described for the 1-D case as a Fickian dispersive process:

$$J_{disp} = -\theta D . \frac{\partial C_i^r}{\partial z} \qquad (9)$$

where D [L2 T-1], the hydrodynamic dispersion coefficient. Crucial for applying Eq. [8] is the independence of advective particle displacements. This implies that a solute particle samples a representative set of the advection velocity distribution on its trajectory. The time needed for a particle to sample a representative set of velocities at the macroscopic scale is the characteristic mixing time τ^*. Although the hydrodynamic dispersion has an analogous effect on solute transport as diffusion, i.e. it tends to spread out the concentration differences, hydrodynamic dispersion is strongly influenced by the advective velocity $\overline{V}$. The exact relationships between D and $\overline{V}$ can only be obtained from theoretical considerations for simple or hypothetical geometrical pore systems. For small values of $\overline{V}$, chemical diffusion is the dominant process, and D is independent of $\overline{V}$ and reaches a value slightly lower than D0 (due to matrix tortuosity effects on the diffusion process). In Taylor's dispersion theory for solutes in a cylindrical tube with a laminar flow field, $D = r^2 . \overline{V}^2 / 48 . D_0$ (Taylor, G.I. (1953)), with r [L], the radius of the tube. In this case, mixing is caused by lateral diffusion so that τ^* does not change with $\overline{V}$. According to the Hagen-Poiseuille relationship, the variance of velocities in a capillary with laminar flow increases quadratically with $\overline{V}$ and so does D. This regime applies for lower flow rates when the mixing caused by local diffusion is faster than the mixing caused by advection of particles into regions with a different particle velocity. For sufficiently high $\overline{V}$, advective mixing due to divergence and convergence of streamlines becomes the dominant mixing process. When the geometry of the water filled pore space remains constant with $\overline{V}$, i.e. in saturated media, and mixing is dominated by advection, the mixing time τ^* is inverse proportional to $\overline{V}$ whereas the variance of velocities increases quadratically with $\overline{V}$. As a consequence, D increases linearly with $\overline{V}$: $D = \lambda . \overline{V}$, with λ [L] a soil characteristic parameter called hydrodynamic dispersivity. Saffman, P.G. (1959) derived the theoretical relationship between D and $\overline{V}$ for saturated networks of pores. At the larger scale, e.g. the scale of an aquifer, the variance of advection velocities is determined by the spatial variability of the hydraulic conductivity. The relation between D and $\overline{V}$ at the larger scale depends, similarly to the relation at the pore scale, on the dominant mixing process: diffusion or advection.

Averaging the pore scale transport process over the REV and assigning the average properties to the centroid of the REV results in continuous functions in space of the hydrodynamic properties and state variables. As for the flow equation (1), differential calculus can be applied to establish mass and momentum balance equations for infinitesimal small soil volume and time increments. For the case of inert solute transport in a macroscopic homogeneous soil, the general continuity equation applies:

$$\frac{\partial C_t^r}{\partial t} = -\frac{\partial Js}{\partial z}. \qquad (10)$$

Considering in this case that the solutes only resides in the liquid phase (i.e. $C_t^r = \theta.C_l^r$) and decomposing the total solute flux in a convective and dispersive component yields the classical convection dispersion model (CDE):

$$\theta\frac{\partial C_l^r}{\partial t} + \bar{v}.\frac{\partial C_l^r}{\partial z} - \frac{\partial}{\partial z}\left(\theta D \frac{\partial C_l^r}{\partial z}\right) = 0 \qquad (11)$$

The classical CDE equation has been intensively used to model solute transport in saturated porous media (1989). In partially saturated soils, the CDE model is usually considered to be more appropriate to describe transport in repacked or non-structured soil (Radcliffe, D.E. et al. (1998); Wilson, G.V. et al. (1998)), but has also successfully been used to describe transport in real structured soils (Bejat, L. et al. (2000); Comegna, V. et al. (1999)).

4. Solvers for the Governing Flow and Transport Equations

For relatively simple initial and boundary conditions and for simplified representations of the heterogeneity of natural porous media, simple analytical and semi-analytical solutions for (1) and (11) exist (e.g. Wooding, R.A. (1968); Philip, J.R. (1969); Simunek, J. et al. (1999); Simunek et al., 1999), resulting in explicit expressions of moisture content, pressure head or concentrations as a function of space and time. Analytical solutions are often based on the transformation of the partial differential equations in the Laplace or Fourier domain to separate variables or the application of Green's function (e.g. Leij, F.J. et al. (2000)). When compared to numerical solutions, analytical solutions are mathematically more rigorous and exact but also much faster to implement. Analytical models are therefore often proposed in nutrient and pesticide management studies, especially when a large number of simulations needs to be performed.

For more complicated descriptions of the variability of the soil properties and of the flow and transport boundary conditions, different numerical methods are used such as finite difference or finite element integration (van Genuchten, M.Th. and Simunek, J. (1996)). The flexibility with which the boundary conditions and the soil variability can be described makes numerical models particularly attractive tools in chemical management. In a simplified form, only 1-D vertical transport in the field is considered and the 1-D forms of eqs. (1) and (11) are numerically integrated.

For a detailed register of agro-system models the reader is referred to dedicated web sites such as the pf-models site (http://www.pfmodels.org), the CAMASE site (http://library.wur.nl/camase/), the REM site (http://eco.wiz.uni-kassel.de/model_db/), and many others.

It should be noted however that the numerical integration introduces an artificial error referred to as numerical dispersion. This numerical error will depend on the quality of the adopted numerical scheme. Recently, Vanderborght J.et al.(2004) reviewed different numerical models which solves governing flow equations (1) and (11) and compared them to some analytical benchmarks as to quantify this numerical dispersion. An exemple of such benchmark results is given in Fig. 7.

To deal with larger scale applications and as such with the specific horizontal variability of soil and land use processes, the quasi 3-D spatial distributed model is often implemented. In its simplest form, this is done by linking a multiple set of 1-D solutions in a spatially distributed modeling approach. In such an approach, there is no interaction between the vertical columns representing the unsaturated zone. In these approaches, it is not possible to distinguish between drainage fluxes to local surface waters and leaching fluxes into deeper aquifers. Examples of this approach are published by Tiktak, A. et al. (2002); Tiktak, A. et al. (2004). In a more sophisticated form, the vertical columns are linked with a regional-scale hydrological model, assuring a proper description of the lower boundary conditions for the 1-D model. The most sophisticated models offer a solution to the full 2-D or 3-D forms of eqs. (1) and (11). This approach is not often used in large-scale contaminant transport modelling, because it involves powerful numerical algorithms (Feyen, J. et al. (1998)).

5. Conceptual Model Limitations with Classical Flow and Transport Models

Although easily established from a conceptual point of view, it is important to realize that the governing flow and transport equations (1) and (11) rely on a series of simplifying assumptions such as i) the existence of a REV; ii) Darcy's law is valid for the soil porous system; iii) the osmotic, geo-static, and electrochemical gradients in the soil water potential are insignificant, iv) the fluid density is independent of solute concentration and temperature; v) the matrix and fluid compressibilities are small; vi) the effective phenomenological properties like the hydraulic conductivity relationship k(h) can be defined; and vii) equilibrium in water pressures and solute concentrations for a Darcian scale REV.

Modeling errors at the conceptual level therefore arise when processes are inappropriately described in the given model or when process descriptions are used in an application for which they were not initially conceived.

We address here two conceptual problems: i) the modelling of preferential flow problems with classical matrix flow concept; and ii) the modelling with the classical CDE concept.

5.1. LIMITATIONS OF THE MATRIX FLOW CONCEPT

For flow, the ignorance of preferential flow is a major point of concern. Preferential flow is a process for which a consensus exists that it is extremely relevant for describing chemical transport in soils (Flühler, H. et al. (2001)), yet in many management models it remains as an example of incomplete conceptualization.

In the context of transport, the presence of macropores plays a particular role. Macropores are large pores, which form at the macroscopic level an obviously distinguished pore system from the soil matrix pore system. Macropores constitute sometimes a separate and/or continuous network in which particle velocities may deviate systematically from those in the soil matrix. As a result, the solutes released in the macropore network will be subjected to a preferential flow as compared to flow in the matrix system and will not completely mix with the total pore water volume at short time intervals. Preferential flow through macropores is considered here as a macroscopic process since concentrations in the macropores cannot easily be determined separately from the concentrations in the micropore system.

Besides preferential flow through macropores, preferential flow may as well occur through certain parts of the soil matrix. Fingered flow e.g. will be caused by wetting front instability (de Rooij, G.H. (2000)) by air entrapment in the matrix (Peck, A.J. (1964)), by water repellency of the solid phase (Ritsema, C.J. and Dekker, L.W. (1994)), or by the increase of the soil hydraulic conductivity with depth (Raats, P.A.C. (1973)). An exemple of fingered flow as observed during infiltration in a sandy soil is given in the Fig. 8.

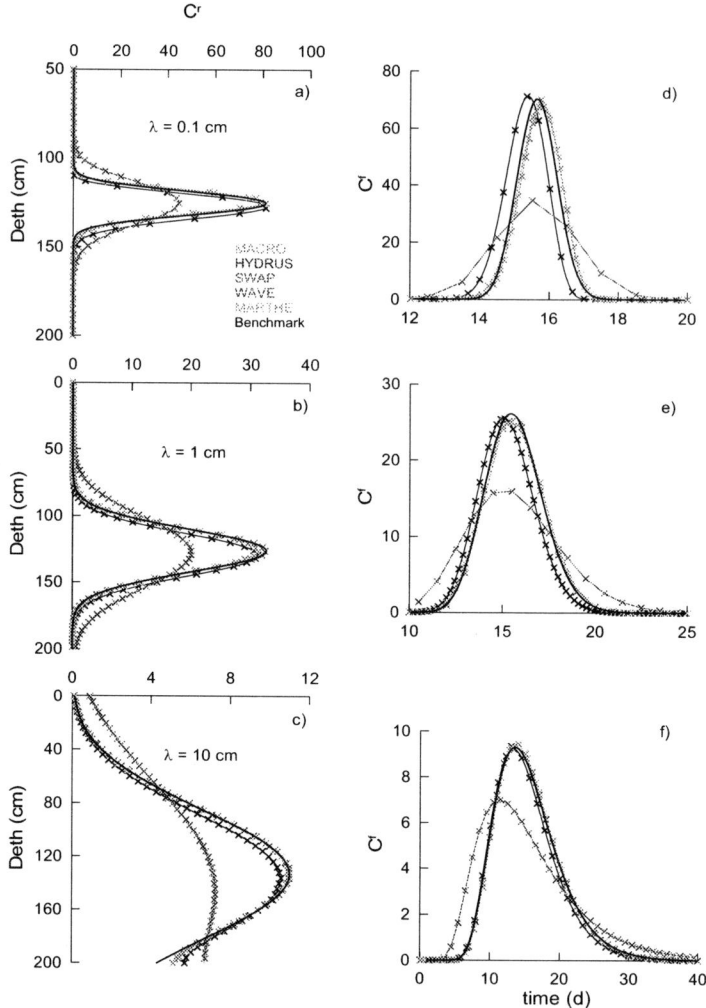

FIGURE 7. Simulated resident concentration, C^r, depth profiles (a,b, and c) and flux concentration, C^f, breakthrough curves at 2 m depth (d,e, and f) for different dispersivities, (Black lines are analytical benchmarks, coloured lines are different numerical solutions). Source : Vanderborght J. et al. (2004).

Funneled flow will typically develop along the inclined bottom of a fine layer overlaying a coarse sub-layer (Kung, K.-J.S. (1990);Poletika, N.N. et al. (1992); Quisenberry, V.L. et al. (1994)). An exemple of observed funnel flow is given in the Fig. 9.

Resistance to using preferential flow models in chemical management has been attributed to the lack of a generally accepted model concept, and the lack of robust techniques that allow prediction of preferential flow. Better observations of preferential flow, made by means of new techniques such as soil tomography, dye tracing and hydro-geophysics such as TDR or GPR (Lambot, S. et al. (2003)), will definitely lead to better understanding of the processes and patterns of preferential flow and will be the basis for improved modelling approaches (Deurer, M. et al. (2003)).

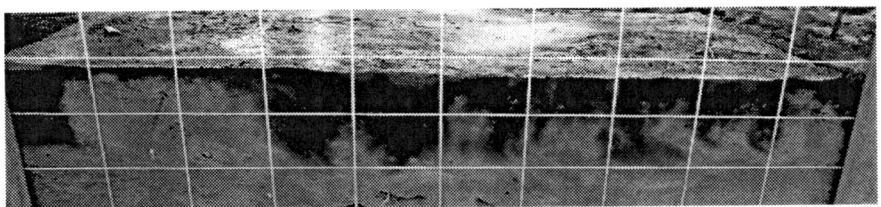

FIGURE 8. Example of observed preferential infiltration in an initial dry homogeneous sand (Source: Bielders C., personal communication).

Indeed, many preferential flow models are bedeviled by an excess of parameters that defy measurement (Gerke, H.H. and van Genuchten, M.Th. (1993)). Preferential flow models are usually applied in an a-posteriori parameter identification approach which has so far limited the range of potential applications as a management tool (Flühler, H. et al. (2001)). However, the recently developed FOCUS scenarios for pesticide exposure assessments for surface water represent one example of how preferential flow models may be used in a management context FOCUS (2001). The preferential flow model MACRO (Larsbo, M. and Jarvis, N. (2003)) is used to calculate pesticide inputs to surface water by subsurface drainage systems for six scenarios representative of drained land in the EU. Four of these scenarios are pre-calibrated, for example with respect to site hydrology and non-reactive solute transport, which should considerably reduce the inherent predictive uncertainty when using preferential flow models.

5.2. LIMITATIONS OF THE CDE

The CDE process model is probably the most popular solute transport model that has been used in soil hydrological research so far. The model considers the soil as a homogeneous matrix in which the pores are well connected, and in which solute mix perfectly laterally when travelling through the soil. The differential description of the process is given in (11) and simplifies for a homogeneous soil profile and steady state conditions to:

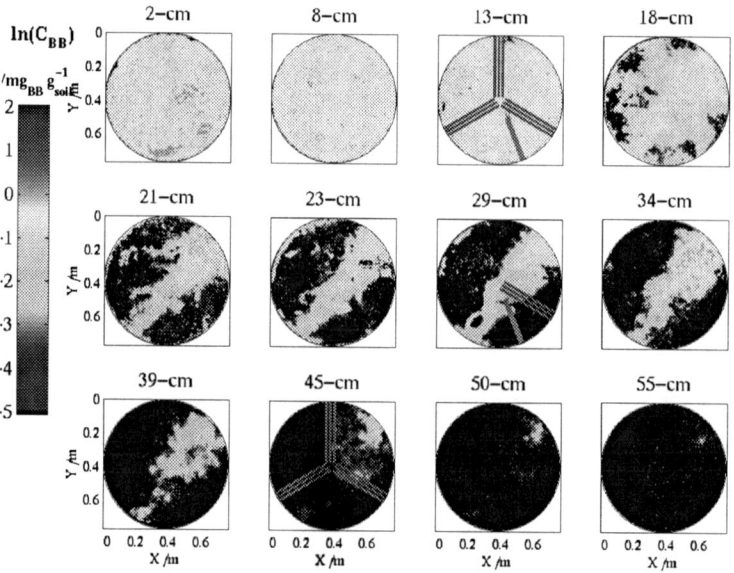

FIGURE 9. Observed preferential flow in an undisturbed sandy lysimeter. Results of a brilliant blue tracer experiment is shown. A discontinuous clay layer is situated at a depth of 20–30 cm causing the brilliant blue to funnel in a preferential flow path. (Source: Javaux et al., 2005).

$$\frac{\partial C_i^r}{\partial t} + \overline{v}\frac{\partial C_i^r}{\partial z} - D\frac{\partial^2 C_i^r}{\partial z^2} = 0 \qquad (12)$$

A characteristic of the CDE travel time distribution is that the variance of the travel times grows linearly with travel distance z. This is equivalent to the particle location distribution, which grows linearly with time for a Brownian motion process. As such, it is essential in the derivation of (12) that the hydrodynamic dispersion can be described as a diffusion process, i.e. on average, all solute particles are subjected to the same forces and the transport time is sufficiently large so that the incremental microscopic particle displacements are no longer statistically correlated. As a corollary, the CDE process cannot be valid for small soil volumes where the travel times are too small as compared to the mixing time, or to describe transport close to interfaces.

A comparison between the particle travel time and the mixing time during a transport experiment forms the basis for discrimination between different 'transport processes' or 'mixing regimes' (Simmons, C.S. (1982)). If the 'mixing' time $\tau*$ is much smaller than the solute travel time, the dispersion of a solute plume can be described as a Fickian, concentration gradient-driven process (1989).

For stationary flow conditions, D and $\bar{v}$ are independent parameters describing the transport process. In transient conditions, however, the relationship between D and $\bar{v}$ must be taken into account. Experimental evidences show that for transport in homogeneous saturated porous media, D is a monotoneous function of $\bar{v}$. In unsaturated media, this relation becomes extremely complicated since the transport volume θ_{eff} changes with the water flux. Therefore, the structure of the water filled pore-space and, hence, the flow field depends on the saturation degree (Flury, M. et al. (1995)) so that the variance of local velocities and the mixing time cannot be simply related to the mean advection velocity. As a consequence, no validated theoretical models exist to calculate the relationship between D and $\bar{v}$ for unsaturated soils and the dispersivity λ cannot be considered to be a material constant, i.e. independent of θ.

It has been observed that the variance of local advection velocities increase dramatically with increasing flow rate, especially when macropores are activated (White, R.E. et al. (1984); Dyson, J.S. and White, R.E. (1987),Bouma, J. (1991)). On the other hand, a decrease in water saturation may result in larger tortuosity of the solute trajectories, a disconnection of continuous flow paths and a physical non-equilibrium due to a slow diffusive exchange of solutes between mobile and immobile pore regions. Therefore, in some experimental studies (e.g.Corey, J.C. et al. (1963); De Smedt, F. et al. (1986); Maraqa, M.A. et al. (1997); Padilla, I.Y. et al. (1999)), a higher solute dispersion was found for unsaturated than for saturated flow conditions.

Flow processes affected by the heterogeneous structural properties of real soils prevailing at different spatial scales, contribute to an incomplete mixing of the solutes during the transport in real soils. At the microscopic scale, structural heterogeneity of the pore liquid system is characterized by the complex geometry of the pore liquid phase and irregular distribution of the carrying liquid water phase in the soil. The heterogeneity of the soil pore systems is expressed by the presence of micro-aggregates, aggregates and pedons in real soil horizons. Therefore, under such conditions, CDE fails to describe the observed dispersion (e.g. Butters, G.L. and Jury, W.A. (1989); Khan, A.U.H. and Jury, W.A. (1990)). Evidence of non-CDE behavior is obtained when the variance of the solute travel time does not grow linearly with depth, or, similarly when the variance of the solute travel distance does not grow linearly with time.

Scale-variable dispersivity was observed in saturated and unsaturated porous media at the laboratory and the field scale, as well in homogeneous as in heterogeneous formations (Pickens, J.F. and Grisak, G.E. (1981); Huang, K. et al. (1995); Vanclooster, M. et al. (1995); Pang, L. and Hunt, B. (2001); Javaux, M. and Vanclooster, M. (2003); Javaux and Vanclooster, 2003)Abbasi, F. et al. (2003). Studying solute transport through two undisturbed Spodosols monolithes (0.8 m i.d., 1 m deep), Seuntjens, P. et al. (2001) obtained opposite depth scaling behaviors of the dispersivity and related this to the different morphological characteristics of the two pedons.

At the field scale, the soil hydraulic properties such as the soil hydraulic conductivity and water retention curve are variable in space (Jury, W.A. et al. (1987); Mallants, D. et al. (1996a)). The macroscopic variability of the flow and transport properties at the field scale will be expressed in a variable flow field, thereby deviating locally the direction of the macroscopic particle velocity vectors. The mixing of solutes from e.g. low to high conductive zones at these larger scales will only be fully established if again the process is evaluated at sufficiently large transport volumes or transport times (Flühler, H. et al. (1996)). A full characterization of the dispersion process at the field scale would therefore need a full description of the variability of the flow properties. This, however, remains a complicated task, which for the time being is only realizable in a research oriented context.

Van Wesenbeeck, I.J. and Kachanoski, R.G. (1995), Kasteel, R. et al. (2000) and Vereecken, H. et al. (1991) predicted solute dispersion observed in field-scale leaching experiments from the spatial variability of measured soil hydraulic properties. Vanderborght, J. et al. (1997) concluded that the mixing regime could be fairly well reproduced for smaller flow rates if small correlation scales of the hydraulic properties were assumed. For higher flow rates, the dispersion was largely underestimated owing to a large variability of particle velocities at the pore-scale when macropores were activated. Based on a summary of leaching experiments that were carried out in a range of soils at different scales and for different leaching rates Vanderborght et al. (2001) linked the solute transport regime to morphological and hydraulic properties. The change of dispersivity with leaching rate was linked to the unsaturated hydraulic conductivity using a multi-domain conceptualization of the pore space as was introduced by Steenhuis, T.S. et al. (1990). The mixing regime was further related to soil morphological features, such as vertical tongues, stratification, macropores, and a water repellent layer. In all investigated soils, the hydrodynamic dispersion increased more than linearly with increasing leaching rate confirming that the dispersivity is not an intrinsic soil characteristic for unsaturated flow conditions at this scale of observation.

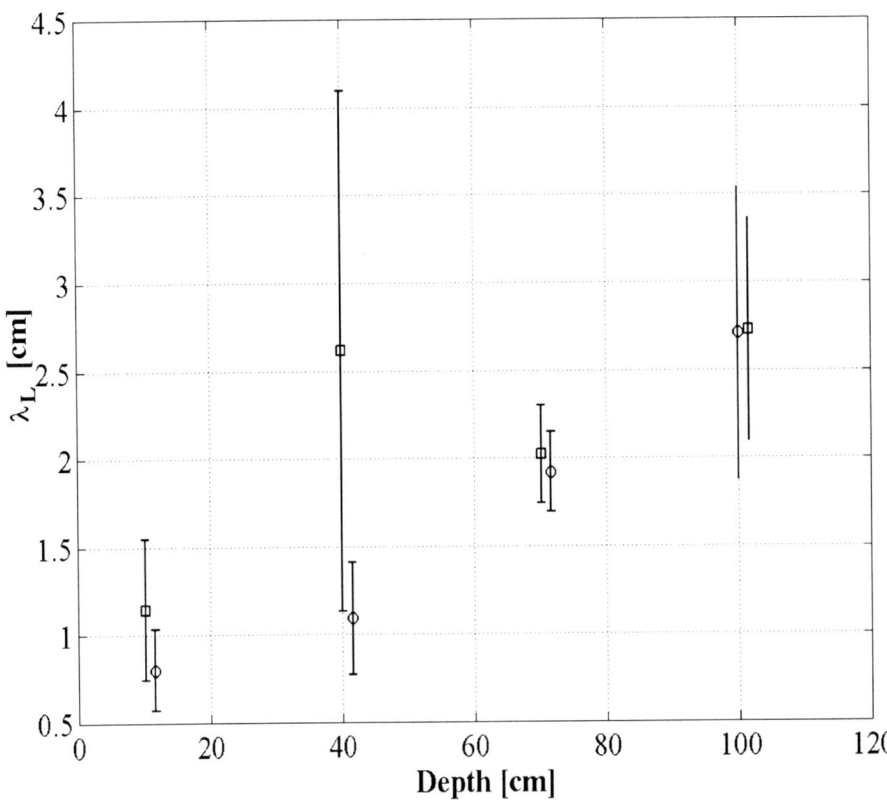

FIGURE 10. Example of depth scaling of meso-scale hydrodynamic dispersivity. Results are represented for transport experiments performed at the scale of an undisturbed sandy soil monolith (Source: Javaux, M. and Vanclooster, M. (2003)).

Vanderborght, J. et al. (1997) concluded that the mixing regime could be fairly well reproduced for smaller flow rates if small correlation scales of the hydraulic properties were assumed. For higher flow rates, the dispersion was largely underestimated owing to a large variability of particle velocities at the pore-scale when macropores were activated.

6. Parameter Estimation Problem

Extensive literature is now available to illustrate the often large variability in space and time of the material properties affecting chemical transport in soils. The variability is present at different spatial scales ranging from the pore scale (Cislerova, M. (1999)), the core scale (Vanderborght, J. et al. (1999)), the field scale (Mallants, D. et al. (1996b); Jacques, D. et al. (1998); Ritsema, C.J. et al. (1993); Ritsema et al., 1998;), to the landscape and regional scales (Roth, K. et al. (1999)). The temporal variability of material properties has often been ignored but is also clearly present. The impact of mechanical stress on the soil hydraulic properties is well illustrated in the literature. However, other factors may also cause temporal variability in transport characteristics. Moutier, M. et al. (1999) and Toride, N. (1999), for instance, illustrated the temporal change of the unsaturated hydraulic properties as a function of water quality parameters. Vanderborght, J. et al. (1997) and Javaux, M. and Vanclooster, M. (2003) illustrated the temporal variability of the solute dispersion length in terms of the governing flow regime. Input and parameter generation problems arise when modelling data are not available to deal with the extreme spatio-temporal variability of the system within the management application exercise.

Despite the difficulty of spatio-temporal variability in the key properties that determine flow and transport through soil, new measurement technologies are being developed to provide direct measures of the parameters required for modeling chemical transport through field soils. The hydraulic properties of topsoil can be measured, reasonably easily, in-situ, using disc permeameters, or as they might be known 'tension infiltrometers' (Perroux, K.M. and White, I. (1988)). With these devices, the spatial and temporal changes in the topsoil's transport properties can be quickly resolved (Messing, I. and Jarvis, N. (1993)). Furthermore, using tracers in disc permeameters, the preferential flow properties of the mobile water fraction in structured soils can now be resolved (Clothier, B.E. et al. (1995); Jaynes, D.B. et al. (1995)) to provide direct parameter input into preferential-flow models By using multiple tracers, including reactive compounds, the adsorption isotherm of invading solutes that exchange with the soil's matrix can now be determined. Measurement technologies for direct parameterization of the soil's hydraulic and transport properties are now available, albeit there is a certain degree of effort required to establish the spatio-temporal pattern in these parameters at the pedon scale. Direct measurement techniques are extensively reviewed by (2005) and (2002).

Another alternative consist of inverse modelling approaches. Inverse modelling consist in estimating the flow and transport parameters by fitting solutions of the flow and transport problems on observed flow and transport. Reviews of inverse modelling are given by Hopmans, J.W. et al. (2002); Durner, W. et al. (1999) and Lambot, S. et al. (2005). An illustration of combined advanced identification techniques consisting of the combination of the inversion of an electromagnetic model and a hydrodynamic model is given in Lambot, S. et al. (2004).

At the regional scale, other means of parameterization are required. A typical example is the evaluation of large-scale non-point source pollution with spatially distributed modeling approaches, which very often rely on the availability of the soil's physico-chemical properties at the scale of each grid of a constructed soil information system. Unfortunately, only limited hard data are available in most soil information systems. Grid scale modeling parameters need to be generated by interpolation, extrapolation, geo-statistics, pedo-transfer functions and the like. Pedotransfer functions will play an important role in this context.

Pedotransfer functions allow us to bridge this gap by translating the basic soil data into functional data. Most available pedotransfer functions, however, have been developed using data from small-scale samples. Given the scaling problem, it would be unsound to consider a pedotransfer approach as a way to obtain effective functional model parameters of, for example, a grid in a spatially distributed model. The role of pedotransfer functions is not to give an exact effective parameter, but rather to generate a realistic *a-priori* parameter that constrains the parameter space in a more generic Bayesian type parameter estimation framework. Many pedotransfer functions have been described in the literature. Good reviews of the use of pedotransfer functions in hydrology are given by Pachepsky, Y. et al. (1999); Pachepsky, Y.A. and Rawls, W.J. (2003) and in a series of papers in (1999). Vanclooster, M. et al. (2004) reviewed approaches for improving pedotransfer functions for solute transport.

In the up-scaling procedures, one is confronted with the core of the scale problem, i.e. the uniqueness in time and space of a transport event and the non-linearity of the transport process. Indeed, each chemical transport event is unique in place and time and a perfect repetition of this event can never occur. Hence, a model inferred from an observation in a given space and time framework can never be tested since this observation is unique. The non-linearity of the transport process further suggests that the transport parameters cannot simply be averaged at the grid or time step scale (Jarvis, N. et al. (2000)).

7. Model User Experience and Good Modelling Practice

A lack of good modelling practice restrains the advanced use of modelling for soil management. In the past, most modelling work was performed within an academic context. The modelling codes were often poorly documented and limited in pre- and post-processing capabilities. The lack of appropriate interfacing, and advanced pre- and post-processing capabilities introduced an additional and often insurmountable burden for the soil manager. The complexity of the modelling process and the lack of operational modelling guidelines further hampers the introduction of advanced transport codes in operational nutrient and pesticide management. This also introduces an additional risk of modelling error due to user subjectivity, as was clearly illustrated by Brown, C.D. et al. ; Jarvis, N. et al. (2000). Boesten, J.J.T.I. compared the estimation of the pesticide half life at 10 °C of ethoprophos and bentazone as estimated by 20 model users using a common laboratory degradation study at reference temperature.

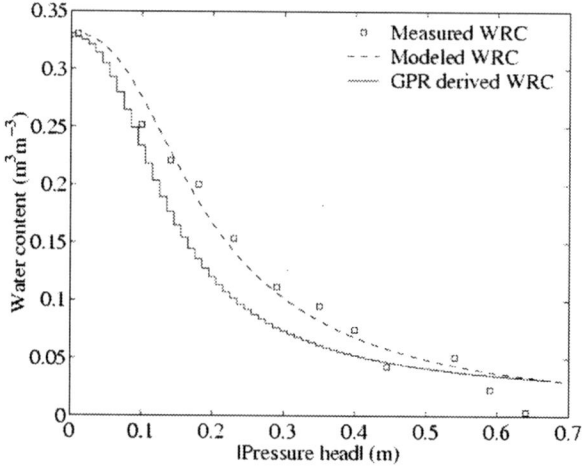

FIGURE 11. Estimated soil moisture retention characteristic from a combination of electromagnetic inversion and flow inversion (Source:Lambot, S. et al. (2004)).

The coefficient of variation of estimated half life were 29% and 46% for ethoprophos and bentazone, respectively. The principal cause of this important user subjective variability was the lack of guidance on the transformation rate dependency on soil temperature. Brown, C.D. et al. compared the outputs from three models operated by five modellers. Differences between the output data from the five modellers using the same model were of a similar magnitude to the variation associated with field measurements. They concluded that model development should seek to reduce subjectivity in the selection of input parameters, and improve the guidance available to users where subjectivity cannot be eliminated.

The appropriate use of transport models for risk management is only possible if the different actors in the management process receive appropriate training. Often the surprisingly poor modelling leads to a user inter-comparison ring test, which clearly elucidates the need for advanced education and training in chemical transport modellng, as well as the implementation of strict guidelines for good modelling practice. The model user is responsible for understanding the model and its appropriate use. He is also responsible for estimating the model parameters and the input for the selected scenarios. He must further keep in touch with the evolution of the different model versions and documentation. He is further responsible for developing modelling reports that contain sufficient and reliable information. Most of the state-of-the-art modelling approaches are developed by the research community and need to be disseminated to soil professionals. In many cases, a tremendous gap still exists between models available in the research community and those used in management applications. However, some progress is being made. The FOCUS modeling tools and scenarios developed for EU-harmonized pesticide registration (e.g. FOCUS (2002)) represent a good example of how this gap can be bridged. Models originally developed in the research arena have been equipped with user-friendly interfaces to improve their useability for this specific purpose. Therefore a plea is made to further upgrade existing scientific models into useful engineering tools, to improve the training of the potential model user, and to implement strictly the concepts of 'Good Modeling Practice'. The latter idea is to make the modelling process completely transparent by documenting each step of the modelling process such that it can be independently executed by any other model user.

References

Abbasi, F., Jacques, D., Simunek, J., Feyen, J., and Van Genuchten, M.T. (2003) Inverse Estimation of Soil Hydraulic and Solute Transport Parameters From Transient Field Experiments: Heterogeneous Soil. *Transactions of the Asae* 46 (4), 1097–1111.

Alvarez-Benedi, J. and Munoz-Carpena, R. (2005) *Soil water solute process characterization: an intergrated approach.* 778 pp, CRC , Boca Raton, London, New York, Washington.

Bejat, L., Perfect, E., Quisenberry, V.L., Coyne, M.S., and Haszler, G.R. Solute transport as related to soil structure in unsaturated intact soil blocks. Soil-Science-Society-of-America-Journal 64[3], 818–826. 2000.

Boesten, J.J.T.I. Modeller subjectivity in estimating pesticide parameters for leaching models using the same laboratory data set. *Agricultural Water Management* 44 389–409.

Bouma, J. (1991) Influence of soil macroporosity on environmental quality. *Advances in Agronomy* 46 (1–37).

Brown, C.D., Baer, U., Günther, P., Trevisan, M., and Walker A. Ring test with models LEACHP, PRZM-2, VARLEACH: Variability between model users in prediction of pesticide leaching using a standard data set. *Pesticide Science* 47 249–258.

Butters, G.L. and Jury, W.A. Field scale transport of bromide in unsaturated soil. 2. Dispersion modeling. Water Resources Research 25[7], 1583–1589. 1989.

Cislerova, M. (1999) Characterization of pore geometry. In: Feyen J. and Wiyo, K. (eds), pp. 103–117, Wageningen Pers, Wageningen, the Netherlands.

Clothier, B.E., Heng, L., Magesan, G.N., and Vogeler, I. The Measured Mobile-Water Content of an Unsaturated Soil as a Function of Hydraulic Regime. Australian-Journal-of-Soil-Research 33[3], 397–414. 1995.

Comegna, V., Coppola, A., and Sommella, A. Nonreactive solute transport in variously structured soil materials as determined by laboratory-based time domain reflectometry (TDR). Geoderma 92[3–4], 167–184. 1999.

Corey, J.C., Nielsen, D.R., and Biggar, J.W. (1963) Miscible displacement in saturated and unsaturated sandstone. *Soil Sci. Soc. Am. Proc.* 27 258–262.

Dagan, G. (1989) *Flow and transport in porous formations*. Springer Verlag, Berlin.

Dane, H.J. and Topp, G.C. (2002) *Method of soil analysis. Part 4. Physical methods*. Soil Science Society of America, Inc, Madison, WI, USA.

de Rooij, G.H. Modeling fingered flow of water in soils owing to wetting front instability: a review. Journal of Hydrology 231–232, 277–294. 2000.

De Smedt, F., Wauters, F., and Sevilla, J. Study of tracer movement through unsaturated sand. Journal-of-Hydrology 85, 196–181. 1986.

Deurer, M., Green, S.R., Clothier, B.E., Böttcher, J., and Duijnisveld W.H.M. Drainage networks in soils. A concept to describe bypass-flow pathways. Journal of Hydrology 272, 148–162. 2003.

Durner, W., Schultze, B., and Zurmul, T. State-of-the-art in inverse modeling of inflow/outflow experiments. M. T. van Genuchten and F. J. Leij and L. Wu. Proceedings of the International Workshop on Characterization and Measurement of the Hydraulic Properties of Unsaturated Porous Media. 661–681. 1999. University of California Press, Riverside, CA.

Dyson, J.S. and White, R.E. A comparison of the convection-dispesion aquation and transfer function model for predicting chloride leaching through an undisturbed, structured clay soil. Journal of soil science 38, 157–172. 1987.

Feyen, J., Jacques, D., Timmerman, A., and Vanderborght, J. Modelling water flow and solute transport in heterogeneous soils: A review of recent approaches. Journal-of-Agricultural-Engineering-Research 70[3], 231–256. 1998.

Flühler, H., Dürner, W., and Flury, M. Lateral solute mixing processes - A key for understanding field-scale transport of water and solutes. Geoderma 70, 165–183. 1996.

Flury, M., Leuenberger, J., Studer, B., and Flühler, H. Transport of anions and herbicides in a loamy and a sandy field soil. Water Resources Research 31[4], 823–835. 1995.

Flühler, H., Ursino, N., Bundt, M., Zimmerman, U., and Stamm, C. The preferential flow syndrome a buzzword or a scientific problem ? D.C. Flanagan and Ascough, J. C. *Soil erosion research for the 21st century symposium and 2nd international symposium on preferential flow*. 2001. Hawai, USA, ASAE.

FOCUS. FOCUS surface water scenarios in the EU evaluation process under 91/414/EEC. 2001. EC-DG Sanco.

FOCUS. Generic guidance for FOCUS groundwater scenarios. Version 1.1. 2002. EC.

Gerke, H.H. and van Genuchten, M.Th. (1993) A dual porosity model for simulating the preferential movement of water and solutes in structured porous media. *Water Resources Research* 29 305–319.

Hopmans, J.W., Simůnek, J., Romano, N., and Durner, W. (2002) Inverse methods. In: Dane, J.A. and Topp, G.C. (eds), Soil Sci. of Am. Book Series 5, Madison, Wisconsin, USA.

Huang, K., Toride, N., and van Genuchten, M.T. Experimental investigation of solute transport in large, homogeneous and heterogeneous, saturated soil columns. Transport-in-Porous-Media 18[3], 283–302. 1995.

Jacques, D., Kim, D.J., Diels, J., Vanderborght, J., Vereecken, H., and Feyen, J. (1998) Analysis of Steady State Chloride Transport Through Two Heterogeneous Field Soils. *Water Resources Research* 34 (10), 2539–2550.

Jarvis, N., Brown, C.D., and Granitza, E. (2000) Sources of error in model predictions of pesticide leaching: a case study using the MACRO model. *Agricultural Water Management* 44 247–262.

Javaux, M. and Vanclooster, M. (2003) Scale- and rate- dependent solute transport within an unsaturated sandy monolith. *Soil Science Society of America Journal* 67 (5), 1334–1343.

Jaynes, D.B., Logsdon, S.D., and Horton, R. (1995) Field method for measuring the mobile/immobile water and solute transfer rate coefficient. *Soil Science Society of America Journal* 59 352–356.

Jury, W.A., Russo, D., Sposito, G., and Elabd, H. The spatial variability of water and solute transport properties in unsaturated soil. I. Analysisof propertyuvariations and spatial structure with statistical models. Hilgardia 55[4], 1–32. 1987.

Kasteel, R., Vogel, H.J., and Roth, K. (2000) From Local Hydraulic Properties to Effective Transport in Soil. *European Journal of Soil Science* 51 (1), 81–91.

Khan, A.U.H. and Jury, W.A. A laboratory study of the dispersion scale effect in column outflow experiments. Journal-of-Contaminant-Hydrology 5, 119–131. 1990.

Kung, K.-J. S. Preferential flow in a sandy vadose zone: 1. Field observation. Geoderma 46, 51–58. 1990.

Lambot, S., Antoine, M., van den Bosch, I., Slob, E.C., and Vanclooster, M. (2004) Electromagnetic inversion of {GPR} signals and subsequent hydrodynamic inversion to estimate effective vadose zone hydraulic properties. *Vadose Zone Journal* 3 1072–1081.

Lambot, S., Javaux, M., Hupet, F., and Vanclooster, M. (2005) Inverse modeling techniques to characterize transport processes in the soil-crop continuum. In: Alvarez-Benedi, J. and Munoz-Carpena, R. (eds), pp. 693–713, CRC Press, Boca Raton, London, New York, Washington.

Lambot, S., Slob, E.C., van den Bosch, I., Stockbroeckx, B., Scheers, B., and Vanclooster, M. GPR design and modeling for identifying the shallow subsurface dielectric properties. A. Yarovoy. Proceedings of the 2nd International Workshop on Advanced Ground Penetrating Radar. 130–135. 2003. Delft, The Netherlands, Delft University of Technology.

Larsbo, M. and Jarvis, N. MACRO5.0: a model of water flow and solute transport in macroporous soil. Technical description. 2003. Uppsala, Sweden, SLU, Department of soil sciences.

Leij, F.J., Priesack, E., and Schaap, M. (2000) Solute transport modeled with green's functions with application to persistent solute sources. *Journal of Contaminant Hydrology* 41 155–173.

Leij, F.J. and van Genuchten, M.T. Characterization and measurement of the hydraulic properties of unsaturated porous media. M.T. van Genuchten and F.J. Leij and L. Wu. {Proceedings of the International Workshop on Characterization and Measurement of the Hydraulic Properties of Unsaturated Porous Media}. 1–12. 1999. University of California Press, Riverside, CA.

Mallants, D., Mohanty, B.P., Jacques, D., and Feyen, J. (1996a) Spatial Variability of Hydraulic Properties in a Multi-Layered Soil Profile. *Soil Science* 161 (3), 167–181.

Mallants, D., Vanclooster, M., and Feyen, J. Transect study on solute transport in a macroporous soil. Hydrological-Processes 10[1], 55–70. 1996.

Maraqa, M.A., Wallace, R.B., and Voice, T.C. Effects of degree of water saturation on dispersivity and immobile water in sandy soil columns. Journal-of-Contaminant-Hydrology 25[3–4], 199–218. 1997.

Messing, I. and Jarvis, N. (1993) Temporal variation in the hydraulic conductivity of a tilled clay soil as measured by tension infiltrometers. *Journal of soil science* 44, 11–24.

Moutier, M., Degand, E., and De Backer, L.W. (1999) Temporal changes in unsaturated hydraulic properties due to clay migration and water quality. In: van Genuchten, M.Th., Leij, F.L., and Wu, L. (eds), pp. 517–526, University of California, Riverside, USA.

Pachepsky, Y., Rawls, W.J., and Timlin, D.J. (1999) The current status of pedotransfer functions: their accuracy, reliability and utility in field and regional scale modeling. In: Corwin, D.L., Loague, K., and Ellsworth, T.R. (eds), pp. 223–234, AGU.

Pachepsky, Y.A. and Rawls, W.J. (2003) Soil Structure and Pedotransfer Functions. *European Journal of Soil Science* 54 (3), 443–451.

Padilla, I.Y., Yeh, T.C.J., and Conklin, M.H. The effect of water content on solute transport in unsaturated porous media. Water-Resources-Research 35[11], 3303–3313. 1999.

Pang, L. and Hunt, B. Solutions and verification of a scale-dependent dispersion model. Journal-of-Contaminant-Hydrology 53, 21–39. 2001.

Peck, A.J. (1964) Moisture profile development and air compression during water uptake by bounded porous bodies: 3. Vertical columns. *Soil Science* 100 (1), 44–51.

Perroux, K.M. and White, I. (1988) Designs for disc permeameters. *Soil Science Society of America Journal* 52 1205–1215.

Philip, J.R. (1969) Theory of infiltration. *Advances in Hydrosciences* 5 215–296.

Pickens, J.F. and Grisak, G.E. Scale-dependent dispersion in stratified granular aquifer. Water Resources Research 17[4], 1191–1211. 1981.

Poletika, N.N., Roth, K., and Jury, W.A. (1992) Interpretation of solute transport data obtained with fiberglass wick soil solution samplers. *Soil Science Society of America Journal* 56 1751–1753.

Quisenberry, V.L., Phillips, R.E., and Zeleznik, J.M. Spatial distribution of water and chloride macropore flow in a well-structured soil. Soil-Science-Society-of-America-Journal 58, 1294–1300. 1994.

Raats, P.A.C. Unstable wetting fronts in uniform and nonuniform soils. Soil Science Society of America Journal 37, 681–685. 1973.

Radcliffe, D.E., Gupte, S.M., and Box, J.E. Solute transport at the pedon and polypedon scales. Nutrient-Cycling-in-Agroecosystems 50[1–3], 77–84. 1998.

Ritsema, C.J. and Dekker, L.W. How water moves in a water repellent sandy soil. 2. Dynamics of fingered flow. Water Resources Research 30[9], 2519–2531. 1994.

Ritsema, C.J., Dekker, L.W., Hendrickx, J.M.H., and Hamminga, W. (1993) Preferential Flow Mechanism in a Water Repellent Sandy Soil. *Water Resources Research* 29 (7), 2183–2193.

Roth, K., Vogel, H.J., and Kasteel, R. The scaleway: a conceptual framework for upscaling soil properties in modelling of transport processes in soils at various time and space. International workshop of EurAgEng's field of interest on soil and water. 477–490. 1999.

Saffman, P.G. A theory of dispersion in porous medium. J. Fluid. Mech. 6, 312–349. 1959.

Seuntjens, P., Mallants, D., Toride, N., Cornelis, Ch., and Geuzens, P. Grid lysimeter study of steady-state chloride transport in two Spodosol types using TDR and wick samplers. Journal-of-Contaminant-Hydrology 51, 13–39. 2001.

Simmons, C.S. A stochastic-convective transport representative of dispersion in one-dimensionnal porous media systems. Water-Resources-Research 18, 1193–1214. 1982.

Simunek, J., van Genuchten, M. Th., Sejna, M., Toride, N., and Leij, F.J. The STANMOD computer software for evaluating solute transport in porous media using analytical solutions of the convection-dispersion equation, Version1/0 and 2.0. 1999. Colorado, USA, International Ground Water Modeling Center, Colorado School of Mines.

Steenhuis, T.S., Parlange, J.-Y., and Andreini, M.S. (1990) A numerical model for preferential solute movement in structured soils. *Geoderma* 46 193–208.

Taylor, G.I. The dispersion of matter in solvent flowing slowly through a tube. Proc. London Math. Soc., Ser. A 219, 189–203. 1953.

Tiktak, A., De Nie, D.S., Pineros-Garcet, J.D., Jones, A2.0., and Vanclooster, M. (2004) Assessing the pesticide leaching risk at the pan European level: the EuroPEARL approach. *Journal of Hydrology* 289 222–238.

Tiktak, A., De Nie, D.S., van der Linden, A.M.A., and Kruijne, R. (2002) Modeling the leaching and drainage of pesticides in the Netherlands: The GeoPEARL model. *Agronomie* 22, 373–387.

Toride, N. (1999) Clay dispersion and hydraulic conductivity of low high swelling smectites under sodic conditions. In: van Genuchten, M.Th., Leij, F.J., and Wu, L. (eds), pp. 507–516, University of California, Riverside, USA.

Van Genuchten, M.Th., Leij, F.J., and Wu, L. (1999) *Characterization and measurement of the hydraulic properties of unsaturated porous media*. University of California, Riverside, USA.

Van Genuchten, M.Th. and Simunek, J. (1996) Evaluation of pollutant transport in the unsaturated zone. In: Rijtema, P.E. and Eliáš, V. (eds), pp. 139–172, Kluwer Academic Publishers, Dordrecht, the Netherlands.

Van Wesenbeeck, I.J. and Kachanoski, R.G. (1995) Predicting field-scale solute transport using in-situ measurements of soil hydraulic properties. *Soil Science Society of America Journal* 59, 734–742.

Vanclooster, M., Boesten, J., Tiktak, A., Jarvis, N., Kroes, J., Clothier, B.E., and Green, S. (2004) On the use of unsaturated flow and transport models in nutrient and pesticide management. In: Feddes, R., De Rooij, G., and Van Dam, J. (eds), pp. 331–361, Kluwer Academic Publishers.

Vanclooster, M., Mallants, D., Vanderborght, J., Diels, J., Van Orshoven, J., and Feyen, J. (1995) Monitoring solute transport in a multi-layered sandy lysimeter using time domain reflectometry. *Soil Science Society of America Journal* 59, 337–344.

Vanderborght, J., Gonzalez, C., Vanclooster, M., Mallants, D., and Feyen, J. Effects of soil type and water flux on solute transport. Soil-Science-Society-of-America-Journal 61[2], 372–389. 1997.

Vanderborght, J., Gähwiller, P., Bujuova, S., and Fluhler, H. (1999) Identification of transport processes from concentration patterns and breakthrough curves of fluorescent tracers. In: Feyen, J. and Wiyo, K. (eds), Wageningen Pers, Wageningen, the Netherlands.

Vanderborght, J., Jacques, D., Mallants, D., Tseng, P.-H., and Feyen, J. Comparison between field measurements and numerical simulation of steady-state solute transport in a heterogeneous soil profile. Hydrology and Earth System Sciences 4, 853–871. 1997.

Vanderborght, J., Vanclooster, M., Timmerman, A., Seuntjens, P., Mallants, D., Kim, D.J., Jacques, D., Hubrechts, L., Gonzalez, C., Feyen, J., Diels, J., and Deckers, J. Overview of inert tracer experiment in key belgian soil types: relation between transport, and soil morphological and hydraulic properties. Water Resources Research 37[12], 2873–2888. 2001.

Vanderborght J., R.K.M.H.M.J.D.T.M.V.C.M.a.H.V. (2004) A test of numerical models for simulating for simulating flow and transport in soil using analytical solutions. *Vadose Zone Journal* 4 206–221.

Vereecken, H., Vanclooster, M., Swerts, M., and Diels, J. (1991) Simulating nitrogen behaviour in soil cropped with winter wheat. *Fertilizer Research* 27 (233–243).

White, R.E., Thomas, G.W., and Smith, M.S. Modelling water flow through undisturbed soil cores using a transfer function model derived from ^{3}HOH and Cl transport. Journal of soil science 35, 159–168. 1984.

Wilson, G.V., Yunsheng, L., Selim, H.M., Essington, M.E., and Tyler, D.D. Tillage and cover crop effects on saturated and unsaturated transport of fluometuron. Soil Science Society of America Journal 62[1], 46–55. 1998.

Wooding, R.A. (1968) Steady infiltration from large shallow circular pond. *Water Resources Research* 4 (1259–1273).

FATE OF CHEMICALS IN THE AQUATIC ENVIRONMENT

BOGDANA KOUMANOVA
University of Chemical Technology and Metallurgy
Kliment Ohridsky, 8
1756 Sofia, Bulgaria
e-mail: bogdana@uctm.edu

Abstract. Three quarters of the earth's surface is covered by water and it is not surprising that water serves as ultimate sink for most anthropogenic chemicals. Study of the sources, reactions, transport, effects and fates of chemicals in the water is very important for the water resources protection. In this paper the main properties of water as a medium for many chemicals as well as the most important chemical processes that occur are discussed. The behavior of pesticides, petroleum and related hydrocarbons in water is also discussed.

Keywords: aquatic chemistry; aquatic chemical processes; water pollutants.

1. Aquatic Chemistry

Water has a number of unique properties that are essential to life, many of which are due to water's ability to form hydrogen bonds (Manahan, 1992). The most important properties of water are:

- Excellent solvent making biological processes possible because of the transport of nutrients and pollutants;
- Highest dielectric constant of any common liquid that means high solubility of ionic substances and their ionization;
- Highest surface tension than any other liquid;
- Colorless, allowing light required for photosynthesis to reach considerable depths in water body;

- Higher heat of evaporation determining transfer of heat and water molecules between the atmosphere and water body;
- Maximum density as a liquid at 4°C;
- Higher heat capacity than any other liquid except ammonia that is significant for the stabilization of temperatures of organisms and geographical regions;
- Higher latent heat of fusion (temperature stabilized at the freezing point of water).

Many chemical processes occur in water. They are influenced by the action of algae and bacteria in water. For example, algal **photosynthesis** fixes inorganic carbon from HCO_3^- ion in the form of biomass in a process that also produces CO_3^{2-}:

$$2\ HCO_3^- + h\nu \xrightarrow{Photosynthesis} \{CH_2O\} + O_2 + CO_3^{2-}$$

Carbonate undergoes an acid-base reaction:

$$CO_3^{2-} + H_2O \xrightarrow{Acid-base} HCO_3^- + OH^-$$

or it reacts with Ca^{2+} to precipitate

$$Ca^{2+} + CO_3^{2-} \xrightarrow{Precipitation} CaCO_{3\ (s)}$$

Oxidation-reduction reactions in water involve the transfer of electrons between chemical species. Metal ions in water are always bonded to water molecules in the form of hydrated ions represented by the general formula, $M(H_2O)_x^{n+}$ Metals in water may be bound to organic chelating agents. Hardness is due to the presence of calcium ion and to a lesser extent to magnesium ion.

Many oxidation-reduction reactions are catalyzed by bacteria. For example, bacteria convert inorganic nitrogen to ammonium ion, NH_4^+, in the oxygen-deficient lower layers of water body. Near the surface, where O_2 is available, bacteria convert inorganic nitrogen to nitrate ion, NO_3^-.

In water bacteria use oxidation-reduction reactions to obtain the energy that they need for their own growth and reproduction. Some bacteria require oxygen for their metabolic needs and are called aerobic bacteria. Other anaerobic bacteria extract their oxygen from sources such as NO_3^-, SO_4^{2-} and other matter represented as $\{CH_2O\}$. The most common bacterially mediated reaction in water is the oxidation of organic matter:

$$O_2 + \{CH_2O\} \longrightarrow CO_2(g) + H_2O \quad \text{(aerobic respiration)}$$

Bacteria interact with environmental chemicals in a number of ways. Many organic compounds are partially or completely degraded by bacteria. In some cases the organic products are even more toxic than the original pollutants.

One of the important characteristics of unpolluted water is gas solubility. Gases are exchanged with the atmosphere and various solutes are exchanged between water and sediments. Since it is required to support aquatic life and maintain water quality, oxygen is the most important dissolved gas in water. One of the most important sources of dissolved oxygen in water is oxygen in the atmosphere which dissolves in the water mass at the water surface. Losses of dissolved oxygen are associated with BOD due to organic discharges.

The solubility of molecular oxygen in water is affected by the:
- partial pressure of oxygen gas in contact with the water;
- temperature;
- salinity.

The most important class of complexing agents that occur naturally are the **humic substances** (Manahan, 1989). These are materials formed during the decomposition of vegetation. They are commonly classified on the basis of solubility. If a material containing humic substances is extracted with strong base, and the resulting solution is acidified, the products are:
- a nonextractable plant residue called **humin**;
- a material that precipitates from the acidified extract, called **humic acid**;
- an organic material that remains in the acidified solution, called **fulvic acid**.

These names do not refer to single compounds but to a wide range of compounds of generally similar origin. Because of their acid-base, sorptive, and complexing properties, both the soluble and insoluble humic substances have a strong effect upon the properties of water. In general, fulvic acid dissolves in water and exerts its effects as the soluble species. Humin and humic acid remain insoluble and affect water quality through exchange of species, such as cations or organic materials, with water.

The binding of metal ions by humic substances is one of the most important environmental qualities of humic substances. This binding can occur as chelation between a carboxyl group and a phenolic hydroxyl group, as **chelation** between two carboxyl groups, or as **complexation** with a carboxyl group (Fig. 1).

Another major type of metal species consists of **organometallic compounds**. They differ from complexes and chelates in that the organic portion is bonded to the metal by a carbon-metal bond and the organic ligand is frequently not capable of existing as a stable species.

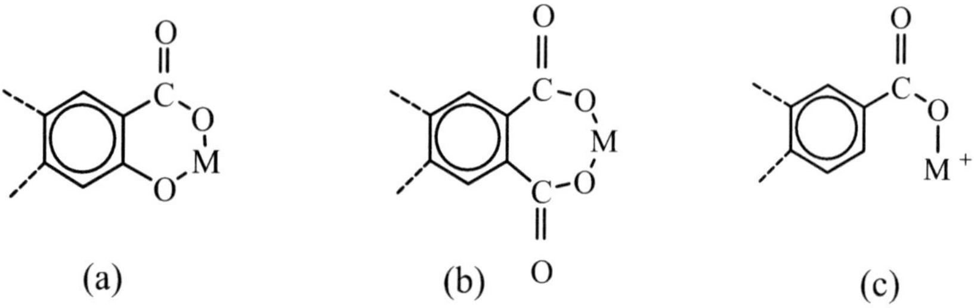

FIGURE 1. Binding of metals (a) between a carboxyl group and a phenolic hydroxyl group; (b) between two carboxyl groups; (c) complexation with a carboxyl group.

Typical examples of organometallic compound species are:

$HgCH_3^+$ $Hg(CH_3)_2$
Monomethylmercury ion   Dimethylmercury

Organometallic compounds may enter the environment directly as pollutant industrial chemicals and some are synthesized biologically by bacteria. Some of these compounds are toxic because of their mobility in living systems and ability to cross cell membranes.

Important chemical phenomena associated with water may occur through interaction of solutes in water with other phases. So, the oxidation-reduction reactions catalyzed by bacteria occur in bacterial cells.

2. Water Pollutants

Wastes rich in organic carbon are commonly discharged into waterways in the form of sewage, food processing wastes from the fruit, meat, diary and sugar industries, as well as wastes from paper manufacturing, and a wide variety of other industries. Organic pollution can also result from urban runoff giving nonpoint discharges to waterways. Primary, secondary, and tertiary wastewater treatment can reduce these problems. The discharge resulting from these treatment techniques can still result in a significant demand on oxygen in a waterway.

Many organic hazardous wastes are carried through water as emulsions of very small particles suspended in water. Some hazardous wastes are deposited in sediments of the water bodies from which they may enter the water through chemical or physical processes and cause severe pollution effects. For example, films of organic compounds, such as hydrocarbons, may be present on the surface of water. Exposed to sunlight these compounds are subject to photochemical reactions (Wayne, 1988). Gases such as O_2, CO_2, CH_4, and H_2S are exchanged with the atmosphere. Particles contributing to the turbidity of water may originate by physical processes, including the erosion. Chemical processes, such as the formation of solid $CaCO_3$, may also form particles in water. Sediments are sinks for many hazardous organic compounds and heavy metals that have gotten into water. A significant problem arises when untreated industrial wastewater is discharged into the water bodies. Study on the surface water quality in an industrial region near Black Sea shows the sources of pollution causing sediment formation (Dimitrova et al., 1998).

Colloids have a strong influence on aquatic chemistry. Colloidal particles are very small particles ranging from 0.001 μm to 1 μm in diameter. They are small enough to remain suspended in water. Toxic substances in colloidal form are much more available to organisms in water than are such substances in bulk form. The ability to undergo ion exchange processes is an important characteristic of some solids in contact with water.

The oxygen-demanding processes usually are measured by the biochemical oxygen demand (BOD). This test measures the uptake of oxygen by water samples incubated over a period of several days (usually 5 days) under standard conditions. Since the BOD measure has some disadvantages such as lack of precision and length of measurement time, other measures as chemical oxygen demand (COD) and total organic carbon (TOC) are often used.

Many polluted waterways have very little plant biomass but in some cases conditions can be suitable for the growth of aquatic plants. Large rivers and estuaries have significant populations of phytoplankton but many small and shallow streams are dominated by rooted aquatic plants and attached algae. Plants can have a very important influence on dissolved oxygen content through photosynthesis and respiration. Considerable increase in the amounts of the inorganic forms of nitrogen and phosphorus in the lakes near Black Sea and accelerated eutrophication due to the pollution with fertilizers have been established (Dimitrova et al., 2004).

Many environmental factors vary seasonally. For example, the incidence of sunlight has an important effect on photosynthesis and resultant dissolved oxygen concentrations in many aquatic areas. So, the abundant phytoplankton blooming in the spring- summer period is related to the seasonal oxygen oversaturation in the surface water of two studied lakes (Stoyneva, 1997).

Turbulent water conditions give well-mixed waters in which vertical profiles of dissolved oxygen are constant from the surface to the bottom waters. However, thermal stratification, due to solar heating of the surface waters, leads to isolation of the bottom waters. If the bottom sediments are enriched with organic matter the bottom waters may become depleted in dissolved oxygen while the surface waters remain unaffected. In this manner, a vertical profile showing considerable variation in dissolved oxygen concentration can occur.

Some important categories of water pollutants are:
- alkalinity, acidity, salinity (HCO_3^-, H_2SO_4, $NaCl$);
- BOD (organic matter that consumes oxygen when degraded);
- Carcinogens (aflatoxins, nitrosamines, polycyclic aromatic hydrocarbons (PAHs));
- Detergents (surface active agents and their builders);
- Fertilizers (phosphates, K^+, NO_3^-); cause excessive algal growth;
- Inorganic pollutants (CN^-, NH_3, SO_2; toxicity, esthetics);
- Pathogens (disease-causing bacteria and viruses);
- Pesticides (toxic to humans and wildlife);
- Petroleum wastes, "oil spills";
- Radionuclides;
- Sediments;
- Sewage (sanitary and other wastes discharged into sewer systems);
- Metal species bound with organics (heavy metals chelates, organometallic compounds);
- Trace elements and heavy metals (toxic to wildlife and humans);
- Trace organics (aromatics, organohalides).

Among them the transformations concerning the pesticides and petroleum wastes will be discussed.

Pesticides

The environmental impact of pesticide use is related to several fundamental properties essential to their effectiveness as pesticides (Connell et al., 1984). Firstly, pesticides are toxicants capable of affecting all taxonomic groups of biota, including nontarget organisms, to varying degrees dependent on physiological and ecological factors. Secondly, many pesticides need to be resistant to environmental degradation so that they persist in treated areas and thus their effectiveness is enhanced. This property also promotes long-term effects in natural ecosystems.

Pesticides include a vast array of natural and synthetic substances of widely different chemical nature. According to function the pesticides of major ecological significance are the insecticides, herbicides, and fungicides.

Commonly used insecticides are the chlorinated hydrocarbons, or organochlorine compounds, and the organophosphorus compounds. As examples the chemical structures of some of them are given on Fig. 2.

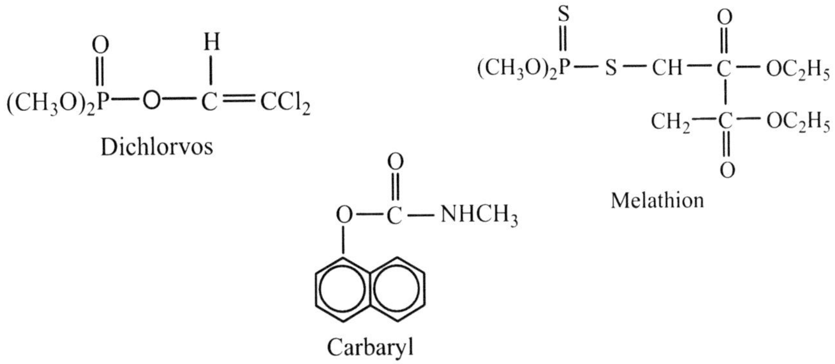

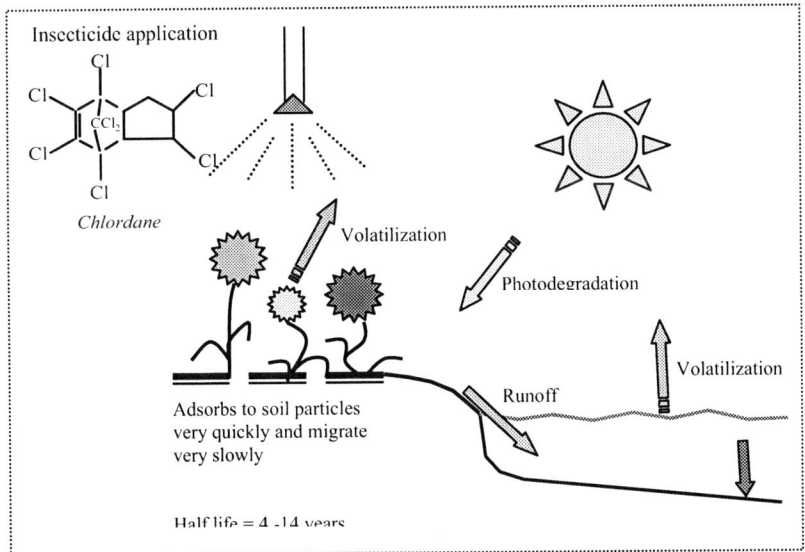

FIGURE 2. Chemical structures of commonly-used insecticides.

For example, the environmental fate of chlordane as summarized in Fig. 3

FIGURE 3. Environmental fate of Chlordane.

A wide variety of substances have been used as herbicides. Some chemical structures are illustrated in Fig. 4.

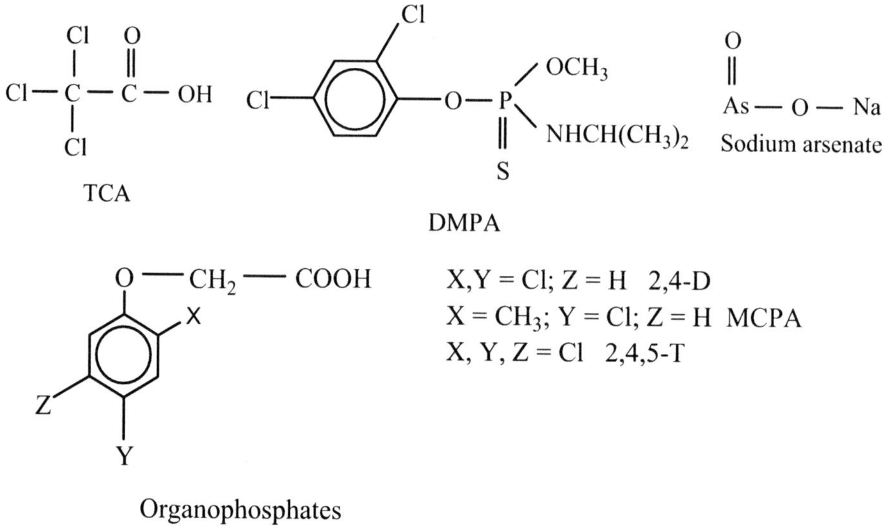

FIGURE 4. Chemical structures of herbicides.

Pesticides enter into the hydrosphere via many pathways including:
- Direct application for pest and disease vector control;
- Urban and industrial wastewater discharges;
- Runoff from nonpoint sources including agricultural soils;
- Leaching through soil;
- Aerosol and particulated deposition, rainfall;
- Absorption from the vapor phase at the air-water interface.

The relative inputs from these sources are difficult to assess. But generally, water bodies associated with urban regions receive substantial pesticide inputs from industrial and domestic effluents. The major input probably originates from agricultural and forestry practices.

Many environmentally significant pesticides are highly lipophilic and therefore relatively insoluble in water. Removal of pesticides from the hydrosphere may occur by: volatilization, absorption by aquatic organisms, settling of particles to which pesticides are adsorbed. Removal by degradation processes occurs by transformation and ultimately mineralization. Photochemical decomposition of pesticides may also proceed via series of photolysis reactions – for example, photooxidation, photonucleophilic hydrolysis, and reductive dechlorination – influenced by factors such as natural photosensitizers, pH conditions, and availability of dissolved oxygen. Once in the sediment, a pesticide may be re-released in the water; absorbed by organisms; transformed or degraded by microorganisms; permanently buried. In certain cases (e.g., DDT, lindane), degradation may proceed more favorably under anaerobic conditions. In Bulgaria aldrin, DDT, dieldrin and endrin have been banned since 1969, and heptachlor since 1991, but they still exist in the environment (Kaloyanova-Simeonova F. et al., 2001). Modeling of environmental chemical exposure and risk is used in the risk assessment procedures (Kambourova et al., 1998).

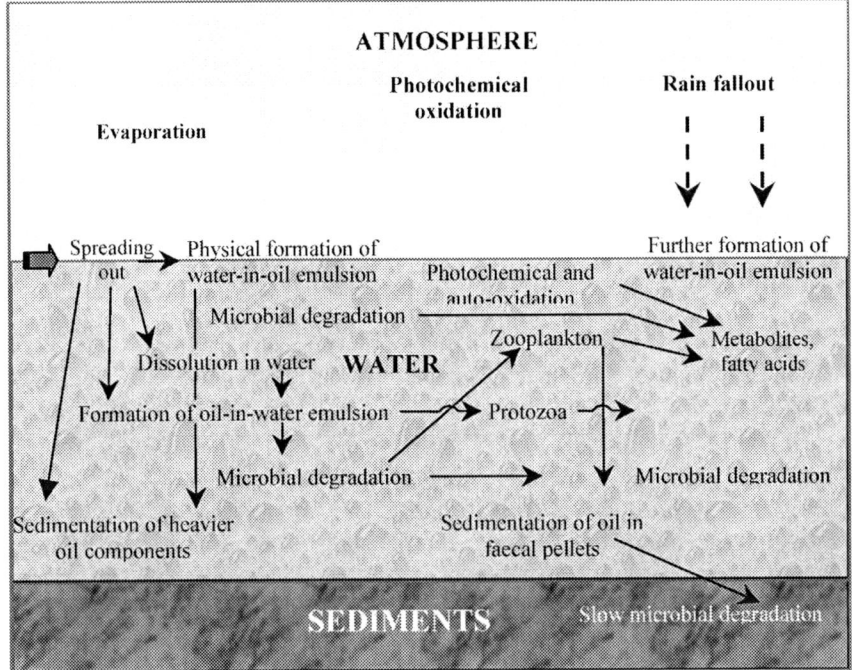

FIGURE 5. Processes and transformations of crude oil and fuels in aquatic environment.

Study on the occurrence and levels of pesticides residues in the river Danube has established the presence of organochlorine pesticides (HCH isomers, HCB and DDT), atrazine and desethylatrazine. Simazine and chlorinated phenols (2,4-dichlorophenol, 2,4,6-trichlorophenol and pentachlorophenol) were also detected (Bratanova Z. et al., 1998). Relatively high levels of DDT and metabolites in the water samples after incidents with pesticide stocks are reported (Kambourova et al., 2005). Atrazine and lindane also were found (Bratanova Z. et al., 2000).

Petroleum and related hydrocarbons

Crude petroleum contains complex mixtures of hydrocarbons as well as relatively small amounts of nitrogen-, sulfur-, and oxygen-containing organic compounds, asphaltenes, and various trace metals (uncomplexed and complexed forms). The hydrocarbons can be divided into two classes related to their chemical structure: the alkanes (normal, branched, and cyclo) and aromatic compounds (mono-, di-, and poly-, i.e., PAH).

Crude petroleum can be converted by physical and chemical processes into a wide range of refined products including gasoline, kerosene, heating oils, diesel oils, lubricating oils, waxes, and asphalts.

The hydrocarbons are low soluble in water and are strongly lipophilic. They are rapidly adsorbed onto particulate matter in aquatic areas as well as onto the bottom sediments. In addition, a variety of other processes can occur as illustrated above (Fig. 5).

Photochemical transformations of crude oil and fuels involve:

- Formation of reactive radicals;
- Formation of potentially toxic intermediate peroxides and hydroperoxides;
- Formation of oxidation products such as carboxylic acids, esters, oxygenated aromatics, carbonyl compounds and carbon dioxide.

Many marine microorganisms such as bacteria and fungi can metabolize either completely or partially the hydrocarbons from petroleum oils and fractions.

The chemical industry has grown enormously in recent decades and has now reached the stage where it is vital to modern society since it provides petroleum fuels, antibiotics, plastics, pesticides, food preservatives, etc. The assessment of hazard to the natural environment by a chemical must be preceded by an evaluation of the fate of the chemical after discharge to the environment.

References

Bratanova Z., Kovacicova J., Gopina G., (1998), A Review of Existing Data on the Occurrence of Pesticides in Water of the River Danube and Its Tributaries, Fresenius Env. Bull. 7, 495–501.

Bratanova Z., Vassilev K., (2000), Pesticides Residues in Ground and Surface Waters in Bulgaria, Hygiene and Public Health, v. xviii, 16–18, (in Bulgarian).

Connell Des W., Miller G.J., (1984), Chemistry and Ecotoxicology of Pollution, John Wiley & Sons, New York.

Dimitrova I., Kosturkov J., Vatralova A., (1998), Industrial Surface Water Pollution in the Region of Devnya, Wat. Sci.&Technol., v.37, No 8, 45–53.

Dimitrova I., Hainadjieva V., (2004), Water Pollution in the Shabla and Ezerets Lakes, Bulgaria, 3rd European Conference on River Restoration, Zagreb, Croatia, 17–21 May, 2004.

Kaloyanova-Simeonova F., Bratanova Z., Bratinova S., Dura G., Simeonov J., Kambourova V., Rizov T., (2001), Human Exposure and Risk Assessment of Soil Pollution with Persistent Organochlorine Compounds in Bulgaria, Centr. Eur. J. Occup. Environ. Med., 7, 3–4, 263–275.

Kambourova V., Simeonov L., Dura G., Vassilev K., Tasheva M., (1998), Comparative hazard assessment of pesticides to aquatic life using estimated concentrations, Centr. Eur. J. Occup. Environ. Med., 4, 343–353.

Kambourova V., Manova J., Bratanova Z., (2005), Environmental Pollution with Organochlorine Pesticides by Small Scale Incidents, Fresenius Env. Bull., 14, 3, 181–184.

Manahan S., (1992), Toxicological Chemistry, Lewis Publishers, Boca Raton.

Manahan S., (1989), Humic Substances and the Fates of Hazardous Waste Chemicals, Chapter 6 in *Influence of Aquatic Humic Substances on Fate and Treatment of Pollutants*, Advances in Chemistry Series 219, American Chemical Society, Washington, DC.

Stoyneva M., (1997), Investigation on the Phytoplankton in the Shabla and Ezerets Lakes and in the Shabla Touzla, Report on the Project "Northern Coastal Wetlands", Varna, 80–89, (in Bulgarian).

Wayne R.P., (1988), Principles and Applications of Photochemistry, Oxford University Press, New York.

VULNERABILITY, RESILIENCE, AND ADAPTATION: RESPONSE MECHANISMS IN AN ENVIRONMENTAL EMERGENCY – THE ASIAN TSUNAMI IN THAILAND AND HURRICANE KATRINA IN THE UNITED STATES

HEIKE SCHROEDER* AND DAYNA YOCUM
Donald Bren School of Environmental Science and Management
University of California, Santa Barbara
Santa Barbara, CA 93106, USA

*To whom correspondence should be addressed: schroeder@bren.ucsb.edu

Abstract. This chapter applies the concepts of vulnerability, resilience, and adaptation to two case studies of natural disasters, the Asian Tsunami in Thailand and Hurricane Katrina in New Orleans. In seeking to unveil opportunities for institutional learning to better cope with similar hazards in the future, the chapter explores the emergency response mechanisms in place in each case, examining the roles of the Thai and US governments and of the non-governmental organizations offering their services in the relief effort. It concludes that the best remedy for both countries is to rebuild natural ecosystems to limit the consequence of such extreme hazards as Hurricane Katrina and the Asian Tsunami, to identify factors of vulnerability within a system, and limit those factors' influence by adopting best practice models for effective disaster mitigation.

Keywords: vulnerability; resilience; adaptation; institutional learning; response mechanisms; environmental emergency; Asian Tsunami; Thailand; Hurricane Katrina; United States; New Orleans

1. Introduction

"Thousands drowned in the murky brew that was soon contaminated by sewage and industrial waste. Thousands more who survived the flood later perished from dehydration and disease as they waited to be rescued. It took

two months to pump the city dry, and by then the Big Easy was buried under a blanket of putrid sediment, a million people were homeless, and 50,000 were dead. It was the worst natural disaster in the history of the United States. When did this calamity happen? It hasn't—yet", writes the National Geographic in its October 2004 issue. (Bourne 2004) Hurricane Katrina has come awfully close.

Hurricane Katrina was among the most powerful storms of all time, occurring in the world's most powerful economy. Widespread failure of emergency response mechanisms aggravated the hazards, death and flooding tolls. Hurricane Katrina struck Florida and the Central Gulf Coast with significant force in the days before it hit New Orleans, giving weather authorities sufficient time to analyze its strength – chances of a direct hit to New Orleans were forecast at nearly 90 percent. Upon landfall on 29 August 2005, the Category 4 hurricane breached New Orleans' levee system – its man-made barrier against the waters of Lake Pontchartrain, made to withstand a Category 4 hurricane – flooding most of the city and damaging the coastal regions of Louisiana, Mississippi, and Alabama. Hurricane Katrina was the most destructive and expensive natural disaster in the history of the United States. The official death toll stands at almost 1,300 (Washington Post October 19, 2005) and property damage at over $200 billion (Everett, 2005). The environmental toll was also significant: multiple oil spills were caused by Hurricane Katrina, and more were added by Hurricane Rita, which devastated the coast off Texas and Louisiana only weeks later. Eleven oil spills occurred in Louisiana, totaling a discharge of 7.4 million gallons of oil. (Appel, 2005) Dangerous levels of lead and coliform bacteria also developed, and around 100 wastewater systems and 1,000 drinking water systems were contaminated and debilitated. (Marris, 2005) The world's largest economy was not able to successfully implement quick recovery from this natural disaster – or was it in fact a human disaster?

Nine months prior to New Orleans' inundation, a massive tsunami unrivaled by any in recorded history devastated coastal nations in the Indian Ocean. Again, many lives could have been saved had these states been better prepared. An estimated 225,000 deaths occurred in the eight countries most affected by the Asian Tsunami that struck on 26 December 2004; Thailand alone reported 5393 dead, 8457 injured and 3062 missing. Waves up to 10 meters high severely impacted the six coastal provinces along Thailand's Andaman Peninsula: Ranong, Phang Hga, Phuket, Krabi, Trang, and Satun. (UNEP 2005) No warning was given to Thailand's residents or visiting tourists in the communities affected, despite the fact that Thai officials received a message from the Pacific Tsunami Warning System authorities notifying them that an earthquake with a magnitude of 9.5 had occurred in the Indian Ocean. When the ocean encroached upon the coastal villages and cities, panic ensued as water flooded into homes and family members watched while loved ones were swept out to sea. This type of disaster was unimaginable, as there are no tsunamis in recorded history along the coast of Thailand. The tsunami flooded coastal areas up to two to three kilometers inland. Of 30 water bodies sampled a month after the tsunami struck, 29 faced significant contamination and can no longer be used as they were prior to the tsunami. Contamination of well water in some areas is significant – roughly a third of wells were unsafe due to coliform bacteria contamination. (UNEP 2005) Communities and local officials were not prepared for this type of environmental disaster, but the Thai government did their best to free up resources for emergency relief.

By investigating these two recent large-scale environmental disasters, this chapter seeks to offer some lessons learned about successful – and unsuccessful–emergency response mechanisms to large-scale environmental disasters. It will look at both governmental (including national and sub-national, bilateral and international) and non-governmental (including transnational and grassroots) emergency response structures in analyzing how well these institutions are adept at learning and implementing lessons learned in subsequent disasters.

2. Institutional Learning: Vulnerability, Resilience, and Adaptation

A growing body of literature points to the fact that the magnitude of a disaster results not only from the magnitude of the natural hazard itself, but to a great extent from the level of vulnerability the affected socio-ecological system exhibits, as well as its resilience and its adaptive capacity. (Gunderson and Holling 2002; Carpenter et al. 2001) Although this body of literature lacks a common understanding of definitions and delineations of the three concepts, the literature does offer a useful means of framing the relationships between the natural disaster itself and the socio-ecological system it impacts, between the socio-economic and ecological systems, and between human beings and nature.

Defining **vulnerability** is not easy or straightforward, as it is used differently among distinct groups of scientists. Perhaps the most prevalent definition is one introduced by Robert Chambers, defining vulnerability as "defenselessness, insecurity, and exposure to risk, shocks, and stresses;" he emphasizes that it is not the same as poverty. (Chambers, 1990, cited in Schoon, 2005) Others, such as Piers Blaikie, define vulnerability as "the characteristics of a person or group in terms of their capacity to anticipate, cope with, resist and recover from the impacts of natural hazard." (Blaikie, et al. 1994, cited in Schoon, 2005) This implies that systems with progressively lower capacities to anticipate, cope with, resist, and recover, have progressively higher vulnerabilities. Hence, whereas Chambers defines vulnerability as status quo and passive, Blaikie's definition is more dynamic and relational.

Piers Blaikie developed a 'Pressure and Release Model' – an analytical framework that identifies disasters as the outcome of both natural hazards and a progression of driving forces that shapes the degree of people's vulnerability to these hazards. (Blaikie, et al., 2005) The model differentiates between three types of driving forces. First, vulnerability is derived from specific socio-economic causes that are structured within sets of more pervasive, widespread, and generalized 'root causes,' such as poverty, war or poor governance. Second, vulnerability is subjected to 'dynamic pressures' which operate in specific ways within the root causes, such as demographic changes, corruption, press censorship, or political under-representation of ethnic minorities. The third category of driving forces is 'on site threats,' which are the most location- and socially-specific: unprotected buildings or lack of disaster preparedness at particular places and at particular times are examples of such threats. (Blaikie, et al., 2005)

Definitions of **resilience** abound, but some of those receiving the most citations will be singled out here. C.S. Holling defines resilience as the buffer capacity or the ability of a system to absorb perturbations, or the *magnitude of disturbance* that can be absorbed before a system changes its structure by changing the variables and processes that control behavior. (Holling, et al., 1995, cited in Schoon, 2005) Stuart Pimm makes the point that resilience is a measure of the *speed* of a system's return to equilibrium following a perturbation. (Schoon, 2005) So whereas Holling stresses the magnitude of disturbance from an original state, Pimm highlights the speed of return to an original state that a system can accomplish. Yet another definition, by Holling and Gunderson, emphasizes the "capacity of a system to experience disturbance and still maintain its ongoing functions and controls." (Holling and Gunderson, 2002).

According to Berkes and Folke, responsible disaster management units or political figures may respond either 'with experience' or 'without experience' and accumulated ecological knowledge. Response without experience may result in a series of policy responses that are not based on knowledge, or it can lead to institutional transformational learning. If the crisis is a true surprise, such as that in Thailand, or of a greater magnitude than anything experienced before, such as that in New Orleans, then the group will have no previous experience with it. The mechanism of institutional learning is trial and error; if it takes place as active learning and deliberately uses management policies as experiments from which to learn, then we see the basics of adaptive management. International experience with large-scale environmental management agencies shows that there often is institutional learning following a crisis, although much of the behavior of these agencies hardly fits the model or an ideal adaptive management approach. Steps essential to learning from experience are: documenting decisions, evaluating results, and responding to evaluation – although publications, data records, and computer databases are often not adequate to serve the institutional memory, and are therefore insufficient. (Berkes and Folke, 2002, p 143) Incorporation of lessons learned into an organization's policies, practices, and protocols are imperative in retaining these lessons.

Response with experience is a situation in which the agent has previous experience with a crisis of that nature and access to management policy that was used in the previous occasion. Characteristic of fully developed traditional ecological knowledge systems, response with experience relies on taking advantage of multigenerational, culturally transmitted knowledge about local and regional resources; this is referred to as institutional memory. Developing mechanisms for the retention of institutional memory may be the key to arriving at responses with experience. (Berkes and Folke, 2002, p 142-143).

Adaptation or adaptive capacity is perhaps the vaguest of the three concepts. In the context of climate change, where much of this literature originates, Smit et al. for example purport that "adaptation involves adjustments to enhance the viability of social and economic activities and to reduce their vulnerability." (Smit, et al., 2000, cited in Schoon, 2005) Others contrast adaptation with mitigation. For some, adaptation is the antithesis of resilience, for others it is not. (Schoon, 2005) According to Wilcox and Horwitz, adaptation refers to the capacity, willingness, and preparedness to heed learned advice and lessons from past events and the growing body of knowledge on human-natural system interactions. While 'control strategies' are structural and built around notions of resisting events of a predictable magnitude, 'prevention' seeks to intervene in the sequence of events that leads to the extreme situation, and 'adaptation' is a strategy that uses learning to avoid or cope with the consequences of the extreme. (Wilcox and Horwitz, 2005).

3. Emergency Response Mechanisms

This section will investigate the vulnerability of coastal communities in Thailand and the United States by applying Blaikie's Pressure and Release Model, examining the root causes of the two countries' vulnerabilities, the dynamic pressures that aggravate these root causes, and the on site threats that were uncovered by Hurricane Katrina in New Orleans and the Asian Tsunami in the affected coastal regions in Thailand. Using Berkes and Folke's formula of analytically differentiating 'response with experience' from 'response without experience,' and by looking into each system's level of experience with the respective hazards and resultant learning, this section will investigate the level of resilience of these communities to large-scale environmental hazards. The aim is to better understand the communities' capacity to maintain – or swiftly retain – their ongoing functions and controls, which, following Holling and Gunderson, is an indicator of their level of resilience. Lastly, this segment will examine how these communities use learning to avoid or cope with the consequences of the tsunamis and hurricanes, as prescribed by Wilcox and Horwitz. In doing so, this section will compare and contrast both governmental and non-governmental emergency responses in Thailand and the United States. The aim here is to investigate what role non-governmental entities play in complementing governmental efforts, and determine if they would be able to fill in the gap should the government be ill-prepared to respond effectively in an emergency.

3.1. GOVERNMENTAL EMERGENCY RESPONSE IN THAILAND

The tsunami may have hit vulnerable communities in Thailand by surprise, but the Thai government was quick to respond. The government demonstrated a high level of resilience (capacity to maintain or swiftly retain ongoing functions and controls), implementing various actions quickly and largely unhindered. Taking initiative, it immediately sent out national troops to the destructed coastal communities. An initial allocation of $2.5 million from the emergency fund plus the government's annual budget for disaster mitigation paid for most early relief operations. Response was carried out by government agencies with support from the private sector and numerous individual volunteers. The Department of Disaster Prevention and Mitigation, operating within the Ministry of the Interior, which has the responsibility to manage coordination of all governmental relief efforts in tsunami-affected provinces, established a database to support decision-making, management, monitoring, and follow-up programs. Additionally, the Thai Prime Minister allocated specific mandates to various governmental agencies to manage the disaster; four national committees focused on emergency response, natural resources, livelihoods, and disaster prevention through an early warning system that was established to coordinate the response of the government on those priority issues. (UNEP 2005).

Initial response at the local level was less than ideal. Weaknesses in the local levels of government became clear, as local authorities did not have quick access to funds, and were not able to initiate quick recovery. Local Thai health authorities were overwhelmed by the extensive deaths, and threw the bodies into mass graves instead of identifying them. Massive backlash from the international community occurred, especially from those countries with a large number of visiting tourists, and the Thai Prime

Minister mandated that all bodies be identified – one of many actions for which the Thai government can be praised.

Another successful response was in providing healthcare for regions in need. The Thai Ministry of Public Health (MOPH) responded within hours of the tsunami with rapid mobilization of local and non-local clinicians, public health practitioners, and medical supplies, assessment of healthcare needs, identification of the dead, injured, and missing, and active surveillance of syndromatic illness. These efforts were supported by the World Health Organization (WHO) and the US Armed Forces Research Institute of Medical Science (AFRIMS). Temporary clinics were established and provided with domestic medical supplies, which lowered the risk of administering out-of-date or poorly labeled foreign drugs, a common problem in international relief operations. The Bureau of Epidemiology set up a real-time surveillance system for tracking and responding to possible outbreaks or bacterial meningitis, pneumonia, cholera, etc. Emergency food, water, and medical supplies were quickly provided to the worst-affected areas, (Watts, 2005) and the Thai Prime Minister ordered banks to lend to the rebuilding effort. (*Economist*, January 6, 2005) Due to the government's rapid and thorough response in providing the required healthcare, a 'second wave' of deaths from disease and starvation never materialized.

Did poor environmental management and urban planning contribute to the devastation caused by the tsunami in Thailand? Thailand has an Environmental Protection and Quality Act (1975) and a more comprehensive and detailed Enhancement and Conservation of National Environmental Quality Act (1992) in place. The latter established a National Environmental Board that is mandated to prepare a policy and plan for the enhancement and conservation of the environment. It provides criteria and procedures to conduct environmental impact assessments (EIAs), as well as detailing the EIA reporting requirements. Some 500 EIA reports are prepared yearly of which around 20 percent are approved without conditions. (UNEP 2005; Weisman, 2005)

Although Thailand shows a commitment to environmental protection through its numerous acts (the National Park Act of 1961, the Forest Act of 1941, the National Reserve Forest Act of 1961, the Fishery Act of 1941, the Harbor Acts of 1913 and 1992, and the Wildlife Preservation and Protection Act of 1992), there is a need to review and streamline these legislations in order to support an integrated coastal zone management approach. (UNEP 2005; Weisman 2005) Coastal areas where mangrove forests had been removed were more affected by the tsunami than those still forested. (Wilcox and Horwitz, 2005) Thailand's near shore brackish shrimp farming industry has grown exponentially, and rapid expansion into mangrove forest areas has resulted in its widespread decimation. (MMWR Weekly, 2005) The level of devastation in the six affected provinces varies depending upon the presence of these natural barriers. Studies have shown that preserving these valuable ecosystems greatly reduces the level of vulnerability of a community. (ICEM, 2003).

The coastal communities hit by the tsunami were indeed vulnerable, but assistance from Bangkok arrived to some cities and villages shortly after the wave marked its path. Although no national preparedness plan for the tsunami phenomenon was in existence prior to 26 December 2004, the legal framework and resulting structural arrangements for disaster management were (and remain) relatively clear in Thailand. Weaknesses lie at the local level of Thailand's government, but the armed forces and national government's actions offset the initial shortcomings. With support from the UN, formal schooling continued practically uninterrupted in the affected areas, despite destruction of infrastructure and loss of life among the personnel. (Government of Thailand 2005) Although this disaster was clearly the largest loss of life that the country has suffered for many years, the Thai government performed initial emergency response commendably.

3.2. GOVERNMENTAL EMERGENCY RESPONSE IN THE UNITED STATES

After 9/11, the United States created a Department of Homeland Security to assume primary responsibility in the event of a terrorist attack, natural disaster, or other large-scale emergency for ensuring that emergency response professionals are prepared for any situation (Department of Homeland Security October 15, 2005) – still they were not prepared for a Category 4 hurricane. The department and its Federal Emergency Management Agency (FEMA) have been criticized for not having responded to Hurricane Katrina quickly enough. America watched in horror as thousands of refugees lived for days in unimaginable conditions at the Superdome 'refuge center.' In New Orleans' Evacuation Plan, this center was meant to be used as a refuge for the small percentage of residents that were not able to evacuate on the city-provided buses. Due in part to FEMA's lack of preparation, some people died, and many more suffered, of thirst, hunger, heat exhaustion, and drowning.

FEMA is not the only culprit; a breakdown occurred on all levels of government. According to the National Response Plan that establishes a comprehensive all-hazards approach to enhance the ability of the United States to manage domestic incidents, local, state, and federal authorities are all designated specific responsibilities and duties. The mayor is responsible for coordinating local resources to address the full spectrum of actions to prevent, prepare for, and recover from major disasters, accidents or acts of terrorism. He/she requests state and, if necessary, federal assistance through the governor of the state when the jurisdiction's capabilities have been exhausted. In turn, the governor is responsible for coordinating state resources and requesting federal assistance in the case that the state does not have sufficient resources to handle the emergency. Once the request is made, the Secretary of Homeland Security coordinates federal efforts in conjunction with state, local and private-sector entities, summoning FEMA. The president's responsibilities under the National Response Plan involve leading the nation in responding effectively, and ensuring that necessary resources are applied quickly and efficiently to all incidents of national significance. (Department of Homeland Security 2004)

Where did they go wrong? In answering this question, much finger-pointing has ensued. For FEMA to successfully act in accordance with the National Response Plan, officials at local and state governments must coordinate efforts and officially request assistance. Michael Brown, defending his actions as FEMA Director, suggested that the blame can be attributed to Kathleen Blanco, Governor of Louisiana, and Ray Nagin, Mayor of New Orleans, who could not come to consensus on how to manage the problem, and FEMA was restricted from acting until the agency was called to action by Nagin and Blanco. Nagin's mandatory evacuation order was issued only 20 hours before the storm struck the Louisiana coast, less than half the time researchers determined would be needed to get residents out, and FEMA in. This allowed for enough time for some, but those who failed to leave town typically did so because they had no means of transport. Two-thirds of the city's population is African-American and some 35 percent of those households did not own a car, compared with 15 percent of white households. Neither the city of New Orleans nor the State of Louisiana evacuated residents from the hurricane zone to begin with, as the state's Emergency Plan dictates, despite the availability of 550 school buses for this purpose. (*Economist*, September 13, 2005) This poverty and disorganization is representative of one of Blaikie's 'root causes' to vulnerability, and greater precautions should be taken in these situations.

FEMA demonstrated coordination problems once it was called to action by refusing entry to Wal-Mart delivery trucks of water although demand was incessant, cutting all emergency communication lines, and deterring Red Cross from sending supplies and medical personnel into New Orleans. The agency attempted to justify the latter action by claiming that Red Cross' presence would deter residents from evacuating the city and encourage others to return to the city, despite the fact that the city-coordinated evacuation never took place. Overestimating its control of the situation, FEMA denied an offer from Chicago's mayor to send manpower days before the hurricane hit. FEMA traditionally relies on the private sector to provide the manpower and logistical help necessary to deal with a major emergency, but there were major gaps in the arrangements it had made. Many of the contracts were poorly executed because of miscommunication and lack of planning. (Merle and Witte, 2005) In addition, the FEMA Director failed to organize the proper distribution of food and water to desperate victims, and he did not actively manage the situation in the overcrowded Superdome, where impoverished refugees vied over floor space and suffered in unsanitary conditions. Evidently, the United States' protocol for acting in the face of an emergency is embedded with superfluous levels of responsibility, inhibiting the agency from successfully coordinating its actions. This lack of coordination between different levels of government, or within the agency, clearly show that the Disaster Response system in the United States is not adequately structured to manage disasters of this magnitude and thus makes the system more vulnerable.

Although initial emergency response to Hurricane Katrina was rather disorganized, the United States government did take initiative on many aspects of recovery. FEMA distributed more than $1.5 billion in federal aid to more than 717,000 households and established a Housing Area Command to oversee all temporary housing operations. The US Army Corps of Engineers is performing a detailed assessment of about 350 miles of hurricane levees. The Corps is developing a comprehensive, prioritized plan to repair the levees and the pumping stations that support New Orleans and surrounding areas. (Department of Homeland Security January 28, 2005; MMWR Weekly 2005) The Department of Housing and Urban Development has partnered with the US Conference of Mayors (USCM) and the National Association of Counties (NACo) to identify available homes to temporarily house displaced families in the wake of Hurricane Katrina. (Department of Homeland Security October 3, 2005; MMWR Weekly 2005).

To ensure proper healthcare for victims, Health and Human Services declared a public health emergency. Suddenly America's traditionally limited healthcare safety net expanded for hundreds of thousands of uninsured Gulf Coast residents. Some patients' long-standing illnesses were finally diagnosed and treated – a welcome side effect for those without health insurance. (*Economist*, September 13, 2005) However, emergency assistance was held up by red tape: government officials did not coordinate the relief effort effectively enough to ensure a swift medical response. For example, a North Carolina mobile hospital developed by the Office of Homeland Security after 9/11 and designed specifically to handle disasters with mass casualties was stranded in Mississippi for several days because Louisiana officials were not able to get authorization for the vehicle to enter the city. (CNN 2005).

The US Senate initially approved funding of $10.4 billion, and after receiving reports detailing the extent of the damage, authorized another $52 billion. With the extensive financial resources of the United States, one would expect disaster mitigation to happen rapidly and effectively.

However, the story is the same throughout. With the wrong policies in place, or policy malfunction, failures occur throughout the system.

3.3. NON-GOVERNMENTAL EMERGENCY RESPONSE IN THAILAND

Although widespread goodwill was directed towards relief for the victims and their families in Thailand, many of the funds were misdirected, and the communication between the government and non-governmental organizations (NGOs) was faulty. By June 2005, six months after the tsunami, governments, companies, and individuals from numerous nations had pledged $6.7 billion to the tsunami effort, but only $2.85 billion had been turned into specific commitments of paid contributions, according to the United Nations Office for the Coordination of Human Affairs (OCHA). (Inderfurth, Fabrycky, and Cohen 2005) French companies Bouygues, Carrefour, and Electricité de France equipped new schools in Baan Nam Khem – an area where half of the 5,000 residents died – with computers and a banquet room, to the envy of the community's inland neighbors. (Mathes 2005) NGOs have raised $1.48 billion in private funds and in-kind gifts, $254 million of which was spent in the three months immediately following the tsunami. (Inderfurth, Fabrycky, and Cohen 2005) Thailand did not accept external financial assistance, but requested support in terms of technical expertise, equipment, and capacity building, in particular in the areas of environmental rehabilitation and community livelihood recovery. (UNEP 2005; Department of Homeland Security October 3, 2005) To this end, PricewaterhouseCoopers offered $1 billion for 800 hours of pro bono advisory services in the form of loaned professional expertise for tsunami-specific UN projects. (UN News Service 2005) The Thai Red Cross played a vital role in coordinating the delivery of public donations to affected communities, but communication and information exchange between foreign and local NGOs was inadequate. (UNEP 2005).

Despite the abundant creativity in helping tsunami victims, coordination was lacking and projects were partly nonsensical, such as the Slum Network Foundation that built 30 houses in Phang Nga on land not owned by the villagers it was trying to help. Without taking into account local preferences or cultural traditions, the American Refugee Committee offered fiberglass boats to fishing communities, rather than traditional wooden boats which they would have preferred. (Trouillaud 2005) Although a disjointed effort, NGOs did provide basic goods like food and water to victims and significantly contributed to the emergency response efforts of the Thai government. (Government of Thailand 2005).

3.4. NON-GOVERNMENTAL EMERGENCY RESPONSE IN THE UNITED STATES

The major NGO assistance during the initial weeks of the hurricane relief efforts were directed by the American Red Cross, which was complemented by countless other NGOs that extended a helping hand to the victims. Elected to take over the organization of providing water, food, and essential supplies to the Superdome as soon as the storm calmed, the American Red Cross implemented the largest mobilization of resources in its history for a single natural disaster. More than two hundred Red Cross shelters housed, fed, and cared for thousands of displaced residents from the day the hurricane struck. (American Red Cross August 29, 2005) The Red Cross received more than $1.3 billion in gifts and pledges. (American Red Cross November 2, 2005) The Salvation Army, funded by corporate, private foundation, and individual donations, also provided emergency assistance; the Army served more than 4.2 million meals, 50,000 boxes of food and 30,000 clean-up kits in the three weeks following the hurricane to residents of Alabama, Louisiana Mississippi, Texas, Arkansas and Oklahoma. The Salvation Army also managed 145,000 social service cases aiming to provide long-term assistance to victims, including housing and rebuilding. (The Salvation Army September 22, 2005) Countless other organizations mobilized, such as the American Society of Civil Engineers, which held emergency roundtables to identify workforces on housing, transportation, environmental health, and reconstruction. Community Wireless Networking experts from throughout the United States headed to the New Orleans region to help rebuild their telecommunications infrastructure in evacuee camps. (Bird and Lubkowski, 2005).

Despite this huge response, there were still shortcomings, primarily regarding coordination with other organizations, primarily the government. For example, the Red Cross was prohibited from entering the city prior to Hurricane Katrina making landfall by the Louisiana State Department of Homeland Security, under the direction of Louisiana Governor Kathleen Blanco. (MMWR Weekly 2005) This type of widespread miscommunication deeply hindered effective recovery efforts.

4. Recommendations for Institutional Learning

This section will identify characteristics of the coastal communities in Thailand and New Orleans that affected their respective levels of vulnerability. It will describe major shortcomings of each country's response mechanisms, give recommendations for policy change, and acknowledge efforts to correct these deficiencies. In critiquing the response mechanisms in each case study, the chapter will identify areas where institutional learning would significantly reduce the two systems' vulnerabilities and enhance their resilience and adaptive capacities in coping with large-scale environmental emergencies.

4.1. THAILAND

An obvious major shortcoming of Thailand's disaster management system was the absence of an accurate tsunami early warning system monitoring the Indian Ocean. (Bird and Lubkowski, 2005) A Pacific Tsunami Warning System did exist, of which Thailand was and is a member state. Although Thai officials were duly notified by the warning system's staff that an earthquake had occurred, this information was not transmitted adequately to local authorities – for fear that it would harm tourism! As a consequence, an evacuation alarm was not sounded, and the tsunami floods entered surprised and unprepared communities. Loss of livelihood may have been unavoidable even if the information had been transmitted optimally, but the death toll would have certainly been lower. (Wilcox and Horwitz 2005; UNEP 2005; Weisman, 2005).

Is there a reasonable expectation that Thailand should have been better prepared? During the past 50 years five major tsunamis were triggered by earthquakes, and although they never hit Thailand directly, several occurred in Southeast Asia. In 1998, a tsunami hit Papua New Guinea caused by an earthquake of 7.1 on the Richter scale that had occurred 15 miles off the coast. Waves reaching 12 meters hit the coast within 10 minutes, killing 2,200 people. In 1976, an earthquake occurring in Moro Gulf, a few miles away from the coast of the Philippine Island of Mindanao, set off a tsunami that devastated 700 km of coastline and killed more than 5,000 people. (Ringsurf, 2005) Some may argue the disasters that occurred in neighboring areas should be impetus enough for the governments in Southeast Asia to take precautions. However, as Berkes and Folge (2002) would argue, the best learning occurs through experience.

Following the tsunami, the Thai government made efforts to produce effective programs that will help avoid future disasters. Sixteen of 27 countries in Southeast Asia have identified national tsunami focal points, and have committed to work with each other and the United Nations on coordinating a regional network of communication. (Inderfurth, Fabrycky, and Cohen, 2005) The Thai government collaborated with others to implement a "Cobra Gold" drill, an exercise aimed to test the disaster response on the scale of the December tsunami. Troops from the United States, Thailand, Singapore, and Japan participated in the mock disaster that focused on previously identified weaknesses, and were accompanied by UN and NGO representatives, practicing interaction, sharpening training, and coordinating communication. (Mathes 2005) This is a commendable first push towards improving response mechanisms in Thailand, and a continued effort will certainly lead to higher resilience to large-scale tsunamis.

An issue that still needs to be addressed is the impact on and recovery of coastal communities. Villages protected by mangrove forests, such as Khura Buri in Phang Nga Province, were practically unaffected by the tsunami, a factor that needs to be considered when rebuilding the less fortunate communities. Also, there is now a window of opportunity to introduce more sustainable fishing practices, as a large number of the lost fishing gear, such as stake traps, are illegal. (UNEP 2005) The recovery effort needs to focus on enhancing human livelihoods while reducing both human and ecosystem vulnerability to future disasters and on protecting ecological diversity. The conservation of forests, mangrove systems, and natural resources will certainly present a large challenge, especially as families illegally harvest resources to rebuild their own livelihoods, and private interest groups try to secure bio-resources in the vulnerable economy. (Blaikie, Mainka, and McNeely 2005) It is necessary to devise and implement a strategy to procure the material to rebuild houses and repair vessels without exploiting local resources. (UNEP 2005) Additionally, it is advisable to negotiate and consult with local people to ensure best practices of resource management. Local people usually have institutional means and knowledge for recovery, which should not be undermined or pre-empted by top-down blueprint planning.

To improve utilization of resources in emergencies, a national disaster plan should include a market analysis to identify procurement options in the various provinces that can be quickly accessed in case of need. Measures should be put in place to allow access to governmental financial resources, which are generally available for local disaster response activities, to be quickly utilized by governors and local civil defense administrators. (Government of Thailand 2005; Watts, 2005) Reform to reduce vulnerability in the future could include establishing management structures that encourage and facilitate rapid response to changing situations. (Blaikie, Mainka, and McNeely, 2005).

Without having experience and past lessons to draw upon, the Thai government pulled its country out of crisis remarkably well. Additionally, it addressed a major flaw in its system: Thailand has since become part of an Early Warning System, which will serve as the country's first line of defense to future tsunamis. Now that the crisis is over, there are valuable opportunities for institutional learning. The local level, however, did not have the tools to be adequately responsive and effective. This may be a systemic problem, but it resulted in capacity foregone that the local level could have played in the emergency response efforts. In seeking best practices of emergency management, Thailand should seek to better prepare its local officials for future disasters.

4.2. THE UNITED STATES

The roots of the disaster unleashed by Hurricane Katrina date back to 1718, when Jean Baptiste le Moyne de Bienville, a French colonist, built his settlement on a hurricane-prone patch of swampland surrounded by three huge pools of water: the Mississippi Delta, the Gulf of Mexico, and Lake Pontchartrain. (*Economist*, September 3, 2005) Much of this settlement, now the city of New Orleans, is 6-10 feet below sea level, and it has grown into a densely-packed city of almost 500,000 inhabitants.

The location of New Orleans inherently contributes to its vulnerability. When Hurricane Katrina struck, Lake Pontchartrain's levees, the only barrier between downtown New Orleans and the lake, were in a state of disrepair far below the Category 3 standard that was chosen in 1965 as the storm intensity level for the levees to defend against. Loss of coastal marshes that dampened earlier storm surges puts the city at increasing risk to hurricanes. Eighty years of substantial levee construction has prevented spring flood deposition of new layers of sediment into the marshes, and a similarly lengthy period of marsh excavation activities related to oil and gas exploration and transportation canals for the petrochemical industry have threatened the wetland's integrity. (Laska 2004; Bird and Lubkowski, 2005) With the Mississippi Delta sinking due to these unsustainable management practices, the most fundamental recovery activity outside of the city's borders is arguably the restoration of the wetlands. Simply rebuilding New Orleans will not make the city less vulnerable to major storms such as Hurricane Katrina. (Schwartz and Revkin, 2005).

Should New Orleans have been more prepared? Yes. Hurricanes appear on our radars days before they hit land, and allow for long preparation phases. Also, several hurricanes hit the countries bordering the Gulf of Mexico every year, each time threatening to tear apart cities and ruin lives. The United States has plenty of experience with hurricanes – what distinguishes Hurricane Katrina from previous storms hitting the region is its sheer magnitude. But experts have accurately predicted what could happen if a powerful hurricane hits, and have used this research to make protocol to address it. Just last year, Hurricane Ivan attacked New Orleans with fury. Although the storm passed without any major damage, the near miss led the Natural Hazards Center to write a report anticipating the damages that could have occurred: massive floods due to breaks in the levees, the destruction of sanitation systems, the perishing of many people who could not evacuate New Orleans, and a toxic mess that New Orleans would inherit after the water cleared out. (Laska, 2004) These predictions are now a reality – a reality that could look far less bleak if protocol had been implemented and effective. The United States has been granted many trial runs to determine and identify best practices in managing a disaster the magnitude of Hurricane Katrina. Although officials responded with some level of experience that could have bestowed institutional learning in the form of protocol, the government did not effectively implement these policies, and thus suffered the consequences.

Immediate goals that need to be addressed are cleanup of debris and toxic chemicals and locating temporary or permanent housing for victims. The most effective result will come about if local, state, and other stakeholders participate in the decision-making process and develop a shared vision for rebuilding the city and surrounding area. Toxic chemical exposures must be minimized; consequences of contaminated water and land may have the widest degree of uncertainty, leaving a legacy of persistent illnesses. (US Environment Protection Agency 2005).

In considering long-term plans to reduce vulnerability, city planners should prioritize their task list according to the three indicators of Blaikie's dynamic 'Pressure and Release Model'. A combination of socio-economic 'root causes,' 'dynamic pressures,' and 'on site threats' contribute to the level of vulnerability, and New Orleans exhibits all three, identified below. A major root cause present in New Orleans at the time of the hurricane, with the potential to be highly exacerbated, is poverty, typically stronger among African-American neighborhoods. In the process of rebuilding, there are fears of gentrification and segregation of races, which should be avoided by city planners to ensure the best welfare for all. This bottom-up approach to rebuilding would address the political under-representation of ethnic minorities, the major dynamic pressure associated with New Orleans. And finally, to address New Orleans' most pressing location-specific on site threat vulnerability, some have suggested relocating the city to a more stable ecological zone, despite the complications involved in this process. Others believe it is simply a matter of improved levee engineering. With some luck, this question will resolve itself with at least some portion of the displaced residents relocating elsewhere.

Addressing the political emergency response system and improving governance and communication among the different levels of government (city, state, and federal) is another requirement. The capacity of the government to maintain or quickly restore ongoing functions and controls, including provision of water, food, transportation, and social order, was unexpectedly low. It took too long for these response structures to get activated, resulting in many lost lives, unnecessary suffering in the Superdome refugee center, and residents trapped in their homes with no means of transportation to escape. Additionally, changes need to be made in the disaster plan implementation process. Since the plan was never implemented, it cannot be known if emergency protocol would have satisfied the situation's demands. Evaluation of this current plan should be initiated, and an assessment of the institutional fit, i.e., the match between the prevailing institutional arrangement and the properties of the biophysical or socio-economic systems to which they relate, can be made according to the actual conditions of the Katrina disaster. (Young, 2002) The federal government should incorporate any lessons learned into the National Response Plan, and pass on these lessons to state and local governments so that they can infuse new protocol into their own charters. Lastly, it is arguable that the rigid process of requiring state and local officials to take certain actions before FEMA is able to interfere and act stands in the way of a streamlined process. Alternatively, an improved ability of local and state officials to recognize when they can benefit from Federal aid is needed, complemented with better communication between the federal and state governments. Yet another possible solution is to develop a branch of the armed services with responsibility to perform rescue and relief efforts when called upon, mimicking that of Thailand's actions. This could avoid the situation where FEMA depends upon private resources which may or may not be available when needed. The United States Department of Homeland Security should find a way to facilitate quicker action with more communication and coordination with the local and state governments.

Adaptation, i.e., how New Orleans officials and the United States government at large use learning to avoid or cope with the consequences of such hazards, will depend on what choices are made in the rebuilding process. If the city is simply rebuilt as before, as President Bush proposes, then the same calamity is likely to occur again at some point in the future. Now is the time to rethink current urban planning approaches, and to resort back to natural protection structures, such as healthy ecosystems. This parallels one of the lessons that coastal countries in Southeast Asia learned from the Asian Tsunami: natural ecosystems, such as mangrove forests, proved the most effective protection against the ravaging waves, and communities where this barrier was destroyed suffered disproportionately.

5. Conclusion

When comparing the emergency response mechanisms of Thailand and the United States, a number of similarities, but also differences, are striking. We have identified three determining factors in how a system manages a disaster: vulnerability, adaptation, and resilience.

The large number of casualties from both disasters already hints at the fact that both coastal regions exhibited a significant degree of vulnerability. Recalling Blaikie's 'Pressure and Release Model', we can identify the level of vulnerability of the affected communities in Thailand and New Orleans. It is clear that Thailand had some significant shortcomings in its infrastructure that led to high levels of vulnerability: widespread poverty in some of the coastal communities ensured that weak houses would easily topple with the force of the tsunami. New Orleans also had to grapple with poverty – those worst affected by the hurricane were African-Americans who lived in poor neighborhoods and had no means of transportation to escape the hurricane. In Thailand, dynamic pressures exerted by large numbers of tourists who added to the death toll also increased the vulnerability of the system, although this may have led to additional international donations that were essential for recovery. Furthermore, without considering the consequences, many of the coastal communities have torn down their coastal mangrove forests to enjoy the income of shrimp fisheries. This destruction of valuable ecological barriers contributed to the vulnerability of the system by adding an unnecessary on site threat to those communities. The political marginalization of minorities, especially of African-Americans, was a significant dynamic pressure that aggravated the devastation caused by Hurricane Katrina while the sub-optimal levee system that was built to endure only Category 3 storms was its most cataclysmic on site threat.

The level of resilience in Thailand and the United States was surprisingly similar; the main contrast between the two, however, is that the United States did have a reasonable expectation of a powerful hurricane and extensive experience with hurricane mediation, whereas the tsunami was the first of its kind experienced in Thailand. One would expect that the system 'with experience' would also act in a more efficient manner, drawing on lessons from the previous experiences. (Berkes and Folke, 2002) However, this can only happen if lessons are learned within the institution. In a highly organized society like the United States, one may also expect the system to have highly developed protocol in place for each type of imaginable situation. In practice, it seems that the United States government has distributed responsibilities to such an extent within the hierarchy of power between local, state, and federal leadership that no agency has ultimate control over when and how resources are distributed. This shortcoming has significant effects on coastal communities' resilience to hurricanes, as Hurricane Katrina has blatantly demonstrated.

Examining Thailand's methods of disaster management, we can see an example of how a more streamlined approach, although less formally organized, may function better in the case of a national emergency. The study of its disaster mitigation actions shows that although the local level authorities lacked capacity to manage the emergency, the immediate action of Thailand's national government rapidly secured the situation.

The disaster management systems in Thailand and the United States have some commonalities in the potential for enhancing their level of adaptation to their natural hazards: both systems could benefit by preparing local authorities to mitigate the disaster while national authorities are organizing their response. Thailand's lack of financial resources contributes to its ill preparation, and its vulnerability. The United States, with strong economic resources, may have an effective process written into the New Orleans Evacuation Plan, but since this plan was not implemented, there is no way to evaluate it.

Another parallel between the two systems is the contribution of NGO support. Essentially, both governments profited enormously from NGO support, especially for first-response actions. The Thai Red Cross and other international NGOs provided the necessary supplies for victims in affected coastal communities in Thailand. In New Orleans, many would have gone for days without food and water if it were not for the American Red Cross and the Salvation Army. Additionally, NGOs continue to support victims in both locations, either by donating funds to community re-development or supplying the necessary manpower to rebuild. This high level of NGO involvement is a critical complement to the disaster management system. Still, in many reported cases, relief assistance was significantly obstructed by government entities. Coordination and communication should be more intimately incorporated into the government program during the early stages of a disaster.

Will Thailand and the United States increase their adaptive capacity by incorporating their lessons learned into new policies and protocols determining disaster management? The Japanese language carries a much clearer connotation for the word 'crisis' than English; it is understood to carry both 'danger' and 'opportunity.' It can only be hoped that these opportunities for institutional learning will indeed be taken seriously. The reconstruction phase in both regions will hopefully not be a purely top-down approach and misled by control strategies, which may be built around notions of resisting such events, such as building even higher levees in New Orleans. Furthermore, it can be hoped that they will not seek simply to intervene in the sequence of events leading to the extreme situation of a hurricane or tsunami, such as building more authoritative emergency response mechanisms. The ideal result is that they will adapt to these hazards by rebuilding natural ecosystems to limit the consequence of such extreme hazards as Hurricane Katrina and the Asian Tsunami, and to adopt best practices for effective disaster mitigation.

ACKNOWLEDGEMENT

We thank Oran R. Young for substantive comments and the US National Science Foundation for funding this research as part of the project on the

Institutional Dimensions of Global Environmental Change (IDGEC) under Grant Number BCS-0324981.

References

American Red Cross, November 2, 2005 Facts at a Glance: American Red Cross Response to Hurricane Katrina and Rita, http://www.redcross.org/news/ds/hurricanes/katrina_facts.html.

American Red Cross, August 29, 2005 American Red Cross Launches Largest Mobilization Effort in History for Hurricane Katrina, Press Release, http://www.redcrossncm.org/news/pr/08292005.html.

Appel, A., September 30, 2005 Gulf Wracked by Katrina's Latest Legacy – Disease, Poisons, Mold, *National Geographic*.

Associated Press, September 17, 2005 Uninsured get Medical Care in Katrina's Wake, www.msnbc.com.

Berkes, F. and Folke, C., 2002 Back to the Future: Ecosystem Dynamics and Local Knowledge, in: L.H. Gunderson and C.S. Holling (eds) *Panarcy: Understanding Transformation in Human and Natural Systems*, Island Press, Washington, DC, p 121–146.

Bird, J. and Lubkowski, Z., January 22, 2005 Controlling meticillin-resistant Staphylococcus Aureus, aka "Superbug", *The Lancet*.

Blaikie, P., Mainka, S., and McNeely, J., February 2005, The Indian Ocean Tsunami – Reducing Risk and Vulnerability to Future National Disasters and Loss of Ecosystems Services, The World Conservation Union IUCN Information Paper, http://www.iucn.org/tsunami/docs/ip-tsunami-risks-and-services-2.pdf.

Blaikie P. et al., 1994 *At Risk: Natural Hazards, People's Vulnerability, and Disasters*, Routledge, London.

Bourne, J.K., 2004 Louisiana Wetlands: The Big Uneasy, National Geographic, October, http://magma.nationalgeographic.com/ngm/0410/feature5/index.html?fs=www7.nationalgeographic.com.

Carpenter, S. et al., 2001 From Metaphor to Measurement: Resilience of What to What? *Ecosystems* 4, p 765–781.

Chambers, R., 1990 Editorial Introduction: Vulnerability, Coping, and Policy, *IDS Bulletin* 20, 2: p 1–7.

CNN, September 5, 2005 Katrina Medical Help Held up by Red Tape, CNN.com.

Department of Homeland Security, October 3, 2005 Hurricane Katrina: What Government is Doing, http://www.dhs.gov/interweb/assetlibrary/katrina.htm.

Department of Homeland Security, October 15, 2005 Emergencies and Disasters, http://www.dhs.gov/dhspublic/theme_home2.jsp.

Department of Homeland Security, December 2004 *National Response Plan*, http://www.dhs.gov/interweb/assetlibrary/NRP_FullText.pdf.

Economist, January 8, 2005 After the Deluge.

Economist, September 17, 2005 Now the Rebuilding Begins.

Economist, September 10, 2005 When Government Fails.

Economist, September 3, 2005 A City Silenced.

Everett, T., September 19, 2005 Washington Responds to Katrina Victims, http://wwwc.house.gov/ everett/news/columns/col_091905.asp.

Government of Thailand, May 2005 Report and Summary of Main Conclusions, National Workshop on Tsunami Lessons Learned and Best Practices in Thailand, http://www.reliefweb.int/library/documents/2005/ocha-tha-31may.pdf.

Gunderson, L.H. and Holling, C.S. eds., 2002 *Panarchy: Understanding Transformation in Human and Natural Systems*, Island Press, Washington, DC.

Holling, C.S. and Gunderson, L.H., 2002 Resilience and Adaptive Cycles, in: L.H. Gunderson and C.S. Holling (eds) *Panarcy: Understanding Transformation in Human and Natural Systems*, Island Press, Washington, DC, p 25–62.

Holling, C.S. et al., 1995 What Barriers? What Bridges? in L. Gunderson, C.S. Holling, and S.S. Light (eds) *Barriers and Bridges to the Renewal of Ecosystems and Institutions*, Columbia University Press, New York.

ICEM, 2003 *Thailand National Report on Protected Areas and Development,* Review of Protected Areas and Development in the Lower Mekong River Region, Indooroopilly, Queensland,Australia,http://www.mekong-protected-areas.org/thailand/docs/thailand_nr.pdf.

Inderfurth, K.F., Fabrycky, D., and Cohen, S., June 2005 The 2004 Indian Ocean Tsunami: Six Month Report, The Sigur Center Asis Papers.

Laska, S., November 2, 2004 What if Hurricane Ivan Had Not Missed New Orleans?, Natural Hazards Observer, Natural Hazards Center.

Marris, E., September 15, 2005 First tests show flood waters high in bacteria and lead, *Nature* 437, 7057, p 301.

Mathes, M., May 12, 2005 Mock Earthquake, Tsunami Tackled in Multinational Drills in Thailand, *Agence France Presse*.

Merle, R. and Witte, G., October 10, 2005 Lack of Contracts Hampered FEMA Dealing With Disaster on the Fly Proved Costly, *Washington Post*, p A01.

MMWR Weekly, January 28, 2005 Rapid Health Response, Assessment, and Surveillance After a Tsunami ---Thailand, 2004-2005.

Ringsurf, October 10, 2005 Tsunamis Through History, http://www.ringsurf.com/info/News/Tsunami/Tsunamis_through_history.

Schoon, M., 2005 A Short Historical Overview of the Concepts of Resilience, Vulnerability, and Adaptation, Workshop in Political Theory and Policy Analysis, Indiana University, Working Paper W05-4.

Smit, B. et al., 2000 An Anatomy of Adaptation to Climate Change and Variability, *Climatic Change* 45, p 223–251.

Schwartz, J. and Revkin, A.C., September 30, 2005 Levee Reconstruction will Restore, but Not Improve, Defenses in New Orleans, *New York Times*.

The Salvation Army, September 22, 2005 Salvation Army Expands Southern U.S. Relief Efforts in Anticipation of Hurricane Rita, Press Release, http://www.salvationarmyusa.org/usn/www_usn.nsf/vw-news/95161FE5D13FB35580257 085007AD038?opendocument.

The Infrastructure Security Partnership, Attendee Roster, Roundtable on Non-Governmental Organization Activities in Response to Hurricane Katrina. www.tisp.org/events/eventdetails.cfm?&eventID=745.

Trouillaud, P., June 20, 2005 Some Aid Schemes Botched in Rush to Help Tsunami-hit Thailand, *Agence France Presse*, http://www.reliefweb.int/rw/RWB.NSF/db900SID SODA-6DJ7UY?OpenDocument.

UN News Service, March 14, 2005 UN Announces Pioneering Agreement to Accept Donated Services to Augment Tsunami Relief Resources, http://www.un.org/apps/news/story.asp?NewsID=13680&Cr=tsunami&Cr1=.

UNEP, February 22, 2005 After the Tsunami: Rapid Environmental Assessment, United Nations Environmental Programme, www.unep.org/tsunami/tsunami-rpt.asp.

US Environmental Protection Agency, September 17, 2005 Environmental Health Needs and Habitability Assessment, http://www.epa.gov/katrina/reports/envneeds_hab_ assessment.html.

Washington Post, October 19, 2005 Katrina Death Toll now at 1,281, http://www.washingtonpost.com/wp-dyn/content/article/2005/10/19/AR2005101901438.html.

Watts, J., January 22, 2005 Thailand Shows the World It Can Cope Alone, *The Lancet*.

Weisman, J., September 16, 2005 Critics Fear Trailer 'Ghettos', *Washington Post*, p A18.

Wilcox, B.A. and Horwitz, P., 2005 The Tsunami: Rethinking Disasters, *EcoHealth* 2, p 89–90.

Young, O., 2002 *The Institutional Dimensions of Environmental Change: Fit, Interplay, and Scale*, MIT Press, Cambridge, MA.

ONE APPROACH FOR COGNITIVE ASSESSMENT ON GROUP OF SCHOOL CHILDREN EXPOSED TO LEAD IN THE COMMUNITY OF VELES

MIHAIL KOCHUBOVSKI
Republic Institute for Health Protection - Skopje
50 Divizija No. 6, 1000 Skopje, Republic of Macedonia
E-mail: kocubov58@yahoo.com

Abstract. The objective of the study performed in 2003 was to investigate and analyze the cognitive functions in children exposed to lead emissions in the city of Veles. Measurements of blood lead levels were executed, cognitive psychological tests (Raven), and investigation of graph motor ability (Bender-Geshtalt) were performed on school children (n = 31) randomly selected from those dwelling near by the Lead Smelter Plant. Most of the tested children have shown slightly increased blood lead levels (average = 16.51 μg/dl), being between the prescribed limit (10 μg/dl) and the critical level (25 μg/dl). This was correspondingly reflected in the measured levels of intelligence and graph motor ability.

Keywords: lead, cognitive abilities, IQ, child population, graph motor ability

1. Introduction

It is well known (Needleman et al., 1979; Thomson et al., 1989; Silva et al., 1988) that lead is most studied pollutant in the environment which corresponds with cognitive functions. Lead's exposure leads to lowering of IQ (Intelligence Quotient). In accordance with that, exposure to lead is related to disordered school's behavior, as well to attention deficit disorder.

As shown by Landrygan et al. (1994) lead causes CNS's damage manifested by delayed development, lowered intelligence and disordered behavior. In young children, this effect appears on blood lead levels in the range of 10-20 μg/dl. Centre of Disease Control and Prevention has recommended that blood lead level of 10 μg/dl, or higher should be considered as an evidence for increased lead's absorption, and National Academy has agreed upon this recommendation.

It is well known (Kristoforovic 1998) that in children exposed to lead it is possible to appear induced neuropsychological deficit: lowered IQ, worsened eye-hand coordination, muscle's tonus disorder and attention deficit hyperactivity disorder.

Lidsky et al. (2003) point out that cognitive development of children and effects of behavior as a consequence of long-term exposure on low levels of blood in the environment is known, but there are evidences that certain genetic and ecological factors can increase the adverse effects of lead to neurological development, so certain part of children are more vulnerable to lead's neurotoxicity.

2. Material and Methods

In a cross sectional study representative sample of randomly selected 31 children with average age of 12.8 years and standard deviation of ± 0.47 years, from the Veles-polluted area were examined in June 2003. The examinees were living or learning in the vicinity of the Lead Smelter Plant in Veles, Republic of Macedonia.

Examinations have been performed on children in the facilities of the Institute for Health Protection-Veles. For the purpose of the investigation has been got agreement from the Ministry of Education and Science, and children's parents have been introduced with the study design, and they have given their consent for the investigation.

Lead levels were analyzed in venous blood samples. 2 ml samples were obtained in sterile vacutainer, reconstituted with 1.5 mg K_2EDTA/ml blood, stored at $+4^0C$ and transported during the same day. Using AAS (atomic absorption spectrometer) with a Perkin Elmer 4100 HGA 700 graphite furnace fitted with an AS-70 brand Auto sampler. Lead was extracted into a mixture of nitric (HNO_3) and hydrochloric acid (HCl) under pressure with microwave digestion. A PAAR PHYSICA - PERKIN ELMER microwave furnace designed for laboratory practice was used. The research was approved by the ethics committee, as well by the parents of the examined children. The parents of the children have been informed about the results of the blood lead levels.

For evaluation of the health effects caused by lead has been performed group psychological tests of pupils:
 a) examination of the intelligence - cognitive psychological tests (Raven - progressive matrices;
 b) graph motor ability (Bender-Geshtalt test and test tree-house-man).

In order to examine the intelligence developments have been chosen Raven - progressive matrices in color for children (1966). By Raven's test is measured the level of general mental abilities of children (global IQ). It shows whether some examinee is capable to compare and think by analogy, and how much is capable to organize its special perceptions in systematic settled completeness.

Bender-Geshtalt test is used as intelligence's test, as a test for diagnosing of brain's damages and as a projective test for emotional problems. Bender's test point out on disturbances in visual-motoric function. Visual-motoric function is integrative function of person as a whole, that control cerebral cortex. Similar goal has the test tree-house-man.

3. Aim

The aim of the study was to find out potential causal-relationship between lead's exposure and:

- IQ reduction in the examined children;
- lowered graph motor ability in the examined children.

4. Results and Discussion

The city of Veles, with its geographical position, atmospheric characteristics, urban and industrial concentration, and its improperly located the Lead and Zinc Smelter Plant upstream to the north wind opposite to the Wind Rose (north wind is dominant and dispersion of pollutants is going directly to the city), multiplied by inconvenient climate-meteorological, hydro-topographic factors has a huge and continuous air pollution problem. The Lead Smelter Plant, built in Veles in 1973, is located on the north of the city only 200-300 m away from first households.

A group of 31 pupils of average age 12.8 years living near the Lead Smelter Plant (1-3 km distance on the southwest) were examined in June 2003. Increased blood lead levels were registered as a result of the lead exposure from the Lead Smelter Plant ($\xi = 16.51$ µg/dl with a range of 8.1-32.9 µg/dl) (Table 1). This average value was significanthly higher than the recommended value of 10 µg/dl proposed by Guidelines for Air Quality (World Health Organization 2000).

There was found statistical significant difference (Student t-test $t = 2.72$; $p < 0.05$) in blood lead levels, between the examinees with good and very good score in school (Table 2). On the other hand there was found low significance with Pearson's test of correlation between the values of IQ and

TABLE 1. Distributions of blood lead levels (µg/dl) in examinees from Veles (June 2003).

Settlement	Number of Examinees	Average	Min	Max	SD
Veles	31	16.51	8.1	32.9	11.56

* WHO recommended value = 10 µg/dl

TABLE 2. Score of schoolchildren and intelligence quotient related to blood lead levels.

Score of Children in School	Number	IQ	µG/DL
Sufficient	1	88	18
Good	5	97.4	20.96
Very good	9	99.55	13.37
Excellent	16	110.25	16.78
Average		104.35	16.51

blood lead levels in the examinees with good score ($r = 0.22$; $p<0.05$) and pupils with excellent score ($r = -0.27$, $p < 0.05$).

With a goal to assess is there any relation between the examinees' blood lead levels, and level of graph motor ability, has been performed a classification of pupils according to recorded level of graph motor ability.

Almost a third of the examinees were with the average level of graph motor ability ($n = 19$), and only three pupils were with bed one (Table 3). In one of pupils with bed level of graph motor ability, has been registered high blood lead level (31.4 µg/dl). There was not found statistical significant difference between the blood lead levels of the examinees with bed and average level of graph motor ability (Student t-test $t = 0.65$; $p > 0.05$), between bed and above the average level ($t = 0.37$; $p > 0.05$), and between the average and above the average level ones ($t = -0.19$; $p > 0.05$). Despite the fact that there was not found a statistical significant difference, it is clear that in group of pupils with bed level of graph motor ability, the average blood lead level (18.76 µg/dl) had been higher compared to other two groups.

Minder et al. (1994) have investigated relation between lead's exposure ane attention in children. There was 43 children (male), at age 8-12, that have learned in a special school for pupils with educational and/or learning problems. Results have shown that children with relatively high lead in the hair, have reacted slower during simple tasks reaction-time, compared to children with relatively low concentration of hair's lead. As well, the first group of the examinees have been less flexible for changing of the attention focus.

TABLE 3. Distribution of blood lead levels - μg/dl in the examinees according to graph motor ability.

Level of graph motor ability	number	average	min	max	SD
	31	16.51	8.1	32.9	6.74
Bed	3	18.76	12.3	31.4	10.94
Average	19	16.09	8.4	32.1	5.84
Above the average	9	16.62	8.1	32.9	7.83

While, most number of studies point out that lead's concentration in the hair are not safe indicator for exposure to lead.

With the aim to assess the risk to the central nervous system has been performed examination of the intelligence - cognitive psychological tests (Raven - progressive matrices in selected number of the examinees (n = 31). Pupils with the average intelligence have been found 5.2 fold more, compared to the group of children with the high above average and above average intelligence. Difference is much higher for the group of the examinees with under average intelligence 10.4 fold less compared to the group of pupils with above average intelligence (Table 4). There was not registered any pupil with very high above average intelligence, as well with limited intelligence and retardation, that corresponds to data from other studies [8].

There was homogenous contribution of children with high above average and above average intelligence in the examined group of pupils in Veles (12.9%). There was not found statistical significant difference between male and female for group of pupils with high above average intelligence (Test of notted differences of proportions $p = 0.3715$; $p > 0.05$). Female were presented with 22.2%, and male with 9.1%. There was not found statistical significant difference between male and female for group of pupils with above average intelligence (Test of notted differences of proportions $p = 0.8517$; $p > 0.05$). Male were presented with 13.6%, and female with 11.1%. There was not found statistical significant difference between male and female for group of pupils with average intelligence (Test of notted differences of proportions $p = 0.3629$; $p > 0.05$). Male were presented with 72.7%, and female with 55.6%. There was not found statistical significant difference between male and female for group of pupils with under average intelligence (Test of notted differences of proportions $p = 0.5157$; $p > 0.05$). Female were presented with 11.1%, and male with 4.6%. One of male examinees had moderate blood lead level - 18 μg/dl, and one female a little higher than the WHO's recommended value - 11.5 μg/dl.

There was not found statistical significant difference between above average and average intelligence (Test of notted differences of proportions p = 0.0518; p > 0.05), as well between average and under average (p = 0.1025; p > 0.05), and between under average and above average intelligence (p = 0.8356; p > 0.05).

TABLE 4. Distribution of the examined children's intelligence.

Rank of Intelligence	Veles		Male		Female	
	number	%	No.	%	No.	%
High above average	4	12,9	2	9,1	2	22,2
Above average	4	12,9	3	13,6	1	11,1
Average	21	67,7	16	72,7	5	55,6
Under average	2	6,5	1	4,6	1	11,1
Total	31	100	22	100	9	100

Distribution of the IQ in the examined group of children from Veles is shown on Table 5. There was not found statistical significant difference (Student t-test t = -0.64; p > 0.05), with slightly lower IQ in male examinees, compared to female ones. There have been registered higher IQ in female's examinees (up to 2.6 points). There was found low significance by Pearson's test of correlation between values of IQ and blood lead levels (r = -0.12; p < 0.05), values of ALAD[1] (r = -0.11; p < 0.05), and number of basophilic stippled erythrocytes (r = -0.10; p < 0.05). [1] ALAD - Aminolaevulinic acid dehydratase

Wang et al. (1998) have performed a study in a Plant for battery recycling, and found out that 31 from 64 workers have suffered by lead poisoning. Children that were going to the kindergarten in the vicinity have shown higher blood lead levels (15-25 μg/dl), and moderate decreasing of IQ (Binet-Simon scale), compared to children that were not exposed from socio-economic comparable kindergarten. Indoor air samples from the kindergarten have recorded average value more then 10 μg/m^3.

TABLE 5. Distribution of the intelligence quotient in the group of the examined children.

Veles	Number	Average	Min	Max	SD
Total	31	104.35	87	122	10.14
Male	22	103.59	88	122	10.14
Female	9	106.22	87	120	10.52

Samples of soil have recorded 400 fold higher content of lead, which was lowering if sample has been taken from 15-30 cm (depth), or 350 m away from the Plant for battery recycling. Follow-up investigation 2.5 years later, after moving out of children away from the source of pollution, has been shown considerable decreasing of blood lead levels of children and partial recovery of their IQ. With a goal to assess possible relation between blood lead levels of the examinees and their school's score, has been performed a classification of pupils according their registered score. Most of the examined children have had excellent score (n = 16), and only 1 pupil had sufficient score (Table 6). There was found decreasing of the average blood lead levels from 20.96 µg/dl in pupils with good score, to 16.79 µg/dl in children with excellent score. The same has been registered for minimal blood lead levels from 18 µg/dl in pupil with sufficient score, to 8.1 in pupils with excellent score. Considering recorded maximal blood lead levels there have been found variations that were not expected.

TABLE 6. Distribution of the blood lead levels-µg/dl in the group of the examined children according to their school's score.

Score	Number	Average	Min	Max	SD
Veles	31	16.51	8.1	32.9	6.74
Sufficient	1	18	18	18	0
Good	5	20.96	13.2	31.4	7.13
Very good	9	13.37	9.38	19.6	3.43
Excellent	16	16.78	8.1	32.9	7.64

There was found statistical significant difference (Student t-test t = 2.72; $p < 0.05$) in blood lead levels, between the examinees with good and very good score. There was found moderate significance by Pearson's test of correlation between values of IQ and blood lead levels in children with good score ($r = 0.22$; $p < 0.05$), and pupils with excellent score ($r = -0.27$; $p < 0.05$). Pearson's test of correlation between values of IQ and blood lead levels in children with very good score was $r = -0.07$; $p < 0.05$.

In order to assess possible relation between the examinee's IQ and their school's score, has been made classification of pupils according their registered score in the school. Highest average IQ -110.25 has been registered in most of pupils with excellent score (n = 16), and lowest in 1 pupil with sufficient score-88 (Table 7). It was found an increasing trend of maximal values of the IQ, from 88 in pupil with sufficient score to 122 in children with excellent score. Trend of increasing has been registered for minimal values of the IQ, from 88 in pupil with sufficient score to 98 in children with excellent score.

There was found statistical significant difference (Student t-test t = 2.58; p < 0.05) in the IQ between the examinees with sufficient and excellent score, as well between the examinees with good and excellent score (Student t-test t = -3.12; p < 0.05). For the evaluation of graph motor ability of the examined children has been performed Bender-Geshtalt test and test tree-house-man. The highest frequency has been found for children with average graph motor ability (61.3%), and lowest in the examinees with bed one (9.7%) (Table 8). Female's examinees (33.3%) with above average graph motor ability have taken higher contribution then male (27.2%).

TABLE 7. Distribution of the IQ in the examinees according to their school's score.

Score	Number	Average	Min	Max	SD
Veles	31	104.35	87	122	10.14
Sufficient	1	88	88	88	0
Good	5	97.4	91	108	665
Very good	9	99,55	87	109	8.60
Excellent	16	110.25	98	122	8.35

There was not found statistical significant difference between male and female for group of pupils with above average graph motor ability (Test of notted differences of proportions p = 0.7364; p > 0.05). There was not found statistical significant difference between male and female for group of pupils with average graph motor ability (Test of notted differences of proportions p = 0.6962; p > 0.05), with higher contribution of female (6.7%), compared to male (59.1%). All three examinees were male, and one pupil had high blood lead level (31.4 µg/dl), while two others have had moderate increased blood lead levels (12.6-12.3 µg/dl). According to the investigation there has been found small difference with better graph motor ability in female. There was not found statistical significant difference between pupils with bed and average graph motor ability (Test of notted differences of proportions p = 0.1132; p > 0.05), between bed and above average graph motor ability (Test of notted differences of proportions p = 0.5211; p > 0.05), and between average and above average graph motor ability (Test of notted differences of proportions p = 0.1258; p > 0.05).

TABLE 8. Distribution of the examinees' graph motor ability.

Level of graph motor ability	Veles		Male		Female	
	number	%	No.	%	No.	%
Bed	3	9.7	3	13.7	/	/
Average	19	61.3	13	59.1	6	66.7
Above average	9	29	6	27.2	3	33.3
Total	31	100	22	100	9	100

With a goal to assess possible relation between blood lead levels of the examinees and their graph motor ability, has been performed classification of pupils according their registered level of graph motor ability. Have been registered mostly pupils with average level of graph motor ability (n= 19), and only 3 children have had bed level of graph motor ability (Table 9). There has been found lowering of the average blood lead levels, from 18.76 µg/dl in pupils with bed level of graph motor ability, to 16.62 µg/dl in children with above average of graph motor ability. Trend of decreasing has been registered in minimal blood lead levels - from 12.3 µg/dl in children with bed level of graph motor ability, to 8.1 µg/dl in children with above average of graph motor ability.

TABLE 9. Distribution of the blood lead levels-µg/dl in the group of the examined children according their graph motor ability.

Level of graph motor ability	Number	Average	Min	Max	SD
	31	16.51	8.1	32.9	6.74
Bed	3	18.76	12.3	31.4	10.94
Average	19	16.09	8.4	32.1	5.84
Above average	9	16.62	8.1	32.9	7.83

There was not found statistical significant difference in the blood lead levels between the examinees with bed and average level of graph motor ability (Student t-test $t = 0.65$; $p > 0.05$), between bed and above average (Student t-test $t = 0.37$; $p > 0.05$), and between average and above average level of graph motor ability (Student t-test $t = -0.19$; $p > 0.05$).

5. Conclusion

It can be concluded that by increasing blood lead levels in the examinees from Veles, there was non significant statistical difference of lowering the IQ, and level of graph motor ability. That point out to potential minimal adverse impact of lead to cognitive function, (intelligence quotient), central and peripheral nervous system (level of graph motor ability) in children. But, it must be stressed that this study was performed on a relatively small group of school children, and it can not be generalized to whole population in the targeted area.

General conclusion is that there was evident environmental health risk in Veles with enough evidences of higher children's blood lead level, as well non significant minimal disorders in cognitive development (IQ), and graph motor ability in the examined children. It can be assumed that all registered outcomes in the examined children could be considered as reversible.

With a goal to prevent adverse health effects of vulnerable group of population have been recommended environmental health measures and program for recovery. Program for monitoring of the environmental pollution by heavy metals and recovery, and environmental health impact assessment has been started since 2004, and it continues till the end of 2005. Ministry of Health and Ministry of Environment and Physical Planning in cooperation with the Republic Institute for Health Protection-Skopje, Institute for Health Protection-Veles, Institute for Occupational Diseases-Skopje, Clinical Centre-Skopje, Medical Centre-Veles and Local Authorities are working in finding the best solutions in order to reduce pollution from the past, and to prevent any further contamination of the region.

References

Kristoforovic, M. 1998. *Komunalna higijena*. Prometej Publishers, Novi Sad; 269–70.

Landrygan, P.J., and Todd, A. 1994. Lead poisoning. *Western Journal of Medicine*. 161; No. 2; 153–9.

Lidsky, T., and Schneider J. 2003. Lead neurotoxicity in children: basic mechanisms and clinical correlates. *Brain*, 126 (Pt 1); 5–19.

Minder, B., Das-Smaal, E.A., Brand, E.F., and Orlebeke J.F. 1994. Exposure to lead and specific attention problems in schoolchildren. *J Learn Disabil*. 27; 393–399.

Needleman, H.L., Gunnoe C., Leviton A., Persie H., Maher C., and Barret P., 1979. Deficits in psychological and classroom performance of children with elevated dentine lead levels. *N Engl J Med*. 300; 689–695.

Silva, P.A., Hughes, P., Williams, S., and Faed J.M. 1988. Blood lead intelligence, reading attainment, and behavior in eleven year old children in Dunedin, New Zealand. *J Child Psychol Psychiatry*. 29; 43–52.

Thomson, G.O., Raab, G.M., Hepburn, W.S., Hunter, R., Fulton, M., and Laxen D.P., 1989. Blood-lead levels and children's behavior-results from the Edinburgh Lead Study. *J Child Psychol Psychiatry*. 30; 515–528.

Wang, J., Soong, W., Chao, K., Hwanh, Y., and Yang C. 1998.Occupational and environmental lead poisoning: case study of a battery recycling smelter in Taiwan. *J Toxicol Sci*. 23 Suppl (2); 241–5.

WHO. *Guidelines for Air Quality*. 2000. Regional Publications, European Series No. 91; Copenhagen; 40–2.

ECOLOGICAL AND PROFESSIONAL CHEMICAL DANGERS FOR REPRODUCTIVE HEALTH

OLGA SIVOCHALOVA* AND EDUARD DENISOV
*RAMS Institute of Occupational Health,
31 Prospect Budennogo, 105275, Moscow, Russia*

*To whom correspondence should be addressed. olga@ixv.comcor.ru

Abstract. Scientific data show the association of population health status with environmental situation of residential area and working conditions. For ecologically induced or work-related pathology it is typical not only prevalence of chronic diseases including reproductive health disorders, such as pregnancy complications, but also health disorders in off-spring such as birth defects, diseases of newborns and children of early age due to parental exposures, etc. So risk assessment and management is essential for environmental and occupational hazards especially for chemicals for adequate reproductive health protection.

Keywords: ecology, harmful factors, reproductive age, the pregnant woman, a fetus, newborn, risk factors, chemical substances.

1. Introduction

Chemical safety of a person nowadays is one of priority problems. By present time there are proofs of direct dependence of health and quality of life on environmental conditions including that of work and household. The modern condition of an environment has put before a society a problem of dependence of health of a person on state of general and working environment.

Environmental hazards for health are basically due to emissions of the industrial enterprises and motor transport vehicles. In a daily life on a person as a rule acts the complex of adverse environmental factors. It results in development of pathological changes in different body systems, manifesting in functional, morphological and genetic shifts in an organism. As result ecologically or occupationally induced diseases of respiratory, cardiovascular, endocrine, reproductive and other systems are formed.

Reproductive health is defined as a condition of full physical, intellectual and social well-being in all questions concerning reproductive system, its functions and processes, including reproduction of posterity and harmony in psycho-sexual attitudes in a family (the United Nations, Cairo, 1994).

Disorders of reproductive and sexual health contribute to a global burden of diseases 20% for women and 14% for men (WHO, 2004).

Among adverse factors one distinguishes the exogenous factors (factors of the surrounding and working environment, social and economic indices of a life, quality of health services, etc.) and endogenous factors (heredity, parents health status and, especially mother's health during pregnancy, a functional status of an organism, etc.) which can cause disorders of reproductive health. In human health the reproductive function is the major biological and socially significant side of life.

2. Reproductive Toxicants

Chemical substances are the most dangerous for human organism via processes of reproduction. Some of them carry in itself reproductive toxicity and are capable to cause gonadotropic (damaging action on gonads and system of their regulation), embriotropic and teratogenic action (damages of development and formation of anomalies of a fetus).

In modern times synthesis of new chemical compounds is carried out and constantly release in environment of poisonous by-products increases. According to US EPA approximately 2,000 new substances are introduced on market every year, thus there is a huge discrepancy between number of used chemical substances (approximately 84,000) and the number of substances for which the influence on reproduction was assessed (4,000). Now in biosphere circulates about 4 million chemical compounds, from them more than 1000 possesses the properties of reproductive toxicity.

In the European Community, according to the Directive 92/32/EEC (7-th addition to the Directive 67/548/EEC), the definition «toxic for reproduction» was introduced which takes into account harmful influence on sexual function and fertility of adult men and women, and also influence on development of posterity.

Reproductive toxicity includes two classes: *action on reproductive ability* and *action on a developing organism*. The woman can transmit a chemical substance *during breast-feeding the child.* If the substance penetrates in milk and can be toxic for the child, it is also referred as toxic for reproduction. It is out-of-category class of chemicals to which the substances influencing on or through lactation belong.

In various countries there are lists of chemical substances hazardous for reproduction and development. These lists contain substances as "proved enough" and with probable action on reproductive system. In the countries of the European Community (EUR 14991 EN 1993) and in Russia there are such lists of chemical substances affecting reproductive system.

3. Chemical Hazards for Reproduction

The state of reproductive health is reflected by the health of adults as well as health of newborns and children of the first years of a life. The degree of damaging influence of an environment depends on vicinity of an industrial complex, its turn-over, polluting elements and their concentration (as compared to maximum permissible values), duration of exposure, input ways to an organism, ability of chemicals to synergism and to potentiation, and also specificity of action. In particular, frequency of gestational complications in ecologically polluted regions in 2001 in Russia manifesting as gestosis and the anemia of pregnant women were diagnosed 1.5-2 times more often, than in clean areas.

Real threat of toxicity of the surrounding and working environment is the long term effects which can manifest at any moment of life. The most expressed and socially important manifestations of toxicity are death of a developing organism, structural anomalies, infringements of growth and functional frustration. Chemical substances in spite of the fact that they have various physical and chemical properties, molecular weight, size and a level of ionization of molecules, etc., cause the same gonad defeats. For example, disorders of specific functions in women are clinically shown by infertility, pathology of menstruation functions and pregnancy, infringements of fetal development of a fetus and newborn, and development of a pathological climax.

At the same time the general rules of input, distribution, metabolism and removal of chemicals from an organism, expressiveness of adverse effects of polluted environments and the limit of stability to them of reproductive system of the person depends on a phenotype, heredity, age, sex, individual sensitivity, somatic health status, conditions of work and life, alcohol abuse, tobacco smoking habit, etc. For example, there are proofs that women are more sensitive in comparison with men to a number of the chemical substances acting through a skin, especially during the periods of menstruation, pregnancy, climax that is caused by anatomic and physiological features of their skin. In connection with the high maintenance of a fat in woman's organism, some poisons (metilmetacrilat, styrene) are capable to be dissolved in fats and can be deposited and change processes of metabolism.

According to the theory of critical stages of intra-uterine development of a fetus, its response to surrounding damaging factors, including working environment, is determined by stage of embryogenesis. The intra-uterine period is one of the basic periods of human life, determining the health of individual in future. Thus surrounding conditions in which there was a woman during pregnancy, under other conditions (heredity, social and economic position, etc.) also determine the quality of health of a person.

4. Urban Pollution and Reproductive Health Disorders

The comparative research on role of environmental pollution on reproductive health of the women living in area, polluted by industrial emissions to more than three times in comparison with conditionally clean one was conducted recently (Golovaniova, 2002). In pollutions there were following chemical substances: acryl nitril, acetone, gasoline, benzene, xilol, lead and its combinations, styrene, carbon oxide, formaldehyde, epichlorohydrin, etc.

As a result of environmental contamination by hazardous chemical substances there was an infringement of such integrated reproductive function of women, as menstrual one. In a basis of disorders of this function lay the changes of the hormonal status. Due to specific action of chemical factor the development of destructive changes, both in central, and in peripheral parts of reproductive system were different. As compared to physical factors chemical influence is accompanied by less distinct reaction phase which expressiveness in many respects is defined by poison concentration. Chemical substances in a combination with other factors change a functional condition of a bark of the brain, specific cells hypothalamus, coordinating endocrine functions of an organism. Duration of the period of indemnification due to overstrain of adaptive mechanisms depends on stability of an organism and has individual character.

It is revealed also, that high frequency of gestational complications (95.4 % of cases) depends on technogenic load. From complications of pregnancy the threat of pregnancy interruption, development of gestosis and anemia of pregnant women was observed more often. Gestosis was predominant in women living in ecologically polluted areas, and also having occupational risk, it was in the early beginning - 20-24 weeks of pregnancy (Table 1). High technogenic load caused in 51 % of examined a sensitization to formaldehyde, lead and nickel, and in 69 % of pregnant women changes of gumoral immunity are established. The analysis has shown that high morbidity levels in children of first year of life are associated with frequency of infringements of mother's immune system.

It is established, that 15 of 100 newborn babies (as compared to 5 of 100 in clean territory) were born with low estimates of health status, 20 babies (as compared to 12 of 100) developed in conditions of chronic hypoxia, 11 babies were hypotrophic (body mass less than 2500 grams), and 15 babies (as compared to 9 of 100) had disturbances of brain blood circulation of hypoxic genesis. An important problem is congenital anomalies development in children which were found out approximately in 4% of newborns.

TABLE 1. Pregnancy complications rate depending on living site in Moscow (per 1000 women)

Indices	West (p±m)	South-East (p±m)	Odds ratio OR	Confidence interval 95%CI
Early miscarriage	24,4±2,0	33,3±3,2* P<0,05	1,62	1,11-2,37
Late miscarriage	5,1±1,1	24,3±2,9 P<0,01	6,01	3,43-10,59
Gestosis	24,8±2,1	33,5±2,7 P<0,05,	1,53	1,03-2,23
Anemia	36,5±2,3	80,2±2,2 P<0,01	13,0	7,88-21,61

In territories of ecological pollution toxic load on organism of pregnant woman is an essential risk factor of congenital pathology formation in children. It is known, that through placenta from mother to fetus can penetrate more than 600 chemical substances capable to some extent negatively influence its development. Changes in placenta can be considered as original display of adaptation to external factors that allows to use hormonal parameters as test - object for biological monitoring of an environment. The absolute proof of harmful effects of chemical substance on development of this or that system of an organism and development of a fetus is the detection of the poison or its metabolites in biological tissues of parental organism.

5. Aluminum Plant Pollution and Reproductive Health Disorders

Studies that have been carried out in nearby zone of an aluminum factory (Odinayeva, 2002) have shown that fluorine and its combinations are leaders in formation of infringements of function of a liver not only in pregnant woman, but also its fetus during intra-uterine development. Children of the women living in territory polluted with fluorine, had 3.5 times more often intra-uterine development disorders - each 8-th child, and at presence of liver pathology in mother during pregnancy every 3-rd child had low estimation on Apgar scale (6 points and less) and for each 3-rd child was diagnosed post hypotoxic CNS defeat (31.2 % as compared to 6.5 % in controls). Thus it is important to note, that aggravating factor for health of the child is breast feeding since the fluoric derivates determined in milk of mother, pass to the child during breast feeding.

The long term effects of exposure to fluorides and their combinations were shown during puberty: hypothalamic syndrome was observed in 38.6 % of cases, the delay of sexual development was established in 18.8 % of cases, and infringements of menstrual cycle were found in 44.2 % of girls.

6. Placenta Permeability

For forecast and prevention of possible disorders during pregnancy the knowledge of condition of fetus and placenta complex and, in particular, placenta is of great value.

High levels of pregnancy complications in women exposed to harmful chemical substances testify to early formation of infringements in system mother – placenta – fetus and, in particular, infringement of its permeability. The index of permeability of placenta (IPP) that enables to assess barrier functions of placenta has been calculated, to predict the possible long term risks for newborn health based on the knowledge of specific character of chemical substance.

The studies that have been carried out in 1970-80-th (Sivochalova, 1976) and also later (Odinayeva, 2002; Zaytseva, 2004) have shown various values of an index of permeability for different chemicals in contact with pregnant woman during gestation. Special research is needed in questions of infringement of reproduction processes of occupational etiology and their long term effect since they are extremely difficult in diagnostics.

7. Russian System of Occupational Risk Assessment

In Russia the system of occupational risk assessment is devised. The system in based on Federal laws #52-FL (1999) on sanitary and epidemiologic well-being of the population, #181-FL (1999) on labor protection, and #125-FL (1998) on obligatory social insurance of occupational accidents and diseases. It also accounts for WHO and ILO concepts, EU directives, and ILO conventions ratified by Russia (e.g. #148 on working environment). The state standard GOST R 12.0.006-2002 on OHS management in organization based on OHSAS 18001-99 and ILO-OSH 2001 covers both workplace certification and risk assessment.

The system includes two guides approved by the Russian Health Ministry for *a priori* and *a posteriori* risk assessment.

In Guide R2.2.2006-05 (2005) on hygienic criteria for assessment of working environment and labor load, conditions of work are classified on a 7-point scale (optimal, permissible, four hazardous classes, and extreme) depending on multiple of OEL's exceeding for a factor.

The Guide R2.2.1766-03 (2003) that addresses principles and criteria of occupational risk assessment for workers' health introduces the weight of evidence criteria (known, presumed, suspected) according to GHS system (UN, 2003).

As a supplement to R2.2.1766-03, specific recommendations are planned on methods and criteria assessment (scales, models, etc) of workers' health (morbidity, mortality, etc), workplace factors (noise, vibration, dust, chemicals, etc), work load (physical load & nervous tension), and reproductive health risk assessment. Relevant data have been published in handbook on occupational risk (Izmerov & Denisov, 2003).

In Table 2 the working conditions classes and risk categories according to Guide R 2.2.2006-05 are presented.

Table 2 is based on qualitative estimates of occupational morbidity index (Izmerov & Denisov, 2003) at logarithmic scale with its values from 0 to 1. The borderline between classes 3.4 - 4 relatively to classes 1 - 2 corresponds to following risk values: a) somatic diseases and mutagenic disorders – RR≥5; b) ageing acceleration – 10 years or more; c) work-related mortality – standardized relative risk SRR≥7. These indices characterize different sides of health and quality of life, including risk for off-spring health.

8. Qualitative Risk Assessment of Reproductive Health Disorders

The women working in harmful environments demand special measures of protection to perform the motherhood function. Many occupational factors are harmful for reproductive health. Thus character, a degree and frequency of reproductive health disorders depend on a class of harm of working conditions and length of service in these conditions. The higher is degree of harm (or class of working conditions 3.1, 3.2, 3.3 etc.), the more frequent and prominent are the clinical features, and more are relative risk and etiological fraction of the contribution of occupational factors in reproductive health disorders.

TABLE 2. Working conditions classes and risk categories (Guide R 2.2.2006-05).

Working conditions classes	Occupational morbidity index	Risk category
Optimal 1	-	No risk
Permissible 2	<0.05	Negligible (tolerable)
Harmful 3.1	0.05-0.11	Low (moderate)
Harmful 3.2	0.12-0.24	Middle (significant)
Harmful 3.3	0.25-0.49	High (intolerable)
Harmful 3.4	0.5-1.0	Very high (intolerable)
Hazardous (extreme) 4	>1.0	Extremely high risk for health and life

In Table 3 the relative risk and etiological fraction of occupational risk factors (heating microclimate and physical loads handling) in reproductive health disorders is shown depending on class of working conditions.

TABLE 3. Relative risk and etiological fraction of work-related reproductive health disorders of female workers.

Disorders	Class of working conditions (according to Guide R2.2.2006-05)			
	3.2 (n=272)		3.3 (n=211)	
	Relative risk	Etiological fraction, %	Relative risk	Etiological fraction, %
Menstrual cycle disorders	1.19	16	2.76	64
Infertility	1.75	43	3.58	72

From Table 3 it follows that etiological fraction of occupational risk factors contribution in frequency of menstrual cycle disturbance increases 4 times depending on harm of working conditions (class 3.2 - 16%; class 3.3 - 64%). It is known that menstrual function disturbance promotes the growth of obstetric morbidity.

It is proved also, that the longer the woman works in adverse conditions, the higher are the parameters of reproductive health disorders.

Influence of occupational hazards on mother was clinically shown by high prevalence of pregnancy complications: the late toxicosis was diagnosed for each 3-rd pregnant women, threat of abortion at every 10-th. Levels of perinatal outcomes depend on parameters of pregnancy pathology: more often complications in the gestation period were diagnosed; the bigger was the contribution of occupational factors to infringements of perinatal period. Thus it was established, that the pathology of a fetus and newborn is significantly more work-related since etiological fraction or contribution of workplace harmful factors in development of health infringements of newborns was more than 50 % (Table 4).

TABLE 4. Qualitative risk assessment for health of mothers and new-born babies depending on working conditions classes of mothers before pregnancy.

Mother's working conditions class before pregnancy	Mothers (according to indices of pregnancy course)		New-born babies	
	Relative risk RR*	Etiological fraction EF, %	Relative risk RR*	Etiological fraction EF, %
3.1	1.5	33.3	1.7	41.2
3.2	1.7	41.2	2.7	63.0
3.3	2.0	50.0	3.8	73.7

* RMS values of the 3 largest values out of 5 and 7 indices for mothers and new-born babies, respectively.

From positions of evidence-based medicine it is possible to approve, that etiological fraction of contribution of harmful workplace factors in health disorders is substantially higher for developing fetus, than for mother's organism and this fraction increases with growth of harmful working conditions of the woman, i.e. the risk of infringements of development of a fetus considerably exceeds risk for a parent organism.

9. Pregnancy Planning from Occupational Health Point of View

The scientific data testify to poor security of an intra-uterine fetus in organism of mother from adverse influence of environmental factors. In this connection the principle *of pregnancy planning from occupational health point of view* is important. This includes the choice by the woman of optimum conception time in view of medical recommendations. Thus favorable working conditions and the way of life excluding or reducing the influence of harmful occupational and other factors on woman's organism during conception, in early terms of intra uterine development of a fetus and during breast-feeding.

The women of fertile age working with harmful occupational factors should be treated as "vulnerable" groups in relation to disorders of development of a fetus and newborn's health. They demand careful medical clinical and laboratory monitoring and surveillance by an expert in occupational health. It is because these women frequently suffer by early beginning of toxicoses of 2-th half of pregnancy, anemia of pregnant women, changes of the immune status, deterioration of development and functional immaturity of a fetus and newborn, development of fetus and placental and hormonal insufficiency.

10. Men's Health

For today it is necessary to take into account a role of health of the man in formation of health of the future child. Worldwide there is a tendency for deterioration of men's reproductive health, e.g. rates of spermatozoa concentration decrease are in the USA - 3 % per year, in the Western Europe and Australia – 1.5 %, in France - 2 % per year.

A number of occupational factors affect reproductive health of men: chemical, physical, biological, stresses, etc. Therefore it is necessary to give special attention to working environment of the men exposed to these occupational hazards, especially chemicals (lead, mercury, manganese, phosphorus, carbon disulfide, ammonia, benzene, granosan, organic peroxides, chlororganic substances, etc.). Occupational exposure to such factors can lead to defeat of rather sensitive testicle epithelium that causes deterioration of sperm, infringement of endocrine status, etc. Similar disorders can be caused by thermal factors.

It is known, that chloride cadmium causes selective defeat of vessels of testicle, and also infringement of permeability of other structures which are carrying out function of testicular barrier.

In men, occupationally exposed to fluoric intoxication, was observed an easing of sexual function (infringement of libido, erection and ejaculation). Laboratory parameters of these men were characterized by reduction of ejaculate volume, spermatozoa concentration in it, and increase in motionless and degenerated forms of spermatozoa. These changes were 3-4 times more frequent, than in control group.

In ejaculate of men exposed to chloroprene the clinically significant infringements were found and also were shown spontaneous miscarriage in their wives.

At work with vinyl chloride in concentration, basically not exceeding maximum permissible concentration, at sexological observation has revealed almost infringements of copulative cycle in 60 % of workers.

There is an opinion, that mechanisms of adverse effects on fetus development include transmission of chemical substances from the father to an embryo through a seed liquid, so-called pass-through "infection" of mother to an embryo - the substance brought by the man from a workplace or in domestic conditions. Besides, harmful chemical substances at man's workplace can induce genetic mutations.

11. R-Phrases for Reproductive Toxicity and Risk Management

For development of preventive measures for work-related reproductive health disorders, pathology of a fetus and newborn, it is expedient to examine a degree of hazard for health depending on character of action of chemical substances because they can produce the gonadotropic and embriotropic effects, etc. Thus it is necessary to take into account standard indices or so-called phrases of risk (R-phrase) for chemical substances (EU Directive 92/32/EEC). To mark the chemicals with specific kind of action in European Union they apply the R-phrases for classification. Such designation of chemical substances helps to receive information on reproductive risks for the substances used on workplaces.

Labeling of substances toxic for reproduction (Commission Directive 2001/59/EC) is as follows:

- R40 (R68): Possible risk of irreversible effects.
- R46: May cause heritable genetic damage.
- R47 (obsolete): May cause deformities.
- R60: May impair fertility.
- R61: May cause harm to the unborn child
- R62: Possible risk of impaired fertility
- R63: Possible risk of harm to the unborn child
- R64: May cause harm to breastfed babies

For practical purposes main phrases were categorized as shown in Table 5.

For occupational risks management during different periods of women's reproductive life the criteria for work-related reproductive health disorders were developed. It is important to establish possible occupational hazards for one or both parents and evaluate the structure and degree of risk.

TABLE 5. Risk phrases (EU Directive 92/32/EEC) of reproductive toxicants for different categories of women.

Categories of women	R-phrases
Women under 18 years of age	R60, R62
Pregnancy planning women	R61, R63
Pregnant women	R61, R63
Breast-feeding women	R64

So, chemical substances to which female or male worker are exposed can cause:

- deterioration of somatic health of the worker;
- damage to gonads with subsequent frustration of hormonal and sexual functions;
- affect ovule during ovulation, impregnated ovule at the moment of implantation and in different stages of embryonic and the germinal periods;
- damage to sexual and somatic cells with induction of mutations which can be seen in the remote period (in subsequent generations).

It is essentially important to develop and implement the Programs of health protection and health promotion for mother and father with preventive measures taking into account both environmental and occupational hazards of both parents to ensure good health and well-being of future child.

Russian experience demonstrates that for the purpose it would be preferable to have two compatible and overlapping systems of chemical safety: a scientifically-based rather sophisticated system to be used for occupational and environmental health monitoring and intended for use by health professionals and a simplified system for small and medium business like Control Banding that is under development by ILO and WHO. The international cooperation in the field seems to be very useful.

References

Golovaniova, G.V., 2002. Health of pregnant women in cities with different technogenic loads. *Meditsina truda i promyshlennaya ecologiya, 8,* 5–7 (in Russian).

Health of women of Russia. 1998. Analytical report (economic, social, ecological, legal and medical aspects). Presidential commission on questions of women, families and demography. Moscow: Joint-Stock Company Informatics, 96 p. (in Russian).

Izmerov, N.F., Sivochalova, O.V., Denisov, E.I., 2001. Medical and social problems of reproductive health protection and decision making in this field. Actual problems of reproductive health in conditions of anthropogenic pollution. Proceedings to the International Symposium. Kazan. P. 5–13 (in Russian).

Occupational risk for workers' health (Handbook) / Ed. by N.F. Izmerov and E.I. Denisov. - Moscow, Trovant, 2003. - 448 pp. (in Russian, abstract and contents in English).

Odinayeva, N.F., 2002. Course of pregnancy, labour and new-born status with account for liver functional activity in women residing near aluminum factory. Diss. Cand. Med. Sci. Ivanovo (in Russian).

Sivochalova, O.V., Feytshans, I.L., Denisov, E.I., and Golovaneva, G.V., 1999. The right of workers on reproductive health: priorities today and in the 21st century. Human rights in Russia: declarations, norms and life. Proceeding to the International Conference. Moscow. P. 214–217 (in Russian).

Workers' reproductive health protection. Main terms and notions / Ed. by N.F. Izmerov and O.V. Sivochalova. - Moscow: Russian Health Ministry, 2003. - 20 pp. (in Russian).

TECHNOLOGICAL TRANSFER. MINIATURE LASER MASS SPECTROMETER FOR EXPRESS ANALYSIS OF ENVIRONMENTAL SAMPLES

LUBOMIR SIMEONOV[1]* AND GEORGY MANAGADZE[2]
[1]*Solar-Terrestrial Influences Laboratory*
Bulgarian Academy of Sciences
Acad.G.Bonchev Str.Bl. 29, 1113 Sofia, Bulgaria
[2]*Space Research Institute, Russian Academy of Sciences*
Profsoiuznaia 84/32, GSP 7, 117997 Moscow, Russia

*To whom correspondence should be addressed. simeonov@bas.bg

Abstract. A miniature laser mass spectrometer, called LASMA, was built around the concept of the first space-born laser time-of-flight mass-spectrometer, the core analyzer of the LIMA-D complex (Laser Ion Mass Analyzer – Distant), developed for the space mission PHOBOS. The new device is a close analog of its predecessor, but its size and mass are by an order of magnitude lower. In concern to space research, the analyzer is suitable for investigation in-situ of the elemental and isotopic composition of the regolith of small bodies and cometary nuclei. The earth-based modification of the instrument is convenient for environmental applications and is capable to perform an express analysis of samples of soil, water and plant origin. The paper presents moments of instrumental design, estimation of the analytical characteristics of the analyzer and the development of its measurement methodology in connection to specific environmental applications.

Keywords: environmental samples; elemental and isotopic analysis; laser mass spectrometry; technology transfer

1. Introduction

Laser time-of-flight mass analysis is a preferred method for solution of difficult analytical problems, because of its basic physical characteristics, such as the ability to achieve a parallel registration of all elements in the mass spectra in a practically unlimited mass range. The high analytical accuracy of the method combines with the simple design of the instrument, which does not pose high requirements for precision of its building elements and mounting procedure. In principle, modern laser sources generate a stable ion production in the whole mass range with equal probability without elemental and isotopic constrains and the yield of all elements is due to even evaporation of the target without fractions.

A convenient way to investigate the elemental and isotopic composition of the upper surface layers of small bodies of the Solar system is to use small-size laser mass spectrometers, mounted on a lander. Optimal for the purpose is a reflectron type time-of-flight mass analyzer with laser evaporation and ionization of the target, called LASMA (**LAS**er **M**ass **A**nalyser). It was created on the basis of a laboratory prototype, initially developed for the LIMA-D laser mass spectrometric system on the PHOBOS experiment, a space mission to the Mars satellite. A spin-off modification of the LASMA device was further developed for the special purposes of environmental research.

2. Instrumental Design

In the process of device creation different instrumental designs has been checked for analytical performance with the main criteria to keep the overall size as small as possible and in conformity to the principal requirements for space-born devices for minimal mass, size and energy consumption. The application of these limits in earth-based instrumentation provides possibilities for its implementation in studies, where an express elemental analysis is necessary, i.e. in cases of accidental or intentional environmental pollution. The accepted final version of the analyzer, whose functional scheme-view is shown on Fig. 1, is that of a linear mass-reflector type. This configuration, along with the general basic advantages of the time-of-flight mass-reflector design, features several specific characteristics. On the first place, the linear and full axial symmetry of the analyzer compartment with coaxial beam impact, orthogonal to the target surface is certainly the most compact instrumental configuration. On the other hand, it achieves high reproducibility of spectra and unlimited layer-by-layer analysis due to elimination of shadowing effects. The alignment of the laser focusing system with the targeting microscope allows the establishment of precision control of the spot position on the target. The time-of-flight of ions having same mass does not depend on the laser beam spot on the target due to the free flight regime without pre-acceleration. The multi-stage reflector with nonlinear electrostatic field preserves a relatively high mass resolving power of the instrument in a narrow ion energy window mode of operation. Further optional upgrades comprise: a "quasi-ideal" field distribution of the reflector to narrow the mass spectrum; a multi-spot laser action on target using a beam splitter to achieve high representativness of probe in a combination with higher sensitivity, repeatability and reproducibility; variable neutral optical attenuator to provide laser power distribution without distortion, resulting in more stable spectra and elimination of multi-charged ion production, etc.

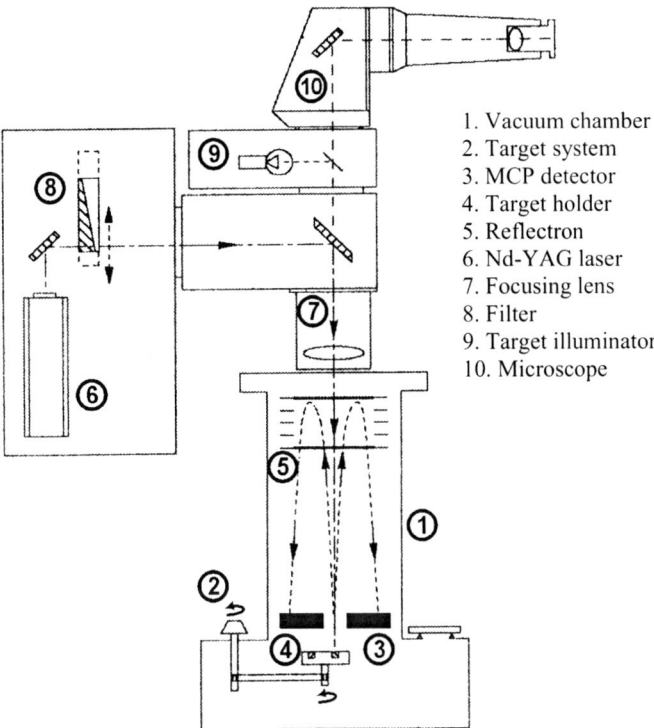

1. Vacuum chamber
2. Target system
3. MCP detector
4. Target holder
5. Reflectron
6. Nd-YAG laser
7. Focusing lens
8. Filter
9. Target illuminator
10. Microscope

FIGURE 1. Functional scheme-view of the laser mass spectrometer LASMA without vacuum, power supply and steering electronics units.

LASMA device consists of the following main functional units: a laser source 6 with a beam-focusing system 7, which enables to concentrate energy into spot with diameter of 30÷50 µm on the surface of the sample in vacuum chamber 1. Ions, emitted as a result of an infrared irradiation influence, are reflected in the field of electrostatic reflector 5 and directed to the detector 3. The detector is a special development of a standard chevron assembly of two micro-channel plates. The micro-channel plates have a central hole with a diameter big enough to let the laser beam down to the target and the flow of ejected ions up towards the time-of-flight compartment of the analyzer. After their reflection the ions of one and a same mass form expanding ring packets on their flight-back to the ring detector. With that form the mass packets arrive at the detector. The high-speed analog-to-digital converter loads a spectrum to a computer for further signal processing. The instrument is equipped with microscope 10 and miniature TV monitor that allows observing intentional movements of laser influence area during the investigation of surface heterogeneity. The samples are introduced in the vacuum chamber through a lock – chamber, which makes it possible to replace a holder with samples on the target platform wheel 4 within one minute without deterioration of the vacuum in the main chamber 1. Each sample holder contains up to ten samples situated on a metal, Teflon or a glass volumes depending on the nature of the sample material and its consistence. The instrument is capable to perform elemental and isotopic analysis of environmental samples of soil, water and plant origin.

The possibility to mount the whole system of the instrument with the pump unit, the power supply and the personal computer on a rack and operate it on a medium-size pick-up van allows the process of sample-taking, target preparation and actual analysis to be accomplished in-field.

The experimentally established analytical and the physical characteristics of the laser mass-reflectron LASMA are presented in Table I.

3. Evaluation of the Analytical Characteristics of the Laser Mass Spectrometer

The analytical characteristics of the laser mass spectrometer, as shown above in Table 1, were confirmed experimentally with the help of laser targets of different materials, alloys, sediments and other samples. They were approved in calibration procedures with standard reference materials with certified presence of elements and isotopes and checked by analysis of targets with preliminary unknown elemental composition.

To demonstrate the mass resolving power of the instrument, a target of a two-component soldering alloy was analyzed. Figure 2 and Fig. 3 show two parts of the mass spectra. The mass resolution is correspondently 600 (at FWHM) for Pb and 500 (at FWHM) for Sn. We have to mention that such good values for the mass resolution are not possible to be achieved with every laser shot and every target. The main factor that limits the mass resolution are not the analytical capabilities of the analyzer itself, but rather the physical properties of the target and the quantities of evaporated and ionized material. The mass resolution strongly depends on the material homogeneity of hard samples.

TABLE 1. Experimental and physical characteristics of LASMA.

Mass range	1 to 250 a.m.u
Mass resolution at FWHM	≥ 400
Mass resolution of individual spectra of a series up to	800
Relative sensitivity of one analysis	10^{-6}
Relative sensitivity in the spectrum accumulation mode for 10 shots at frequency of 0.3 Hz	$5 \cdot 10^{-7}$ to 10^{-7}
Absolute detection threshold in terms of mass in one analysis	$5 \cdot 10^{-14}$ g
Absolute detection threshold in mass detection in the spectrum accumulation mode	$5 \cdot 10^{-16}$ g
Diameter of the sampling area	20 to 50 µm
Depth resolution during layer-by-layer analysis depending on the target material	0.2 to 3 mm
Reproducibility of Cu isotopes ratio	1%
Type of laser	NdYAG
Wavelength (IR)	1064 nm
Specific power density of the laser emission	10^9 W/cm^2
Laser pulse duration	5 ns
Instrument speed of response to 1 a.m.u.	200 ns
Magnification of the microscope	60 X
Scanning resolution on the target surface	2 µm
Size of vacuum chamber of the mass spectrometer	35 x 18 x 18
Overall size with of the instrument with pumping unit and power supply	50 x 50 x 50
Power consumption	600 W

The reproducibility of measurement that the laser mass analyzer LASMA could reach was approved with a sample of high homogeneity. The target was a $Ni_{60}Cu_{63}$ metal folio, used for the production of contacts for integrated circuits. On the target were produced 56 laser shots in a sequence and the spectra data for each single shot was introduced automatically in the computer. The reproducibility of measurement that the laser mass analyzer LASMA could reach was approved with a sample of high homogeneity. The target was a $Ni_{60}Cu_{63}$ metal folio, used for the production of contacts for integrated circuits. On the target were produced 56 laser shots in a sequence and the spectra data for each single shot was introduced automatically in the computer.

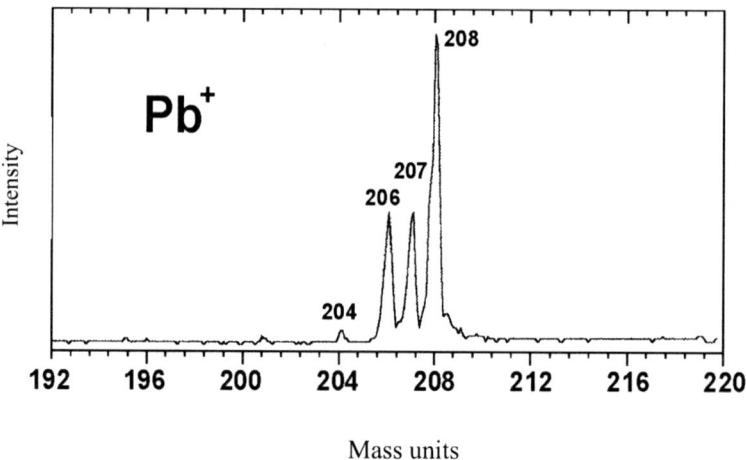

FIGURE 2. Part of a mass spectra made with the laser mass analyzer with the Pb isotopes of a soldering alloy target, featuring a mass resolution M/ΔM 600 (FWHM).

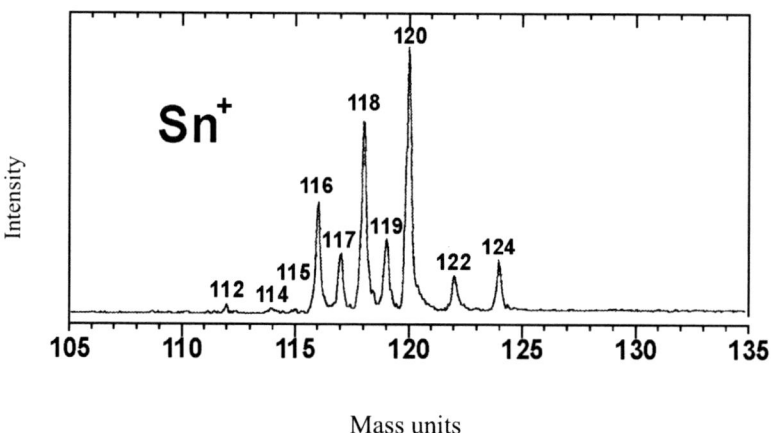

FIGURE 3. Part of a mass spectra made with the laser mass analyzer with the Sn isotopes of a soldering alloy target, featuring a mass resolution M/ΔM 500 (FWHM).

The reproducibility of measurement that the laser mass analyzer LASMA could reach was approved with a sample of high homogeneity. The target was a $Ni_{60}Cu_{63}$ metal folio, used for the production of contacts for integrated circuits. On the target were produced 56 laser shots in a sequence and the spectra data for each single shot was introduced automatically in the computer. The spectra are produced in equal time intervals, defined and controlled in order to avoid introduction of false data record from a possible unintentional laser self-shot. In the following data analysis the dependence of the computed by the program for mass identification ratios of areas of Ni_{60}/Cu_{63} are plotted against the distribution of these ratios from the 56 laser shots. The Gauss-approximation shows a deviation of 12%. Assuming, that the sample is of high homogeneity of distribution of Ni and Cu, it can be stated, that the measurement results have a statistical error of 12%. This error is caused by the statistical fluctuations in the production of Ni and Cu ions in dependence to the self fluctuations of the laser beam power density, which is not lower, than 8%.

The spectra are produced in equal time intervals, defined and controlled in order to avoid introduction of false data record from a possible unintentional laser self-shot. In the following data analysis the dependence of the computed by the program for mass identification ratios of areas of Ni_{60}/Cu_{63} are plotted against the distribution of these ratios from the 56 laser shots. The Gauss-approximation shows a deviation of 12%. Assuming, that the sample is of high homogeneity of distribution of Ni and Cu, it can be stated, that the measurement results have a statistical error of 12%. This error is caused by the statistical fluctuations in the production of Ni and Cu ions in dependence to the self fluctuations of the laser beam power density, which is not lower, than 8%.

The reproducibility of mass spectra was checked also with the help of a standard Cu wire, determining the characteristic isotopic ratio of copper Cu_{63}/Cu_{65}. The laser produced totally consecutive 165 shots and the isotope ratio was calculated for every shot by the areas under the mass peaks. The average value of the so calculated isotope ratios was estimated to be 2.24 ± 6%, which fully corresponds to the natural abundance distribution of 2.24. Similar measurements with the same instrument with a standard reference material (SRM 610) of NBS showed an averaged value of 2.24 ± 1% from 50 consecutive laser shots. It is obvious that the differences in the two measurements are a function of different homogeneity of the targets.

The calibration of every laser mass spectrometer in principle is performed with the help of special standard reference materials with high homogeneity and certified elemental and isotopic composition. The industrially produced laser mass spectrometers undergo this obligatory procedure because of possible performance differences in each instrument, introduced by its main building elements. This approach is of much greater importance and must be performed with every instrumental design, because besides the preliminary evaluation of the analytical characteristics, the checking with standardized materials is in fact an evaluation of the new ideas, introduced in the instrument [R, R]. One of the standard types, used for the calibration of the LASMA laboratory prototype was the target material series NIST SRM 310-312, provided by the NBS (National Bureau of Standards). According to the attached certificate, the standard is produced and certified to facilitate the development of trace analytical methods, and is one of a series of four. The nominal trace element concentration is 500 ppm for each of 62 elements that have been added to the glass support matrix. The material is prepared in a rod form and is sliced into wafers. A normal wafer is a 3 mm thick glass-plate matrix with implanted 62 trace elements from hydrogen to uranium with high homogeneity, certification guaranteed with accuracy under 2%. The elements with masses from 60 to 238 a.m.u. have approximately the same presence in the standard, equal to ~ 50 ppm.

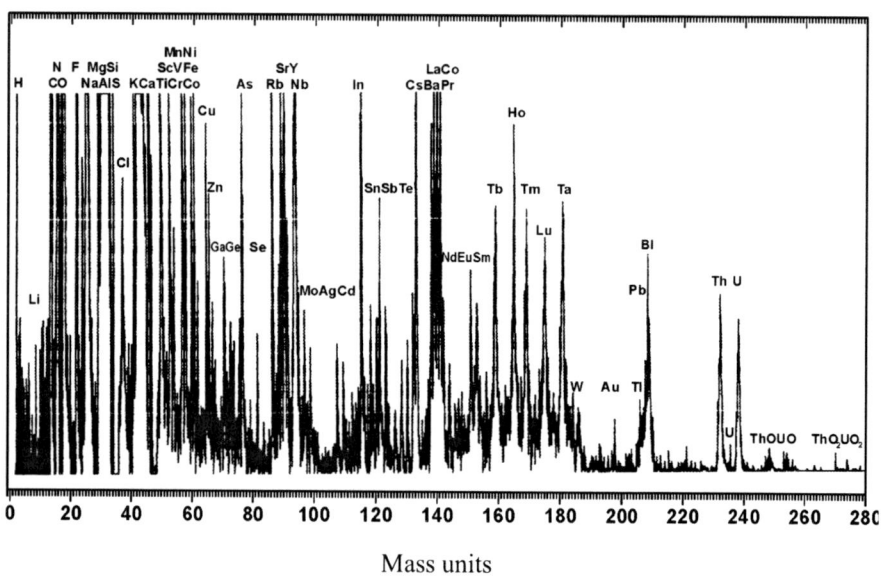

FIGURE 4. A calibration mass spectrum from a single laser shot on a standard target SRM 610.

In order to match the dimensions of the sample holding unit of the spectrometer a smaller in size set of sizes was prepared by careful cutting of the initial glass matrix plate. The cutting was accomplished with a copper-bonded diamond wheel. The debris was removed and the surfaces of the slices were cleaned with alcohol and with diluted 1/10 nitric acid in order to avoid a possible etching and to remove the expected copper contamination. On Fig. 4 is presented a single laser shot spectrum of SRM 610 with the presence of all certified elements with good analytical accuracy, as the elements after 60 a.m.u. are with amplified appearance. A more detailed observation of the part of the spectrum, which includes the heavier elements Th and $_{2387}$U could be noticed several additional mass peaks, which correspond to UO, ThO, UO$_2$ and ThO$_2$. The evaluation of the presence of the heavier Pb, Th and $_{238}$U was used to check the reproducibility of the laser analyzer and the analysis included 19 consecutive laser shots, each on a different spot of the matrix. The areas under every mass spike were calculated and the ratios of Th/U and Th/Pb were estimated. It was found, that the spreads were respectively 13% and 26%.

Further demonstration of the analytical capabilities of the miniature laser mass analyzer LASMA was accomplished with an alloy target, called IMAGE 2, with preliminary unknown elemental and isotopic composition. In principle, such kind of laser targets are not considered as reference standards, although they are produced especially for the tests and their certification is carried out in at least two independent laboratories. The check of the analytical capacity of the prototype of the mass-spectrometer LIMA-D, which served initially as a technological basis for the design of the miniature LASMA device, was accomplished with similar target material. The results of the tests with LASMA are presented on Table 2, featuring very good compatibility of the measured to the passport elemental abundances in weight percentage.

TABLE 2. Laser mass analysis with LASMA of a standardless Target with preliminary unknown elemental composition.

Element	Measured in weight %	Passport in weight %
Au	84.68	84.50
Pt	7.12	6.90
Pd	4.30	5.00
In	1.22	1.75
Ag	1.10	1.00
Re	0.65	0.1
Fe	0.49	0.70
H	0.314	-
Zn	0.13	0.15
C	0.0025	0.002

The above presented results confirm the applicability of the miniature laser mass analyzer LASMA in different earth-based areas, where an express elemental and isotopic analysis is needed. The instrument is especially suitable for in-field environmental studies and meets the basic requirements such as: small dimensions and mass; high mobility; high degree of autonomy; operation in various working conditions; minimal consumption of energy and consumables; easy operation and maintenance; easy exchange of the main functional blocks; fully-automated analysis cycle, registration and identification of the mass spectra.

4. Laser Mass Analysis of Environmental Samples

The application of laser mass spectrometry in environmental analysis is connected to the problem of anthropogenic pollution with heavy metals in soil, water and higher plants. The evaluation of heavy metal concentrations is in the first order a task of analytical chemistry. Generally, there are several tenths of different methods and instrumental modifications that people use in elemental and isotopic analysis of samples of different origin and state of matter. A great number of these techniques are task-orientated or they have been developed to solve particular and often very narrow problems. Only a few of them are applied more widely in environmental chemical analysis, like for example: the atomic absorption spectrometry (AAS); the inductively-coupled plasma/emission spectrometry (ICP/AAS); the neutron activation analysis (INAA and RNAA); the X-ray fluorescence analysis (XRFA); the voltametry; laser mass spectrometry, etc. Without the intention to analyze precisely the advantages and disadvantages of the available methods and devices and because of the great complexity of objects of investigation, related to environmental analytics, it becomes evident that a combination of different instrumental and methodology approaches is needed. The general requirements of environmental research are formulated below in concern to the principal analytical characteristics of laser mass spectrometry and with regard to the qualities of the miniature laser mass analyzer LASMA in order to define its proper application areas.

1. *Ability to perform multielement analysis* – this is one of the main advantages of time-of-flight mass spectrometry. In principle, there is a possibility to obtain analytical information about unlimited mass range from a single laser shot/ mass spectrum. The heavy metals that are interesting for environmental research are in the mass range up to 250 a.m.u. (U) and a mass resolution (FWHM) of 400 $M/\Delta M$ (LASMA typical) would be sufficient enough to obtain good-resolved mass spectra.

2. *Express and easy sample preparation* – the homogeneity of the sample is of primary importance in laser mass spectrometry. The preparation of environmental samples includes pure physical manipulation. In the case of:

- soil samples – cleaning, if necessary, drying and homogenization (grinding and pulverization);
- water samples – in a method for laser target preparation from water samples and solutions, developed for LASMA [Simeonov et al., 1996], the water content of approximately 0.1 ml is carefully evaporated on a glass holder to produce a thin and homogeneous sedimentary film. The preliminary insertion of a reference metal (Atomic standard solution) provides a basis for quantification of the measurement. This method

could be used with soil and biological samples after decomposition in a high pressure microwave oven;

- plant and bio–indicator moss and earthworm samples – cleaning if necessary, slicing, compartment procedure and pressing with careful drying at 40 to 50 °C [Markert, 1993A; Markert, 1993B].

3. *Analysis without destruction of the object (in plant monitoring)* – the application of laser micro-analysis allows analytical information to be obtained without destruction of the whole object of investigation. For example, while monitoring of heavy metal uptake, translocation and accumulation in higher plants by analyzing tiny parts from the root and leaf system.

4. *High determination rate* – the determination rate is defined as the number of investigated samples for a period of time. It depends on the medium time it takes to investigate one sample while the sample preparation time and the data processing of mass spectra are not included. The investigation rate of LASMA is nearly 40 samples per hour with a sample change duration of nearly 1 minute without affecting the vacuum in the system.

5. *High determination sensitivity* – the detection limit for laser mass spectrometry is within the 1 to 100 ppm range, depending on the target characteristics [4] (LASMA typical is 10 to 100 ppm). A comparative survey of the Tables of concentration levels of heavy metals in soil for Holland, Germany and Bulgaria shows that only the detection of Cd maximum permissible limit of 5 ppm should be a problem [Klocke, 1990; Van Lidth de Jeude, 1983; Maximum permissible levels, 1979]. The LASMA detection limits for water samples are in the 0.5 to 1 ppm range.

6. *Accuracy and precision* – in respect to LASMA's implementation in environmental screening and monitoring of heavy metal contamination, the requirement for measurement precision is not decisive. The accuracy depends on the quality of the reference material. Two approaches are possible – to use commercial reference samples (Atomic standard solutions of metals and powder standards) or to prepare sets of reference samples with elemental compositions, not available on the market. Another possibility is to use the chemical matrix of clearly defined soil types [Zimmermann, 1989].

7. *Analytical control and justification of the measurement with parallel methods* – in literature, a great number of publications discuss the question of verification of analytical data. Many authors point to the reality to obtain false results even with highly sensitive devices and precision measurement procedures [Lieth and Markert, 1988]. The verification of the measurement accuracy of laser mass spectrometry (LASMA device in particular) could

be checked by independent analytical methods (for example by AAS – graphite furnace and flame), periodically or in cases of uncertainty.

8. *Reproducibility of measurement* – the best results for laser mass spectrometry, reported in literature, point standard deviations better than 8%. In respect to environmental analysis, the decisive factor is the homogeneity of the sample and the quality of the reference material.

9. *Relative quantification (determination of the elemental concentration) as a minimum* – since laser mass-spectrometry is not an absolute method, the quantitative analysis depends highly on the sample and reference homogeneity. A semi-quantitative analysis is possible only with water samples after an averaging of the spectral data for every single sample. The analysis of biological samples is practically qualitative, but with the possibility of a relative quantification, because the determination of the elemental concentration ratios is often enough for the task of investigation.

10. *Automatical analysis procedure and data analysis in near-to-real time* – the automatic data analysis is achieved with the application of computer programs, especially written for the particular instrument. The monitoring of the processes of extraction and accumulation of heavy metals in plants and a screening of a phytoremediation process in a contaminated media would be effective only, if the spectral data could be analyzed in near-to-real time, i.e. the elemental identification of the mass spectrum and the semi- or –relative quantification is performed in a short time after the data recording. The computer analysis time of LASDAT, a program for mass-spectra analysis of LASMA instrument, which includes a print-out of the integral elemental concentration, in percentage and absolute, is approximately 1-2 minutes, depending on the computer speed.

11. *Moderate to low investment and exploitation costs (energy, material, additional costs)* – the only consumable of LASMA laser mass-spectrometer, if we do not count the materials for sample preparation, is a moderate electric power, not higher than 1 kW for the whole system, including the mass spectrometer itself, the vacuum pump facility and the recording system with the transient recorder, the computer and the printer.

12. *Compactness, transportation, easy installation and operation* – the overall size and mass of the instrument with the vacuum and data recording systems allow an easy transportation, require no special room for its installation and permits mounting on a small laboratory desk. The operation of the instrument is easy and does not require any special qualifications, although knowledge of basic chemistry and understanding of the problems of contemporary environmental research would be appropriate.

5. Conclusion

The creation of the miniature laser mass spectrometer LASMA is a proof of successful technological transfer of instrumental designs and a measurement methodology initially developed for the purposes of space research and is an example of justification of the huge costs, which are spending in this area of human activity.

References

Simeonov L., C. Schmitt and K. Sheuermann, 1996 Scnelle semiquantitative Elementanalyse wässeriger Lösungen mit dem Laser-Massenanalysator LASMA, TerraTech, 6, pp. 29–31.

[A]Markert B., Instrumental analysis of plants, in Plants as biomonitors, B. Markert, Ed. , VCN: Weinheim, 1993, pp. 81–83.

[B]Markert B., Instrumetelle Multielementanalyse von Pflanzenproben, Weinheim, VCN: Weinheim, 1993, pp. 241–265.

Moenke-Blankenburg L., Laser Micro Analysis, John Wiley & Sons, Chemical Analysis, vol. 105, p. 227, 1989.

Van Lidth de Jeude J.W., Leidraad bodemsaniering. Staatuitgeverij. S-Gravenhage, 1983.

Klocke A., Richtwerte'80. Orientierungsdaten für tolerierbare Gesamtgehalte einiger Elemente in Kulturböden, Mitt. VDLUFA, H. 1–3, 1990.

Maximum permissible levels of toxic substances in soil. Ministry of environment. Regulation Nr. 3, Bulgarian State Journal, 39, 1979.

Zimmermann, R. 1989 Erste ergebnisse einer Ringanalyse zur Erstellung eines internen Buchenblatt Referenzmaterials für Ökosystemuntersuchungen", Fresius Z.Anal. Chem., 334, pp. 323–325.

Lieth H., and B. Markert, Aufstellung und Auswertung ökosystemarer Element-Konzentrations-Kataster, Sprigerwerlag. Berlin. Heidelberg, 1988.

INCIDENTAL AND ACCIDENTAL POLLUTION IN ITALY

MICHELE ARIENZO
Dipartimento di Scienze del Suolo, della Pianta e
dell'Ambiente
100 VIa Università, 80055 Portici,Italy
E-mail: michele.arienzo@unina.it

Abstract. Three different scenarios of pollution of the environment in the Campania Region, south Italy, were studied. The first study (a) deals with the spatial and temporal distribution of the radioactive levels of ^{134}Cs, ^{137}Cs, ^{103}Ru in the unpolluted area of the Vesuvius National Park after the nuclear accident of Chernobyl in 1986. The effectiveness of *S. vesuvianus* as biomonitor and the effective half time of Ru and Cs at each height and time were also determined. Results indicated that *Stereocaulon vesuvianum* is a valid biological indicator of environmental contamination by radionuclides, capable to detect the radioactive contamination even though there are no more effects on other vegetal community. The estimated removal half-life of Cs in lichen indicated the strong retention by lichen. The second study (b) assessed the total content of Cu, Cr, Pb and Zn in surface and sub-surface soils of the city of Naples and described metals spatial distribution. It also defined the chemical and mineralogical forms of metals in soil and compared current data with those of a 1974 sampling. Many surface soils from Naples urban area contain Cu, Pb and Zn levels largely above the limits set by the Italian Ministry of Environment. Cu apparently accumulates in soil contiguous to railway lines and tramways; Pb in soil on the border of motorway and high traffic flow streets; Zn in soil influenced by industrial activities. Cu and Cr exist in soil mainly in organic forms, Pb essentially as residual mineral phases, Zn is present in more readily available pools Cu, Pb and Zn levels have greatly increased since 1974, especially in soil from roadside fields.

Keywords: incidental and accidental pollution; radionuclides, heavy metals; biomonitor

1. Introduction (a)

The use of lichens as biomonitors for radio contamination assay rather than direct soil measurements is advantageous because they show high radionuclide concentrations. After Chernobyl, a great number of studies has been undertaken to monitor radioactivity levels in lichens from several European countries. Relatively few studies have been carried out in Italy (Triulzi et al., 1996). Residence time of radiocesium in lichens is very variable, depending on the species and on geographic situation and microclimatic variability (Nimis, 1996). The dependence of lichens radioactivity concentration on altitude has been the subject of much study in the past years with evidence often controversial. The main purpose of this work was to assess the variation of the activity concentrations of ^{134}Cs, ^{137}Cs, ^{103}Ru and ^{106}Ru in the thallus of *Stereocaulon vesuvianum Pers.* growing on the slopes of Mt. Vesuvius, Campania Region (south Italy) from October 1986 through May 1999 at different quotes.

2. Materials and Methods (a)

Sampling was done on different sectors and elevations of the slopes of Mt. Vesuvius: 370 m a.s.l. west; 490 m a.s.l. north-east; 580 m a.s.l. north-west; 780 m a.s.l. north-west; 960 m a.s.l. north. Lichens were collected on lava flows on October 28 1986, December 5 1986 and October 5 1987 and a further sampling was carried out on May 20 1999. lichens were separated from the substratum, dried at 30°C to constant weight. The activity of all radioisotopes was measured by gamma-ray spectroscopy.

3. Results and Discussion (a)

Overall, and within the first two sampling times, about 7 months from the Chernobyl accident, significant (P=0.05) highest average activities of radionuclides were observed at the quote of 960 m (Table 1). This attest for the transport to higher altitudes and hence deposition of radionuclide contaminants at higher quotes. A discontinuity in this trend was observed in the case of specimens collected at 780 m. this could be related to the placement in the studied area of the so-called 'boundary layer' between 700 and 800 m. local environmental factors such as orographic effect and wind canalization determined differences of the fallout pattern as well as of local growth conditions. The contents of radiocesium found in *S. vesuvianum* at all four sampling dates are generally higher or in a few cases similar to those measured by Roca et al. (1989) immediately after the Chernobyl accident in various crop from the Campania Region. This attested the high retention capacity of lichens and the long persistency of Cs isotopes in their thallus. Compared with higher plants, lichens act as a substantial reservoir of ^{137}Cs radionuclide for long period of time.

TABLE 1. Specific activity of radionuclides in *Stereocaulon vesuvianum* from slopes of Mt. Vesuvius (Bq kg^{-1} dry weight).

Nuclide	Elevation (m)				
	370	490	580	780	960
28 October 1986					
^{134}Cs	458	695	873.2	555	1021
^{137}Cs	1325	2035	2201	1465	2500
^{103}Ru	88.8	85.1	159.1	111	196
^{106}Ru	360	455.1	580.9	381	714
5 december 1986					
^{134}Cs	370	558	732	632	891
^{137}Cs	1200	1665	2146	1702	2480
^{103}Ru	40.7	44.4	70.3	48.1	74
^{106}Ru	322	418	680	518	736
5 October 1987					
^{134}Cs	322	455	629	359	584
^{137}Cs	1306	1683	2149	1343	1970
^{103}Ru	-	-	-	-	-
^{106}Ru	178	230	333	248	340
20 May 1999					
^{134}Cs	-	-	-	-	-
^{137}Cs	266	251	618	385	573
^{103}Ru	-	-	-	-	-
^{106}Ru	-	-	-	-	-

The slow growth rate and hydrolability of thallus along with longevity and relative absence of abscission processes are considered the main factors inducing the high radionuclide concentrations in lichens (Bargagli, 1998). Overall, the range of ^{137}Cs, 250-2500 Bq kg^{-1} was of lower order of magnitude than that determined by Triulzi et al. (1996) in Parma in the period 1990-1993 in various lichen species, 118-11084 Bq kg^{-1} dw. The estimated removal half-life of radiocesium in *S. vesuvianum* on Mt. Vesuvius (6.1 years) was comparable to the majority of the values reported in literature (Nimis, 1996) and indicative of the strong retention of the Cs element by lichens. The high radionuclide activities found after Chernobyl confirm in situ-lichens as a useful, sensitive and inexpensive monitoring method of radionuclides fallout patterns in full agreement with more traditional methods.

References (a)

Bargagli, R., 1998., Trace elements in terrestrial plants. An ecophysiological approach to biomomitoring and biorecovery. Springer, Berlin.

Nimis, P.L., 1996. Radiocesium in plants of forest ecosystems. Studia Geobotanica, 15, 3–49.

Roca, V., Napolitano, M., Speranza, P.R., Gialanella, G., 1989. Analysis of radioactivity levels in soils and crops from the Campania Region (South Italy) after the Chernobyl Accident. J. Environ. Radioact. 9, 117–129.

Triulzi, C., Nonnis Marzano, F., Vaghi, M., 1996. Important alpha, beta and gamma emitting radionuclides in lichens and mosses collected in different world areas. Annali di Chimica, 86, 699–705.

4. Introduction (b)

Heavy metals reaching the soil remain present in the pedosphere for many years even after removing of the pollution sources and increased amounts of heavy metals in soils of urban areas have been reported (Chen et al., 1997; Pichtel et al., 1997). However, it is known that the severity of pollution not only depends by heavy metal total content in soil, but more by the amount of their mobile and bioavailable forms, which are generally controlled by the texture as well physicochemical properties of soils. Therefore, to define hazards and to propose treatments and eventually new more appropriate utilisation of soils in urban areas the speciation, geochemistry and behaviour of heavy metals in soil have to be investigated.

In Italy research on urban areas has been low. Napoli is a very densely populated city, located near the sea and included in one of the largest urbanised and densely populated areas in the world. In particular, the presence and the activities in the eastern and western districts of the city, respectively, of various oil refineries (Q8, AGIP, ESSO) along with combustible deposits and of steelworks (ILVA, Cementir) and chemical industrial plants (Eternit, Montedison) have produced widespread trace metal contamination of the surroundings green areas.

Data from the Agencies for the control of air pollution in Campania indicate that the urban area of Napoli is seriously affected by nitrogen, sulphur and carbon oxides, ozone, dust, hydrocarbons and trace elements contamination from the atmosphere.

Early studies in Napoli urban area indicated increased concentrations of Cu, Pb and Zn in soils surrounding industrial plants and streets (Basile et al., 1974). Limited information exist concerning the spatial distribution and availability of heavy metals in soil. Furthermore, do not exist data regarding temporal changes of metal concentrations in contaminated soils.

The objective of this study was to assess spatial and temporal changes of surface soil metals in Napoli urban area. Specifically, (i) the total content of Cu, Cr, Pb and Zn in surface and sub-surface soils of the city of Napoli was measured, (ii) the chemical and mineralogical forms of metals in soil were determined using a sequential fractionation procedure in order to assess mobility and availability to plants and (iii) current data were compared with those of a 1974 sampling, to define the accumulation or removal of metals.

5. Materials and Methods (b)

5.1. THE STUDY AREA AND SOIL SAMPLING

The study concentrated on the urban core area including the northern and eastern districts of the city, where, respectively, the motorway and the old oil refineries along with combustible deposits are located. For the sampling the study area was divided in regular grids of 500x500 m. One sample of surface soil (0-2 cm depth) was collected in 1999 in each grid in gardens, parks, roadside fields and industrial sites. A total of 173 surface soil samples were collected. In 36 of the selected sites soil samples at 10, 20 and 30 cm of depth were also collected.

5.2. CHEMICAL ANALYSIS

After collection soil samples were air dried, 2 mm sieved and kept in PP bottles. Separation of the coarse sand (2-0.2 mm) and clay (< 2μm) fractions was achieved respectively, by wet sieving and centrifugation. The total heavy metal contents were determined treating 5 g of soil in HCl/HNO_3 (3:1 solution) at 100 °C for 1 h. The four-step chemical extraction procedure (0.11 M HOAc, 0.1 M NH_2OH HCl, H_2O_2/1 M NH_4OAc, HF/HNO_3), developed by the Measurement and Testing Programme of the European Commission, was used to fractionate heavy metals chemical forms in the fifteen most contaminated soil samples. In order to compare current data with those collected by Basile et al. (1974), extractions of Pb with Na_2EDTA (0.05 mol l^{-1}), at a ratio sample:extractant of 1:10 with shaking for 2 h, and of Cu and Zn with $(NH_4)_2EDTA$ (0.05 mol l^{-1} at pH 9), at a ratio sample:extractant of 1:10 with shaking for 24 h, were carried out on the soil samples from the same twelve site locations used in the 1974 assay. The concentration of Cu, Cr, Pb and Zn in the extracts was determined by atomic absorption spectrometer.

5.3. STATISTICAL ANALYSIS AND SPATIAL DISTRIBUTION MAPS

Data were submitted to descriptive statistical analysis to define their frequency distributions. Simple correlation analysis was used to examine the relationship between the different analysed heavy metals. The evaluation of the degree of association among variables, assumed as normally distributed, was based on calculating the value of the correlation coefficient (r) and testing it $[|\text{t-test}| = |r| \sqrt{(n-2)/(1-r^2)} > t_{(n-2;\,\alpha/2)}]$ for 5% significance level. Scatter diagrams of all associated variables were drawn to check the uniform dispersion of plots and to help in the interpretation of significant and non-significant correlation coefficients. The total concentrations of heavy metals in the surface soil samples were used to construct contour maps using the program SURFER (Golden Software Inc., Colorado).

6. Results and Discussion (b)

6.1. HEAVY METAL TOTAL CONTENT

Copper total content ranges from 6.2 to 286 mg kg^{-1}, with a median of 54 mg kg^{-1}. Approximately 85% of the analyzed soils contained from 6.2 to 120 mg kg^{-1} total Cu, and only 15% exceeded the maximum concentration (120 mg kg^{-1}) established for soils of public, residential and private areas by the Italian Ministry of Environment. The relative number of contaminated soils increases to 36% when only the 35 samples collected in the industrial eastern area of the city are considered. Chromium content ranges from 1.7 to 73 mg kg^{-1} with a median of 8.4 mg kg^{-1} never resulting above the regulatory limit (120 mg kg^{-1}). Soil Pb total content ranges between 4 and 3420 mg kg^{-1}, with a median of 184 mg kg^{-1}. A large number of soils (~76%) were polluted by Pb displaying concentrations in excess of 100 mg kg^{-1} set for public, residential and private areas. Zn total content ranges from 30 to 2550 mg kg^{-1} with a median of 180 mg kg^{-1}. The percentage of contaminated soil (Zn > 150 mg kg^{-1}) increases from 53 to 82% when all soils and the soils from the industrial eastern city area are considered, respectively. 14% of the overall analyzed soils show levels of Cu, Pb and Zn above the regulatory limits. Considering only the soils from the eastern area of the city this percentage arises to 34%. In conclusion, the surface soils of the Napoli urban area appear to be polluted in the order by Pb > Zn > Cu >> Cr. This sequence changes as Zn ≥ Pb > Cu >> Cr when only the soils from the eastern area are considered. In all the soil samples collected from 36 selected sites at 10, 20 and 30 cm of depth Cu, Cr, Pb and Zn total content decrease with depth, indicating extensive soil contamination by aerial heavy metal fallout. The trend is particularly steep for Zn (80-185 mg kg^{-1}) and Pb (170-278 mg kg^{-1}). Also Cr not enriched in the studied soils shows a gentle decline moving from surface to 30 cm depth. The depth of penetration may give indications about the mobility of anthropogenic Pb in soil. Taking into account that the majority of Pb deposition in Napoli urban area is from automobile exhausts and that the mobility of the element in the soils reaches on average 8-10 mm/year the levels measured at greater depth would reflect past inputs. However, the rate of Pb decrease with depth does not appear to be uniform. The spatial distributions of Cu, Cr, Pb and Zn are shown in the maps of Figures 1. For copper, chromium and zinc all the data are included; for lead, the two largest values measured in soil samples collected on the edges of the city motorway are excluded in order to improve the visualization of the geochemical anomalies. High levels of all metals are located in sites of the eastern part of the city, corresponding with areas of heavy industry and where various oil refineries (Q8, AGIP, ESSO) along with combustible deposits occur. Copper apparently accumulates in soils contiguous to railway lines and tramway mainly present along the south eastern coastal line. Lead concentrations fluctuate throughout the city. However, from the city centre and in industrial eastern and western areas, Pb contamination of soils is substantial. The most contaminated soils are located in proximity of the motorway and of streets with high traffic flows. Only soils on the northwest part of the city, which is characterised by greater elevation (~ 150 m a.s.l.), contain low Pb levels. Despite the sharp increase of unleaded fuel utilization, followed by a rapid decline of Pb levels in the atmosphere, the content of Pb in urban soil still remain high with consequent associated risk for children via the soil-hand-mouth pathway (Bargagli, 1998).

6.2. HEAVY METAL SPECIATION

Copper and Cr are prevalently held in the oxidizable (~68%) and residual (~24%) fractions, with other forms making up much less than 10%. Pb is primarily held in the residual mineral fraction (77%) with smaller amounts held in the oxidizable (18%) and reducible (3%) forms. Zn is uniformly distributed between easily extractable (23%), reducible (24%) and residual (49%) fractions, with very small percentage occurring in oxidizable forms (4%). Thus, only Zn appears to occur in the studied soils in consistent amounts as bioavailable and leachable forms. Copper and Cr, which are mainly associated with organic matter, for which both metals have high affinity are characterised by a lower mobility, unless the occurrence of oxidising conditions may induce their release in solution. Lead, and in lower proportion also Cu and Cr, extracted mainly by the fourth step, result relatively immobile.

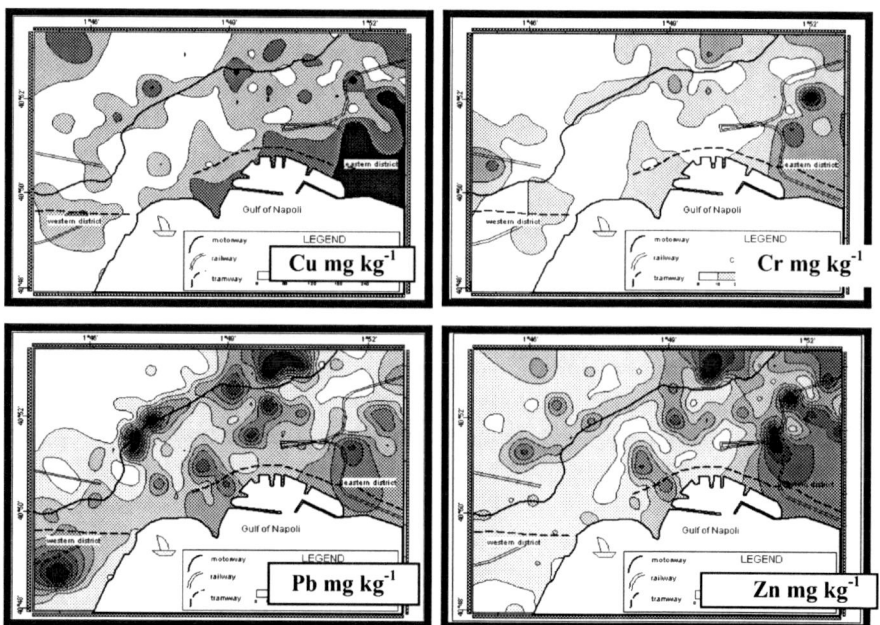

FIGURE 1. Kriged maps of Cu, Cr, Pb and Zn content in surface soils of Napoli urban area.

6.3. COMPARISON BETWEEN PRESENT AND HISTORICAL DATA

For comparison between present and historical data emphasis has been placed on selecting soil sampling locations and methodologies of metal extraction that had been previously used. Twelve sites were selected for comparison: three urban parks (P), five square gardens (G) and four roadside fields (R). Generally and according with 1974 results higher content of heavy metals characterize soils from roadside fields and square gardens compared with soils from parks, which, as expected, result more protected from pollution. Comparison between the 1999 data and those of 1974 reveals a significant increase of Cu, Pb and Zn content in the surface layer of all Napoli urban soils. In comparison to 1974, Cu content in soil increases from 41% (site G5) to 878 % (site R3), with mean variation of 134%. Lead content increases from 17% (site G1) to 876% (site R1), with mean variation of 182%. This trend is observed also for Zn, whose levels in soil at sites G1 and R1 are characterised, respectively, by the lowest (22) and highest (765) percentages of increase. Among the analysed metals Zn levels in urban soils show higher percentages of increase compared with 1974 data. Presumably, the higher increases of Zn in urban soil can be related to a continuous input of the metal in the urban environment principally as a consequence of vehicle emissions, tire and brake abrasion. This would confirm the urban traffic as the main anthropogenic source of airborne trace elements in Napoli urban area. By contrary, the utilization of unleaded fuel could explain the lower increases of Pb element content in urban soils.

References (b)

Bargagli R., 1998. Trace Elements in Terrestrial Plants: an Ecophysiological Approach to Biomonitoring and Biorecovery. Springer-Verlag, Berlin, Germany.

Basile, G., Palmieri, F., Violante, P., 1974. Inquinamento da zinco, rame e piombo nel suolo dell'area urbana ed industriale di Napoli. Annali della Facoltà di Scienze Agrarie dell'Università di Napoli, 8, 3–12. A.G. Della Torre, Portici.

Chen, T.B., Wong, J.W.C., Zhou, H.Y., Wong, M.H., 1997. Assessment of trace metal distribution and contamination in surface soils of Hong Kong. Environmental Pollution 96, 61–68.

Pichtel, J., Sawyerr, H.T., Czarnowska, K., 1997. Spatial and temporal distribution of metals in soils in Warsaw, Poland. Environ. Poll. 98, 169–174.

PESTICIDES AS GLOBAL ENVIRONMENTAL POLLUTANTS

KOSTA VASSILEV* AND VESKA KAMBOUROVA
National Center of Public Health Protection
15 Ivan Ev. Geshov Blvd, 1341 Sofia, Bulgaria

*To whom correspondence should be addressed: k.vasilev@ncphp.government.bg

Abstract. The use of the chemicals for vector control in various regions of the globe is commented in the paper. The current trends in the World Health Organization policy for the application of diverse groups of pesticides against the vectors are discussed. The information on the residues of persistent organochlorine pesticides in the environmental media - air, freshwater, seawater, sediments and soil as well as the tendencies for their elimination in regional and global aspect is analyzed. The adverse effects for the living organisms as a result of bioaccumulation and biomagnification of those substances in the tissues are outlined. The conditions and the principles of the long range transport of the persistent pollutants in the environmental compartments are also addressed. Special attention is paid to the international agreements. The related conventions are discussed including the Stockholm Convention on Persistent Organic Pollutants. Retrospective data on residuals of persistent pesticides in water and soil in Bulgaria are presented. It is concluded that there is a clear tendency for decline of the residual levels of ogranochlorine pesticides in the natural waters. Currently the relative rate of the positive samples for organochlorine pesticides is about 10% in comparison to the 70-s when almost 100% of the analyzed water samples in Bulgaria were positive. The decline rate of soil residuals is much slower as the expected one.

Keywords: organochlorine pesticides; DDT, lindane; atrazine; POPs; global pollutants

1. Introduction

The industrial production and usage of pesticides has its relatively short but controversial history. The modern development of agriculture would be impossible and present status of public health would be on far lower level without pesticides application. In the two main areas of pesticide current utilization - agriculture and public health - there are two international organizations of the United Nations which work independently and in cooperation on different aspects of regulations of usage and minimizing of pesticide effects on environment and human health – these are WHO and FAO. WHO makes big efforts to reduce the mass infectious diseases transmitted by carriers or vectors – mosquitoes and other insects. The main instrument for control of these diseases in the basic centers of infections is still pesticide application. WHO is the only international organization with its own current programme that promotes and coordinates the testing and evaluation of pesticides intended for public health use. The experts of WHO on the base of regularly gathered information from member states make periodical review of the status and derive the trends on pesticide use in vector control in the so called WHO regions.

2. Pesticide Use for Vector Control

Table 1 shows the main infectious diseases transmitted by insects and their regional distribution according the report of WHO (2001).

The African region has the highest morbidity of tropical diseases - between 80 to 90 % of the total number of clinical malaria cases (300-500 mln. people), with 1.5–2.7 mln deaths, and 55 mln people at risk of trypanosomiasis as well.

The use of chemicals against vectors remains the most important method for vector and pest control in this region. Here all classes of pesticides: organochlorine, organophosphorus, carbamates, pyrethroids and bacterial larvicides are applied. The most widely used ones are organophosphorus followed by pyrethroid pesticides. The pesticides against malaria have the biggest relative rate – 80 to 90 % of the total amount of used pesticides against all diseases. Generally DDT remains a basic chemical used in programmes for vector control of malaria, although some countries replace it by pyrethroids for indoor treatment and mosquito nests treatment. The lack of regulations for pesticide use, lack of effective system for registration and monitoring, as well as quality control of used pesticides is a big problem for African countries with several exemptions.

Malaria and dengue are the most important infectious diseases for Latin America. Integrated methods are used here more widely against pests – improvement of sanitation and house construction are applied parallel with use of insecticides to exclude the contact with vectors. The use of DDT during the last decade has been generally reduced. For example during the 70s, DDT was used for malaria control in 21 countries, but in 1997 only 5 countries (Argentina, Ecuador, Guyana, Mexico and Venezuela) reported use of DDT. DDT was replaced by organophosporus pesticides (malathion, fenitrothion, pirimiphos-methyl) still in 70s. Carbamates are widely used in some countries. During the last years the use of pyrethroids (deltamethrin, gamma cyhalothrin, cypermethrin) was increased, but the relative part of organophosphorus pesticides and carbamates was decreased.

TABLE 1. The most important vector born diseases in WHO regions.

Disease	Vector	WHO region				
		Africa	Latin America	Eastern Mediterranean region (Egypt, Iran, Iraq, Somalia, Saudi Arabia, Afghanistan)	South-East Asia region (India, Bangladesh, Thailand, Sri Lanka, Indonesia)	Western Pacific region (China, Malaysia Vietnam)
Malaria (parasite)	Mosquitoes	+	+	+	+	+
Trypanosomiasis (parasite)	Tsetse fly	+				
Onchocerciasis (helminth)	Fly	+		+		
Leishmaniasis (parasite) Kala - Azar	Insects (Phlebotomus)	+	+	+	+	
Shistosomiasis (helminth)	Molluscs	+		+		+
Lymphatic filariasis (helminth)		+	+	+	+	+
Dengue / Dengue hemorrhagic fever (virus)	Mosquitoes (Aedes aegypti, Aedes albopictus, Aedes hebridens)		+	+	+	+

The most important vector borne diseases for Eastern Mediterranean Region are malaria, onchocerciasis, leishmaniasis, schistosomiasis and dengue. About 60% of the population are at risk of malaria. Most countries have developed a strategy for integrated vector control, but actually most of them continue to rely on chemicals. Dramatic decrease in the use of organochlorine and increased usage of pyrethroids was reported, personal protective means were introduced – insect repellents, mosquito coils, mats and aerosols, as well as elevated use of insect growth regulators and bacterial larvicides. For example the total amount of DDT used for indoors residual spray decreased from 168 tones for 1989 to 68 tones for 1997. Other reason for the decreased usage of chemicals is the development of resistance – 14 malaria vector species have been reported as resistant to organochlorine pesticides, 8 to organophosphorus and 3 to carbamates and 3 to pyrethroid permethrin.

Vector borne diseases are the major problem in South East Asia region too. About 1 billion people are living in malaria endemic areas with about 3 mln cases of malaria annually. Epidemics are common. All basic chemicals are used for vector control. In 1990, DDT was used in 7 countries, but after 1997 only 3 continue to use it against malaria, while Bangladesh reserved DDT only against kala-azar.

Malaria is the most important vector borne disease for the region of Western Pacific, followed by dengue. It is a great concern that 107 mln. people are living at risk of malaria in this region. Basic method for control is the insecticide treatment of mosquito nests, mainly with pyrethroids. The use of household insecticides is the major component of public health use of pesticides in the developed countries from the region. The biggest market of these products is China.

Generally the modern strategy for vector control includes limited use of organochlorine, organophosphorus and carbamate pesticides on the account of increased relative rate of synthetic pyrethroids and bacterial larvicides. This action is in the right direction, but as the experts conclude, the work in searching for new effective chemicals for vector control should be intensified in order to avoid the problem with resistance to pyrethroids which tend to replace the persistent chemicals. Serious attention should be paid to the information for increased diseases prevalence in some regions where the quantity of persistent pesticides for vector control had been reduced. WHO commission on vector biology and control pointed the following issues of importance related to public health use of pesticides:

- Vector born diseases continue to be of global public health importance and there is continued reliance of chemical control of vectors.
- Regulations for the control of pesticides exist in most countries but enforcement of the regulations is frequently ineffective. Monitoring of the quality of pesticides used in public health programs is limited.

- The use of public health and agricultural pesticides have to be monitored at national level. In the same time countries should involve adequate regulation for registration, purchising, quality control of pesticides, etc and to ensure these regulations are enforced.

3. Residues of Persistent Pesticides in Environmental Media and Biota

Except WHO and FAO during the last years UNEP developed a wide range of activities in the frame of its programme for reducing and elimination of emissions and discharges of persistent toxic substances. In the frame of this programme an assessment has been made by the experts of the global dispersion of the toxic organic substances among the regions. The results of this first by his nature and range analysis are presented in UNEP Global report 2003 (2003). Some basic data and conclusions concerning the global dispersion of persistent organic pesticides – subject of the present report will be presented below. Where in the present report analytical data not included in the Global report 2003 are cited, the particular sources are especially mentioned. According to the experts in all regions it is likely to detect persistent toxic chemicals including organochlorine pesticides in all environmental compartments – air, water, soil, living organisms and humans as well. Certain evidence for the global dispersion of pesticides is their detection in the air and depositions from all parts of the globe from the equator to the poles which shows that the atmosphere is one of the most important means for pesticide transport in the nature. The biggest amount of analytical data exist for DDT and HCHs. The most elevated air concentrations are found in tropical countries, particularly in India, countries from Asia and Africa, but the lowest concentrations are detected in polar regions. The levels vary up to 992 ng/cub.m for DDT and for HCHs up to 780 ng/cub.m. The concentrations of DDT and lindane, the most reported pesticides in depositions vary from parts of nanogram up to hundreds ng/sq.m/d. During the last years there is a general trend of decrease in the amount detected of all persistent pesticides especially for these chemicals that had been still banned.

Pollution of rivers and fresh water lakes with pesticides is strongly influenced by the hydrogeological regime of water courses and the presence of suspended solids. It is reasonable to analyze water samples paralelly for dissolved content and content on suspended solids. Very often after stormy rains and floods high levels are registered because of elevated amount of suspended soil particles. The highest mean levels of organochlorine pesticide in rivers and lakes are detected in developing countries. DDT and its metabolites, as well as HCHs in concentrations up to and about 0.1

mkg/l are the most frequently found substances. In the developing countries of Africa, Sought America, South Asia detectable amount of wide spectrum of pesticides like endosulfan, dieldrin, heptachlor, chlordan, HCB, atrazine has been found. It is not surprising having in mind their popular use in the past, but may be still in present days in some regions. The presence of DDT and lindane are most often reported for European countries. Thirty five percent positive samples for HCHs and levels above 0.1 mkg/l are registered in the waters of low Danube. The data reported are usually from single samples or scientific reports and only single countries have their own monitoring systems – for example France during 80s and 90s. In conclusion, except in cases of accidents, the highest levels in fresh waters are reported from the regions of Africa, Sought America, India and South-East Asia where some persistent chemicals are still in use for vector control and probably in agriculture. The global trend is declining of registered values in natural waters compared with 70s and 80s.

Seawaters are the final receivers of land based pollution sources and through the sea currents is realised the global transport of some pollutants in the world ocean. Generally the content of pollutants in coastal waters is much greater than this in open sea waters. The coastal waters are polluted by discharges of sewage and industrial effluents and rivers, whereas the major sources of pollution in open sea are atmosheric depositions. The presence of persistant and hydrophobic chemicals in sea waters is important because of the possibilities for bioacummulatiom and biomagnification through food web. The highest levels of DDTs in costal waters were reported near Sought America. Detected levels for total DDTs and HCHs in costal waters vary between tenths to hundreds of mkg/l. In open sea the amount of organochlorine compounds is very low down to several pg/l and lower. According to some reports HCHs and lindane which are more water soluble than most of the other chlorinated pesticides show a trend for higher concentrations in open sea of Arctic regions compared with some south regions.

This effect is not observed for the rest of organochlorine chemicals that are lower soluble. The explanation of this fact is given by the so called "cold condensation effect" that is typical for more volatile organic chemicals.

Soils are natural sinks for lipophilic organic compounds which adsorb to soil organic carbon and remain relatively immobile. DDT and the rest of persistent organochlorine pesticides are found in soil samples from all regions of the globe. The registered values for these compounds vary from not detectable in some remote locations up to several mg/kg dw soil. Only in some heavily polluted areas levels up to 100 mg/kg dw soil are reported. HCHs also vary in wide ranges depending on the distance from pollution sources. The main trend is decreasing of the soil concentrations but not so quickly as in waters and air.

Sediments concentrate organochlorine pesticides released in water bodies because of their low water solubility. The highest concentrations of DDT – thousands mkg/kg/dw were found in river sediments in India, Africa and Sought America. Much lower levels – several mkg/kg/dw were determined in river sediments for other representatives of this group: cyclodien pesticides, HCB, heptachlor, chlordan. Common trend is decreasing of registered amounts compared with the 80s. The predominant part of pesticides entering into sea waters through rivers sedimentate in coastal areas, and only very small part deposit in sediments in open sea. In costal sediments from different regions content of DDT of tenths of mkg/kg dw were found, while in deep basins the amount of DDT and its derivates is about 1 ng/kg/dw.

The most strong evidence for the global despersion of persistent organic pesticides in the environment is their accumulation in living organisms - plants, water organisms, terrestrial animals. Very high concentrations of DDT were found in some terrestrial and water plants like pine needles (Eriksson, G. et al., 1989) and water hyacinth – tenths and hundreds mkg/kg dw in African countries during 80s, which is a consequence of application of DDT in the agriculture and public health sector. Accumulation of these chemicals in tissues through the food chain has much more negative effects on birds and animals. Among most investigated terrestrial animals are rein deer, polar fox, and among sea mammalians – mediterranean dolphin, whales, arctic seals (Reijnders, P.J.H.,1986) etc. The most popular examples are birds of pray, which accumulate great amount of persistent compounds through the food chain, or so called biomagnification, as well as fish eating birds, which are the top predators on the aquatic food chain.

The most typical event illustrating the negative effects of these processes is the strong reduction of the population of birds of pray and fish eating birds in some regions. Accumulation of DDT and other persistent organochlorine chemicals such as aldrin, heptachlor, HCB, PCBs and Dioxines (Mocarelli, P., et al., 1996) causes adverse effect on endocrine system leading to imbalance in calcium metabolism resulting in shell thinning and behaviour abnormalities as well. These effects were observed among different species – Australian falcon, Norwegian sea eagle, cormorans in Belgium and Netherlands from the estuaries of Rhine and Meuse, sea gulls in the Great lakes in USA and Mediterranean region. Shell thinning occurs, increasing loss of eggs and modified behaviour of birds. For example male sea gulls ignore nesting, and females may pair and nest together. It was established that these substances can also block androgen receptor mediated processes acting as androgen receptor antagonists. These events lead to reduction of reproduction capacity resulting in total decrease of birds population. These events started during the 50s, reached maximum in the 60s and 70s, which coincides with the maximum intensity of aplication of organochlorine pesticides.

Humans have highest position in terrestrial and aquatic food chains and as a result consume food with relatively high concentration of persistent organic compounds. They accumulate in the adipose tissue of the organism and are found in almost equal quantities if calculated on a lipid basis in adipose tissue, serum and breast milk. Practically the monitoring is easily made in maternal milk. Most investigations conducted during the last years have proven mainly the presence of pp DDE that indicates limited entering of these compounds from new sources. The studies have shown relatively higher concentrations of persistent organic compounds in breast milk among the population in Sub Saharan Africa, South America and India, especially in regions with intensive agriculture and availability of vector control programs. Most frequently the content of total DDT, HCHs and HCB is examined. The values found vary largely, and for total DDT in breast milk they are between hundred mkg/kg and several mg/kg lipid basis. Similar and lower levels are registered for HCHs, HCB and other organochlorine compounds. Levels in breast milk increased significantly with maternal age. Higher values of DDT and metabolites were detected among the population of North Canadian regions compared to the Southern parts, that is associated with the higher fish availability in the diet. In general, during the last decades there is a decreasing tendency in the amounts of organochlorine compounds detected in human tissues.

The main part (about 90%) of organochlorine compounds in the human intake is due to the diet, including commercial milk (Kara, H., et al.,1999). Therefore the risk assessment is carried out by comparing dietary intake with established tolerable daily intakes (TDI) for separate substances. The adverse effects of the different pesticides most frequently registered in animal studies and more rarely in humans are presented in Table 2.

It is generally accepted that in comparison to the 70s and 80s the content of organochlorine pesticides in the environment and in the living organisms tends to decline. It could be explained by the reduction of the inputs into environment. Nevertheless these compounds may still be a problem in regions, like Sub Saharan Africa, Asia region, and South America, where agriculture and vector control requires application of these substances. One exception from this tendency are some Arctic parts where elevated contents of these substances were found in adipose tissue of humans living in remote regions. It is well known that the environmental conditions in these regions favor a higher persistence.

3.1. LONG RANGE TRANSPORT

The global dissemination of the persistent organochlorine pesticides is determined by their global use and the big variety of long range transport possibilities in the different environmental media as well. The investigation of the transport mechanisms of these substances on long distances is necessary in order to select the strategy of their elimination from the environment. For the understanding of this process knowledge of substance properties and some physical, geographical and climate characteristics of the regions where these are applied or transported are necessary.

TABLE 2. Potential effects of individual PTSs (selected, Global Report 2003).

Type of effects	Aldrin Dieldrin	Chlordane	DDT	Toxaphene	Mirex	HCB	HCH
Reproduction and/or development	+	+	+	+	+	+	+
Cytochrome P450 system	+	+		+	+	+	+
Porphyria						+	
Immune system	+	+	+	+	+	+	+
Adrenal effects			+	+			+
Thyroid and retinol effects			+	+		+	+
Carcinogenic effects	+	+		+	+	+	+
Skeletal effects			+				

Common property of all persistent organic substances is their persistence in the environment. This is a necessary but not sufficient condition for realization of long range transport in the atmosphere or hydrosphere.

The substances have to have such characteristics that allow them to move relatively easily among different environmental compartments - air, water and soil. These compounds are not polar i.e. they are sparingly soluble in water and have low to intermediate volatility. Substances with such characteristics cannot be present exclusively neither in water nor in air, i.e. they are not exclusively water or air pollutants. Because of their properties to be presented simultaneously in different environmental media they are often referred to as intermedia pollutants. The Long Range Transport (LRT) of persistent substances could be realized in different ways:

- As a vapor dissolved in cloud water or adsorbed to suspended particles in the atmosphere;
- Dissolved in water or adsorbed to suspended particles in rivers, lakes, oceans;
- Concentrated in the tissues of migratory birds and animals.

The relative part of each of these ways of transport depends on the specific properties of the compounds. According to their volatility and water solubility the substances are divided in four main groups (Table 3). As a measure of the affinity of the substances to certain phase (air, water) the partition coefficients octanol/air (log Koa) and air/water (log Kaw) are used. The octanol is used as an equivalent of the organic matter in the environmental media. The substances with low Koa values are volatile, i.e. they have affinity to the air phase, and in the opposite - substances with high coefficient levels have low volatility and do not tend to the air phase. Substances with low Kaw coefficient have higher water solubility and affinity to the water phase, whereas those with higher values of the coefficient have lower solubility in water.

Chemical substances with low Koa and high Kaw coefficients belong to category A ("no hop"). Because of their high volatility and affinity to the air phase these substances do not tend to bond on the earth surface, and therefore till the moment of their degradation they are presented mainly in the atmosphere.

Substances with low Kaw value (relatively higher water solubility) and low Koa belong to category D ("no hop required"). The substances of this group are water soluble enough in order to ensure long range transport on considerable distances with the surface and ground waters.

Substances with very low volatility (high values of the Koa coefficient) and very low water solubility (high Kaw coefficient) are added on group C ("single hop"). These substances are so non-volatile and not water soluble that they can be transported only bond to the suspended particles in the atmosphere or in the water. Because of this they contaminate relatively small areas around the pollution sources.

Substances with moderate expressed volatility (Koa) and water solubility (Kaw) belong to category B ("multi-hop"). Depending on the environmental temperature the chemical substances of category B move more often from gaseous state to condense state (i.e. hop) and so carried by the drifts in the atmosphere they are transported on longer distances.

No substance from the POPs in the Stockholm Convention is sufficiently volatile in order to be attached to group A. It has to be taken into account that the boundaries between the different categories are not sharp, and depending on the external conditions one substance may belong to more than one group.

TABLE 3. Categorization of organic chemicals in term of their LRT behaviour. (Global Report 2003)

Category		Characterization	Examples
A	no hop	Chemicals that are so volatile that they do not deposit substantially to the Earth's surface and therefore remain in the atmosphere	Chlorofluorocarbons
B	multi-hop	Chemicals that readily shift their distribution between gas phase and condensed phase (soil, vegetation, water) in response to changes in environmental temperature and phase composition, and therefore can travel long distances in repeated cycles of evaporation and deposition	HCB Toxaphene Dieldrin Chlordane Endosulphane Lighter PSBs Lighter PCDD/PCDF
C	single-hop	Chemicals that are so non-volatile and so water insoluble that they can undergo LRT only by piggy-backing on suspended solids in air and water	Mirex Benzo-a-pyrene Heavier PCDD/PCDF
D	no hop required	Chemicals that are sufficiently water soluble to undergo LRT by being dissolved in the water phase	HCHs PCP Atrazine

Beside this, the products used in the practice are often mixtures of several substances with different partition coefficients, that is why single representatives fall into different categories. For instance the lighter Dioxins and Furans are in category B (multi hop), but the heavy Dioxins and Furans get into category C ("single hop").

Because the partition coefficients are temperature depending values the volatility and the water solubility of the substances changes with the change of the environmental temperature. At lower temperatures the deposition on the earth surface, but at higher the process of volatilization is prevailing. The behavior changes of the substances in the environment follow the twenty-four-hour and the seasonal variations of the temperature. Due to the differences in the temperature and the continuous movement of the air masses in the atmosphere the transport of the pollutants from warmer to colder regions of the globe is facilitated, i.e. from lower to higher latitudes and respectively altitudes.

4. Transport of Persistent Pesticides in Tissues of Migratory Birds and Animals

The tissues of the migratory birds and animals often contain high concentrations of pollutants because of the high bioaccumaulation potential of the organochlorine pesticides. Usually, in order to avoid the worsening of the weather conditions, the birds and the animals migrate from north to south and vice versa. In spite that the amounts of the pollutants in the tissues are not very high the consequences could be considerable, especially for other classes of living organisms or animals who consume migratory birds and animals. This concerns also people who hunt and consume those birds. Often the birds of passage, from Arctic for example acquire the bulk of their pollutants load in their wintering locations.

5. Residues in Bulgarian Environment

The development of programs for pesticide monitoring in environmental compartments started in Bulgaria only in the last decade. As in many other countries the data available are collected mainly by scientific investigations or at necessity to resolve certain local problems.

There is a greater data basis from the 70s in Bulgaria when the concentration of lindane, DDT and diene pesticides in natural waters - surface, ground and sea water were studied (Gitsova, S., 1975). In that period almost 100% of the analysed samples from all kind of waters - river, ground, sea and rain water showed presence of lindane and DDT (with a maximum value of pp DDT). As a rule diene pesticides (aldrin, dieldrin, endrin) were not found in natural waters. Alfa and gamma HCH were detected in all kind of natural waters in concentrations of 0.002 – 0.063 mkg/l, while DDT was in the range of 0.023 - 0.408 mkg/l. The DDE/DDT ratio determined in the same period was 13% indicating the prevailing introducing of DDT in the water compartment.

The lowest concentrations of lindane were found in sea water samples - 0.003 mkg/l and the highest in rain water (mean value of 0.039 mkg/l). The average content of DDT was the lowest one in ground waters – 0.046 mkg/l, but higher in rain and sea waters (0.101 and 0.128 mkg/l respectively). DDT concentrations up to several mkg/l were detected only in single samples. A considerable part of the organochlorine pesticides adsorbs on the solids suspended in the water. The adsorption capacity of DDT is relatively greater, and that of the lindane - lower.

Though non-systematic, the investigations carried out by the end of the 90s demonstrated considerable decline of the number of positive samples for organochlorine pesticides. Relatively low concentrations (0.01 - 0.06

mkg/l) of lindane were found only in 10% of the analysed surface and ground waters (Bratanova, Z. and K.Vassilev, 2001). DDT was not detected, and its derivative DDE was determined only in single water samples. Other representatives of persistent pesticides – heptachlor and HCB were detected also in single samples in low concentrations (0.004 and 0.02 mkg/l respectively).

Higher concentrations - 0.14 mkg/l of lindane, 1.63 mkg/l of DDT, and 0.57 mkg/l of DDE were found only in cases of accidental pollution associated with illegal introduction of pesticides in water bodies. There is a clear tendency of decreasing the content of the classic organochlorine pesticides in the Bulgarian hydrosphere. Whereas before three decades DDT and lindane were found in near 100% of the samples, at present this percentage is about 10. This trend is an expected consequence because the import and the usage of DDT and other organochlorine pesticides were banned decades ago in Bulgaria. Like other European countries Bulgaria introduces step by step import and usage restrictions for the most toxic and environmentally persistent pesticides. The pesticides banned for import and use in Bulgaria are listed in Table 4.

Simtriazine pesticides (atrazin, simazin) are not part of the classical POPs according to the Stockholm convention, but they are a concern in many countries, especially because of their presence in groundwaters. Ripparbelli et al. (1996) records for 11500 positive samples from a total of 113000 samples for atrazin in Italy (above 0.1 mkg/l). Authors from different countries (USA, Italy, Germany, UK) report high frequency of triazine contamination of groundwater as well as drinking water, in regions with intensive agriculture activities (Pionke, H.B. et al., 1988; Deladedova, P., 1996; Grohman, A., 1994; Carter, A.D., 1993). The presence of triazine pesticides was showed very often in the tested natural waters by the end of the 90-s. The main representative of the group – atrazin demonstrates highest frequency of the positive samples (about 13% according Bratanova, Z. and K.Vassilev, 2001) in ground waters, including drinking waters. The concentration detected is usually below 0.1 mkg/l, that is the maximum value stated in the EU Directive 98/83/EC (1998) on the quality of water intended for human consumption. The usual degradation time (DT_{50}) for atrazin in soil is 1-2 months, but in ground waters it takes more than 200 days (The Pesticide Manual, 1994), that practically covers the persistence criteria of 6 months, set by the Stockholm Convention.

Nowadays the soil pollution with persistent organochlorine compounds in Bulgaria is mostly because of their use in the past, result of local environmental accidents, associated with improper storage or illegal use of obsolete pesticide stocks.

At present, the pesticide soil monitoring is carried out by the Ministry of Environment and Waters (MoEW), has started in 1997 and covers the whole country. In the scope of the monitoring programme are the following persistent organochlorine pesticides: aldrin, DDT, dieldrin, endrin, heptachlor, hexachlorbenzene (HCB), methoxychlor, cis-heptachlorepoxide, alfa-HCH and lindane.

TABLE 4. List of pesticides banned for import and use in Bulgaria.

No	Persistent organochlorine pesticides under the Stockholm Convention *	No	Other pesticides
1	Aldrin	1	Azinfos-methyl
2	Dieldrin	2	Aldicarb
3	Endrine	3	Amitrol
4	DDT	4	Endosulfan
5	Chlordane	5	Mercury containing fungicides
6	Heptachlor	6	Captafol
7	Toxaphene	7	Crimidin
		8	Lindane
		9	Metamidofos
		10	Methidathion
		11	Methyl-parathion
		12	Monocrotofos
		13	Nitrofen
		14	Organotin fungicides
		15	Paraquat
		16	Parathion
		17	Pyrasophos
		18	Fenvalerate
		19	HCH (alfa and beta)
		20	Cyhexatin

* Mirex has never been registered in Bulgaria

Of all organochlorine pesticides analyzed during the period of 1997-1999 at 313 sites, only DDT and its metabolites were detected in almost all sites under study, with geometric mean values varying from 0.005 to 0.119 mg/kg. Regions having no DDT residues have not been observed (Kaloyanova-Simeonova, F., et al., 2001).

The first large investigation on soil contamination with organochlorine pesticides was performed by the end of 70s by Kujumdzieva (1976) and Kujumdzieva et al. (1979) who analyzed 650 soil samples for pesticide residues in 14 regions of the country. At that time, the proportion of positive findings for DDT total was almost the same in all sites, with a geometric mean of the concentration of 0.54 mg/kg.

Compared to the levels observed in the seventies, the soil contamination with DDT has dropped, but less than expected. It is very likely that the actual values at that time were higher, but also the recent levels need more detailed confirmation, nevertheless the monitoring is supported by QA/QC program.

In general, the percentage of (p'p-DDE) from the presented DDT-total, varies from 31 to 66 % for the different districts, with an average of 44%. According to Kujumdzieva (1976) in the seventies this percentage was lower - 33% in an average (variation from 17 to 44%). This finding as well as the increased value of the main metabolite ppDDE indicates a certain trend of breakdown. The maximum single-site values registered in a number of regions (Sofia, Smolian, Montana, Plovdiv, and Pazardzik) exceeded the MAC value of 1.5 mg/kg, pointing to possible unauthorized use of DDT.

Among the other pesticides studied the most frequently detected is HCB (68%) in relatively low concentrations of 0.004 mg/kg.

Potential source for environmental pollution for Bulgaria is also the relatively high number of existing stockpiles of obsolete pesticides. According to Kambourova (2004) the total amount of identified organochlorine pesticides by 2000 was 58 tonns, stored in 100 storage places on the territory of 22 districts.

6. Conventions

Because of the global distribution of the persistent organic compounds in the environment and their negative impact on the living organisms the international community has developed several agreements aimed at strengthening the control on their distribution and minimizing the negative consequences for the people and the environment.

Four international conventions are developed at present by means of which the governments execute actions for controlling the dissemination and usage of the environmentally persistent pesticides on international scale.

Those are as follows:

1. Basel Convention on the Control of Transboundary Movement of Hazard Wastes and Their Disposal – adopted in 1989 and entered into force in 1992. This convention regulates transboundary movements of hazardous waste across national boundaries in such way that these wastes are not a threat for the environment.

2. The Rotterdam Convention on the Prior Informed Consent Procedure for Certain Hazardous Chemicals and Pesticides in International Trade. The convention was adopted in 1998. The convention enables for monitoring and control the international trade in very dangerous substances. According to the convention export of a chemical can be realized only with the prior informed consent of the importing country. The convention includes a list of 22 pesticides including aldrin, dieldrin, DDT, chlordane, heptachlore, HCB, and 5 industrial chemicals.

3. International Convention for the prevention of pollution from ships (Marpol 73/78). The convention combines 2 treaties adopted in 1973 and 1978 respectively. The agreements include all aspects of pollution from ships – harmful substances, chemicals, garbage, sewage, oil, with the exception of the disposal of waste into sea by dumping.

4. Stockholm Convention on Persistent Organic Pollutants (POPs) – adopted 2001. The Stockholm Convention is a global treaty to protect human health and Environment from persistent organic pollutants. POPs are defined as chemicals that remain intact in the environment for long periods, become widely distributed geographically, accumulate in the fatty tissues of living organisms and are toxic to humans and wildlife. With the convention governments will take measures to eliminate or reduce the release of POPs into the environment (Table 5).

The most broad is the scope of the Stockholm Convention. Its objective is to protect human health and the environment from persistent organic pollutants. In order to meet this goal the parties agree by means of prohibitions and restrictions in the production, use and control of the import to achieve elimination or maximum restriction of 12 persistent organic compounds in the environment. Among them are 9 organochlorine pesticides, Dioxins, Furans and PCBs. The 12 POPs are divided in 3 groups. In Annex A (elimination) are listed the compounds which production is prohibited (Aldrin, Dieldrin, Endrin, Chlordane, Heptachlor, Hexachlorobenzene, Mirex, Toxaphene and PCBs). The production and use of DDT is especially regulated in Annex B, and in Annex C are included Dioxins, Furans and PCBs and the antropogenic sources that lead to their formation like incinerators, thermal processes in the metallurgical industry, open burning etc. The Convention foresees that every country could produce or use some of these substances if necessary, but only after registration in the special Register of the Convention concerning the specific exemptions that are shown in Annexes A and B of the Convention. The parties are obliged to undertake all necessary measures to reduce or eliminate releases of POPs from stockpiles and wastes by means of a proper and environmentally sound manner of management.

TABLE 5. POPs according Stockholm Convention.

No	chemical	CAS No	production	use	sources of environmental pollution	
					Stockpiles	Unintentional production
1	Aldrin	309-00-2	none	local ectoparasiticide insecticide	+	
2	Dieldrin	60-57-1	none	in agricultural operations	+	
3	Endrin	72-20-8	none	none	+	
4	Toxaphene	8001-35-2	none	none	+	
5	Heptachlor	76-44-8	none	termiticide termiticide in structures of houses termiticide (subterranean) wood treatment in use in underground cable boxes	+	
6	Chlordane	57-74-9	As allowed for the parties listed in the Register	local ectoparasiticide insecticide termiticide termiticide in building and dams termiticide in roads additive in polywood adhesives	+	
7	Mirex	2385-85-5	As allowed for the parties listed in the Register	termiticide	+	
8	HCB	118-74-1	As allowed for the parties listed in the Register	intermediate solvent in pesticide closed system site limited intermediate	+	+
9	DDT	50-29-3	Acceptable purpose: Disease vector	Acceptable purpose: disease vector	+	

		control use in accordance with part II of this Annex	control in accordance with part II of this Annex	
		Specific exemption: Intermediate in production of dicofol Intermediate	Specific exemption: Production of dicofol Intermediate	
10	PCDD			+
11	PSDF			+
12	PCBs			+

7. Conclusions

1. The main vector borne diseases remain an exclusively important factor for the public health therefore the chemical vector control is still a need. It is necessary to continue the work for improvement of all activities in the field of the registration, import, trade, quality control and application of the pesticides for agricultural and public health use.
2. It is a need to develop monitoring schemes on national level in order to ensure systematic data collection of the residues of persistent organic compounds, including pesticides in various environmental media, living organisms and humans.
3. It is necessary to build up the capacity of the different countries and regions for investigation and modeling of the conditions favoring the transport of persistent pollutants on regional and global level.
4. On the basis of already existing international agreements, and especially the Stockholm Convention, the primary task of all parties of the convention is the development of National Implementation Plans for the management of persistent organic pollutants.

References

Basel Convention on the Control of Transboundary Movement of Hazard Wastes and Thear Disposal – adopted 1989, into force 1992.

Bratanova, Z. and K. Vassilev, 2001 Pesticide residues in ground and surface water in Bulgaria *Fresenius Environ. Bull., 10,4,401–4.*

Carter, A.D., 1993 Pesticides in soil and water., Pesticide outlook.,4,23–28.

Council Directive 98/83/EC of 3 November 1998 on the quality of water intended for human consumption. Official Journal of the European Communities, L 330, 32–54.

Deladedova, P., Sesana, G., Bersani, M., Riparbelli, C., Maroni, M., 1996 Atrazine groundwater contamination in an intensive agricultural area west of Milan – Italy., Period 1986–1994. Proceedings of the X Symposium Pesticide chemistry, Piacenza, 669–73.

Eriksson, G., Jensen, S., Kylin, H. and Stracham, W.M.J., 1989 The pine needle as a monitor of atmospheric pollution, *Nature*, 341,42–44.

Gitsova, S., 1975 Natural water pollution in Bulgaria by residues of oganochlorine insecticides and methods for their detection, PhD thesis, Sofia.

Grohman, A., Winter, W., Ottenwalder, H., 1994 Mogliche Beeintrachtigungen des Trinkwassers in den neuen Landern durch Pflanzenschutzmittel., Bundesgesundhbl., 12, 496–502.

International Convention for the prevention of pollution from ships (Marpol 73/78)

Kaloyanova-Simeonova, F., Bratanova, Z., Bratinova, St., Dura, Gy., Simeonov, J., Kambourova, V., Rizov, T., 2001. Human Exposure and Risk Assessment of Soil Pollution with Persistent Organochlorine Compounds in Bulgaria. Central European Journal of Occupational and Environmental Medicine, vol. 7, No. 3–4; 263–275.

Kambourova, V., 2004. Impact of Obsolete Pesticides on Rural Environment. Journal of Balkan Ecology, vol. 7, No 4; 422–427.

Kara, H., Aktumsek A. and Nizamlioglu, 1999 Some organochlorine pesticide residues in commercial milk konya-region/Turkey. *Fresenius Environ. Bull.*, 8, 257–263.

Kujumdzieva, T., 1976. Determination of organochlorine pesticides in soil and their levels in some regions in Bulgaria. Ph.D Thesis, Sofia.

Kujumdzieva, T., Rizov, N., Danova, S., 1979. Content of organochlorine pesticides in soil down the river Marica, Information Bulletin – Hygiene of soil and water, VIII, 2; 15–31.

Mocarelli, P., Bambilla, P., Gertoux, P.M., Patterson, D.G. and Needham, L.L., 1996 Change in sex ratio with exposure to dioxin, *The Lancet*, 348, 409.

Pionke, H.B., Glofelty, D.E., Lucas, A.D., Urban, J.B., 1988 Pesticide contamination of Groundwater in the Mahantango Greec Watershed, Pennsylvania, USA, J.Environ.Qual., 17, 76–84.

Reijnders, P.J.H., 1986 Reproductive failure in common seals feeding on fish from polluted coastal waters, *Nature*, 324, 456–7.

Ripparbelli, C. Scalvini, C., Bersani, M., Auteri. D., Azimonti, G., Maroni, M., Salamana, M.,Carreri, V., 1996 Groundwater contamination from herbicides in the region of Lombardy – Italy., Period 1983–1993., Proceedings of the X symposium Pesticide chemistry, Piacenza, 559–63.

Stockholm Convention on Persistent Organic Pollutants (POPs) – adopted 2001

The Pesticide Manual, 1994 Tenth edition, BCPC, 51–2.

The Roterdam Convention on the Prior Informed Concent Procedure for Certain Hazardous Chemicals and Pesticides in International Trade. Adopted 1998.

UNEP, 2003 Global Report 2003, Regionally based assessment of PTS, UN, UNEP, Geneva.

WHO, 2001 Technical report series 899 - Chemistry and specifications of pesticides, WHO, Geneva.

REGULATORY USE OF PESTICIDE CONCENTRATIONS IN THE ENVIRONMENT IN THE EU

JAN LINDERS B.H.J.
RIVM-SEC,
P.O.Box 1HL-3720 BA Bilthoven, Netherlands,
E-mail: Jan.Linders@RIVM.NL, http://www.rivm.nl/csr/

Abstract. Pesticides used in agriculture are deliberately introduced into the environment to prevent harmful organisms and pests from damaging edible crops and non-edible other plants intended for human consumption or use. Therefore, unintended side effects of these agrochemicals should be avoided or minimized. In the European Union methods have been developed to estimate the environmental concentration in different environmental compartments, like soil, groundwater, surface water and air for use in the evaluation process for the registration of pesticides. The evaluation systems developed are highlighted shortly as well as the way the concentrations established may be use by registration authorities in the decision making process for registration. Also the decision making criteria are dealt with and possible mitigation measures to minimize the environmental hazards.

Keywords: pesticides; registration; predicted environmental concentrations; modeling; decision making; tiered approach; effect assessment; risk assessment

1. Introduction

As in many countries and federations the registration or authorization of agrochemicals (plant protection products (PPPs)) is an active process of carrying out risk assessments for this type of substances that involves several aspects and a lot of data to be taken into account. In the European Union (EU) it is the same of course. It needs for example an active positive decision on the marketing of PPPs from the European Commission considering human health, worker exposure, agricultural efficacy and environmental criteria. (EU, 1991) In this presentation an overview will be given of the environmental aspects taken into consideration and the way

Member States of the EU have decided how they will act if criteria are being breeched or not. Emphasis will be put on the prediction or estimation of environmental concentrations in surface waters which will be used to make a comparison to different kinds of effect concentrations. Depending on the outcome of the comparison a positive or negative decision is taken on that aspect of the evaluation. The registrants have the opportunity to deliver new data if a decision is not favourable for the compound they wish to register.

The guidance that was developed in the EU on the risk assessment of PPPs is dealt with taking into account the relevant data and a tiered system of data evaluation. There is a base set of information needed to register a substance in the EU but additional information is needed if the specific criteria set for side effects are breeched.

2. Environmental Risk Assessment of Pesticides

The risk assessment for PPPs in the EU is always based on sound scientific data, no matter which topic is involved. (EU, 2001) There are 4 main areas of risk to be considered:

- agricultural efficacy;
- human health,
- worker exposure and
- environmental safety.

Each area is analysed separately and will be brought together in a final decision making document published by the EU in accordance with all Members States. The publication date of the decision is determining the moment of the regulation getting into force. There is a possibility for the registrant to disagree with the decision and to ask justice in court. Although a substance may be considered a risk for a specific aspect cost benefit analysis may be used to prepare an exception for a limited time frame. The data in the dossier may be used anywhere in the risk assessment where considered relevant. Generally the distinction is clear enough where the data will be used, some data as e.g. the dosage of the substance will be used in all areas.

In the following scheme an outline is given of the risk assessment procedure. (RIVM, VROM, VWS, 2002).

3. Tiered Approach

Generally the risk assessment is carried out according a tiered approach requiring more advanced data in the next tier if in the current tier the criterion for safe use is breeched. The data needed in Tier 1 is called the base set of information and has to be delivered to the authorities in all normal cases. Sometimes a registrant may ask for certain waivers. Depending on the risk assessment the registrant carries out with his own substances he may decide for higher tier studies and higher tier risk assessments for problematic topics.

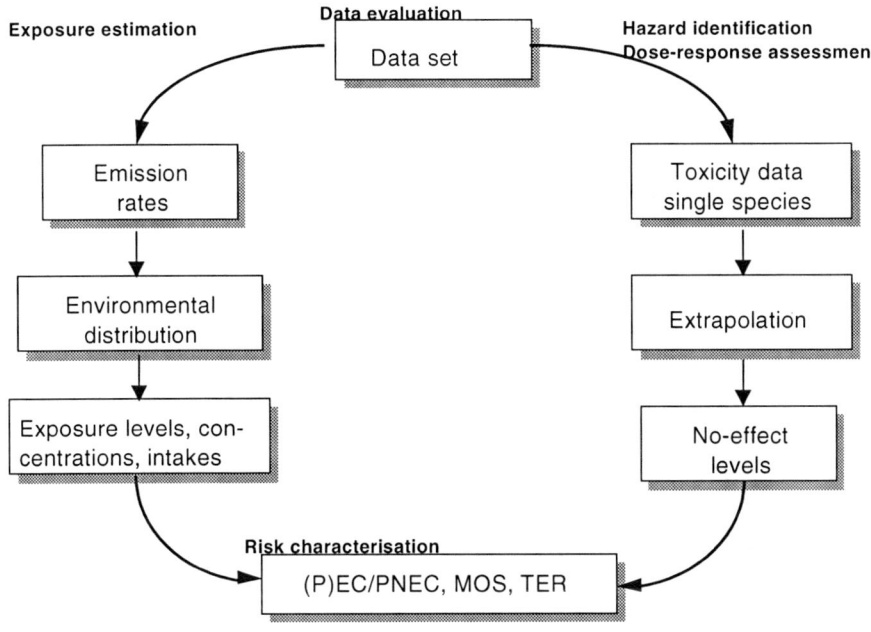

FIGURE 1. Generalised procedure for risk assessments of chemicals.

If the topic stays problematic, e.g. the triggers are breeched in all situations than he may withdraw the substance. Also if the registrant decides the data required exceed the potential profits a withdrawal may be appropriate. All studies have to be carried out under Good Laboratory Practice conditions and according to internationally accepted guidelines.

Base set information consists in any case of identification data, physico-chemical data, fate and behaviour data and ecotoxicological data for different organisms, like birds, fish, daphnids, algae, earthworms, honey bees, etc. Higher tier information may contain field studies to fate and behaviour, like accumulation or lysimeter studies, and micro- or mesocosm studies for aquatic or terrestrial organisms, cage studies for assessing the risk for honey bees, etc.

If a dossier is considered incomplete by the registration authorities the registrant is given the opportunity to provide additional information on the topic requested within a limited time, e.g. 1 year. In principle, the registrant should be able to decide on the registrability of the substance or provide the authorities with sufficient mitigation measures that a safe use is warranted. A substance may be registered in the EU if at least one safe use can be identified.

4. Predicting Environmental Concentrations

An important prerequisite for carrying out a scientifically sound or as sound as possible risk assessment is that all data have been determined without any scientific prejudice. (Mensink et al., 1995) It is not possible to carry out a risk assessment without good quality data. Internationally accepted guidelines, like the OECD test guidelines, should be used in determining relevant parameters for active substances of PPPs.

Predicting environmental concentrations is a process based on certain simplifying assumptions and using up-to-date scientific tools. Generally, mathematical models are developed where all relevant and as much as possible scientific information is included. The models are based on a mass balance approach using the conservation of mass principle, the advection – dispersion equation and contain processes like transport, transformation, sorption and volatilisation. In the EU, the FOCUS organisation has been set up; FOCUS stands for FOrum for the Coordination of pesticide fate models and their use (FOCUS, 2005). For the environmental compartments, soil, groundwater, surface water, including sediment, and air, fate models have been made operational for the use in the EU registration process to determine the PEC in these compartments. Guidance documents are available from a web site located at the Joint Research Centre in Ispra, Italy, from where the tools can be downloaded. (FOCUS, 2005) At the moment models and scenarios are available for groundwater and surface water. Models for determining the PEC in air are currently inventoried and probably be made operational in 2006. A working group on Version Control keeps track of all the changes in models, scenarios and shells.

For the estimation of the predicted environmental concentration in groundwater, there are 4 models and 9 scenarios available. The scenarios are located in several parts of the EU covering a wide variety of soils and climatic regions. The models are PEARL, PELMO, PRZM and MACRO. It should be noted that MACRO is only parameterised for one scenario, whilst the other three models are parameterised for all 9 scenarios. (FOCUS, 2005).

The estimation of the predicted environmental concentration in surface waters is much more complicated compared to groundwater. There are 3 possible input routes: drift, drainage and run-off and in addition there are three different water types identified. Ten scenarios are used to characterise the different hydrological, climatological and soil situations all over the EU, all ten contain drift as input, 6 scenarios are defined for drainage and 4 for runoff. As the inputs of drift and drainage or runoff are feeding into surface water the final fate of the substance is determined by a great variety of influences and processes. In the scheme below the interaction of the models are visualised. The models used are MACRO for drainage, PRZM for runoff and TOXSWA for fate. A specific drift calculator has been developed that combines different drift data gathered in several drift investigations. SWASH is a user friendly shell that guides the user through the successive model steps and keeps the data base updated. (FOCUS, 2001).

The results of the fate calculation are used as the predicted environmental concentrations for the risk assessment by combining the data with the effect concentrations for aquatic organisms. For exposure during longer periods the time weighted average PEC (TWAEC) is determined for different periods varying from 2 to 100 days. (EU, 2001 and EU, 2002).

5. Effect Assessment

Using the aquatic compartment as an example, there are ecotoxicological data available for three different water organisms considered representative for the food chain, algae, Daphnids and fish. Generally, acute (4 day studies, EC50 for algae and Daphnids, LC50 for fish) and chronic (4 day studies for algae, 21 day studies for Daphnids and 28 day studies for fish) data (NOEC in all cases) have to be delivered to the authorities. The most sensitive species of the three will be used in the risk assessment. In the European Union assessment factors of 100 (AF_{acute}) are common for the acute risk assessment and a factor of 10 for the chronic situation ($AF_{chronic}$). The effect concentration of the ecotoxicological study divided by the assessment factor is considered to be the safe concentration for the aquatic ecosystem. This concentration is called the Predicted No effect concentration: PNEC. (EU, 2001 and EU, 2002)

Therefore,

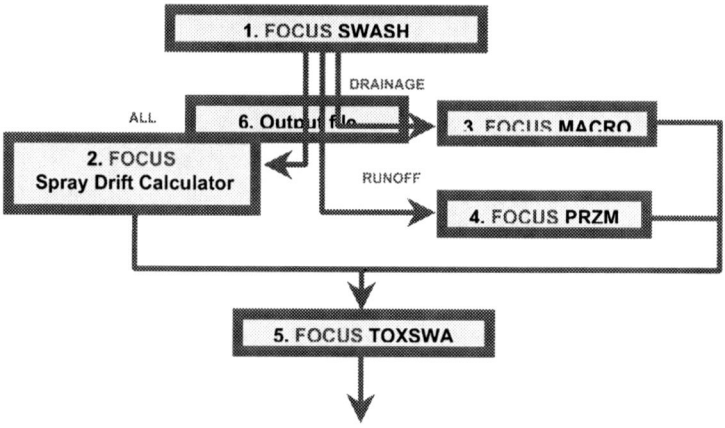

FIGURE 2. Overview of model interactions in FOCUS Surface Water Scenarios.

$$PNEC_{acute} = \frac{L(E)C50}{AF_{acute}} \quad \text{and} \quad PNEC_{chronic} = \frac{NOEC}{AF_{chronic}} \quad (1)$$

are the estimates relevant for the risk assessment as described in the next section.

6. Decision Making

As shown in the risk assessment figure, the decision making takes place at the risk characterisation step, where the results of the environmental distribution, the calculated concentration, and the ecotoxicological data, the effect concentration, come together. The most common way is to divide the predicted environmental concentration by the effect concentration in the acute or chronic situation, called the PEC-PNEC-ratio. In the European Union the toxicological concentration is divided by the predicted environmental concentration revealing the TER: the toxicity-exposure-ratio.

Thus, the decision making criteria are defined as follows:

for the acute situation:

$$\frac{InitialPEC}{PNEC_{acute}} < 1 \quad \text{or} \quad TER_{acute} > AF_{acute} \quad (2)$$

for the chronic situation:

$$\frac{TWAEC}{PNEC_{chronic}} < 1 \quad or \quad TER_{chronic} > AF_{chronic} \quad (3)$$

Rearranging the equations (1), (2) and (3), shows that the following equation is valid in all cases:

$$\frac{PEC}{PNEC} = \frac{1}{TER}.$$

Up to now it did not show possible to harmonise the risk quotient approach internationally!

Depending on the tier under consideration it may be required to perform additional studies. (EU, 2001 and EU, 2002) If at Tier 1 the criterion is breeched then there are different options available for the registrant to solve this problem. On both sides of the risk assessment scheme, the emission and distribution, on one hand and on the other the effect assessment, higher tier studies may result in non-breeching situations. In the fate area, more sophisticated modelling with more realistic assumptions may lower the PEC whilst in the effect area, micro- or mesocosm studies may show higher no-effect concentrations. Both will result in a adjustment of the risk ratio to more favourable outcomes. Finally, monitoring studies should be carried out to show ultimate safety or risk. Another possibility to reduce the risk is risk mitigation as an ultimate risk management tool. Risk mitigation may be applied to all areas of the risk assessment procedure. Examples are no crop zones in the fields applied (drift reduction), limit the use area of the substance, e.g. green house instead of full field applications, increase of toxicity data (reduction of assessment factor), probabilistic risk assessment (increase level of realism).

7. Conclusions and Recommendations

In principle, yes, the environmental risk assessment of plant protection products in the European Union is a straightforward process. The criteria are clearly defined and the methods are well documented and of highly scientific sophistication. Some gaps are still present, like the criterion on persistence, the higher tier evaluation of groundwater contamination and the risk assessment for the total terrestrial ecosystem. However, currently the Directive 91/414/EEC on the marketing of plant protection products is under reconstruction. It is not yet clear how this process will affect the evaluation methodologies available. Furthermore, the organisation of the evaluation process is also under reconsideration. The European Food and

Safety Authority will take responsibility in this area. How this will work out has to be seen in the near future. Nevertheless, the straightforwardness of the approach does not hide the relatively complex methods that are used or have to be used. The models applied in the determination of the PEC are far from simple; they contain the current state-of-the-art of the scientific description of chemical, physical and microbiological processes in the environment. The methods will be kept update and communications with the research institutes and the users of the models will continue. The main aim is to keep having available a system of methods to carry out risk assessments for plant protection products.

Growing knowledge on items of interest will be made applicable in the methodology, e.g. research on drift, reduction of runoff, influence of volatilisation and deposition through the atmosphere, use of geographical information systems, increase of user friendliness, increase of calculation power of computers, all these developments will continue to change the methods of risk assessment but it is also a great scientific challenge to keep track with these developments. Plant protection products still have to play an important role in providing sufficient food to mankind.

References

EU (1991) Directive 91/414/EEC. The authorization Directive, Anon. (1991) Official Journal of the European Communities No L 230, 19-8-1991.

EU (2001) Guidance Document on Aquatic Ecotoxicology in the context of the Directive 91/414/EEC. DOC Sanco/3268/2001 rev.4 (final) Brussels, 17 October 2001, 62 pp.

EU (2002) Guidance Document on Terrestrial Ecotoxicology under Council Directive 91/414/EEC. DOC Sanco/10329/2002 rev.2 (final) Brussels, 17 October 2002, 39 pp.

FOCUS (2001). "FOCUS Surface Water Scenarios in the EU Evaluation Process under 91/414/EEC". Report of the FOCUS Working Group on Surface Water Scenarios, EC Document Reference SANCO/4802/2001-rev.2. 245 pp.

FOCUS (2005) http://viso.ei.jrc.it/focus/index.html

Mensink, B.J.W.G., Montforts, M., Wijkhuizen-Maslankiewicz, L., Tibosch, H., Linders, J.B.H.J. (1995) Manual for Summarizing and Evaluating the Environmental Aspects of Pesticides. RIVM-report 679101022, Bilthoven, The Netherlands, 117pp.

RIVM, VROM, VWS (2002) Uniform System for the Evaluation of Substances 4.0 (USES 4.0). National Institute of Public Health and the Environment (RIVM), Ministry of Housing, Spatial Planning and the Environment (VROM), Ministry of Health, Welfare and Sport (VWS), The Netherlands. RIVM report 601450012.

THREATS ON THE SEASIDE LAKES WATER AND THEIR ACUTE AND LONG-TERM CONSEQUENCES

LIVIU-DANIEL GALATCHI[1], SIMONA DOBRINAS[2] AND ELISABETA CHIRILA[2]*
[1] Department of Biology, University Ovidius of Constanta,
124, Mamaia Blvd, 900527, Constanta, Romania
[2] Department of Chemistry, University Ovidius of Constantza,
124, Mamaia Blvd, 900527, Constanta, Romania

*To whom correspondence should be addressed: cchirila@univ-ovidius.ro

Abstract. Black Sea's pollution by organic and inorganic pollutants occurs frequently due to the fact that in the sea water there are discharged some waste waters (industrial or home originated) and surface waters with a high pollutants concentration. The conservation of surface waters quality in order to prevent and control the Black Sea coast pollution constitutes a problem of national, even international interest. In this purpose data were collected between 1995 – 2004 regarding the Corbu, Tasaul, Siutghiol, Tabacarie, Hagieni, Limanu lakes. From time to time, the Nuntasi, Agigea, Tatlageac, Neptun, Golovita, Zmeica, Razim, Sinoie, and Belona Lakes were also investigated. This lakes are located in the central and southern part of Romanian Black Sea coast and represent ecosystems contaminated by the anthropic influences. Mean values of inorganic constituents (including nutrients, metals etc.) and global organic pollution in correlation with measured biological parameters (primary production) are reported.

Keywords: seaside lakes; eutrophication; water quality parameters; nutrients

1. Pollution Threats

Water is one of the most important environmental factors, that contribute to the quality of live. For this reason the legislation worldwide protect the water resources against any form of pollution. Surface water quality monitoring has the aim to evaluate the "state of the art" for each river, lake et.al. in order to protect aquatic ecosystems, maintaining and improving their quality and natural productivity (Perniu, 2002, Sica et al., 2002).

The economic development of Dobrogea County, particularly of its seacoast area, during the last 4-5 decades had a great impact on the natural terrestrial and aquatic environment. As there has been practically no concern regarding the protection of the environment, this impact has appeared with all the intensity, which has lead to the occurrence of degradation processes that according to their intensity have altered the environment in different ways.

1.1. AGRICULTURE

For instance, agricultural works have lead to the following of a large part of the old Dobrogea steppe, which has increased the erosion process. To this, the administration of chemical fertilizers was added, which, washed down by the rain from the slopes, has led to the silting of the aquatic basins, plus an intensification of the anthropic eutrophication process of the respective basin. Such a situation can be seen in the cases of the Razim and Tasaul Lakes.

1.2. PISCICULTURE

Another form of alteration of the aquatic basins is represented by the establishment of the pisciculture system, which besides the creation of dames and the partioning of the lakes, has resulted in the so-called "fattening" of the basin. This process consists in throwing into the water substantial quantities of manure or nutrients. This is how man with a view to increasing the fish production deliberately accelerates the eutrophication process. Such a situation can be seen at the Tatlageac Lake.

1.3. WASTE DISCHARGE

A third form of accelerating the eutrophication process is achieved by the input of liquid or solid wastes from the localities situated in the close vicinity of the aquatic basins. In human settlements, people throw various wastes: garbage, manure etc. on the shores of the lakes. The rainwaters wash down such waste, which reach the lake, causing the loading of the waters with organic matters, carbohydrates, oils, detergents, household chemicals etc. If there are factories in the riparian localities, their liquide wastes are also disposed of in the nearby lake, or in one of its tributaries. Such situations occurred at Corbu, Tasaul and Siutghiol. The Tabacarie Lake represents an extreme case, where part of the drain waters brought from the northern part of the city of Constanta have been discharged for over three decades. As a result, this is the most degraded lake due to the impacts of anthropic pollution on the Romanian seacoast of the Black Sea.

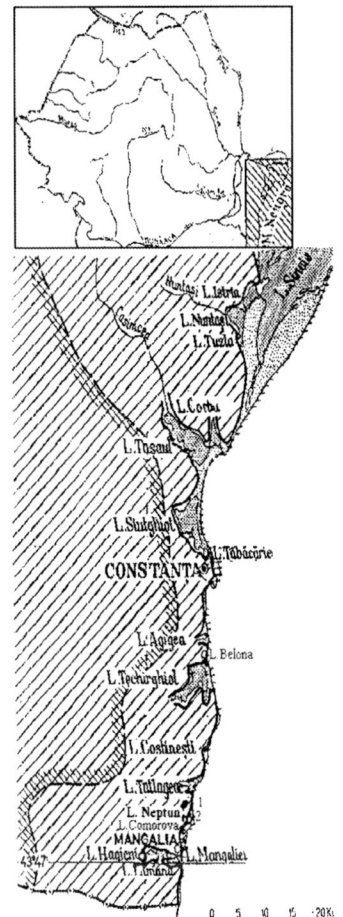

FIGURE 1. The sampling sites from Seaside Lakes.

The investigated Seaside lakes are located in the southern part of Romanian Black Sea coast with reduced anthropic influence (Limanu and Hagieni), followed by the big lakes in the northern part of the seacoast (Razim and Sinoe) which are feeding, directly or indirectly, with water from the Danube River (fig. 1). The other lakes, being situated in agricultural or suburban areas, evolve more rapidly towards eutrophication.

2. Inorganic Constituents of Seaside Lakes Water

2.1. PH, ALKALINITY, HARDNESS, SALINITY

Alkalinity, total and calcium hardness and salinity were the monitored quality parameters using standard titrimetric analytical methods according Romanian regulations as follows:

- the alkalinity was determined by titration of neutralization with HCl 0.1N to pH 8.3 ("p" alkalinity) or 4.5 ("m" alkalinity);
- the total hardness was measured by complexation titration with disodium salt of EDTA at pH 10 in the presence of Eriochrome Black T, and calcium hardness by the titration with the same reagent at pH 12 in the presence of murexid;
- chlorides were determined by precipitation titration with silver nitrate 0,01 in the presence of chromate ions; the salinity was calculated using empiric formula: $S = Cl(g/L) \cdot 1.8 + 0,03$;

Table 1 presents the annual mean values for four quality parameters in Tabacarie lake water: "m" alkalinity, total and calcium hardness and salinity.

TABLE 1. Annual mean values for some water quality parameters for Tabacarie lake.

Year	"m "Alkalinity, meq/L	Hardness, meq/L		Salinity, g/L
		Total	Calcium	
1995	6.15	7.66	2.68	0.27
1996	5.86	7.14	2.14	0.28
1997	6.17	7.00	2.82	0.32
1998	5.08	7.09	3.52	0.33
1999	5.49	8.00	2.95	0.37
2000	5.57	6.66	2.14	0.27
2001	5.18	6.53	2.00	0.25
2002	5.04	5.50	2.10	0.28
2003	6.39	8.27	2.24	0.36

The Razim Lake is feeding, directly or indirectly, with water from the Danube River and so the lake water hardness increased due to the high calcium and magnesium ions concentrations. In Sinoe Lake the ratio between calcium and magnesium ions is keeping constant, due to the seawater influence (Chirila et al., 1998a).

Previous works by authors present studies about these quality parameters from Nuntasi, Agigea, Tatlageac, Neptun and Belona lakes

(Mihaiesi et al., 1992; Chirila et al., 1998a) when was observed a large variety of concentration salinity from 0.52 g/L to 9.32 g/L. In Table 2 are presented the annual mean values for pH and salinity in Corbu, Limanu, Hagieni, Tasaul and Sinoe lakes waters. These two parameters were studied in Razim, Sinoe, Zmeica and Golovita lakes waters only in 2000 year. The salinity for Razim, Zmeica and Golovita lakes waters is 0.5 g/L and for Sinoe lake is 4.3 g/L and pH varies from 7.6 (Razim) to 8.3 (Sinoe).

TABLE 2. Annual mean values for some water quality parameters for some Seaside lakes.

Year	Corbu		Limanu		Hagieni		Tasaul		Siuthiol	
	pH	Salinity, g/l	pH	Salinity, g/l	pH	Salinity, g/l	pH	Salinity, g/l	pH	Salinity, g/l
1997	8.1	0.4	8.20	0.30	8.20	0.28	7.65	0.30	8.30	0.50
1998	8.0	0.5	8.30	0.30	8.30	0.27	7.80	0.40	8.20	0.40
1999	8.0	0.9	8.40	0.49	8.10	0.38	8.00	0.35	8.10	0.37
2000	8.2	0.8	8.10	0.31	8.20	0.28	7.90	0.30	8.10	0.35

The pH was measured with a portable pH-meter and as we know microorganisms present a optima activity at a neutral pH, but exists some species which are evolving normal at a value of pH between 8 and 8.5. The values of pH vary between 7.65 –8.40. So, with mean of 8.13 the value of pH from lakes waters shows an alcalin character. For Tabacarie Lake the values of pH ranged from 7.8 to 8.5 and these values are higher than those encountered in 1992 by Mihaiesi et al. This fact indicates the reduction of CO_2 content, respectively a intensive photosynthetic activity of algae mass.

2.2. NUTRIENTS

Nutrients are usually thought of as compounds of nitrogen or phosphorus, although certainly other elements, such as iron, magnesium, potassium and silicium are also necessary for bacterial and plant growth.

These nutrients are important in natural waters because, in excess, they can cause nuisance growth of algae or aquatic weeds. In wastewater treatment, a deficiency of nutrients can limit the effectiveness of biological treatment processes. In some plants treating industrial wastewaters, ammonia or phosphoric acid must be added as a supplement.

2.1.1. *Ammonium, Nitrate, Nitrites and Total Nitrogen*

Nitrogen occurs primarily in the oxidized forms of nitrates (NO_3^-) and nitrites (NO_2^-) or the reduced forms of ammonia (NH_3) or "organic nitrogen", where the nitrogen is part of an organic compound such as an amino acid, a protein, a nucleic acid, or one of many other compounds. All of these can be used as nutrients, although the organic nitrogen first needs to decompose to a simpler form. The studied nutrients were measured by spectrometric methods.

The evolution of nutrients in Razim, Sinoe, Zmeica and Golovita lakes waters were studied only in 2000 year. Nitrate, nitrite and total nitrogen concentration in effluents discharged in natural waters must be lower than 37mg/L, 2 mg/L, respectively 15 mg/L (HG 188/2002). These parameters have lower concentration values than those imposed in all cases. In Golovita lake there are registered the lowest nitrate (47.0 µg/L), nitrite (6.1 µg/L) and total nitrogen (136.7 µg/L) concentration. According to this parameter, all lakes water can be included in the 1st category of surface waters (STAS 4706/88).

Ammonium ions concentration in effluents discharged in sewage system must be lower than 30 mg/L and in effluents discharged in natural waters lower than 2-3 mg/L (HG 188/2002). Higher concentrations values than those imposed were observed in case of Tabacarie Lake in 1997 and 1999 (table 3). Total nitrogen varies from 9.6 µg/L (Razim lake- figure 2) to 6335, 3 µg/L (Tabacarie lake) with season (higher concentrations were noticed in spring).

2.2.2. Phosphorus

Phosphorus is biologically important in the form of phosphate, the most highly oxidized state of the element. The most biologically available form is dissolved orthophosphate, (PO_4^{-3}). There are also condensed forms of phosphate, with more than one phosphorus atom per ion, such as pyrophosphate and polyphosphates.

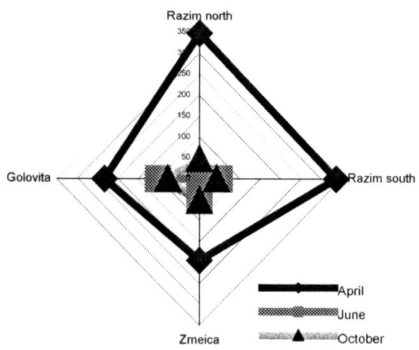

FIGURE 2. Evolution of nitrogen concentration in Razim, Golovita Zmeica lakes during 2000 year.

There are also organic phosphates, and all of these forms can be either dissolved or particulate (i.e., insoluble). The sum of all the forms is known as total phosphorus.

Phosphorus concentration in effluents discharged in sewage system must be lower than 5 mg/L and in effluents discharged in natural waters lower than 2 mg/L (HG 188/2002). Obtained data (Table 3) show lower concentration than those imposed for all studied lakes and maximum concentration values were registered in 1999 for all lakes. It can be noticed higher phosphorus concentration in summer (Fig. 3).

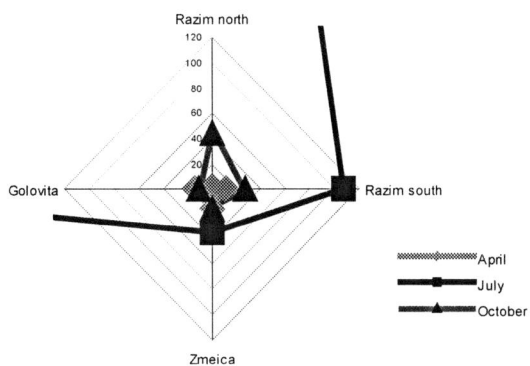

FIGURE 3. Evolution of phosphorous concentration in Razim, Golovita Zmeica lakes during 2000 year.

Razim Lake in north area has very low phosphorus concentration in spring (1.3 μg/L) followed by huge concentrations in summer (749.7 μg/L) contrary to nitrogen concentration, which are huge in spring (see fig.2.). The same effect of phosphorus exces is presented in Golovita lake ($\Sigma N/P < 1$ in summer), but in autumn the ratio $\Sigma N/P$ return to normal ($\Sigma N/P=15$). Zmeica Lake is izolated and this way the influence of phosphorus is not so high. Concern Sinoe Lake the variation of nitrogen and phosphorus concentrations has minor amplitude (in summer ratio $\Sigma N/P=15$ followed by increasing of phosphorus concentration in autumn). In Hagieni lake total nitrogen concentration was huge comparatively with phosphorus concentration ($\Sigma N/P=59.75$). This paradoxical values are absolute adversary to a good reproductivity and are critical for biosynthesis. Also, for Limanu Lake ratio $\Sigma N/P$ is far from optimal value (being higher from normal value).

According to phosphorus maximum admissible concentration in surface waters (0.1 mg/L), only Hagieni, Corbu and Limanu lakes can be included in the 1st category (STAS 4706/88).

TABLE 3. Annual mean values for nutrients in Hagieni, Corbu, Limanu, Tasaul, Tabacarie and Siutghiol lakes waters, during 1997-2000.

Station	Year	$N-NO_2^-$ (µg/L)	$N-NO_3^-$ (µg/L)	$N-NH_4^+$ (µg/L)	Total N (µg/L)	$P-PO_4^{3-}$ (µg/L)	$Si-HSiO_3^-$ (µg/L)
Hagieni Lake	1997	80.7	1022.0	87.4	1180.1	11.4	1308.0
	1998	59.8	1023.2	59.7	1142.7	20.1	2194.0
	1999	102.0	1089.9	126.5	1329.4	16.7	2077.0
	2000	49.4	911.5	65.9	1007.3	14.5	1083.0
Corbu Lake	1997	167.3	1064.8	503.5	1735.6	15.7	1394.0
	1998	50.1	1140.8	64.6	1255.5	21.4	1117.0
	1999	25.0	801.0	113.5	939.5	52.0	2182.0
	2000	100.8	1362.4	112.1	1575.3	11.5	696.0
Limanu Lake	1997	39.0	512.0	36.6	587.6	11.1	1044.0
	1998	20.5	730.2	63.8	813.6	9.7	1402.0
	1999	45.7	904.5	37.8	983.0	26.2	1955.0
	2000	92.1	1026.4	236.5	1355.0	10.9	1914.0
Tasaul Lake	1997	4.4	40.2	17.5	62.1	213.7	1875.0
	1998	10.3	432.4	117.3	560.12	124.5	542.2
	1999	68.2	608.8	335.1	1012.2	269.8	2033.5
	2000	4.9	22.7	8.6	36.2	179.1	1949.5
Tabacarie Lake	1997	83.0	445.5	2943.0	3471.5	215.2	2736.5
	1998	202.2	2192.1	1551.5	3945.8	513.5	3546.2
	1999	198.7	2686.4	3450.2	6335.3	1033.2	4522.2
	2000	140.6	1610.3	1408.0	3158.9	616.6	4106.0
Siutghiol Lake	1997	6.2	40.5	92.4	139.1	24,4	1540.0
	1998	69.1	288.7	89.9	447.7	85,7	529.0
	1999	14.9	147.7	18.7	181.3	156,4	667.7
	2000	7.9	104.7	132.0	244.6	45,1	126.0

2.2.3. Silicium

Silicium is an important element for some primary producers of planktonic synthesizer being an essential factor in primary production. This element was found in similar concentrations in all lakes, with maximum value in Tabacarie Lake (4522,2 µg/L). If silicium is in low concentration in lake water (1.5–3.5 mg/L), the diatomee population decreases (Moss, 1988). The optimal concentration of silicium varies from a primary producer of planktonic synthesizer to another and in water this concentration has to range from 30 to 40 mg silicium/L (Horne and Goldman, 1994).

2.3. METALS

Most metals are discharged to the environment as a consequences of urbanization and industrial activity but the metals themselves, unlike some organic micro pollutants are no biodegradable and are readily accumulated within ecosystems. Metal interaction with biological systems, however, may be of greater concern because of their toxicity and bioaccumulation potential.

The choice of an analytical method depends on its performance characteristics (detection limits, accuracy and precision, speed etc). Other conditions to be reached are the concerned element, the concentration in the sample of interest, the variability of their concentration. The concentration of metal ions in studied Seaside Lakes were determined by flame atomic absorption spectrometry (FAAS) (Chirila et al., 2003a), inductively coupled plasma atomic emission spectrometry (ICP-AES) (Chirila et al., 2002), molecular absorption spectrometry in visible (Chirila and Carazeanu, 2001). These investigations were carried out in the biotope (sediment and water) and biocenosis (different plants and fish) from one ecosystem (Tabacarie Lake) and in water samples from the other Seaside lakes.

From Tabacarie Lake ecosystem significant samples of water, sediment and different fish species were collected during 1997-2002 for Cd, Cr, Cu, Fe, Mn, Ni, Pb, Zn analysis. In 1997 for the first time in Tabacarie Lake sediments were analyzed 18 chemical elements (Al, B, Cd, Cr, Cu, fe, Mg, Mn, Mo, Ni, Pb, Sb, Si, Ti, Zn, Zr) by ICP-AES (Birghila et al., 1997). It was noticed that the analyzed elements have slightly high concentration, which proves their accumulation by physical, chemical and biochemical processes. As shown by Chirila et al. (1998b) the metal capacity of accumulation in sediment increase in order: lead, cadmium, nickel, zinc, chromium and copper. Metal concentrations in plants and fishes varied widely, both within and between species, but in general higher metal burdens were observed for essential (Fe, Mn and Zn) rather than non-essential one (Cd). Variations in essential metals within species were low, but different plant species accumulated variable quantities of these metals (Chirila et al., 2003b).

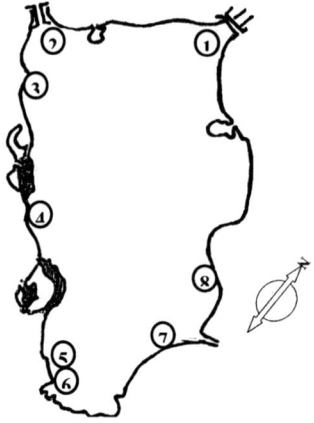

FIGURE 4. The sampling sites from Tabacarie lake.

From the research concerning the application of ICP-AES method (Birghila and Chirila, 1997) and visible molecular absorption spectrometry (Chirila et al., 1997) within Tasaul Lake waters it has been noticed that the micro and oligoelements content is variable from or element or sample to another. Also micro and oligoelements analysis were accomplished in Nuntasi, Corbu, Siutghiol, Agigea, Tatlageac, Neptun and Belona Seaside lakes (Chirila et al., 1998; Birghila et al., 1998).

Iron, chromium and copper concentrations were determined during 2003-2004 from Tabacarie Lake waters. In this purpose eight sampling sites were established around the lake and analyses were carried out weekly, three month per year. The location of the sampling points is presented in the Fig. 4.

TABLE 4. Cu, Cr and Fe concentration in Tabacarie lake water (mg/L).

Element	Year	Sampling site							
		1	2	3	4	5	6	7	8
Cu	2003	1.15	1.80	1.10	2.00	0.58	0.41	0.29	0.68
	2004	1.52	1.01	1.71	1.41	0.51	1.02	1.30	0.77
Cr	2003	0.28	0.26	0.11	0.17	0.15	0.27	0.43	0.20
	2004	0.30	0.40	0.21	0.35	0.11	0.32	0.45	0.29
Fe	2003	0.36	0.33	0.19	0.26	0.20	0.29	0.34	0.11
	2004	1.08	0.27	0.10	0.12	0.15	0.77	0.10	0.009

Chromium concentration in Tabacarie Lake water varies between 0.11 and 0.40 mg/L and all founded concentrations are under the limits for surface waters (0.5 mg/L). Copper concentration varies from 0.29 to 1.80 mg/L and iron from 0.009 to 1.08 mg/L and these concentrations are higher than those imposed in almost sampling sites.

3. Organic Pollution Degree of Seaside Lakes Water

Romanian regimentations about water quality protection show that between the strictly surveyed parameters that can indicate organic pollution degree are: chemical oxygen demand (COD_{Mn}), dissolved oxygen (DO), and biochemical oxygen demand (BOD).

3.1. DISSOLVED OXYGEN (DO)

Dissolved oxygen represents one of the pollution indicators; oxygen is consumed in the dissolved organic compounds' chemical, photochemical or biochemical oxidation processes (Chirila, 2000). Oxygen becomes dissolved in surface waters by diffusion from the atmosphere and from aquatic plant photosynthesis.

DO can be measured by a fairly tricky wet chemical procedure known as the Winkler titration. The DO is first trapped, or "fixed", as an orange-colored oxide of manganese. This is then dissolved with sulfuric acid in the presence of iodide ion, which is converted to iodine by the oxidized manganese. The iodine is titrated using standard sodium thiosulfate. The original dissolved oxygen concentration is calculated from the volume of thiosulfate solution needed.

The mean values of the dissolved oxygen show that all the lakes are well oxigenated (they have mean values of 7.5-10.3 mg/L).

The DO concentration in Razim, Zmeica, Hagieni, Corbu and Siutghiol is lower in July than April and in autumn increases again. Unlike these lakes, in Galovita, Tabacarie, Sinoe and Limanu were observed an increase of DO concentration in summer followed by lower values in October (see Fig. 5). The balance of DO concentration is the most favorable for Siutghiol Lake water because the multi annual average is 8.5 mg/L, value which can assure all forms of biologic and biochemical consumption of organic substances in lake water.

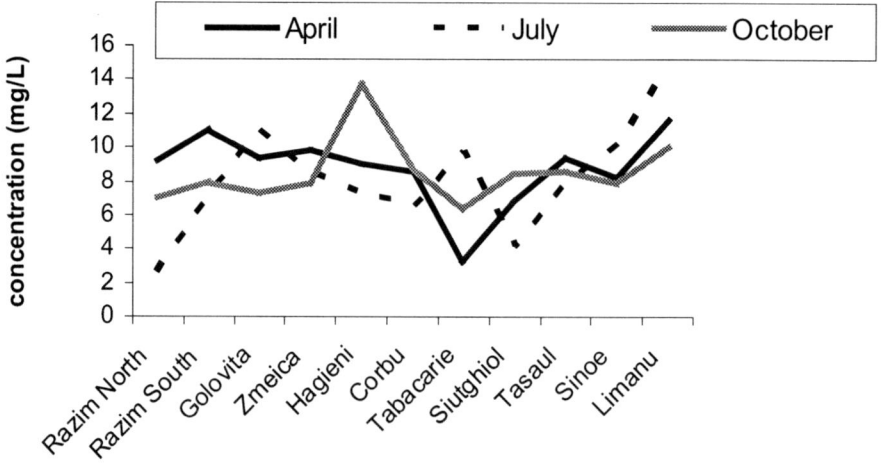

FIGURE 5. The distribution of dissolved oxygen in Seaside Lakes waters with season.

3.2. BIOCHEMICAL OXYGEN DEMAND (BOD)

The biochemical oxygen demand is a test for measuring the amount of biodegradable organic material present in a sample of water. The results are expressed in terms of mg/L of dissolved oxygen which microorganisms, principally bacteria, will consume while degrading these materials.

The BOD test is performed in a specially designed bottle with a flared cap, which forms a water seal to keep out air. The bottle is filled completely with sample, which must be near neutral pH and free of toxic materials. After an initial measurement of the DO, the bottle is sealed and stored in a dark incubator at 20C for five days. The dissolved oxygen is measured again after this incubation period. The difference is the BOD. (The bottles are kept in the dark because algae that may be present in the sample will produce oxygen when exposed to light). Since most wastewaters have BOD's which are much higher than the limited solubility of oxygen in water, it is necessary to make a series of dilutions containing varying amounts of sample in a nutrient-containing, aerated "dilution water." The measured BOD's are then multiplied by the appropriate dilution factors. A variation of this test, called the carbonaceous BOD, adds an inhibitor, which prevents the oxidation of ammonia, so that the test is a truer measure of the amount of biodegradable organic material present. Samples that do not contain enough bacteria to carry out the BOD test can be "seeded" by adding some from another source. Examples of samples which would need seeding are industrial wastewaters which may have been at high temperatures or high or low pH, or samples which have been disinfected. (If there is residual disinfectant present, it must be neutralized before testing.)

The values of B0D vary within narrow limits (table 5). The Razim and Tasaul lakes have lower values of B0D (3.3, respectively 3.9 mg O_2/L), the Limanu and Hagieni lakes have the maximum values of B0D (7.2, respectively 6.7 mg O_2/L). We estimate that this value is the result of the way in which the areas of reed through which the water passes to reach the lake and which retain a large part of the biodegradable organic substances filter the water from the Danube (which feeds the lake).

The accepted values of DO concentration in surface waters vary between 4 - 7 mg/L, of CODMn concentration vary between 10-25 mg O_2/L and BOD concentration vary between 5-12 mg O_2/L (STAS 4706/88). According to these parameters, Seaside lakes water can be included in the 2nd–3rd category of surface waters.

The situation of the Tabacarie Lake is, obviously, different, this being the most polluted lake on the Romanian seacoast of the Black Sea (the quality parameters indicate this fact). Chirila et al. (2005) showed the evolution of annual averages of CODMn in eight sites from Tabacarie Lake during 1995-2003. This evolution indicates that in 2003 the Tabacarie lake quality was the worst and CODMn vary in a large interval, between 3.91 and 15.57 mgO_2/L.

3.3. CHEMICAL OXYGEN DEMAND (COD-MN)

As the BOD test is a fairly long-term bioassay test (5 days), a more rapid one is often used to estimate the BOD and it is known as the COD, or chemical oxygen demand test. CODMn test was done by oxidation-reduction titration.

The analysis of the chemical oxygen demand shows a very wide variation of the values of this parameter, both from one lake to the other, and from one sample taking to another. The maximum values were recorded in the Sinoe Lake, followed by those of the Tabacarie Lake. The minimum values were found in the lakes situated in the southern part of the Romanian seacoast, at Hagieni and Limanu. We estimate that the extremely high values recorded in the Sinoe lake are due to the presence of the humic substances carried over by the water currents at the surface of the benthos and not by excessive pollution. This is confirmed by the values of the biochemical oxygen demand, which are not higher than the values recorded in the other lakes (table 5). The situation of the Tabacarie Lake is, obviously, different, this being the most polluted lake on the Romanian seacoast of the Black Sea.

It can also be noticed that the CODMn values of the lakes north of Constanta are higher than those of the lakes in the southern part of the Romanian seacoast (Limanu, Hagieni), which are less damaged by the human impact.

4. The Biology of Seaside Lake Water

4.1. THICKNESS OF THE EUPHOTIC LAYER

The measurements carried out to determine the thicknesses of the euphotic layer were especially performed because this is where the main biological processes, particularly the primary production of phytoplankton, are achieved. So, the thickness of this layer has a decisive influence on the primary production of the lakes and, implicitly, on the fish productivity. The thickness of the euphotic layer is relatively uniform, 32-61 cm on the average, which indicates that the productivity of these lakes obviously depends on their size and not on their absolute depth. The minimum values of the euphotic layer are found in the Razim (21.6 cm), Sinoe (34.8 cm) and Corbu (16.8 cm) lakes; the mean values are in the Tabacarie Lake, and the maximum values are in the Tasaul (79.2 cm), Siutghiol (88.8 cm), Limanu 9143.4 cm) and Hagieni (123.6 cm) lakes.

The mean transparency in the seacoast lakes varies between 19 and 42 cm in the Razim Lake and 21-112 cm in the Limanu Lake. On the overall, the Razim, Sinoe and Corbu lakes (all of them strongly exposed to dominant winds) have transparencies under 40 cm (usually 26-37 cm), and the others have mean values of 38-51 cm (table 6).

TABLE 5. Multiannual average values (1997-2000) for DO, BOD and CODMn in some lakes waters.

Station	Dissolved oxygen (mg/L)	BOD (mg O_2/L)	CODMn (mg O_2/L)
Razim	7.5	3.3	6.7
Golovita	9.3	5.6	5.1
Zmeica	8.8	4.0	4.7
Sinoe	8.2	5.7	27.9
Corbu	8.9	5.0	3.7
Tasaul	7.8	3.9	5.7
Tabacarie	9.3	7.1	8.2
Hagieni	10.3	6.7	3.5
Limanu	10.0	7.2	2.7
Siuthiol	8.5	4.7	7.8

The limits between the minimum and the maximum transparency vary little (except for the Limanu lake, followed by the Hagieni and Siutghiol lakes).

TABLE 6. Multiannual average values (1997-2000) for transparency and tickness of the euphotic layer in Seaside lakes waters.

Lake	Transparency (cm)	Thickness of the euphotic layer (cm)
Razim	25.0	30.0
Golovita	29.3	35.2
Zmeica	30.6	36.8
Sinoe	35.0	42.6
Corbu	26	31.2
Siutghiol	38.0	46.9
Tabacarie	78.3	93.5
Hagieni	51.0	61.2
Limanu	51.0	61.2
Tasaul	35.7	43.1

It should be noticed that there is a consistency between the values of the mean transparency and those of the thickness of the euphotic layer, which is slightly higher than the first one.

4.2. PHYTOPLANKTON

In order to determine the phytoplankton, samples of 1 liter of water were collected, in which, after treatment with IIK, formaldehyde fixing, a 30-day decantation and the concentration of the material, the components were identified quantitatively and qualitatively. For the determination of the zooplankton, water was collected using a device of the Patalas type and filtered through plankton net with 40: mesh. The quantitative and qualitative determinations were performed with a Kolkvitz chamber in the laboratory.

The examination of the mean values of the number of species of phytoplankton organisms shows that this varies within very wide limits, from 12 species in the Corbu Lake, up to 119 species in the Razim Lake. The aquatic basins most affected by the high input of exogenous organic substances have a lower number of species (there are 52 species in the Tabacarie lake and 35 species in the Corbu lake), whereas the cleaner fish basins have a greater biodiversity (119 species were identified in the Razim lake, 81 species in the Hagieni and Limanu lakes, 105 species in the Siutghiol lake).

The biomass of the phytoplankton (that of the primary producers in the water mass/ euphotic layer) is relatively similar in point of mean values between the lakes, fluctuating between 28.56 g/mc in the Siutghiol Lake and 106.12 g/mc in the Tabacarie Lake. As against the overall mean value of 53.69 g/mc, the Sinoe, Corbu, Siutghiol, Limanu and Hagieni lakes are below this mean value, whereas the Razim, Tasaul, Tabacarie Lakes are above the mean value. It should be noticed that the Tasaul Lake has the highest absolute values, followed by the Razim and Tabacarie lakes. The minimum absolute values are recorded in the Razim, Sinoe and Hagieni lakes, which we consider to be the cleanest lakes on the Romanian coast of the Black Sea.

In the seacoast lakes, the number of taxons of the zooplankton organisms is relatively reduced. It varies between 12 (in the Hagieni lake) and 44 (in the Navodari lake). The green biomass of the zooplankton is very reduced in the Corbu Lake (7.21 mg/mc), the maximum values being recorded in the Tabacarie lake (9028.2 g/mc). In the other lakes the mean biomass is around 300-600 g/mc. The Hagieni and Corbu lakes have values below 100 g/mc. The zooplankton furnishes the lake water with dissolved phosphates, which are then eliminated and are used as foodstuff for algae (Lahman and Scavia, 1982).

5. Conclusions

The following conclusions can be drawn from the research carried out about examined Romanian Seaside Lakes:

1. The general state of the lakes that were studied is extremely different: there are lakes with reduced anthropic influence (Limanu and Hagieni), all of them situated in the southern part of the Romanian sea coast, which are the cleanest; they are followed by the big lakes in the northern part of the sea coast (Razim and Sinoe). The other lakes, being situated in agricultural or suburban areas, evolve more rapidly towards eutrophication. The lake that is in the most advanced stage of eutrophication is the Tabacarie Lake, as it has been under the influence of the discharges of wastewaters from the municipality of Constanta for many years.

2. The organic load of the waters of the seacoast lakes is relatively high (CODMn concentration varies between 10-25 mg O_2/L, BOD concentration varies between 5-12 mg O_2/L and DO concentration varies between 7.5-10.3 mg/L). According to these parameters, Seaside lakes water can be included in the $2^{nd} - 3^{rd}$ category of surface waters. The fluctuations from one lake to another are relatively reduced.

3. The availability of nutrients in Razim and Golovita lakes was influenced seriously by the exces of phosphorus.
4. The values of quality-monitored parameters are variable in quasi-large ranges, depending on the position of the sampling sites and the seasonal characteristics, but except Cu and Fe, all of them are in the limits imposed by the last regulations.
5. Most of the primary plankton production is achieved, as a rule, in the first 50 cm of the mass of water from the surface (in the euphotic area), where the transparency is also the best.
6. The mean transparency of the waters of the lakes is 30-40 cm, with maximum fluctuations between 19 and 112 cm, but which varies from one lake to the other (the highest fluctuations are usually recorded in the cleanest lakes).
7. The data indicate that the most eutrophicated lakes are Tabacarie and Tasaul. The values of the parameters analyzed in the Razim Lake are contradictory as they indicate the existence of an intense process of eutrophication here as well.
8. Based on the studies undertaken during the last decade concerning the lakes on the Romanian sea coast of the Black Sea it is imperious to take urgent steps, specific for every lake, to efficiently fight against the process of eutrophication and organic silting in order to achieve systems for the retention rainfall waters and polluted waters and the extraction of the nutrients which might reach the lakes.

ACKNOWLEDGMENTS

We would like to express our gratitude for the support received from our colleagues Semaghiul Birghila, Mirela Mihaiesi from the "Ovidius" University of Constanta, from the "Danube Delta" National Institute for Research and Development in Tulcea, from Marioara Godeanu, Adriana Comanceanu and Cristina Radulian from the Institute of Applied Ecology in Bucharest who have graciously placed at our disposal the results of the sectorial tests they had performed. Also we want to thanks to the students from "Ovidius" University of Constanta, Ecology, Environmental Protection and Chemistry specialization that did measurements of some quality parameters during their fourth year of study, between 1995 and 2003. Without these, our synthesis would not have been possible.

References

Birghila S. and Chirila E., 1997 The heavy metal content determination within Tasaul lake water by inductively coupled plasma atomic emission spectrometry (IPC-AES), *Analele Universitatii Ovidius Constanta*, VIII, 14–16.
Birghila S., Chirila E. and Capota P., 1997 Analytical characterization of Tabacarie lake sediments by atomic emission spectrometry with inductively coupled plasma (ICP-AES), *Analele Universitatii Ovidius Constanta*, VIII, 17–20.
Birghila S., Chirila E. and Capota P., 1998 Metal determination from seacoast lakes by ICP-AES technique, *Analele Universitatii Ovidius Constanta*, IX, 40–42.
Chirila E., Birghila S., Godeanu S. and Godeanu M., 1997 The analytical characterisation of Tasaul lake water, *Analele Universitatii Ovidius Constanta*, VIII, 9–13.
Chirila E., Birghila S. and Godeanu S., 1998a Analytical characterization of seacoast lakes waters, *Analele Universitatii Ovidius Constanta*, IX, 48–51.
Chirila E., Birghila S., Capota P., Godeanu S. and Stoenescu V., 1998b Occurence of cadmium, copper, chromium, lead, zinc and nickel in different phases of an ecosystem, *Roum. Biotechol. Lett.*, 3, 47–54.
Chirila E., 2000, Study about analytical control in tertiary treatment stage of petrochemical effluents, *Rev.Chim.* (Bucharest), 51, 638–643.
Chirila E., Godeanu S., Godeanu M., Galatchi L.D. and Capota P., Analytical characterization of the Black Seacoast lakes, *Environmental Engineering and Management Journal*, 2, 205–212.
Chirila E. and Carazeanu I., 2001 Cadmium determination in water, fish and sediment of Tabacarie lake, *Ovidius University Annals of Chemistry*, 12, 17–19.
Chirila E., Draghici C. and Carazeanu I., 2003a The chromium determination in seaside lake and mineral ecosystems, *Ovidius University Annals of Chemistry*, 14, 43–46.
Chirila E., Draghici C. and Carazeanu I., 2003b About Cd, Fe, Mn, Zn occurence in Tabacarie lake ecosystem, *Ovidius University Annals of Chemistry*, 14, 47–50.
Chirila E., Dobrinas S., Carazeanu I. and Draghici C., 2005 Tabacarie lake water quality monitoring, *Environmental Engineering and Management Journal*, 4, 169–176.
Horne, A.J. and Goldman, C.R., 1994 *Limnology*, Mc. Graw-Hill, Inc., New York - St. Louis-San Francisco-Auckland-Bogota-Caracas-Lisbon-London-Madrid-Mexico City - Milan-Montreal-New Delphi-San Juan-Singapore-Sidney-Tokyo-Toronto.
Lahman, J.T. and Scavia, D., 1982 Microscale Nutrient Patches Produced by Zooplankton, *Proc. Natl. Acad. Sci.*, USA., 79, 5001–5005.
Mihaesi, M., Chirila, E., Godeanu, S.P. and Galatchi, L.D., 1992 The analythical caracterization of the conventional clean influents from Tabacarie Lake, *Analele Universitatii Ovidius Constanta*, III, 15–18.
Moss, B., 1988 *Ecology of Fresh Water*, Blackwell Scientific Publications, Oxford – London – Edinburgh – Bostos – Palo Alto – Melbourne.
Perniu D., 2002 *Environmental quality and environmental monitoring in Pollution and Environmental Monitoring*, Draghici C., Perniu D. (Eds), ed. Universitatii Transilvania Brasov, 8–9.
Sica M., Toader R., Draghici C., Tica R. and Dragan D., 2002 Barsa river monitoring, *Environmental Engineering and Management Journal*, 1, 347–354.

EVALUATING HEALTH RISKS AND PRIORITISING RESPONSE ACTIONS FOR CONTAMINATED LANDS

THOMAS MCHUGH* AND JOHN CONNOR
Groundwater Services, Inc.
2211 Norfolk, Suite 1000
Houston, Texas, 77098, USA

*To whom correspondence should be addressed: tcmchugh@gsi-net.com

Abstract. Industrial organizations and government entities that manage large numbers of contaminated sites need a systematic management process to ensure that all of sites are managed consistently and effectively. When using a risk-based management approach, individual sites should be managed to i) identify potential mechanisms for exposure to site contaminants, ii) evaluate risks associated with exposure, and iii) identify response actions required to control risk and achieve a consistent standard of care. At the same time, information from the portfolio of contaminated sites managed by the organization should be evaluated in order prioritize response actions among site. By establishing a coordinated system of program management and site management, a portfolio of contaminated sites can be managed in an efficient manner while maximizing the overall risk reduction and protection of health.

Keywords: contaminated land; risk assessment; prioritisation

1. Introduction

Past and present human activities have resulted in the release of contaminants at a large number of sites across Europe and around the world. These activities have results in over 750,000 contaminated sites in Europe alone (Ferguson, et al., 1998). The limitations of available technologies and resources prevent the restoration of all of these sites to background conditions that existed prior to industrialisation. As a result, methods are needed to i) evaluate the likely impact of these sites on human health and ecological resources, ii) manage individual sites in a manner that efficiently utilizes the available resources to prevent or mitigate impacts to human health and ecological resources, and iii) prioritise responses among contaminated sites based on the potential for adverse impacts in the absence of corrective action. This paper presents a process for evaluation and management of contaminated lands covering both management of individual contaminated sites and prioritisation of response actions among contaminated sites.

1.1. KEY STEPS IN RISK MANAGEMENT

To protect human health and ecological resources from effects of spills or other releases, effective risk management consists of three principal steps, *release prevention*, *emergency response*, and *remediation*, as follows:

Release Prevention: The first step in effective risk management is to prevent releases from occurring through spill prevention programs, active preventive maintenance of equipment, and leak detection programmes. Release prevention for industrial sites begins in the facility design phase, with selection of appropriate equipment for containment and transport of products or wastes, and continues throughout facility operation, with routine monitoring, maintenance, and personnel training. To focus release prevention efforts, risk classification tools can be used to identify sites posing the greatest concern with regard to a possible release. Examples include i) sensitive receptor surveys conducted to identify nearby water supply wells, schools, or surface water bodies, and ii) equipment evaluations used to prioritize facility equipment upgrade efforts.

Emergency Response: Upon discovery of a product or waste leak or spill, appropriate regulatory agencies are notified and immediate actions are taken to repair the source of the release and abate any immediate threat to safety, health, or the environment (e.g., fire, explosion, etc.). Such emergency response measures may include site access control, containment diking, product removal, vapour suppression, protection of water resources, and/or contaminated soil and debris removal. The emergency response is complete once the release has been terminated and any associated *acute* hazards (i.e., immediate threats to safety, health, etc.) have been identified and controlled.

Remedial Action: If affected soil, groundwater, or other impacted environmental media remain in place upon completion of the emergency response, additional remedial action may be needed to address potential *chronic* hazards associated with long-term exposure to site constituents.

Chronic health or ecological impacts are often not immediately evident and may therefore require evaluation of long-term, future exposure patterns in order to establish appropriate site cleanup standards. Under a risk-based approach, response actions are prioritized at near-term, high-risk sites, and remedial measures are designed to minimize health/environmental risks by preventing exposure to harmful levels of site constituents.

The risk management process presented in this paper corresponds to the third step of this program, i.e., how to manage remediation sites where an environmental release of a product or waste material has occurred and emergency abatement actions have been completed.

1.2. SITE MANAGEMENT VERSES PROGRAM MANAGEMENT

The effective management of contaminated lands requires both i) the management of individual contaminated sites to prevent or mitigate impacts to human health and ecological resources (i.e., site management) and ii) prioritisation of response actions among contaminated sites based on the potential for adverse impacts at these sites (i.e., program management). Risk control for contaminated lands is achieved through coordination between these two levels of management where by the site managers compile key site information concerning the potential adverse impacts at the individual sites and the program managers compile these data in order to prioritise site remediation and allocate available resources to most effectively prevent and mitigate impacts (see Figure 1). This paper presents the key steps involved in both site management and program management.

2. Risk-Based Program Management

At the program level, risk management objectives for remediation sites include: i) advancing all program sites to the risk management standard, ii) identifying and prioritizing response actions for near-term impacts, and iii) maximizing overall program risk reduction by efficient application of available resources. As illustrated on Fig. 1, this program management process involves the following steps:

Data Compilation: Collect current information from all program sites regarding site classification, cost projection, and status of remediation/-closure process.

Risk Management Analysis: Evaluate program data to establish risk-based priority ranking for all program sites, measure overall risk reduction relative to the prior site classification survey, and identify principal risk drivers for site population.

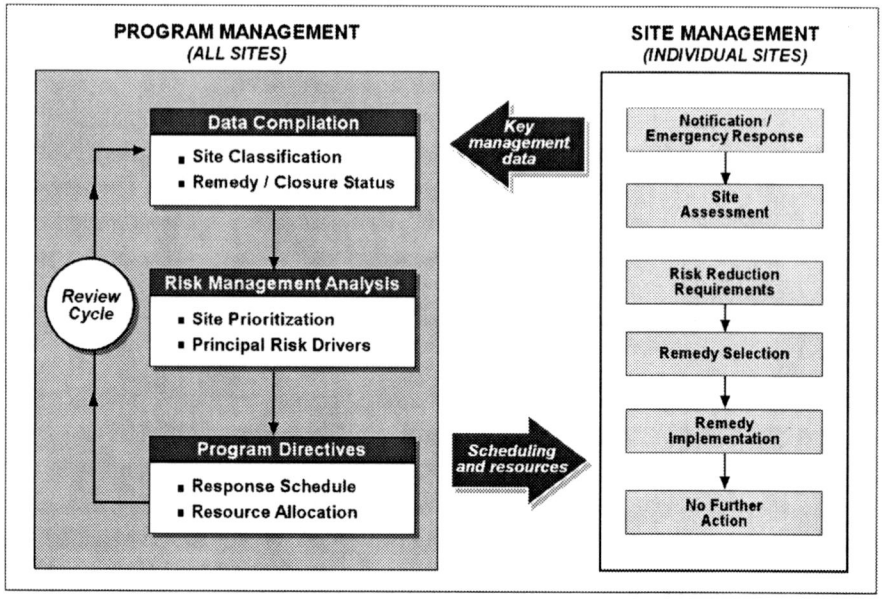

FIGURE 1. Risk Management Process Overview.

Program Directives: Expedite response actions for near-term impacts (e.g., current or imminent exposure above safe chronic level), and target available resources toward principal risk drivers so as to maximize risk reduction.

Review Cycle: Repeat measurement, data analysis, and adjustment cycle on routine basis to achieve standard of care for full site population.

General guidelines for each process step are discussed in further detail below.

2.1. PROGRAM DATA COMPILATION

To support effective program management, key information regarding individual site conditions and remediation activities should be surveyed for individual program sites on a periodic basis. These data can be provided by the site manager. However, quality control oversight will be required for critical data inputs (site classification, costs, etc.). Principal elements of the program data survey are as follows:

Site Classification: Sites can be characterized regarding the nature and immediacy of potential impacts using a simple site classification system based on the ASTM RBCA Site Classification system (ASTM, 1995):

- Class 1: Immediate impact or concern
- Class 2: Near-term impact or concern (within 2 years)
- Class 3: Potential future impact or concern (within 2 to 10 years)
- Class 4: No anticipated impact or concern

This rating system can expanded to address the full range of risk drivers critical to program management decisions (i.e., health/environmental factors and community/regulatory factors):

Primary Risk Factors:
- Groundwater (G)
- Soil (S)
- Air (A)
- Surface water (W)

Secondary Risk Factors:
- Regulatory factors (R)
- Land use (L)
- Off-site impacts (O)
- Community issues (C)

For use in site prioritization, a two-digit rating communicates the type and relative timing of response actions required for each site (e.g., Class 3/1 rating indicates health/environmental concern within a 2 to 10-year timeframe). The specific risk drivers responsible for the primary and secondary factor ratings can be identified by letter codes added to the classification, such as 3G/1L indicating that the site has a Class 3 health/environmental rating due to a groundwater concern (G) and a Class 1 community/regulatory concern due to land use issues (L; e.g., planned property re-development may lead to exposure).

Site Remediation Scope and Cost Projections: The selected remedial action must serve to address all primary and secondary risk drivers associated with the site. To characterize appropriate response actions and associated resource needs, site-specific measures (e.g., site investigation, remedy selection, implementation, etc.) must be defined to address all identified site risks. Based on this work scope, a life-cycle cash flow projection can be established for each site, addressing each phase of the site remediation effort through site closure. Periodic updates of the project scope may be needed as the site evaluation process provides more complete information regarding site conditions and appropriate remedial measures.

Remedy/Closure Status: Periodic site surveys should update information regarding the current project phase and anticipated timeframe for completion of the site remediation effort. Sites meeting the applicable risk objectives should be designated for *monitoring-only* or *no further action* status, subject to applicable regulatory requirements, pending formal agency approval of case closure. The project reports should serve to document that the risk management standard has been achieved. Approved case closures should be removed from the active program site portfolio.

2.2. RISK MANAGEMENT ANALYSES

Data collected from individual remediation projects can be analyzed to prioritize response actions, measure risk reduction progress, and identify the principal program risk drivers to be targeted in the coming year. Suggested procedures for these data analysis tasks are outlined below.

Site Prioritisation: In all cases, emergency conditions (e.g., risk of fire, explosion, or acute exposure) represent the highest priority and are to be handled on a site-specific basis in advance of chronic risk management concerns. Following abatement of emergency conditions, priority can then be given to response actions at sites posing a current or imminent threat of human or environmental exposure in excess of safe chronic (long-term) levels. For sites posing no immediate health or environmental concern, secondary factors, such as community or regulatory issues, may serve as the principal drivers in prioritizing response actions.

Risk Reduction Audit: Measurement of risk reduction progress is critical to assessing the value of on-going site remediation efforts and optimizing future resource allocations. For this purpose, *site classification profiles* can be used to track risk reduction over time for the full program site portfolio. Progress in achieving the risk reduction should be reflected by lower site classification ratings over time for a given site population (e.g., fewer Class 1 or 2 sites, increased Class 3 or 4 sites).

Principal Risk Drivers: For purpose of strategic planning, site classification data may also be analyzed to identify the risk drivers of greatest importance to the total program risk score. For example, if air impacts were identified as the largest contributor to the total risk score, resources could be specifically targeted toward vapour concerns to maximize risk reduction in the coming year.

2.3. PROGRAM DIRECTIVES

For each program review cycle, the scope and schedule of response actions proposed at individual sites should be adjusted to maximize the potential risk reduction to be achieved with available resources. Response actions are prioritized to address near-term human health or environmental concerns. Key elements of this program management plan include:

Priority Response: Remedial measures must be expedited for high-priority health/environmental concerns (i.e., Class 1 ratings). In some cases, interim stabilization actions (e.g., surface cover, vapour cut-off, etc.) may serve to control the potential exposure and reduce the risk rating, pending implementation of final remedial actions. For sites receiving a high-priority classification due to the absence of critical site data (e.g., no groundwater measurement), the response action may involve an expedited site assessment to confirm the presence or absence of an actual concern. In the absence of near-term health/environmental concerns, time-sensitive secondary risk drivers (i.e., Class 1 community concern) will serve as the principal factor in prioritization of response actions.

Targeted Risk Drivers: Based on the analysis of principal risk drivers, site assessment and remediation efforts can be focused on the principal risk driver(s) with the largest contribution to risk across the portfolio of sites. In this manner, limited resources can be leveraged to achieve maximum risk reduction. Again, in all cases, priority is given to control of unsafe exposures for human or environmental receptors.

2.4. REVIEW CYCLE

Periodic reviews should be conducted to monitor and adjust the program management strategy. For this purpose, site data must be updated and re-analyzed to establish appropriate response priorities. In general, such updates are less labour intensive than the initial baseline data surveys, given that site classification and cost data need only be revised for sites for which additional information is available. The review cycle provides useful information once site assessment or remediation efforts have progressed sufficiently to affect the site classification profile. Consequently, the appropriate review frequency may vary for different programs, depending on site activity levels and program planning needs.

3. Risk-Based Site Management

At the site management level, risks to human health and ecological resources are identified on a site-specific basis and appropriate protective measures are implemented in the timeframe necessary to prevent unsafe conditions. Key steps in the risk management effort for each contaminated site include:

Site Assessment: Characterize affected soil, groundwater, surface water, and sediment, identification of applicable receptors, and potential constituent transport mechanisms as needed to support risk-based site evaluation.

Conceptual Model/Pathway Screening: Define applicable risk concerns based on the evaluation of contaminated areas (i.e., sources), transport mechanism and locations where exposure to contamination may occur (i.e., receptors).

Risk Assessment: Determine the potential for adverse impacts to human health or ecological resources based on a site-specific evaluation of contaminant toxicity and the potential for exposure based on applicable transport mechanisms and receptors.

Remedy Selection: Select a site management strategy based on nature and immediacy of anticipated impacts, current and future land use, and site-specific soil, groundwater, and surface water conditions.

3.1. SITE ASSESSMENT

The site assessment must provide sufficient information to identify relevant health and ecological concerns and assess the need for remedial action. To support a risk-based site evaluation, these data must be sufficient to construct a site conceptual exposure model characterizing affected environmental media (i.e., soil, groundwater, surface water, and sediment), applicable exposure pathways and receptors, and potential contaminant transport mechanisms. General guidelines regarding key data requirements and data acquisition procedures are summarized below.

Table 1 itemizes the possible data requirements for risk-based evaluation of contaminated sites. For each individual site, actual data needs will depend on site conditions and applicable exposure pathways, as determined by an environmental professional. Relevant information can be collected in a phased manner as the site management effort proceeds. For example, an initial site assessment may target data requirements for a generic screening level evaluation, including release characterization, receptor information, applicable regulatory standards, and maximum constituent of concern (COC) concentrations in affected media (see Table 1). More complete characterization of the lateral and vertical extent of affected media zones, applicable exposure pathways, and fate-and-transport parameters can be completed as needed to address subsequent steps of the risk management process.

3.2. CONCEPTUAL MODEL/PATHWAY SCREENING

The conceptual model of the site describes the potential for contaminant transport from the affected media source zone to a point of contact with a human or ecological receptor via various *exposure pathways*. For most contaminated sites, the primary exposure pathways of human health concern are i) groundwater ingestion, ii) soil ingestion or dermal contact, iii) inhalation of vapours from contaminated soil or groundwater, or iv) surface water and sediment ingestion, dermal contact, or fish ingestion. Additional exposure pathways may apply based on site conditions and land use (e.g., ecological exposures). To pose a risk, three components of each exposure pathway must be present: an affected source medium (i.e., soil, groundwater, surface water, or sediment), a mechanism for contaminant transport, and a receptor. In practical terms, exposure pathways may therefore be screened from further consideration based on the presence and mobility of the contaminants and the proximity of receptors to the source zone. For example, if an affected groundwater plume is in a stable or diminishing condition, no potential exists for impacts on water supply wells located outside the current plume area, provided groundwater flow conditions remain the same.

To characterize the presence or absence of affected source media, constituent concentrations measured in affected soil, groundwater, surface water, and sediment can be compared to generic screening levels to determine if any contaminants are present at levels sufficient to pose a possible human health or ecological concern. Such screening levels may be specified under the regulatory program governing the contaminated site.

TABLE 1. Possible Site Assessment Data Requirements.

Required Information	Data Source/Methodology
Release Characterization	
Source/Nature of Release: Time, location, material.	*Evaluate historical data.*
Affected Environmental Media: Presence of soil, groundwater, or other environmental media impacted by release.	*Conduct site inspection or field/laboratory measurements.*
Constituents of Concern (COCs): Most prevalent, mobile, toxic constituents in source material.	*Evaluate chemical/toxicological properties of source material constituents.*
Receptor Information	
Land Use: On-site and adjacent development.	*Conduct site inspection, zoning review.*
Human Receptor Survey: Distance to downgradient supply wells, residential or recreational property, buildings, utilities, etc.	*Inspect land use in surrounding area, conduct water supply well survey.*
Ecological Receptor Survey: Distance to sensitive habitats or wetlands. Presence of endangered or threatened species.	*Conduct site inspection. Review published references.*
Surface Water and Groundwater Resources: Distance to streams, lakes, etc., and applicable use classifications. Depth to principal water supply aquifers; use classification of shallow water-bearing units.	*Review topographic map, government water resource references.*
Current Exposures: Presence/absence of actual exposure in excess of applicable exposure limits.	*Evaluate current access to source material or affected media.*
Applicable Regulatory Standards	
Affected Groundwater Zone: Groundwater resource classification and applicable protective limits.	Review applicable environmental regulations.
Affected Soils: Exposure pathways and associated screening limits designated under applicable rules.	Review applicable environmental regulations.
Affected Surface Water/Sediments: Applicable water quality standards for classified surface water bodies.	*Review applicable environmental regulations.*
Affected Media and Migration Pathways	
Affected Groundwater Zone: Lateral and vertical extent of affected groundwater zone. Representative COC concentrations.	*Conduct field/laboratory measurements to delineate groundwater plume. Identify presence/absence of free-phase materials.*

TABLE 1. Possible Site Assessment Data Requirements (Continued).

Required Information	Data Source/Methodology
Affected Soil Zone: Lateral and vertical extent of affected soils. Representative COC concentration.	*Conduct field/laboratory study to delineate affected soils.*
Migration Pathways: Groundwater flow patterns, buried utilities, drainage improvements, exposed affected surface soils.	*Identify based on site characterization and fate-and-transport data (see below).*
Potential Receptor Impacts: Current or potential exposure in excess of applicable exposure limits.	*Re-evaluate initial receptor inventory (see above) based on final delineation results, migration pathway.*
Physiographic Features: Topography, drainage features, flood plain boundary, etc.	*Conduct site inspection. Review topographic maps, floodplain maps.*
Hydrogeology: Site stratigraphy; ground-water flow gradient; hydraulic conductivity (K) and effective porosity (n_e) of water-bearing unit.	*Define stratigraphic profile based on soil borings. Measure flow gradient. Conduct field measurements of K or estimate based on soil type. Base n_e on soil type.*
Soil Properties: Effective porosity, volumetric air and water content, bulk density.	*Estimate based on soil type in transport pathway.*
Soil Leachate: Rainfall infiltration rate through affected surface soil zone.	*Develop site-specific value using i) model estimate or ii) field measurement.*
Plume Condition: Stable, diminishing or expanding plume.	*Evaluate historical monitoring data (e.g., 2-year minimum sampling record).*
Biodegradation Indicators: Either COC decay coefficient or aquifer electron acceptors (dissolved O_2, NO_3, SO_4, Fe, MN^{-2}, methane, CO_2).	*Estimate site-specific decay rate coefficient from historical data and/or measure electron-acceptors in site groundwater.*
Retardation Parameters: Soil fraction organic carbon (foc) content. Soil-water partitioning coefficient (k_d).	*For organic constituents, estimate $k_d = K_{oc} \times foc$. Measure foc on representative background soil samples or use conservative default value per soil type. For metals, run soil leaching test (SPLP) to estimate k_d.*

NOTES:
This table provides a general summary of the possible site investigation data needed to support a site-specific risk-based evaluation. Actual data needs will depend on site conditions and applicable exposure pathways. An example site data questionnaire, addressing all information that may be required for a site-specific risk evaluation, is provided in Appendix B of this guide. Assistance in defining and addressing data requirements can be provided by your Mobil EHS Remediation Services and Government Affairs contacts.

COC = Constituent of Concern. K_{oc} = organic carbon partition coefficient   SPLP = Synthetic Precipitation Leaching Procedure

Alternatively, conservative screening-level concentrations for each contaminant may be derived using the procedures documented by various organizations, including ASTM in the United States, CONCAWE in Europe, and the New Zealand Ministry of the Environment. In general, these generic screening levels are derived in the same manner as described below for Site-Specific Target Levels, based on the assumption that the point of exposure occurs directly in the source area. To use this approach, screening levels must be developed for each contaminant and each pathway of concern. Only those constituents found to be present in excess of generic screening levels and only those media containing these constituents require further assessment. Pathways or contaminants for which screening limits are unavailable must also be retained for further evaluation. If all contaminants are shown to be less than applicable screening limits for each pathway, the site evaluation is complete, and no further response action is required for the site.

The results of the pathway screening analysis are used to develop a conceptual model of the site identifying applicable source media, transport mechanisms, and receptors. The Conceptual Exposure Flowchart, shown on Figure 2, can be used to document site-specific exposure conditions in this manner. Upon completion of the pathway screening analysis, pathways determined to be potentially complete based on the various screening criteria, should be retained for site-specific evaluation. However, if the preliminary screening analysis shows no complete exposure pathways, no further action is required to achieve health/ecological protection.

3.3. RISK ASSESSMENT

To identify necessary response actions, each complete exposure pathway can be analyzed on a site-specific basis to evaluate the potential for exposure in excess of safe levels. For the purpose of this evaluation, Site-Specific Target Levels (SSTLs) can be derived for relevant exposure pathways, receptors, and constituents of concern and compared to measured source media concentrations. Source media exceeding these target levels will require remedial action in a timeframe necessary to control exposure.

Site-Specific Target Level: For a given exposure pathway and contaminant, the SSTL represents a concentration in the affected source medium (soil, groundwater, surface water, or sediment) that is protective of a human or ecological receptor located at a relevant point of exposure (POE). For example, for the human health soil-to-air exposure pathway, the SSTL is the mean concentration in the affected surface soil zone that will prevent unsafe human exposures via vapour or particulate release to air. For each complete exposure pathway and contaminant, SSTL values for the source medium can be back-calculated from safe exposure levels at the POE using the following general expression:

$$SSTL = RBEL \times NAF$$

where:
- RBEL = Risk-based exposure limit at POE for each COC
- NAF = Natural attenuation factor defining natural reduction inconstituent concentrations during transport from source to POE.

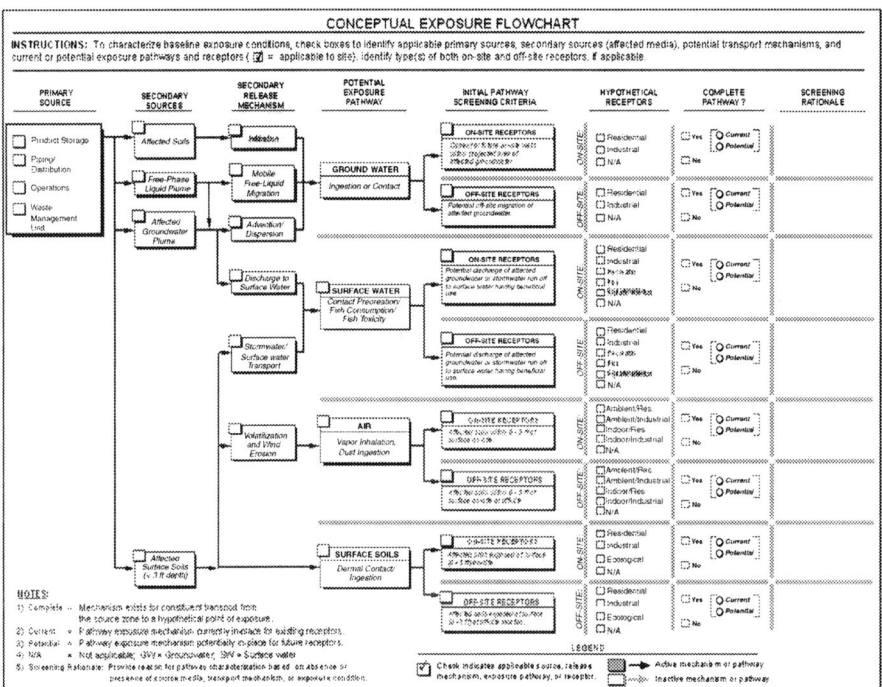

FIGURE 2. Conceptual Exposure Flowchart.

Determination of applicable RBEL and NAF values is addressed below.

Risk-Based Exposure Limits: The Risk-Based Exposure Limit (RBEL) represents the contaminant concentration that is safe for consumption by the receptor, based the toxicity of that contaminant. The RBEL applies at the point of exposure (POE), i.e., the likely point of constituent intake or contact by a human or ecological receptor. For each complete exposure pathway and contaminant, the applicable RBEL is matched to each relevant POE based on the type of exposure medium (air, water, soil) and the type of receptor (resident, commercial/industrial worker, etc.). For certain exposure media, human health-based exposure limits are specified under applicable regulations, such as, in the drinking water standards for drinking water ingestion. In the absence of such standards, human health RBELs can be derived for each constituent and exposure medium (air, water, soil) using the following general expression:

$$\text{RBEL} = \frac{TR \bullet TF}{E}$$

where
- E = effective exposure rate for specified pathway, based on applicable exposure factors (i.e., daily intake rate in mg/day - kg body weight);
- TR = target risk limit effects of individual contaminants (dimensionless);
- TF = toxicity factor for contaminant (mg/kg-day).

Applicable target risk limits (TR) for health protection can be matched to levels specified by the environmental regulatory authority. Toxicological parameters for each contaminant can be determined from published references, such as the U.S. EPA Integrated Risk Information System (IRIS). Exposure rates correspond to the chronic rate of contact or intake of the affected exposure medium (air, water, soil) by the receptor under anticipated land use conditions. As a conservative measure, these rates can be estimated based on standard exposure factors published by the regulatory authority or other source (e.g., American Industrial Health Council) for the anticipated land use at the site (e.g., residential, commercial, etc.).

Quantitative measures for derivation of RBELs for ecological receptors are not well defined. However, if the pathway screening evaluation indicates a reasonable potential for ecological exposure (e.g., surface water/aquatic species), applicable RBELs may be based on published standards or ecological screening criteria (e.g., government surface water quality standard for aquatic life protection, ecological screening limits for terrestrial species, etc.).

Natural Attenuation Factor: For each complete exposure pathway and each separate contaminant, the natural attenuation factor (NAF) represents the sum effect of various partitioning, dilution, and attenuation factors acting to reduce constituent concentrations during transport from source to receptor. These NAF components may involve both cross-media transfer factors (e.g., soil-to-air volatilization) and lateral transport factors (e.g., groundwater advection-dispersion and degradation). For exposure pathways with multiple POEs, separate NAF values must be derived for each POE location (e.g., groundwater ingestion at actual wells and hypothetical future wells as illustrated on Fig. 3).

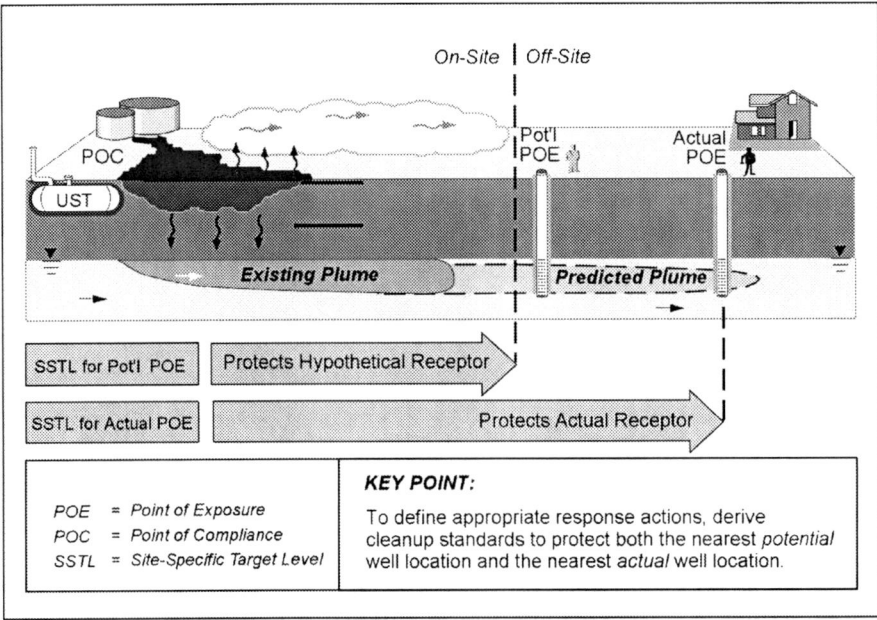

FIGURE 3. SSTL Analyses for Both Potential and Actual Receptors.

For each complete exposure pathway and each COC, the applicable NAF value can be derived based on either: i) the actual measured concentration ratio between the source medium and the POE or ii) fate-and-transport modelling analyses predicting this concentration ratio.

For purpose of simplicity and accuracy, direct field measurements represent the preferred method of NAF estimation, whenever feasible. However, due to temporal variability and sampling difficulties, some of these factors can prove difficult to quantify via direct field measurements (e.g., soil volatilization). In this case, modelling analyses, based on appropriate site-specific data and conservative assumptions, provide a convenient method of estimation. Groundwater dilution attenuation factors are amenable to direct measurement via wells spaced along the spine of the plume. In all cases, time-series groundwater monitoring data should be evaluated to determine if the affected groundwater plume is stable, shrinking, or expanding. Stable or diminishing plumes pose no risk to downgradient receptors located outside the plume area (i.e., NAF = infinite), and, in such case, the groundwater exposure pathway is incomplete and requires no NAF or SSTL calculation. Groundwater modelling analyses are necessary only for plumes for which available data are insufficient to establish the stability condition or indicate an expanding plume.

Various models are available to derive site-specific NAF values, ranging from simple analytical equations with limited data requirements to complex numerical models requiring three-dimensional data resolution. The appropriate modelling tool will depend on the adequacy of the available data, the relative complexity of the site, and the acceptable degree of conservatism.

3.4. REMEDY SELECTION

Remedy selection involves three principal steps: i) development of an overall *exposure control strategy* (i.e., what to do), ii) selection of optimal *remediation technologies* to implement this strategy (i.e., how to do it), and iii) development of appropriate design and construction specifications. The general decision process for remedy selection is described in further detail below.

Exposure Control Strategy: The goal of the risk-based site management effort is to minimize health/ecological risks by preventing exposure to harmful levels of site contaminants. As indicated on Figure 4, risk reduction can be achieved by addressing any component of the exposure pathway: i) removing or treating the source, ii) interrupting contaminant transport mechanisms, or iii) controlling activities at the point of exposure. The remedial action plan may consist of one or more exposure control strategies, including:

Removal/Treatment Action: Removal or treatment of affected source media (i.e., affected soils, groundwater, etc.) to reduce COC concentrations to levels less than or equal to applicable SSTLs (e.g., using technologies such as excavation, soil venting, pump-and-treat, etc.).

Containment: Long-term engineering controls to prevent migration of harmful concentrations of COCs from the source to the POE (e.g., using technologies such as surface cover/capping, barrier walls, soil stabilization, hydraulic containment, etc.).

Natural Attenuation Monitoring: Periodic sampling and analysis to confirm stabilization or reduction of affected media concentrations via natural attenuation processes.

Institutional Controls: Legal or administrative measures to control the nature and frequency of human exposure at the POE (e.g., deed notice, alternative water supply, etc.).

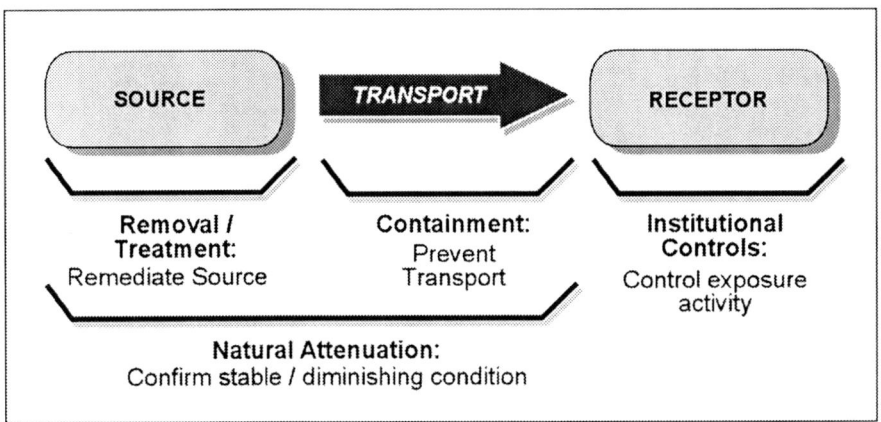

FIGURE 4. Potential Site-Specific Risk Management Strategies.

The exposure control strategy for a given site should be designed to address each of the health/environmental risk reduction requirements identified in the risk-based evaluation. The appropriate exposure control strategy for a given site will depend on the nature of the health/environmental risk reduction requirements.

Remediation Technology: Under many regulatory programs, a remedial action plan must be submitted specifying not only the selected exposure control strategy, but the specific remediation technologies to be used to implement that strategy. Following selection of an appropriate exposure control strategy, further evaluation may be required to define the remediation technology best suited to site conditions. At a given site, the optimal remediation technology will depend on the technical applicability and cost efficiency of each alternative, the degree of constituent reduction or control to be achieved for health/environmental protection, and the secondary risk drivers associated with the site. In some cases, consideration of secondary risk drivers may support selection of technologies that achieve site remediation goals in a shorter timeframe or to a greater degree than required for health/environmental protection alone.

Design Documentation and Review: For the selected remedy, engineering design specifications should be documented in sufficient detail to support safe and economical installation and operation. For remedies involving electrical or mechanical operating equipment or chemical treatment, a safety review must be conducted per applicable regulatory standards. Under some regulatory programs, review and approval of the preliminary design package by the regulatory authority is required prior to implementation.

References

American Society for Testing and Materials, 1995, "Standard Guide for Risk-Based Corrective Action Applied at Petroleum Release Sites," ASTM E-1739, Philadelphia, PA.

Connor, J.A., R.L. Bowers, and T.E. McHugh, "RBCA Toolkit: Comprehensive Risk-Based Modelling System for Soil and Groundwater Cleanup," in Linders, J.B.H.J. (ed.), Modelling of Environmental Chemical Exposure and Risk, Kluwer Academic Publishers, The Netherlands, 2001.

Ferguson, C., Darmendrail, D., Freir, K., Jensen, B.K., Jansen, J., Kasamas, H., Urzelai, A. and Vegter, J. (editors) 1998 *Risk Assessment for Contaminated Sites in Europe.* Volume 1. *Scientific Basis.* LQM Press, Nottingham.

SAFEFOODNET: AN ONGOING PROJECT ON CHEMICAL FOOD SAFETY: NETWORK FOR ENLARGING EUROPE

MARIANNA ELIA*, ANGELO MORETTO AND SARA VISENTIN
Department of Environmental Medicine and Public Health
University of Padua
2 Via Giustiniani,35128 Padova,Italy

*To whom correspondence should be addressed: marianna.elia@tin.it

Abstract. Since the credibility of the information on food safety is very important, the methods how to disseminate verified data on chemical food contaminants to whom it may concern is the SAFEFFOODNET project's principal objective. The Safefoodnet research project connects independent research institutions, universities, government agencies and ministries from across the EU. The objective is to harmonize and integrate the infrastructure and activities of the NMS and ACC in the field of chemical food safety with those of Old Member States and to provide the recently established European Food Safety Authority with an expert network in the field of chemical food safety. The structure of the general work plan has been broken down into 5 work packages (WP) to respond in a practical way to the SAFEFOODNET proposed objectives.

Keywords: contaminant; diet; food; safety; network; analytical

1. Introduction

To the surprise of many people living in modern, technologically advanced societies, food has come to be a problem. Many experts would argue against this notion, but public opinion surveys and other sociological researches clearly demonstrate that people are increasingly confused when trying to organise their food consumption behaviour, because they no longer trust the labels on food products. And what makes the problem worse is that even experts often fail to consent on the notion of safe food. In short, people in contemporary societies often face difficulties in finding reliable and transparent answers about food safety. It is therefore a quite urgent research task to find out what information on safe food is available, how and at what level the information is collected, how reliable the information is, who is responsible for gathering and storing the information and, last but not least, who should provide this information to the public and how.

It is increasingly obvious that large section of the public no longer perceive the existing system of guaranteeing safe food as consistent. The system's credibility is dropping despite the claims by experts that much of the suspicion is not rationally based. The question on how to restore the credibility of the food safety process is therefore of key significance to all the actors, from "farm to fork", and since food production is essential to the quality of life, it is a central issue for societies in general.

The SAFEFOODNET research project connects independent research institutions, universities, government agencies, and ministries from across the EU. The objective is to provide tools to help harmonise and integrate the infrastructures and activities of the New Member Countries and Associated Candidate Countries in the field of chemical food safety with those of Old Member States and to provide the recently established European Food Safety Authority with a network of experts in the field of chemical food safety.

The first step to achieve this objective is to have a picture of the systems of the chemical food safety in the participating countries: Bulgaria, Cyprus, Czech Republic, Estonia, Hungary, Latvia, Lithuania, Malta, Poland, Rumania, Slovakia, Slovenia and Turkey. On the basis of this information, recommendations how to improve chemical food safety management in the participating countries will be prepared. This will be achieved by the activities planned in two project working groups: one aimed at gathering knowledge on sources of dietary data in individual countries and another focussing on strategies/approaches and capabilities of monitoring food chemical contamination. Such information will be organized in "country profiles" where strengths, weaknesses and training needs will be described and discussed, and further steps forward identified. In addition, a network of scientists, researchers, and institutions able to address the several different aspects related the chemical contaminants in food will be available to all stakeholders.

Since the credibility of the information on food safety is very important, the methods how to disseminate verified data on chemical food contaminants to whom it may concern is the project's principal objective. The communication strategy should consider the existing formal institutional framework of chemical food safety systems, but also many other socio-cultural characteristics of the food production in individual countries.

One of the main objectives is to establish a transparent information system, which will enable everybody who is concerned about chemical food safety to obtain reliable information and data on the regulation and control of the food in their daily diet. This means that expert knowledge and expertise should reach the general public, i.e. all interested consumers. To these ends, besides the project web-site (www.safefoodnet.net), several local, regional and European workshops will be convened. These will be organized in 2006 in different European locations with the participation of local authorities and relevant stakeholders in order to discuss the results of the project, investigate areas of cooperation, and propose steps forward a better assessment and management of food safety.

2. Participants in the Project and General Overview

SAFEFOODNET consortium has 20 partners from 17 countries, which include 4 MS (Italy, Denmark, Germany and Belgium), 10 NMS (Hungary, Czech Republic, Slovakia, Poland, Latvia, Lithuania, Estonia, Slovenia, Malta, Cyprus), and 3 ACC (Bulgaria, Romania, Turkey).

Objectives of the project are to gather information that will help to harmonize and integrate Associated Candidate Countries (ACC) and New Member States (NMS) infrastructures and activities in the field of chemical food safety with those of Member States (MS) and to provide the European Food Safety Authority (EFSA) with a network of scientists, researchers, institutions able to address the different aspects of the chemical contaminants in food. In addition, through participation in this project, scientists and research groups will have the opportunity to join mainstream research activities, such as the ones developed in "Food Chemical Safety in Europe" (FOSIE), the NoE "Chemicals as contaminants in the food chain: an NoE for research, risk assessment and education." (CASCADE), the project "Harmonized Environmental Indicators for Pesticide Risk" (HAIR), or to promote new projects in the field of food safety.

Initially, ACC and NMS profiles will be developed with respect to: the definition of analytical capabilities in terms of both infrastructure and knowledge; the identification of the possibilities of dietary assessment for the establishment of a "standard diet" (according for instance to the indications of GEMS/FOOD of WHO) and consumption patterns.

On the basis of country profiles, strengths, weaknesses and training needs will be identified. Consequently, workshops will be organized with local authorities and relevant stakeholders to discuss results of the project, and investigate areas of cooperation.

TABLE 1. List of participants of the project.

Partic. Role	Partic. No.	Participant name	Participant short name	Country
CO	1	International Centre for Pesticides and Health Risk Prevention	ICPS	Italy
CR	2	Dept. of Environmental Medicine and Public Health University of Padova	UNIPD	Italy
CR	3	Istituto Mario Negri	IRFMN	Italy
CR	4	Danish Institute for Food and Veterinary Research	DFVF	Denmark
CR	5	Institute and Outpatient Clinic of Occupational, Social and Environmental Medicine	UNE NRNB	Germany
CR	6	National Centre of Hygiene, Medical Ecology and Nutrition	NCHMEN	Bulgaria
CR	7	Godollo Agribusiness Centre	GAC PUC	Hungary
CR	8	Institute of Agricultural and Food Information Slezska	IAFI	Czech Republic
CR	9	Iuliu Moldovan Institute of Public Health, Cluj-Napoca	IPHCN	Romania
CR	10	Food Research Institute Bratislava	FRI	Slovakia
CR	11	National Food and Nutrition Institute	IZZ	Poland
CR	12	Lavtia University of Agriculture	LUA	Latvia
CR	13	Ministry of Agriculture and Rural Affairs, Ankara Province Control Laboratory Directorate	APCL	Turkey
CR	14	Estonian Environmental research Centre	EERC	Estonia
CR	15	Research and Consultancy Institute Ltd	RCI	Cyprus
CR	16	Agricultural Research Institute	ARI	Cyprus
CR	17	Kaunas University of Technology	KTU	Lithuania
CR	18	University of Ljubljana, Faculty of Social Science	FSS	Slovenia
CR	19	Food Safety Commission	MOH	Malta
CR	20	International Life Science Institute	ILSI Europe	Belgium

*CO = Coordinator
CR = Contractor

A web site will be developed with the information on contaminants in food in the different countries, methods to analyze them, regulations. The network, the project reports and the website will serve EFSA and other interested parties implementing chemical food safety strategies (e.g.: WHO, especially in the GEMS/FOOD program, and, possibly, the International Program on Chemical Safety (IPCS) initiatives related to chemical food safety).

At the end of the project, ACC and NMS will be aware of their capabilities and needs, will have their own network of parties involved (or to be involved, if or when needed) with chemical food safety issues/problems. In addition, within the project, National Co-ordinators will be identified for a proper and efficient co-operation with EFSA, with the aim of a better integration and harmonisation of activities relating to chemical food safety in the new enlarged 25-member European Union. This network of National Co-ordinators and experts will be available on a web-site and will be instrumental in helping interested parties in joining mainstream research Networks of Excellence and other research projects.

3. Work Planning

SAFEFOODNET will last two years. This time-window is necessary to assemble thorough country profiles and to strengthen project's network and dissemination activities. In fact, beside the establishment of the NMS and ACC network of experts, SAFEFOODNET aims at developing a set of long-term activities, to be continued after the end of the project and hopefully expanded to other countries. On the other hand, a longer duration would probably undermine the prompt alignment of the project's results to EFSA and Community needs.

Based on the above-mentioned considerations, it is evident that the final goal of the project will be the harmonization of the approach to food safety among the enlarging European Union. A second significant result will be the creation of a European food Safety network, which will continue its activities after the end of the two-year period of the project. The network will work in close relation with the European Food safety Authority (EFSA) and will disseminate information through the website which will be created and implemented in the frame of SAFEFOODNET activities.

The project can be divided into two phases: in the first phase, selected information will be collected to define the state of the art in NMS and ACC either on availability and reliability of existing sources on food consumption patterns or on strategies for monitoring food contaminants, laboratory infrastructure, analytical procedures and available data.

Data collected will be evaluated and elaborated; in this phase, SAFEFOODNET will highlight strengths and needs in NMS and ACC in order to reach common standard with EU countries and plan future activities.

The second phase of the project will be mainly addressed at dissemination and training. To this aim, a website will be created either to exchange information and documentation among partners, or to support dissemination activities. SAFEFOODNET activities will culminate in the last six months of the project, when the project's results will be disseminated, through the organization of seminars, to national authorities and relevant stakeholders, either at the local or at the international level.

To reach relatively homogeneous groups of countries, SAFEFOODNET seminars will be addressed to four well-defined areas, that is Baltic Countries, Central European Countries, Southern European Countries and islands (Malta and Cyprus).

The structure of the general work plan has been broken down into **5 work packages (WP)** to respond in a practical way to the SAFEFOODNET proposed objectives:

WP1: Management
WP2: Diet
WP3: Chemical residues
WP4: Web-site
WP5: Dissemination and exploitation

The network includes 4 MS, 10 NMS and 3 ACC. Activities carried out in 17 countries need a timely and firm coordination. Therefore, the management of SAFEFOODNET is assigned to a specific work package **(WP1)**. WP1 addresses the co-ordination of administrative and scientific areas of the project. It will assure a well-timed delivery of internal and official documents. WP1 has the task to create and maintain the administrative and technical archive of accounts, contractual record, and progress reports, intermediate and final deliverables. Moreover, the management will include: organisation of project meetings; communication with European Commission, circulation of documents to partners.

WP2 and **WP3** concern the collection of information on available data on diets and analytical methods for detection of food chemical contaminants.

Experts in WP2 and in WP3 prepared standardized forms (questionnaires) for the collection of relevant information at national level on available diets and residues from local and international monitoring programs. Such questionnaires are be encompleted on-line by the co-ordinator or by the institution addressed by the co-ordinator before the end of the first reporting period. Instructions on how to complete the questionnaire are on the web-site (see below) and have been communicated to the Commission Project Officer.

Data – collection will be organized by the area coordinators and results discussed within experts of WP2 and WP3. The latter will provide a homogeneous description of available data together with the judgment about the necessity of a dedicated database and/or the possibility to append available data to existing databases, if not included yet.

Results concerning national diets may influence adequate sampling strategies in planning national monitoring program. The Scientific Manager and members of the Scientific Management team will assure adequate integration between WP2 and WP3 at every stage of the project.

WP2 and WP3 will also describe current Community and National *acquis* and/or current European and International scientific activities in the area. Actually, information on chemical residues in food involves several aspects, such as pre-analytical and analytical procedures, laboratory accreditation, etc.

Based on this information country profiles will be developed and will include suggestions for NMS and ACC alignment to Community *acquis* (body of common rights and obligations which bind all the Member States together within the European Union) and country-specific needs in food chemical safety.

A specific work package (**WP4**) has been assigned to the creation of the web-site because the web-site is a key system to work and communicate with partners and a valid instrument for a broad and timely passive dissemination of activities. Therefore, WP4 expertise will be required to address technicalities related to scientific issues, including judgement about data structure and database organisation. Moreover, experts in WP4 advised on the best way for circulating and using forms for data collection. WP4 maintains the electronic NMS and ACC network, and support dissemination outside the project. The web-site offers a restricted area for project's participants for internal communication and circulation of documents.

WP5 participants will focus on dissemination and exploitation activities of the project. For the sake of practicality, the activities in this work package will be mainly conducted on a regional basis: four areas have been identified (Table 2), having each a co-ordinator, who manages its national co-ordinators. WP5 will focus more on issues related to dissemination and exploitation strategies within and outside the consortium, rather than to scientific topics as such. WP5 will investigate the possibility to join existing mainstream activities especially in NMS and ACC and Europe with a view to implementing dissemination and exploitation activities. Since the start of the project, WP5 will work in close co-operation with area and national co-ordinators to discuss local activities. Dissemination of results will include: publications, participation in international conferences; organisation of workshops.

WP4 and **WP5** will consolidate the capability of this network to join the international community of scientists with particular regard to the mainstream ongoing European projects. NMS and ACC profiles, dissemination of the project's results and the established NMS and ACC network will serve EFSA and the European Commission to plan strategies for improving dietary risk analysis and management in the enlarging Europe. Moreover, the project's outcomes will help relevant national authorities to plan, follow-up and revise national monitoring programmes in the field of food chemical safety.

TABLE 2. Regional and final workshops.

	Month N° from the start of the project	N° Days	Country hosting the event	n. Participants
Workshop in Baltic countries	18	2	Latvia	18 (*)
Workshop in Central Countries	19	2	Slovak Republic	20 (*)
Workshop in Southern countries	20	2	Bulgaria	17 (*)
Workshop in Islands	21	2	Malta	12 (*)
Final Workshop	24	1	Brussels	27

(*) Plus local participants.

4. Potential Impact of SAFEFOODNET Activities on Food Safety Issues in EU

The primary goal of SAFEFOODNET is to promote ACC and NMS capabilities of addressing all aspects relating to chemical food safety, and facilitate the interaction of interested local parties with mainstream research activities, with EFSA and other international and supranational bodies or activities.

4.1. STRATEGIC IMPACT

The impact of this project will be at both National and European level. At the National level, National co-ordinators and networks of scientists/experts in chemical food safety will be identified (WP2 and WP3). Individual countries will be characterised in terms of existing resources, infrastructures, organisational features, data sources, points of strength and weakness, with a view of contributing to improved understanding of gaps and needs, and to highlight actions needed to harmonise processes in ACC and NMS with those of Member States. At the European level, the results of SAFEFOODNET will mainly contribute to a better integration of ACC and NMS institutions involved with food safety with ongoing activities and bodies of the European Union, and to the development of durable links to join mainstream European research in related scientific fields. This project will ultimately lead to societal benefits in terms of consumers' health.

Project components' chart

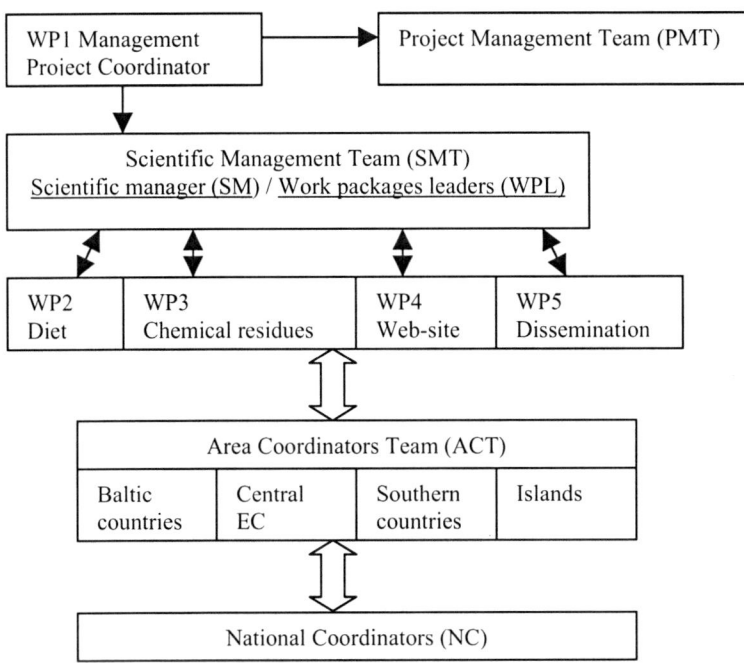

FIGURE 1. Graphical presentation of work packages.

4.2. INTEGRATION TOWARDS THE ENLARGING EU

As the European Union enlarges, it is essential that the ACC and NMS have implemented food safety control systems equivalent to those in place within the Community. This represents a significant challenge to those countries, in terms of:

- adopting the necessary legislation and guidance, and harmonising it with the current EU *acquis,*
- establishing relevant institutions to implement and enforce this legislation,

- upgrading their chemical food safety information and monitoring systems,
- developing a network of scientists, researchers, institutions that will be able to address the several aspects of food chemical contamination and to build the necessary connections with EFSA.

SAFEFOODNET will contribute to address this challenge, mainly by facilitating networking (at both country, regional, and European level) of regulatory bodies, institutions and experts involved with chemical food safety. By the end of the project, the establishment of these networks will ensure durable connections, beyond participation in the support action. SAFEFOODNET will represent a step toward a full integration and cooperation with Member States, a fostered coordination of ACC and NMS policies by devising and implementing country networks, joint measures and systems for exchanging information and experience at European level.

4.3. PREPAREDNESS TO FOOD SAFETY EMERGENCIES

Another relevant point is the ability to react to chemical food safety emergencies (e.g.: outbreaks of chemical food contamination). EFSA has to play a key role in supporting a rapid and effective EU response. In order to do this, EFSA should be able to collect, analyze and distribute relevant information to Member States and to the Commission for an improved planning and handling of crisis situations at the European level. It is clearly evident that such a task can only be accomplished if a proper system is set up in all Member States to collect, analyze and communicate relevant information, as well as to act in response to the indication provided by EFSA. These indications might include rapid responses as well as follow-ups, monitoring and epidemiological surveillance on a local or more general basis. Only if all involved bodies are prepared to the task, will networking be effective. One of the outcomes of SAFEFOODNET will be the identification of problems and needs of ACC and NMS in this respect. Identification of needs is the first step towards the development of a strategy to address specific problems in individual countries.

DEGRADATION OF *N*-NITROSODIMETHYLAMINE (NDMA) IN LANDSCAPE SOILS

M. ARIENZO[1]*, JAY GAN[2], FREDERICK ERNST[2], STEPHEN QIN[2], SVETLANA BONDARENKO[2] AND DAVID SEDLAK[3]

[1]*Dipartimento di Scienze del Suolo, della Pianta e dell'Ambiente*
100 Via Università, 80055 Portici, Italy
[2]*Department of Environmental Sciences, University of California Riverside, CA 9252, USA.*
[3]*Department of Civil and Environmental Engineering, University of California Berkeley, CA 94720, USA*

*To whom correspondence should be addressed: michele.arienzo@unina.it

Abstract. *N*-Nitrosodimethylamine (NDMA) is a potent carcinogen that is often present in municipal wastewater effluents. In a previous field study, it was observed that NDMA did not leach through turfgrass soils following four months of intensive irrigation with NDMA-containing wastewater effluent. To better understand the loss pathways for NDMA in landscape irrigation systems, a mass balance approach was employed using *in situ* lysimeters treated with ^{14}C-NDMA. Negligible leaching was observed in the field study, and suggests volatilization as the only significant loss pathway. Therefore, although NDMA may be present at relatively high levels in treated wastewater, gaseous diffusion and volatilization in unsaturated soils may effectively impede significant leaching of NDMA, minimizing the potential for groundwater contamination from irrigation with treated wastewater.

Keywords: degradation; landscape soils; wastewater; contamination

1. Introduction

The increasing use of treated wastewater to irrigate golf courses and landscaped lands may pose a threat to groundwater quality due to the potential for organic pollutants such as *N*-nitrosodimethylamine (NDMA) to leach into the groundwater (WateReuse Foundation, 2005). NDMA is soluble in water, does not adsorb to soil, and has a moderate to long persistence in soils, suggesting a high potential to leach through a soil profile to reach groundwater (Yang et al., 2005). Under laboratory conditions, NDMA was found to leach through saturated soil columns at nearly the same rate as the inert tracer chloride (Dean-Raymond and Alexander, 1976). NDMA was largely absent in the leachate from mature turfgrass plots that were heavily irrigated with NDMA-containing wastewater effluent (Gan et al., 2005). Incubation experiments conducted with soils from the field site further showed that NDMA was relatively persistent in the turfgrass soils, suggesting that mechanisms other than degradation or adsorption were responsible for the rapid NDMA removal and low leaching in the turfgrass plots.

The purpose of this study was to evaluate the importance of plant uptake and volatilization in NDMA dissipation from turfgrass soils. The second objective of this study was therefore to determine if NDMA persistence changes after soil is irrigated with wastewater for extended periods.

2. Materials and Methods

2.1. *IN SITU* LYSIMETER EXPERIMENT

An *in situ* lysimeter experiment was carried out to understand the loss mechanisms of NDMA from turfgrass soils. About 10 months prior to the *in situ* lysimeters experiment, galvanized aluminum tubing (5 cm i.d. and 30 cm long) was inserted into six turfgrass plots (three plots for each soil type) that were planted with a hybrid Bermuda grass (*Cynodon dactylon X transvaalensis*) and allowed to acclimatize under field conditions. These plots were irrigated with plain water prior to the lysimeter experiment. For treatment, ^{14}C-NDMA was dissolved in water and 50 mL of the solution (166 Bq mL^{-1}, or 200 µg L^{-1} in total NDMA concentration) was uniformly applied to the top of each lysimeter using a glass pipette. The plots with the treated lysimeters were then subjected to standard irrigation practices at three irrigation events per week with plain water. From the time ^{14}C-NDMA was applied until the end of the experiment (14 d or 336 h), the total amount of irrigation water applied was 48 mm, which was 103% of the total ETo (46.4mm) for the site. Triplicate lysimeters from each soil type were removed at 0, 4, 8, 12, 28, 72, 168 and 336 h after the treatment. For extraction, the frozen soil column was cut into 0-5, 5-10, 10-20 and 20-30 cm increments. The whole plant sample or 40 g of wet soil was individually placed in Soxhlet extraction cups, spiked with 0.1 ml of 10 mg L^{-1} d6-NDMA, and extracted with 300 mL dichloromethane for 8 h. The solvent extract was passed through a layer of anhydrous sodium sulfate and condensed to 10 mL on a vacuumed rotary evaporator. Duplicates of

0.5-mL aliquot of the extract were mixed in 6 mL of scintillation cocktail and measured for ^{14}C. Following solvent extraction, ^{14}C associated with the extracted sample matrix was also determined by combusting 50 mg of air-dried plant tissues or 1.0 g of air-dried soil on a Biological Oxidizer. The evolved $^{14}CO_2$ was trapped in a basic solution and the activity was measured by LSC. The fraction of ^{14}C measured in the methylene chloride extract was defined as extractable ^{14}C, while the ^{14}C measured after the extraction was considered non-extractable ^{14}C. The sum of these two fractions was further defined as total ^{14}C. It must be noted that as NDMA was not distinguished from its potential metabolites, the measured ^{14}C activity could overestimate the actual NDMA concentration.

2.2. VOLATILIZATION EXPERIMENTS UNDER CONTROLLED CONDITIONS

Volatilization of NDMA from water and soil was further evaluated in the laboratory using controlled conditions. In the first experiment, ^{14}C-NDMA was spiked into 200 mL of $0.01 M$ $CaCl_2$ solution in wide-mouth glass containers (12.5 cm i.d.) at an initial NDMA concentration of 72.5 µg L^{-1}. Three replicates were prepared, and the containers were placed on magnetic plates (600 rpm) in a fume hood at the ambient temperature (20 ± 1 °C). At 0, 0.5, 1, 2, 4, 8, 24, 48, and 72 h after the treatment, each container was individually weighed to determine water loss from evaporation, and aliquots of 1.0 mL were removed from each container for measurement of ^{14}C radioactivity by LSC. In the second experiment, a sandy loam soil with 0.91% organic carbon content was sieved through 1 mm and air dried to about 6% water content. Two hundred grams of soil (dry weight) was placed in open mouth glass containers (12.5 cm i.d.), and were uniformly treated with 12 mL of an aqueous solution of ^{14}C-NDMA. Three replicates were prepared, and the initial NDMA concentration in the soil was about 3.8 µg kg^{-1}. The treated soil containers were equilibrated in a fume hood at ambient temperature. At 0, 0.5, 1, 2, 4, 8, 24, 48, and 72 h after the treatment, 1.0 g soil was removed and immediately combusted on the biological oxidizer and the released ^{14}C activity was determined by LSC. Losses of NDMA were calculated as % of the initial applied amount and used for analysis of NDMA volatilization kinetics.

2.3. NDMA DEGRADATION IN TURFGRASS SOILS

The long term application of NDMA-containing wastewater effluents may have selectively influenced soil microbial populations capable of degrading NDMA. Immediately following the previous wastewater irrigation study (Gan et al., 2005), soil samples were taken with an auger from turfgrass plots that were previously irrigated with wastewater for approximately four months. The soil cores were divided into 0-10, 10-25, and 25-50 cm layers. The soil samples were passed through a 2-mm sieve after slight air drying. The initial water content of the soil samples was adjusted to 10% (w/w) by adding deionized water. Ten grams (dry weight equivalent) of soil was weighed into 125-ml glass serum bottles and spiked with 0.5 ml of 2 mg L^{-1} NDMA aqueous solution to give an initial NDMA concentration of 100 µg kg^{-1}. The sample bottles were then incubated in the dark at room temperature (20 ± 1°C) and triplicate samples were removed 0, 3, 7, 14, 21, 28 and 56 d after the treatment for analysis of NDMA. For extraction, each soil sample was spiked with 0.1 mL of d6-NDMA in dichloromethane (10 mg L^{-1}) and then extracted with 50 mL dichloromethane by shaking at high speed for 4 h. The solvent extract was filtered through a funnel containing 20 g anhydrous sodium sulfate, and the dried extract was concentrated to a final volume of about 1 mL under a stream of dry nitrogen. An aliquot of the final extract was analyzed by GC-MS. The recovery of NDMA was determined to be 38.5 ± 5.1% for the turfgrass soil. However, as d6-NDMA was used as a surrogate, the results were not corrected for recovery. Detailed analytical conditions used for quantification of NDMA in the final extract were the same as in the proceeding paper (Gan et al., 2005). The detection limit of NDMA using the above protocol was 0.20 µg kg^{-1}.

3. Results and Discussion

3.1. DISSIPATION OF ^{14}C-NDMA IN *IN SITU* LYSIMETERS

The DT$_{50}$ for extractable ^{14}C was estimated to be 2.2 h for the sandy loam soil ($R^2 = 0.99$; P < 0.0001), and slightly longer at 2.7 h for the loamy sand soil ($R^2 = 0.95$; P = 0.0013). The respective DT$_{95}$ values were 9.6 h and 11.8 h for the two soil types. After 72 h, the total ^{14}C in the lysimeters remained at 1.5-3% until the end of the experiment. However, the extractable fraction of ^{14}C in the lysimeters decreased to less than 0.3% of the applied mass at ≥ 72 h after the treatment. This suggests that the non-extractable fraction accounted for the majority of the total ^{14}C activity in the lysimeters after the initial rapid dissipation of NDMA. During the first 24 h after the treatment, the extractable fraction accounted for 72-94% of the total ^{14}C in the sandy loam lysimeters, and 90-98% in the loamy sand lysimeters. However, from 72 to 336 h, the ratio of the non-extractable ^{14}C contributed for 84-95% of the total ^{14}C in the sandy loam lysimeters, and 95-99% in the loamy sand lysimeters. Therefore, even though trace levels of ^{14}C were present in the lysimeters after 72 h, the residue was mostly in the non-extractable form and would have little potential for leaching.

3.2. DISTRIBUTION OF ^{14}C-NDMA IN *IN-SITU* LYSIMETERS

The vertical distribution patterns of total and extractable 14C at different times after the treatment in sandy loam lysimeters is shown in Fig. 1. The initial distribution showed that ^{14}C-NDMA percolated into the 10-20 cm layer, and for the loamy sand soil, a small amount (3.4%) even reached the 20-30 cm segment. In the lysimeters, most of the activity initially resided in the 0-10 cm section, and about 10.2% and 5.1% were found in the 10-20 cm layer for the sandy loam soil and the loamy sand soil, respectively. However, both total and extractable ^{14}C dissipated quickly from all depths for both soil types. At 4 h after the treatment, less than 0.3% of the original activity was detected in the soil below 20 cm. It is also evident that the amount of non-extractable ^{14}C was small compared to the total ^{14}C in the soil during the first 28 h. However, at ≥ 72 h after the treatment, the trace amounts of ^{14}C detected in the soil were predominantly in the form of non-extractable ^{14}C, possibly indicating partial degradation and reduced leaching risk for the small amounts of remaining residue. Overall, uptake of ^{14}C-NDMA by turfgrass appeared to be relatively insignificant, and the absorbed ^{14}C was rapidly converted into the non-extractable form. The significant NDMA volatilization loss was likely a result of both rapid NDMA volatilization at the soil-air interface, and active upward transport of NDMA due to efficient gaseous phase diffusion and negative water potential gradients expected under dry, warm conditions in the soil. Volatilization loss of NDMA from water at 20 °C was rapid under the relatively static conditions, with a first-order half-life of about 25 h. From the measured volatilization kinetics, Henry's Law constant K_H was estimated to be 1.45×10^{-3} using a chemical property estimation method (Lyman et al., 1990). This value was in close agreement with that (1.4×10^{-3} at 25 °C) reported by the International Programme on Chemical Safety (IPCS, 2002). It is known that for organic compounds with $K_H > 10^{-4}$, chemical movement in unsaturated soils is dominated by gaseous phase diffusion and can thus be highly efficient. Rapid volatilization loss of NDMA was further observed to occur from a layer of moist soil under similar laboratory conditions, with a first-order half-life of only 10.3 h.

3.3. EFFECT OF WASTEWATER IRRIGATION ON NDMA PERSISTENCE

In a separate laboratory incubation experiment, the potential effect of long-term wastewater irrigation on NDMA persistence was evaluated using soil samples taken from the turfgrass plots previously irrigated with NDMA-containing wastewater. Irrigation with treated wastewater for four months did not result in significantly enhanced NDMA degradation in the soil when compared to soils that did not receive wastewater irrigation.

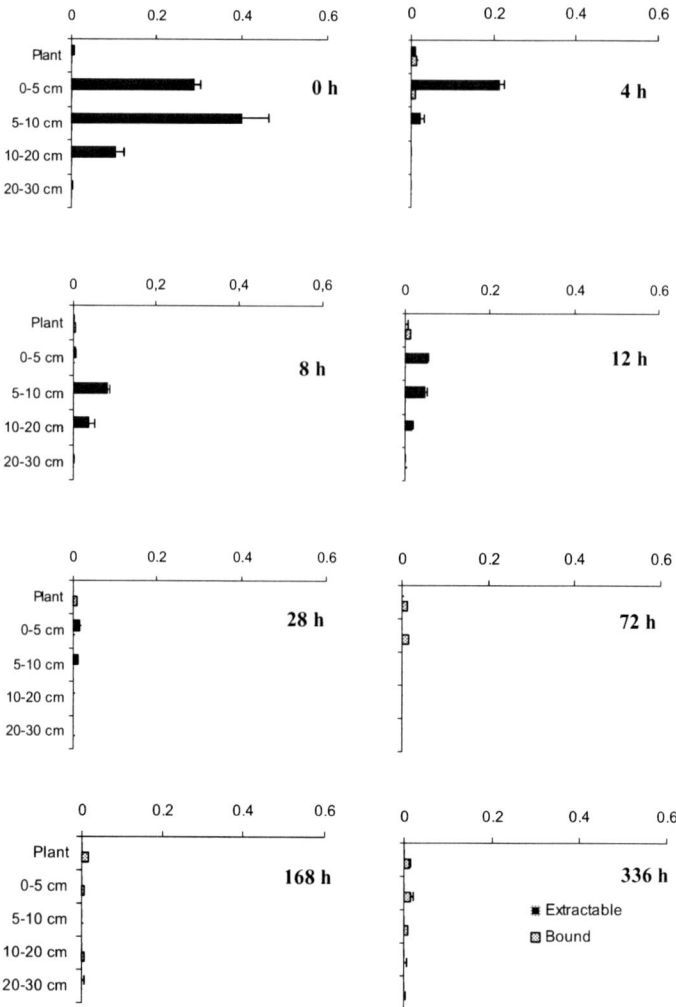

FIGURE 1. Vertical distribution of total and extractable ^{14}C activities at different times in ^{14}C-NDMA treated *in situ* sandy loam lysimeters in turfgrass plots under field conditions. Bars are means of three replicates and lines are standard errors.

In summary, results from this study support the hypothesis that the limited leaching of NDMA observed in the companion field study was mainly attributable to volatilization.

References

Dean-Raymond, D., and M. Alexander. 1976. Plant uptake and leaching of dimethylnitrosamine. Nature 262:394–396.

Gan, J., S. Bondarenko, F. Ernst, W. Yang, S. Bries, and D.L. Sedlak. 2005. Leaching of N-nitrosodimethylamine (NDMA) in turfgrass soils during wastewater irrigation. J. Environ. Qual. (submitted).

ICPS. 2002. N-nitrosodimethylamine. Concise International Chemical Assessment Document 38. Available at http://www.inchem.org/documents/cicads/cicad38.htm

Lyman, W.J., W.F. Reehl, and D.H. Rosenblatt. 1990. Handbook of Chemical Property Estimation Methods: Environmental Behavior of Organic Compounds. American Chemical Society, Washington, D.C. pp. 15:1–36.

WateReuse Foundation. 2005. Final Report on Investigation of N-Nitrosodimethylamine (NDMA) Fate and Transport. WateReuse Foundation, Alexandria, VA 22314.

Yang, W.C., J. Gan, W.P. Liu, and R. Green. 2005. Degradation of N-Nitrosodimethylamine (NDMA) in landscape soils. J. Environ. Qual. 34:336–341.

POLLUTANTS EFFECTS ON HUMAN BODY – TOXICOLOGICAL APPROACH

GHEORGHE COMAN[1], CAMELIA DRAGHICI[1],
ELISABETA CHIRILA[2] AND MIHAELA SICA[1]
[1]*Transilvania University of Brasov, 29 Eroilor Bdv.
500036 – Brasov, Romania*
[2]*Ovidius University Constanta, 124 Mamaia Bdv.
900527 – Constanta, Romania*

*To whom correspondence should be addressed: cchirila@univ-ovidius.ro

Abstract. It is a fact that any compound coming into contact with biological systems will cause some perturbations in that system. These biological responses may not be always toxicologically relevant, and from that point of view it is important to correlate information on the sources, interactions, dose and their toxicity. The interest in toxicology increased continuously in the last years with the synthesis and use of different chemicals. In the environment, humans and other leaving organisms are exposed to support the increasing number of xenobiotics: industrial chemicals, pesticides, drugs, and additives from food or beverage. The effects of some most frequently present and relevant pollutants on human body, presented as typical diseases registered at organisms' level are presented in this paper work.

Keywords: pollutants; toxicology; biotransformation; public health; diseases

1. Introduction

Toxicology is concerned with the study of the noxious effects of chemical substances in living systems. It is a multidisciplinary subject and covers areas of chemistry, biochemistry, biology, toxicology, pharmacology, physiology, pathology, and ecology.

The 'keys' of toxicology are based on all the above-mentioned domains that try to focus on understanding how the toxic chemicals interacts with environment, and what should we do in order to protect harmful effects. For example, chemistry, biochemistry, and biology relieves which types of chemicals that are present in the environment, their quantity and distribution in environmental matrices, such as water, air, soil, and biota. Toxicology, together with physiology, pharmacology, pathology, elucidates the dose response relationship between chemicals present in the environment and organisms with which they interact. The legislative aspects of toxicology try to ensure the adequate protection by law the entire environments and humans. The study of ecology demonstrates how populations, communities and ecosystems respond to toxic contaminants.

We can freely say that anything can be toxic if the dose is high enough. Low toxicant concentrations may produce no observable effects, but as the concentration increases up to a critical level, increasing adverse effects can be observed and finally the death may occur. When two or more pollutants are acting together the synergic effect of pollutants may occur.

Exposure is a concentration of material in the air, water or soil to which an animal is exposed and is different from the dose received by an organism. Dosage is a term that includes the dose, frequency and duration of dosing.

Considering the exposure time to toxicants, there are two distinguished types of toxicity (Vander et al., 1998):

1. **acute toxicity** – observed soon after short-time exposure to the chemical;
2. **chronic toxicity** – resulted after long-term and/or repeated periods of exposure to lower doses of the chemical.

The **dose – response relationship** is given by the response (effect) quantified in a reproducible way, relevant to the toxic processes under investigation. In case of experiments on animal, doses are often expressed in:

- weight;
- molecular units per kilogram of body weight;
- molecular units per square meter of body surface area.

When we establish the relationship between dose and response, it is important to distinguish between the concentration of a chemical in the environment and the actual amount that reaches the target.

It is very important to examine chemicals as contaminants, considering their types, sources, properties, distribution and transformation in environment. Different chemicals generate different effects that depend on many factors like dose, species, health, age, nutrition and sex of the person or animal exposed to the specific exposure conditions, knowing that organisms are reacting different. Toxicants cause different biological responses, at different molecular, biochemical, physiological and behavioural levels. Finally, a response of the whole ecosystem is registered that can be illustrated as cascading from the source through individual organisms to ecosystem, like Figure 1 shows (Trebse, 2004).

If we want to express the toxic effect clearly, a compound has to come into contact with the biological system under consideration, but it also has to penetrate the organism in order to have a systemic effect.

The contact of a toxic compound in the biological system (binding, storing and interacting) may determine and follow four interrelated phases:

- absorption of the contaminant;
- distribution into the organism;
- metabolism;
- excretion of the toxic compounds or metabolites.

There are three main **absorption** routes to enter the body: inhalation into the lungs, ingestion into the gastrointestinal tract, or penetration across the skin, or through placenta in the case of a fetus (Vander et al., 1998). A chemical can also enter the body by direct injection into the bloodstream or under the skin, but these are not ordinary means of entry.

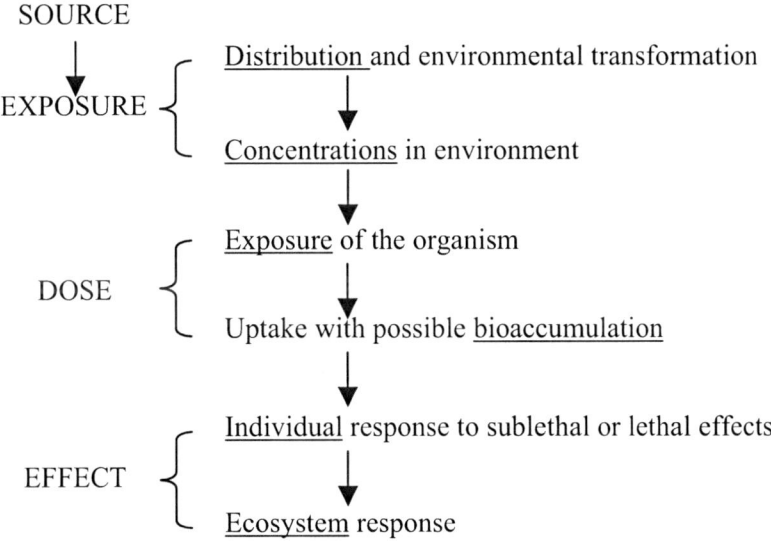

FIGURE 1. The effect of chemicals on the ecosystem.

It is possible that a chemical is toxic only by one route of entry. Such example is formaldehyde, which is a carcinogen only if inhaled. On the contrary, arsenic is toxic by all three routes: skin absorption, ingestion and inhalation. In the context of absorption it is important to mention another term – **bioavailability**. Whenever the chemical enters the body it has to be bioavailable in order to be absorbed into the bloodstream. Chemicals quantity and their absorption velocity are determined by their physical and chemical properties.

Xenobiotics enter in the bloodstream, following one of the described absorption routes, are **distributed** into the body and undertaken by different organs. It is possible that a part of the xenobiotics distributed into the body may be stored in tissues or even in the blood. A word commonly used to refer to storage of a pollutant at higher levels than those found in the environment is **bioaccumulation**, for example, PAHs, PCBs, dioxins and some organometallic forms of metals bioaccumulate in fat, fluoride and lead in bones etc. Sometimes **biomagnification** takes place, as a bioaccumulation process within the trophic chain, in indirect relationship with biota.

Usually, a specific chemical has different effects on different organs. To express a toxic effect, a chemical has to reach a sensitive organ at a high enough dose. As dose increases or time of exposure increases, additional organs may be affected. When the exposure to a chemical is reduced or eliminated, the amount of stored chemicals into the body decreases. It may be released slowly under ordinary conditions.

Metabolism, sometimes also called **biotransformation**, deals with the conversion of absorbed chemical into other substances, process done by living organisms. Biotransformation usually converts xenobiotics into less toxic chemicals, but it is also possible that the resulted substances are more toxic.

Excretion of the toxic compounds from the organism occurs via three main ways:

3. volatiles are partially eliminated in the exhaled breath;
4. some xenobiotics are water-soluble, while others become soluble during the biotransformation processes; all these water-soluble chemicals arrive in urine and than excreted from the body;
5. chemicals that cannot be transformed into water-soluble ones are excreted with faeces; in the case of the nursing mother, water-insoluble chemicals are also excreted in the milk.

2. Biotransformation Processes of Pollutants in the Organisms

The pollutants metabolites resulting from the biotransformation could enter the blood and be eliminated from the body through urine, skin secretion, faeces or breath (Fig. 2). Hydrophilic and lypophilic compounds behave different in the excretion mode, due to the **hydrophile-hydrophobe balance,** knowing that water-soluble compounds have lower toxicity. Hydrophilic pollutants and the water-soluble metabolites are easy excreted by excretory fluids (detoxication) while lipid-soluble pollutants and their water-insoluble metabolites usually accumulate in depots or undergo enzyme-mediated biotransformation, to be finally excreted by faeces. Lipophylic pollutants present in excretory fluids, tend to diffuse into cellular membranes, and are reabsorbed (readily absorbed but poorly excreted).

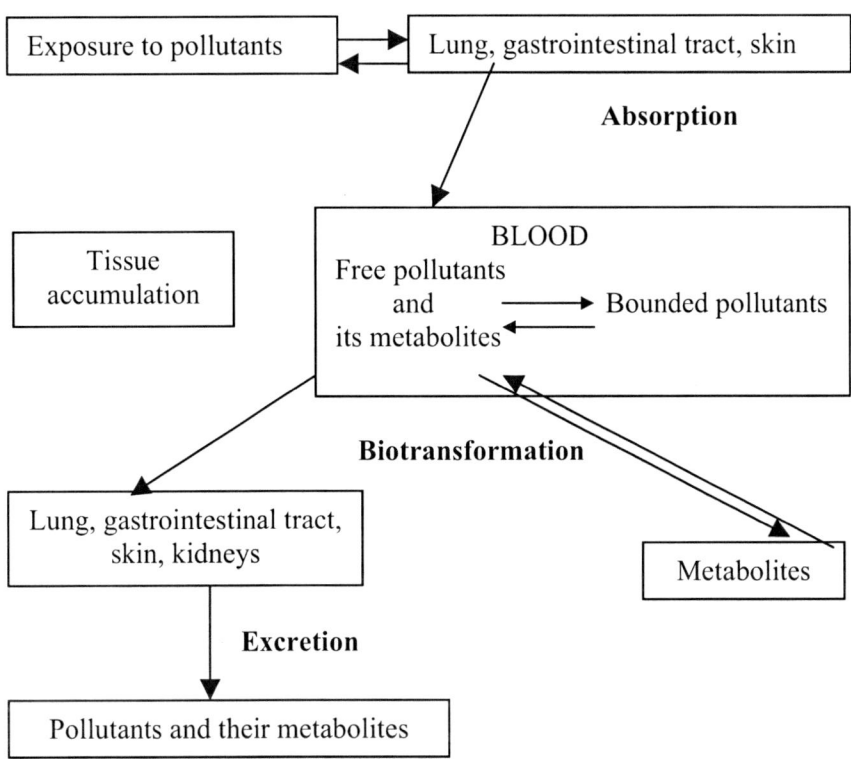

FIGURE 2. Absorption, metabolic alteration, and excretion of the pollutants.

The biotransformation is the sum of the processes by which a pollutant is subject to chemical change in living organisms. During biotransformation a large number of enzymes and pathways are involved and the common activity of most of them is that they transform lypophilic pollutants into more polar metabolites. Biotransformation (metabolic alteration) of pollutant molecules occurs mainly in the liver, but to some extent also in skin, kidney, placenta, plasma, intestine or brain (Nove et al., 2000). Extrahepatic tissues have a limited role in biotransformation of pollutants and their contribution is medium (lung, kidney, intestine) or low (skin, testis, placenta, adrenals).

The primary function of the liver is to receive and process chemicals absorbed from the gastrointestinal tract before they are stored, secreted into bile or released into the general circulation. The microsomal enzyme system (hepatic enzymes) is located mainly in the smooth endoplasmatic reticulum and performs this biotransformation (Olinescu and Greabu, 1990). The most important enzymes involved in biotransformation are the microsomal oxidases, especially cytochrome P-450 monooxygenases. The pollutants act as a substrate for the microsomal enzyme system and the most important feature is that it is easy to induce the biosynthesis of the enzyme (the number of these enzymes can be increased by exposure to pollutants).

Biotransformation is not strictly related to detoxication, because in a number of cases the metabolites are more toxic than the parent pollutants, and in that case, the term of **bioactivation** or toxication is used. Metabolites may have comparable or greater toxicity in organism than the parent pollutant since during the biotransformation process functional groups are inserted into the pollutant by oxidation, reduction or hydrolysis reactions.

The transformation of the water-insoluble pollutant to a water-soluble compound occurs during two steps:

- **biotransformation** – the addition of the reactive functional groups (hydroxyl, amine, carboxylic) to the pollutant, in order to increase the chemical reactivity of the pollutant,
- **conjunction** – some reactive pollutants react with highly polar or ionic species, in order to form a product with higher water solubility therefore can be easily excreted.

The most important mammalian conjugation molecules for pollutants and their metabolites are uronic acids, glutathione, amino acids, taurine, and sulphate. The conjugation is catalyzed by liver enzymes like UDP-glucuronyl-transferases, sulfotransferases, glutathione S-transferases and amino acid N-acyltransferases.

In order to arrive at urine, the pollutant or its metabolites should be either filtered through the renal corpuscle or secreted across the tubular epithelium. If the pollutants could be transformed into more polar molecules, their passive reabsoption from the tubule would be retarded and they would be excreted more readily from urine (Vander, 1981).

TABLE 1. Diseases induced by some inorganic and organic pollutants in human organism.

Pollutants	Nervous system	Lung[2]	Kidney[3]	Liver[4]	Cardiovascular system
As	- cerebral lesions	- pulmonary cancer; chronic bronchitis	-hemoglobinuria; hematuria; kidney cancer	-hepatoportal sclerosis; liver cancer	- anemia; leukopenia; tachycardia
Be, Al, Ni		- pulmonary cancer			
Cd			- nephrolithiasis		-anemia; cardiomyopathy
Co					
Cr		- pulmonary cancer; asthma		-hepatic necrosis	
Hg	- demyelination; cortical and cerebellar atrophy		-acute necrosis; nephritic syndrome; glomerulonephritis		
Mg	- hallucinations				
Pb	- encephalopathy; cerebral edema; demyelination; mental disturbance		-intravascular hemolysis; hyperruricidemia		-anemia; vertigo; tachycardia; hemolitic anemia
Tl	- cerebral edema				
Se				-hepatocellular carcinomas	
White P				-hepatomegaly	

TABLE 1. (Continued).

Si, SiO$_2$		- silicosis			
Inorganic gases		-pulmonary edema			
CO	- demyelination; memory disturbance				- polycithemia; anemia
CS$_2$	- aterosckle-rotic vasculoence phalophaty; optic neuritis		-nephrosclerosis; proteinuria; hematuria		-atherosclerotic vasculo-encelopathy; retinal microaneurism; myocardial infaction
CN					-tachycardia; hypertension
Aromatic hydrocarbons	- encephalopathy; peripheral neuropathy; ataxia; aplastic anemia			-hepatomegaly	--myocarditis; cardiomyopathy; leukemia; methemoglobinemia
Toluidine			-hematuria; cystitis; malignant tumors		
Halogenated derivatives	- myelopathies; depressions; ocular lesions	- asthma; chronic bronchitis	-necrosis; immune reactions	-hepatomegaly; fibrosis; cirrhosis	-thrombocytopenia; methemoglobinemia
Methanol					-metabolic acidosis
Phenol		- pulmonary edema	-nephritis; albuminuria; kidney edema		- methemoglobinemia
Ethylene glycol		-inflammatory reaction; interstitial fibrosis	-interstitial fibrosis; necrosis		-bradicardia

TABLE 1. (Continued).

Nitro, amino derivatives		-pulmonary edema	-urinary carcinogenic effect		- methemoglobinemia; haemolitic anemia
Asbestos		-asbestosis; pulmonary cancer			
Pesticides: organophosphate carbamate, organochlorine	- demyelination; paralysis	-asthma	-hematuria; proteinuria		-leukemia; anemia; multiple myeloma

[1] Neagoe et al., 2004a
[2] Dima, Neagoe, 2004
[3] Neagoe et al., 2004b
[4] Dima et al., 2004
[5] Neagoe et al., 2004c

3. Pollutant Effects on Human Body

All living organisms are dynamic systems, functioning as a result of interdependent (bio)chemical reactions that are permanently maintained in an equilibrium state. The presence of xenobiotic substances in a living system can easily disrupt this balance (inhibiting or activating) by interacting with different components. The biological effects of the pollutants and their metabolites are governed by several factors (Coman, 2004):

- penetration into the organism and their traslocation to the action sites;
- binding, storing and interacting with some biological receptors;
- resisting to the action of degradative enzymes.

An important role in the pollutant penetration is played by the external barriers (bacterial membranes, insect cuticle, and mammalian skin) which have a primary protective function for all living organisms (Koren, 1991). For mammalians, in addition to these external barriers, there are also internal barriers: cell membranes, gastro-intestinal epithelium, and plasma-cerebrospinal fluid. The physical-chemical properties of the pollutants are responsible on the penetration, binding, storage, and interaction with the biological receptors, leading to pathological changes.

The degree of health status depends on the quantity of the pollutant and on the length of exposure time (the dose of the pollutant). When increasing the chance of producing injuries of illnesses, the relationship between two or more pollutants acting together is called synergic effect.

The main diseases induced by some inorganic (heavy metals, oxides and sulphides) and organic (aromatic hydrocarbons, halogenated, hydroxylic and amino derivatives, pesticides) pollutants at different organs level were selected and are presented in Table 1, with no emphasis on acute or chronic effects.

The determinative factor in **acute health effects** is the dose of the pollutant. Most of the acute effects are temporary, but may also cause coma and death. Relatively low concentrations exposures of the human body to pollutants, longer and/or repeated periods may cause **chronic health effects.** The period of time between the first exposure and the development of the disorder (leukemia, cancer, cirrhosis, scarring of the lung or kidneys) is called **latency period**. The same pollutant (i.e. benzene) may cause acute effects (encephalophaty – central nervous system disorders) and/or chronic effects (leukemia – a type of cancer affecting the bone marrow and the blood cells) (Reese, 2003).

Table 1. shows that there are pollutants affecting some or more organs:

- Mg, Tl (nervous system), Be, Al, Ni, Si, SiO_2, inorganic gases (lung), toluidine (kidney), Se, white P (liver), Co, CN^-, methanol (cardiovascular system);
- Cd (kidney, cardiovascular system), Cr (lung, liver), Hg (nervous system, kidney), CO (nervous and cardiovascular systems);
- As, halogenate derivatives and different pesticides – nervous system, lung, kidney, liver and cardiovascular system.

On the other side, there are organs more or less affected by the pollutants presence. For example, cardiovascular system and kidney are considerable affected by most of the discussed pollutants.

Some diseases are induced by different pollutants, such as:

- methemoglobinemia – aromatic hydrocarbons, phenols, halogenated, nitro and amino derivatives;
- cancers, tumors and leukemia – As, Cr, toluidine, aromatic hydrocarbons, pesticies;
- encephalophaty – Pb, aromatic hydrocarbons;
- asthma – Cr, halogenated derivatives, pesticides;
- anemia – As, Cd, CO, nitro and amino derivatives, pesticides.

4. Conclusions

The human body may be exposed to small or substantial quantities of pollutants, sporadically or continuously. Within the human body the pollutants are transported and diffused, diluted-concentrated, or transformed through different complex ways, by physical, chemical, or biochemical processes.

The human body is exposed to an enormous number of non-nutrients chemicals in the environment, many of which could be toxic. All foreign chemicals can easily enter the body because they are present in the air, water or food we use. The presence of some pollutants in the human body may alter cell function and may induce various desieses.

ACKNOWLEDGMENTS

The present paper was prepared with the support of the Socrates/Erasmus program, Curricula Development projects "Bioanalytical Methods – Linking the Environmental Protection and Public Health" (51388-IC-1-2001-1-RO-ERASMUS-PROG-3 and 51388-IC-2-2002-1-RO-ERASMUS-MODUC-3) and of NATO-ASI project "Chemicals as Intentional and Accidental Global Environmental Threats" (CBP.EAP.ASI. 981563).

References

Coman, Gh., 2004, Pollutants and Their Metabolites Transport, Distribution, and Action in Human Body, in Coman, Gh., and Draghici, C., (Eds), *Pollutants and Their Impact on Human Body*, Transilvania University Press, Brasov, 54.

Dima, L. and Neagoe, C., 2004, Induced Lung Diseases and Asthma, in Coman, Gh., and Draghici, C., (Eds), *Pollutants and Their Impact on Human Body*, Transilvania University Press, Brasov, 79–81.

Dima, L., Neagoe, C. and Coman, Gh., 2004, Liver Diseases, in Coman, Gh., and Draghici, C., (Eds), *Pollutants and Their Impact on Human Body*, Transilvania University Press, Brasov, 93–96.

Koren, H., 1991, *Environmental Helth and Safety, I*, Lewis Publishers, London.

Neagoe, C., Dima, L. and Tiut, M., 2004a, Neurological Diseases, in Coman, Gh., and Draghici, C., (Eds), *Pollutants and Their Impact on Human Body*, Transilvania University Press, Brasov, 86–90.

Neagoe, C., Dima, L. and Coman, Gh., 2004b, Kidney and Urinary Tract Diseases, in Coman, Gh., and Draghici, C., (Eds), *Pollutants and Their Impact on Human Body*, Transilvania University Press, Brasov, 81–86.

Neagoe, C., Dima, L. and Tiut, M., 2004c, Cardiovascular and Hematologyc Diseases, in Coman, Gh., and Draghici, C., (Eds), *Pollutants and Their Impact on Human Body*, Transilvania University Press, Brasov, 90–93.

Nove, L.H., Capel, P.D., and Dileanis, P.D., 2000, *Pesticides in Stream Sediment and Aquatic Biota*, R.J. Gillon, London.

Olinescu, R., and Greabu, M., 1990, *Mecanisme de apărare a organismului împotriva poluării chimice* (Romanian), (*Defence Mechanisms of Organism against the Chemical Pollution*), Ed. Tehnica, Bucharest.

Reese, C.D., 2003, *Occupational Health and Safety Management*, Lewis Publishers, London.

Trebse, P., 2004, Toxicological Aspects of Pollutants, in Coman, Gh., and Draghici, C., (Eds), *Pollutants and Their Impact on Human Body*, Transilvania University Press, Brasov, 22–34.

Vander, A.J., 1981, *Nutrition, Stress and Toxic Chemicals*, University of Michigan Press, Michigan.

Vander, A., Sherman, J., and Luciano, D., 1998, *Human Physiology*, WCB McGraw-Hill, Boston.

THE USE OF MODELS FOR THE EVALUATION OF CHEMICAL ATTENUATION IN THE ENVIRONMENT

CHARLES NEWELL AND THOMAS MCHUGH*
Groundwater Services, Inc.
2211 Norfolk, Suite 1000, Houston, Texas, 77098, USA

*To whom correspondence should be addressed. tcmchugh@gsi-net.com

Abstract. Natural attenuation (the natural breakdown of chemicals in the environment) has been demonstrated to be an effective tool for the clean-up of many contaminated sites. This report reviews several free software tools that have been developed to assist in the site-specific determination of whether natural attenuation will be effective for the management of contaminated sites.

Keywords: contaminated land; natural attenuation; model

1. Introduction

Monitored natural attenuation projects address two key questions: How far will a groundwater plume travel? How long will one have to wait until the groundwater is remediated? Several free software tools have been develop that can help site managers address these key questions. These tools include:

BIOSCREEN: A natural attenuation model for hydrocarbon plumes;

BIOCHLOR: A natural attenuation model for chlorinated solvent plumes;

SourceDK: A tool for estimating the lifetime of groundwater source zones;

MAROS: A trend-estimating groundwater database system; and

BIOPLUME IV: A new natural attenuation model for evaluating emerging degradation pathways.

2. Discussion

2.1. BIOSCREEN

BIOSCREEN (Newell et al., 1996) is an easy-to-use screening model that simulates remediation through natural attenuation (RNA) of dissolved hydrocarbons at petroleum fuel release sites. The software, programmed in a spreadsheet environment and based on the Domenico analytical solute transport model, has the ability to simulate advection, dispersion, adsorption, and aerobic decay as well as anaerobic reactions that have been shown to be the dominant biodegradation processes at many petroleum release sites.

BIOSCREEN includes three different model types:
1) Solute transport without decay,
2) Solute transport with biodegradation modelled as a first-order decay process (simple, lumped-parameter approach),
3) Solute transport with biodegradation modelled as an "instantaneous" biodegradation reaction (approach used by BIOPLUME models).

The model is designed to simulate biodegradation by both aerobic and anaerobic reactions.

2.2. BIOCHLOR

BIOCHLOR (Aziz et al., 2000) is a screening model that simulates remediation by natural attenuation of dissolved solvents in groundwater. The software, programmed in the Microsoft® Excel spreadsheet environment and based on the Domenico analytical solute transport model, has the ability to simulate 1-D advection, 3-D dispersion, linear adsorption, and biotransformation via reductive dechlorination (the dominant biotransformation process at most chlorinated solvent sites). Dissolved solvent degradation is assumed to follow a sequential first order decay process.

BIOCHLOR includes three different model types:
1. Solute transport without decay;
2. Solute transport with biotransformation modelled as a sequential first-order decay process;
3. Solute transport with biotransformation modelled as a sequential first-order decay process with 2 different reaction zones (i.e., each zone has a different set of rate coefficient values).

Both BIOSCREEN and BIOCHLOR have simple first-order decay functions to simulate the effects of source decay on downgradient plumes.

2.3. MAROS

The MAROS (Monitoring and Remediation Optimization System) (Aziz et al., 2000) Software is a Microsoft® Access database application developed to assist users with groundwater data trend analysis and long term monitoring optimization at contaminated groundwater sites. This program was developed in accordance with the Long-Term Monitoring Optimization Guide Version 1.1 developed by AFCEE. The Monitoring and Remediation Optimization System (MAROS) methodology provides an optimal monitoring network solution, given the parameters within a complicated groundwater system, which will increase its effectiveness. By applying statistical techniques to existing historical and current site analytical data, as well as considering hydrogeologic factors and the location of potential receptors, the software suggests an optimal plan along with an analysis of individual monitoring wells for the current monitoring system.

The software uses both statistical plume analyses (parametric and nonparametric trend analysis) developed by Groundwater Services, Inc., as well as allowing users to enter External Plume Information (empirical or modelling results) for the site. These analyses allow recommendations as to future sampling frequency, location and density in order to optimize the current site monitoring network while maintaining adequate delineation of the plume as well as knowledge of the plume state over time in order to meet future compliance monitoring goals for their specific site. The User's Guide will walk the user through several typical uses of the software as well as provide screen-by-screen detailed instructions.

2.4. SOURCE DK

SourceDK (Farhat et al., 2004) is a planning-level screening model for estimating groundwater remediation timeframes and the uncertainties associated with the estimated timeframe. While SourceDK is primarily geared for natural attenuation processes, it can also be used to estimate source lifetimes for some flushing-based technologies, primarily groundwater pump-and-treat.

The software, programmed in the Microsoft® Excel spreadsheet environment, gives the user different approaches or Tiers. From easiest to most complex, the three tiers are:

Tier 1 – Extrapolation: Source zones that have extended records of concentration vs. time can be analyzed using the Tier 1 extrapolation tool. With this tool, log concentration vs. time is plotted and then extrapolated to estimate how long it will take to achieve a cleanup goal, assuming the current trend continues. This tool also provides the 90% and 95% confidence level in this estimate of the time to achieve the cleanup goal.

Tier 2 – Box Model: In this tier, the simple box model developed for the BIOSCREEN model (Newell et al., 1996) has been enhanced to include source mass estimation software and other features. The box model estimates source attenuation from a source mass estimate, the mass flux of constituents leaving the source zone, and biodegradation processes in the source zone. The uncertainty in the source lifetime estimate is also provided.

Tier 3 – Process Models: This tier employs more detailed fundamental process-based equations to determine the time and amount of naturally flowing groundwater required to flush out dissolved-phase and NAPL dominated constituents from the source zone.

2.5. BIOPLUME IV

BIOPLUME IV will give users the ability to simulate both biological and non-biological degradation pathways that have recently been identified involving utilization of biologically available iron as an electron acceptor or involving chemical reactions with iron species such as green rust, magnetite, and iron sulfides. BIOPLUME IV is currently under development.

References

Aziz, C.E., C.J. Newell, J.R. Gonzales, P.E. Haas, T.P. Clement, and Y. Sun, 2000. *BIOCHLOR Natural Attenuation Decision Support System, User's Manual Version 1.0*, U.S. EPA, Office of Research and Development, EPA/600/R-00/008, Washington D.C., January, 2000, http://www.epa.gov/ada/csmos/models.html

Aziz, J.J., C.J. Newell, H.S. Rifai, M. Ling, and J.R. Gonzales, 2000. *Monitoring and Remediation Optimization System (MAROS) Software User's Guide*, October 16, 2000. www.gsi-net.com

Farhat, S.K., P.C. de Blanc, C.J. Newell, J.R. Gonzales, and J. Perez, 2004. *SourceDK Remediation Timeframe Decision Support System*, Groundwater Services, Inc., Houston, Texas. www.gsi-net.com

Newell, C.J., J. Gonzales, and R. McLeod, 1996. *BIOSCREEN Natural Attenuation Decision Support System*, U.S. Environmental Protection Agency. EPA/600/R-96/087, August, 1996. http://www.epa.gov/ada/csmos/models.html

MECHANISM OF THE REMEDIATION (DETOXIFICATION) OF CHEMICALS (PESTICIDE, HEAVY METALS, OTHER TOXIC CHEMICAL COMPOUNDS)

Mercury Volatilisation by Immobilized Klebsiella Pneumoniae

YOUSEFF ZEROUAL AND MOHAMED BLAGHEN[*]
Laboratoire de Microbilogie, Biotechnologie et Environnement.
Faculté des Sciences Aïn chock. Université Hassan II.
Km 8 route d'El Jadida, B.P. 5366 Mâarif, Casablanca, Maroco

[*]To whom correspondence should be addressed: blaghen@ facsc-achok.ac.ma

Abstract. *Klebsiella pneumoniae*, mercury resistant bacterial strain which is able to reduce ionic mercury to metallic mercury was isolated from wastewater of Casablanca. This strain exhibit high minimal inhibition concentrations for heavy metals such mercury 2400µM, lead 8000µM, silver 2400µM and cadmium 1000µM. This bacterium was immobilized in alginate, polyacrylamide, vermiculite and cooper beech and use for removing mercury from synthetic water polluted by mercury using a fluidized bead bioreactor. Immobilized bacterial cells of *Klebsiella pneumoniae* could effectively volatilize mercury and detoxify mercury compounds. Moreover, the efficiency of mercury volatilization was much greater than with the native cells. The highest cleanup and volatilization rates were obtained when *Klebsiella pneumoniae* was entrapped in alginate beads with cleanup rate of 100% and volatilization rate of 89%. Immobilized cells in alginate continuously volatilized mercury after 10 days without loss of activity.

Keywords: mercury; immobilization; biotechnology; bacteria; enzyme purification; mercury reductase; volatilization; heavy metals

1. Introduction

Mercury is one of the most toxic heavy metals in the environment (Shaolin and David, 1997). Many investigations have been dedicated to microbial detoxification of mercury salts (Gavis and Ferguson, 1972; Robinson and Tuovinen, 1984; Summers and Silver, 1984; Harry et al., 1985).

To prevent the detrimental effects of heavy metals, many bacterial species have evolved a sophisticated and highly regulated detoxification system in which mercurials and Hg (II) are actively transported into the intracellular space, where ultimate reduction of Hg (II) to the much less toxic Hg (o) leads to its elimination from the cell (Stefan and Miller, 1999).

The crucial two-electron reduction step in this pathway is catalyzed by the flavoprotein mercuric ion reductase. The high vapor pressure of elemental mercury results in the volatilization of mercury from aqueous media (Chang et al., 1993; Ogunseitan 1997; Ogunseitan 1998).

Immobilization of living microorganisms has been described by several investigators (Chjibata I, 1983; Hyde et al., 1991) as being useful in biological wastewater treatment. Immobilization of bacteria, yeast cells, and fungi has been done in a variety of ways. Conventional immobilization methods are generally classified into three categories: covalent binding, physical adsorption and entrapment processes (Monsan, 1982).

It is widely known that immobilized cells offer lot advantages: reusability of the same biocatalyst, control of reactions, and the noncontamination of products (Engasser, 1988).

Being based on the bacterial mercury volatilization mechanism, and exploiting advantages that offer immobilization's techniques, we have planned to remove mercury from a synthetic mercurial water using a bacterial strain that has been isolated, identified and appeared to be resistant to high mercury concentrations, compared to those reported in literature.

This bacterial strain has been entrapped into both alginate and polyacrylamide gels, and immobilized by physical absorption on either vermiculite or cooper beech. The mercury volatilization was studied in a fluidized bed reactor. Cleanup and volatilizing rates obtained were compared.

2. Materials and Methods

Isolation of mercury resistant bacteria

Mercury contaminated sludge was collected from the wastewater outlet of an industrial area in Casablanca. One g of sludge was suspended in 10ml of sterile sodium chloride solution 0.85% (w,v) and mixed thoroughly. The mixture was serially diluted with sterile sodium chloride solution 0.85%(w,v). Aliquots of 0.1ml of 10^{-1}, 10^{-2} and 10^{-3} dilutions were spread onto nutritive agar plates containing 200µM $HgCl_2$. All plates were incubated at 37°C for 2 days. Colonies were picked and sowed onto nutritive agar plates containing 200 µM HgCl$_2$. The plates were again incubated at 37°C for 2 days to confirm their resistance to mercury.

Bacterial identification was done by biochemical analysis according to the standardized micromethod API 20 E and 20 NE (Biomereux), after effectuating Gram coloration tests and cultures on selective media (Kligler, King A and Chapman).

Escherichia coli ATCC 25922 a commercial strain was used as control.

3. Determination of Minimal Inhibitory Concentrations (MICs)

Either the solid media Muller Hinton, or liquid media non amended (controls) or amended with the respective metal element at different from stock solutions were inoculated with 100 µl of cell suspensions from precultures of night diluted to 1%.

The minimal inhibitory concentration (MIC) is defined as the lowest concentration that causes no visible growth.

3.1. IMMOBILIZATION OF BACTERIAL CELLS

Entrapment in Calcium gel: 100ml sterile sodium alginate solution (2% w/v) was mixed with 0.5g of bacteria until homogenous. The mixture was extruded through a needle (2mm i.d). Into 150 mM CaCl$_2$ forming beads of 3.0 mm diameter. The beads were allowed to harden in the CaCl$_2$ solution at room temperature for 30 minutes, and rinsed with 50mM Tris-HCl buffer pH 7.5.

Entrapment in polyacrylamide gel: 0.5g of bacteria were mixed with 78 ml of 50mM Tris-HCl buffer pH 7.5 and 20ml acrylamid-bisacrylamide solution (30-0.8% w/v) and 1ml ammonium persulfate solution (10% w/v). The polymerization was initiated adding 100µl of N,N,N',N'-Tetramethyl-ethylenediamine. Polyacrylamide gel has been then divided in particles of 0.5 cm diameter and rinsed with 50mM Tris-HCl buffer pH 7.5.

Immobilization on vermiculite: 5g of vermiculite were washed with distilled water and placed in a column (1.5cm i.d) with 20ml of bacterial solution (25mg/ml) in 0.1M phosphate buffer pH 7.5. The solution was recycled for 24 hours at room temperature (25°C ±0.1). The supernatant was removed and the actived vermiculite washed with the same buffer.

Immobilization on cooper beech: 10g of fine particles cooper beech were washed with distilled water and placed in a column (1.5cm i.d) with 20ml of bacterial solution (25mg/ml) in 0.1M phosphate buffer pH 7.5. The solution was recycled for 24 hours at room temperature (25°C ±0.1). The supernatant was removed and the actived support washed with the same buffer.

3.2. VOLATILIZATION OF MERCURY IN FLUIDIZED BED REACTOR USING IMMOBILIZED AND FREE CELLS

Two Erlenmeyer flasks constitute the fluidized bed bioreactor. The first one contains 100ml of mercurial solution at 250µM of $HgCl_2$ and immobilized bacteria. The mercury vapors are conveyed by current air generated by a compressor outputting150 L/h of air from the bottom of the mercurial solution flask and directed towards a mercury trap flask containing 100ml of oxidizing solution composed of 1.5 N nitric acid, 4 N sulfuric acid, 1ml Potassium permanganate (5% w/v). Mercury rate was followed according time in both mercurial solution flask and mercury trap flask. The same bioreactor was used for study mercury volatilization with free cells; thus 4g of bacterial pellet obtained by centrifugation of bacterial culture at 8000 g for 15 minutes were suspended in the mercurial solution. Mercury dosage was carried out from both flasks. Samples taken from mercurial solution was centrifuged at 8000 g for 15 minutes and mercury levels is determined in supernatant.

Mercury contents were determined by flameless atomic absorption spectrophotometer M.A.S. 50 (Mercury Analyzer System, Bacharach, USA). Ionic mercury was reduced with $SnCl_2$ (5 g/Liter) to metallic mercury which was volatilized by a vector gas (air) and detected at 253.7 nm by the atomic absorption spectrophotometer.

4. Results and Discussion

4.1. IDENTIFICATION OF ISOLATED BACTERIAL STRAINS

Four bacterial strains which grow on mercurial plates were isolated from mercury contaminated sludge. These mercury resistant bacteria were subjected to Gram stain and all were Gram-negative. The results of identification of these mercury volatilizing bacteria were *Klebsiella pneumoniae, Pseudomonas putida, Enterobacter agglomerans* and *Proteus mirabilis*.

4.2. MINIMAL INHIBITORY CONCENTRATIONS

Minimal inhibitory concentrations (MICs) of heavy metals values registered in solid culture are represented in Table 1. The tests obtained in liquid medium were essentially the same. The results obtained indicate that these strains are potent resistant to heavy metals. Indeed, each strain shows a resistance to different compounds tested with a great MICs compared to those reported by several authors (Devincent, 1990; Blaghen et al., 1993; Spangler, 1983; Filali, 2000).

Klebsiella pneumoniae, reveled itself the most resistant species to all of the metals tested. For example the MIC in solid medium of *Klebsiella pneumoniae* to $HgCl_2$ and $AgNO_3$ is 2400 µM, that of *Pseudomonas putida* to $CdCl_2$ is 4000 µM and that of *Enterobacter agglomerans* to $Cr(NO_3)_3$ is 8000µM.

Klebsiella pneumoniae, which have the highest mercury resistance in both liquid and solid media was selected for study the biovolatilization of mercury in fluidized bed reactor.

TABLE 1. Minimal inhibitory concentration (µM) of some heavy metals.

Metal compounds	K. pneumoniae	Ps. putida	E. agglomerans	P. mirabilis	E. coli ATTC 25922
Hg^{2+}	2400	600	600	300	75
Ag^+	2400	150	75	150	75
Pb^{2+}	8000	8000	8000	4000	ND
Cd^{2+}	1000	4000	1000	1000	500
Cu^{2+}	8000	8000	8000	8000	ND
Ni^{2+}	8000	8000	8000	8000	ND
Co^{2+}	8000	2000	8000	8000	2000
Cr^3	8000	4000	8000	8000	ND
Fe^{2+}	8000	800	8000	8000	6000
Zn^{2+}	8000	8000	8000	8000	6000

ND: Not determined

4.3. VOLATILIZATION OF MERCURY IN FLUIDIZED BED REACTOR

Klebsiella pneumoniae was entrapped on both alginate and polyacrylamide gels and physically adsorbed on vermiculite and particles wood. The concentrations of mercury were measured in both mercurial and oxidizing solutions according times for all immobilized *Klebsiella pneumoniae*, as well as for free cells.

Figure 1a shows a quick fall of the mercury concentration in the mercury solution to the end of one hour of treatment; this could be explained by mercury adsorption on different immobilization supports and also to mercury adsorption on the bacteria cellular envelope due to the peripheral charges and functional groupings present in the envelope and also to absorption phenomena by which these the bacteria accumulates the metals inside the cell by active or passive transport (Jairo-Alberto, 1990). Then we are witnessing a progressive decrease of the load polluting and this for the different effectuated experiences due to mercury detoxification. In fact mercury can bind with cell surface proteins, highly specific transport of Hg^{2+} into the cell in the protein-bound form.

The bound mercury is delivered to the mercuric reductase in the cytosol, which requires intracellular NADPH and thiol as cofactor for enzymatic activity. Mercury reductase reduces toxic Hg^{2+} to less toxic Hg^0 which is volatilized out of the system due to its high vapor pressure (Ghosh, 1996).

The evolution of mercury volatilization according to the time is represented in Figure 1b.

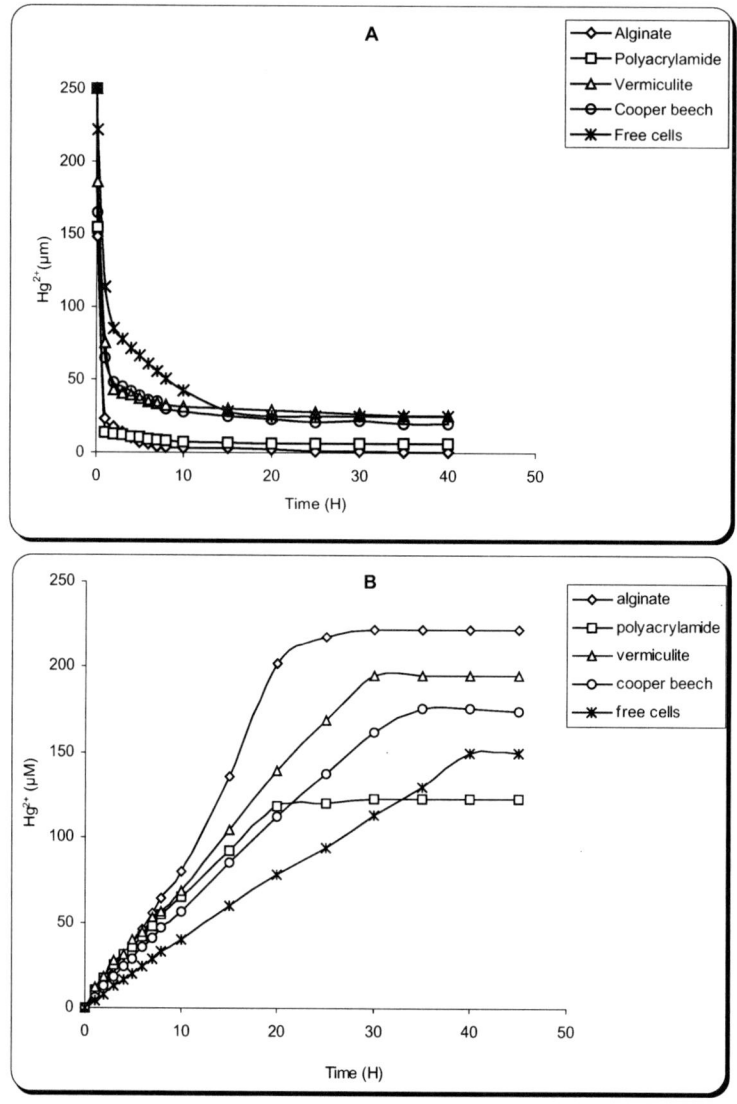

FIGURE 1. Evolution of mercury concentration according to the time.

A: in the solution mercurial

B: in the oxidizing solution

Depollution and volatilization rates were calculated. Results obtained are shown in Table 2.

TABLE 2. Cleanup and volatilization rates obtained for the immobilization of *Klebsiella pneumoniae* by different methods.

	Cleanup rate (%)	Volatilization rate (%)
Alginate	99.8	89.0
Polyacrylamide	97.4	48.8
Vermiculite	91.0	78.0
Wood	92.0	68.0
Free cells	90.0	60.6

It can be observed that *Klebsiella pneumoniae* entrapped in calcium alginate shows the greatest epurifying performances with a quasi-total elimination of mercuric ions and a volatilization rate of 89% (Table 2). All the more that the hydrodynamic behavior and mechanical beads properties of alginate make this polymer a matrix of choice for the utilization in fluidized bed bioreactor (Smid Srod and Skjak-Braek, 1990; Badalo et al., 1991). The immobilization of *Klebsiella pneumoniae* in polyacrylamide gel presents the weakest volatilization rate (48.8%), although cleanup rate of the mercury reach 97.4% (Table 2). This can be explained by the great capacity of the polyacrylamide gel to fix mercury. This limitation of the enzymatic activity is due to the existence of an unfavorable microenvironment inside the gel matrix and the presence of residual monomer that leads to a toxicity of bacterial cells (Mosbach K and Mosbach R, 1996). By fixing *Klebsiella pneumoniae* on vermiculite or on woods, we gets eliminate more 91% of the pollutant charge. Volatilization rate obtained by using activated vermiculite is superior than these obtain with activated woods (Table 2). This is due to the mercury affinity for a woods that fixes to it alone more of 30% of the pollutant charge, while the vermiculite fixes only 6% (data not shown). All the more that researches led by Taoufik (1998) have shown that the vermiculite fixes more bacteria than woods. Utilization of free cell for the volatilization of mercury from waters polluted by mercury allows having cleanup rate comparable to these obtain with vermiculite or woods activated. However volatilization rate remain inferior to these obtain by immobilizing the bacterium.

Using of alginate as entrapment gel in a continuous fluidized bed bioreactor offers a great stability of the bacterial activity and this during more of 10 days without adding nutritional elements (Figure 2). This stability returns to sweet polymerization conditions of the gel and direct role that plays the calcium in the cells conservation (Tamponnet, 1989). After we notes a progressive decrease of the bacterial yield. The adds

278

nutritive elements allows the revitalization of bacteria allowing the maintains of volatilization and cleanup rates to the maximum.

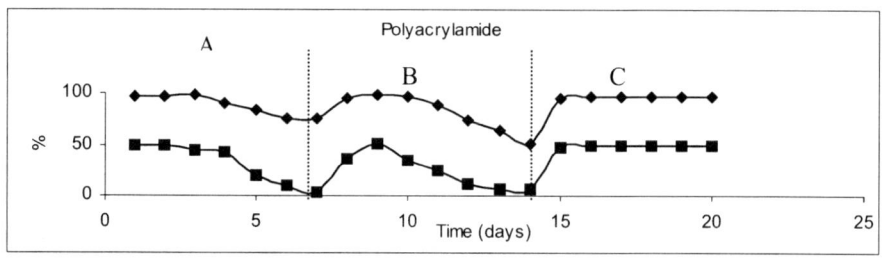

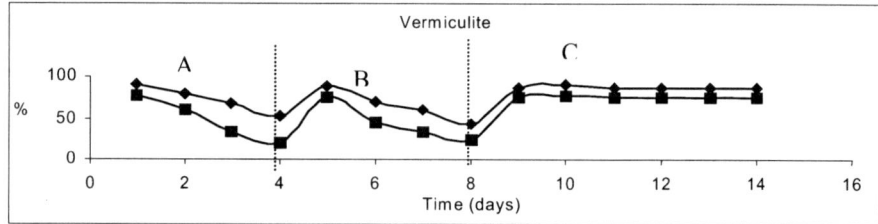

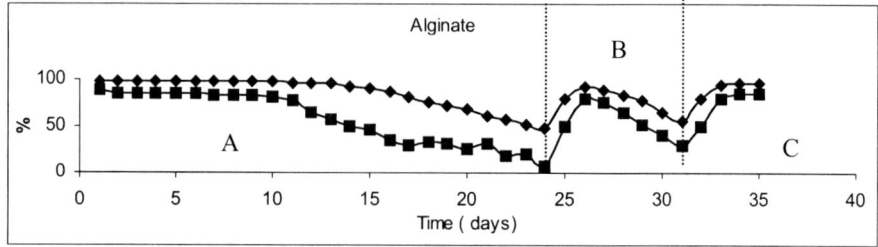

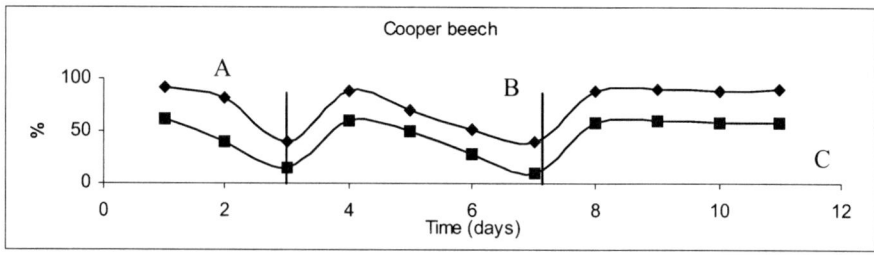

FIGRE 2. Evolution of cleanup and volatilization rates according to the time obtained while immobilizing *Klebsiella pneumoniae* in different supports in continuous fluidized reactor. 250 μM of Hg2+ is added daily.

A: Without adds nutritive elements
B: Adds nutritive elements uniquely the first day
C: Adds daily of the nutritive elements

Cleanup rate ◆
Volatilization rate ■

By immobilizing *Klebsiella pneumoniae* in a polyacrylamide gel the cleanup rates remain constant more 10 days, however the volatilization rates begin diminishing from the 3rd day for practically to cancel out to the end 7 days (Figure 2). Utilization of *Klebsiella pneumoniae* fixed on vermiculite or on wood in continuous fluidized bed bioreactor has proven interesting the moment where bacteria have remained active during the functioning of the bioreactor (figure 2), but addition of nutritional element is obligatory to maintain the same performances.

5. Conclusion

It has been shown that high efficiency removal of mercury from water polluted by mercury using immobilized mercury resistant bacteria in laboratory fluidized bed bioreactor is feasible. The obtained results show that this technology allows specifics and reproducible transformations of mercury with high yield.

The volatilization rate of mercury obtained by fixing *Klebsiella pneumoniae* on vermiculite is superior to those obtained by immobilizing the bacterial stain on cooper beech or in polyacrylamide gel. However the entrapment of *Klebsiella pneumoniae* in alginate offers the greatest epurifing performances as well as the biggest stability of the bioréacteur. All the more the hydrodynamic behavior of the beads, in fluidized bed reactor, is very satisfactory.

The high cleanup and volatilization rates obtained, the simplicity of the immobilization method mean that alginate would be suitable immobilization matrix for immobilization of bacterial stain for removing mercury from wastewater on an industrial scale.

References

Badalo A., Gaez E., Gomez J.L., Bastida J., Maximo M.F. and Diaz F. (1991) Comparison of different methods of ⇓- galactosidase immobilization. Process biochemistry 26: 349–353.

Blaghen M., Vidon D.J.M. and El Kabbaj M.S. (1993) Purification and properties of mercuric reductase from *Yersinia enterolitica* 138 A14. Canadian J Microbiol 39: 193–200.

Chang J., Hong J., Ogunseitan O.A. and Olson B.H. (1993) Interaction of mercuric ions with the bacterial growth medium and its effect on enzymatic reduction of mercury. Biotechnol 9: 526–532.

Chibata I. (1983) In Allaender A.H. and Rogers P. (ed), Basic biology of new developments in biotechnology pp 465–496. Plenum Press New York.

Devincent A., Aviles M., Codima J.C., Borreg J.J. and Romeo P. (1990) Resistance to antibiotic and heavy metals of *Pseodomonas aeruginosa* isolated from natural waters. J App Bacteriol 42: 717–743.

Engasser J.M. (1988) Réacteurs à enzymes et cellules immobilisées. In Biotechnologie. Tec et Doc (eds) pp 468–486.

Filali B.K., Taoufik J., Zeroual Y., Dzairi F.Z., Talbi M., Blaghen M. (2000) Wastewater bacterial isolates resistant to heavy metals and antibiotics. Curr. Microbiol. 41: 151–156.

Gavis J. and Ferguson J.F. (1972) The cycling of mercury through the environment. Water Res 6 :989–1008.

Ghosh S., Sadhukan P.C., Chaudhuri J., Ghosh D.K. and Madal A. (1996) Volatilization of mercury by immobilized mercury-resistant bacterial cells. J. Appl. Bacteriol. 81: 104–108.

Harry A., Harrie G. and Jan van't R. (1985) Detoxification of mercury, cadmium and lead in *Klebsielle aerogenes* NCTC 418 growing in continuous culture. Appl Environ Microbiol 50: 1262–1267.

Hyde F.W., Hunt G.R. and Errede L.A. (1991). Immobilization of bacteria and *saccharomyces cerevisiae* in poly (tetrafluoroethylene) membranes. Appl Environ Microbiol 57: 219–222.

Jairo-Alberto F.A. (1990) Rôle des enveloppes bacteriennes dans l'accumulation des métaux (Cd^{2+}, Ni^{2+}, Cu^{2+}, Zn^{2+}) Thèse de Doctorat en Biologie, Université Metz, France.

Monsan P. (1982) Les enzymes production et utilisations industrielles. In Durand G and Monsan P (Eds) Gauthier-villars, pp 81–118.

Mosbach K. and Mosbach R. (1966) Entrapment of enzymes and microorganisms in synthetic cross-linked polymers and their application in column techniques. Acta Chem Scan 20: 2807.

Ogunseitan O.A. (1997) Direct extraction of catalytic proteins from natural microbiol communities. J. Microbiol Methods 28: 55–63.

Ogunseitan O.A. (1998) Protein method for investigating merciric reductase gene expression in aquatic environments. Appl Environ Microbiol 64: 695–702.

Robinson J.B., Tuovinen O.H. (1984) Mechanisms of microbial resistance and detoxification of mercury and organomercury compounds : physiological, biochemical and genetic analyses. Microbiol Rev 48: 95–124.

Shaolin C. and David B.W. (1997) Construction and caracterization of genetically engineered for bioremediation of Hg^{2+} contaminated environments. Appl Environ Microbiol 63: 2442–2445.

Smid srod O. and Skjak-Braek G. (1990) Alginate as immobilization matrix for cells . Tibtech, 8 :71–78.

Spangler W.T., Spigeralli J.L., Rose J.M., Filippin R.S. and Miller H.H. (1973) Detoxification of methyl mercury by bacteria isolated from environment samples. Appl mocrobiol 25: 488–493.

Stefan E. and Miller S.M. (1999) Alternative routes for entry of HgX_2 into the active site of mercuric ion reductase depend on the nature of the X ligands. Biochemistry 38: 3519–3529.

Summers A.O. and Silver S. (1978) Microbiol transformation of metals. Annu Rev Microbiol 32: 637–672.

Tamponnet C., Matsumara M. and Veliky I.A. (1989) Physical stabilization of *Euglena gracilis* cells by high extracellular calcium (100mM). Appl Microbiol Biotechnol 32: 211–217.

Taoufik J. (1998). Epuration biologique des eaux polluées par le mercure et biodegradation des hydrocarbures aromatiques en bioréacteurs à lit fixe et fluidisé. Thèse de Doctorat en Biologie, Université Hassan II, Maroc.

ENVIRONMENTAL PROTECTION AND PUBLIC HEALTH PROJECTS. EDUCATIONAL AND TRAINING PROGRAMS

CAMELIA DRAGHICI
Transilvania University of Brasov
29 Eroilor Blvd., 500036 Brasov, Romania
E-mail: c.draghici@unitbv.ro

Abstract. Education and training are by far the most important and efficient means for population awareness on environmental changes, pollution effects and public health related concerns. This complex activity is carried out at different levels, starting with institutionalised education by regular curricula in schools and high education institutions, or non-institutionalised training activities. For all educational and training levels there are institutions that introduced and develop programs and are granting projects by partners coming from different countries and very diverse domains. The paper presents the framework of international programs available for education and training projects, covering partnership in Europe, with emphasis on NATO and Socrates programs developed in European countries and in Transilvania University of Brasov, on environmental protection and public health subjects.

Keywords: education and training programs; environmental protection; public health.

1. Introduction

When discussing about learning and training activities we are regularly visualising the actors on this stage, that are represented by two different and well defined groups, strongly linked to each other, in a mutual interaction:

- educators – as teachers, academic staff or trainers are; they are prepared to develop and offer learning programs and teaching materials, in order to transfer particular knowledge and skills, involving all their expertise and experience in this noble activity of education;
- learners – as pupils, students, and/or other professional trainees are; they have different educational/training needs and, therefore, acting like receptors and processors of all these information.

- As a stage, specialized institutions are ready to offer the required framework, logistic and funding for all these actors involved in the educational program: schools, high education institutions (HEI), academy departments, research institutions, public service institutions, governmental bodies, NGOs, or mixtures of all these. In order to act like learning and training systems they have to respond to several compulsory conditions:
- to identify the learning and training interest and needs,
- to have and update their professional and methodological competences,
- to be accredited to develop this specialized activity,
- to be able to manage educational events: study programs, courses, seminars.
- Learning and training activity can be carried out in all these institutions by different means, and at different levels:
- regular curricula in schools and HEI (undergraduate, master and PhD),
- open and distance learning (ODL),
- short or long term postgraduate training courses, as long life learning (LLL),
- even sometimes autodidactic education and training as LLL activity.

During the last decade several international institutions have intensified their interest for educational and training activities, and, therefore, have decided to launch specific programs for financial support, in order to develop an international dimension in education and to encourage cooperation between different education and training institutions. The interest of this paper will mainly be dedicated to those programs granting projects on environmental protection and public health.

2. Financing Institutions and Programs

There are initiatives are coming to give a substantial and complementary aid to the institutionalised education and training systems and nowadays consist in real instruments for international cooperation in the education field. Such frameworks are now available, have very divers financial sources and some of them are subject of an inventory list:

1. bilateral (multilateral) agreements/conventions at governmental, ministerial or regional levels, or directly signed between education/-research institutes,
2. European development, educational & research programs,
3. diverse initiatives (EUREKA, COST, NATO),
4. international organisms (ONU-UNESCO, AUF).

Part of these programs is strictly related to educational purpose (EC Socrates program), while some of them are dedicated to research (EC FP programs) but having always an education component embedded in their activities. It should be agreed that education and research are two specific activities that are rather difficult to be delimited. It is hardly to imagine a research activity without a strong educational background of the researchers, as well as thinking about educational and training programs with no impact on research development. Therefore they are approached and considered as a double-way act and funding institutions are sometimes granting both education and research projects. For high education institutions, like universities and technical institutes are, of interest are the both type of programs, considering the dual activities in which academic staff is involved, education and research.

Of interest for this paper are the NATO programs, as well as the European Commission ones, as funding institutions. Even if the two programs are addressing to different target groups/countries, they have in common the declared interest in promoting collaboration and networking in research and educational programs, as development and progress tools. The NATO "Program for Security through Science" offers support for international collaboration on three priority research topics (http://www.nato.int/science): *defence against terrorism, countering other threats to security, and partner country priorities*. The collaboration supported is between scientists of certain countries of the Euro-Atlantic Partnership Council (NATO-countries and those in eligible Partner countries) and the Mediterranean Dialogue ones, in two main directions:

A. collaborative activities in priority research topics:
- Expert Visits
- Advanced Research Workshops (ARW)
- Advanced Study Institutes (ASI)
- Collaborative Linkage Grants (CLG)
- Science for Peace Research and Development Projects
- Reintegration Grants (RIG)

B. computer networking support for partner countries:
- Networking Infrastructure Grants
- Advanced Networking Workshops

The support is channelled through a range of different mechanisms or activities which promote collaboration, networking and capacity-building, while at the same time catalyzing democratic reform and supporting economic development in partner countries. Among all these type of funding support, some are strictly for research (i.e. CLG, RIG), but some are hybrid projects, for both research and training (i.e. ARW, ASI).

The declared interest for NATO program for security through science and its related projects comes from the interest in collaborative approach of issues like environment and public health protection, as priorities of the "other threats to security" or partner country priorities. This framework offers to the partners from the eligible countries the opportunity to develop common educational and research projects on these two important subjects of interest.

The European Commission (EC) development, educational and research programs are very diverse in terms of goals, priorities, target groups or type of expected outputs, and, therefore, are organised under different units called "Directorate General" (http://www.europa.eu.int). Among them, the Socrates, Leonardo da Vinci and Youth programs are dedicated to education and training activities, covering all the levels of the trained persons: pupils, students and adults (http://www.socleoyouth.be).

In 1995 the EC launched the SOCRATES program designed to educational purposes, aiming to promote the European dimension and the increasing of the educational quality, encouraging cooperation between the European educational institutions (http://www.europa.eu.int/comm/-education/programmes/socrates). The program was developed for two phases: 1995-2000 and 2000-2006, respectively. After the first phase, several European events came to decide the trends in educational policy in Europe and, therefore, also defined the future priorities of the Socrates program (Pyörälä, 2005):

- Bologna declaration (1999) – "promotion of the necessary European dimensions in higher education, particularly with regards to curricular development, inter institutional co-operation, mobility schemes and integrated programmes of study, training and research";
- Lisbon European Council (2000) – "to become the most competitive and dynamic knowledge based economy in the world, by 2010, capable of sustainable economic growth with more and better jobs and greater social cohesion"
- Prague Communiqué (2001) – "development of modules, courses and curricula at all levels with European content, orientation or organisation" – "a way of further strengthening European dimensions of higher education and graduate employability".

The SOCRATES program "Actions" are designated to different participants to the educational process and represent specific types of projects, developed with EC financial support (http://www.socleoyouth.be): COMENIUS (for education in school), ERASMUS and THEMATIC

NETWORKS (for high education), GRUNDVIG (for adult education and other educational pathways), LINGUA (for language teaching and learning), MINERVA (for information and communication technologies in education for Open & Distance Learning – ODL), and DISSEMINATION projects (for any type of the above mentioned Actions).

Following the HEI educational needs, the Socrates ERASMUS Action supports the following activities:

- student and teaching staff organised mobility (OM);
- European Credit Transfer System (ECTS) – supports the activities related to the introduction, implementation and/or extension of ECTS;
- Intensive programs (IP) – for the organization of short courses;
- Curriculum Development projects (CD) – aiming to develop an entire study program or a modular one, for undergraduate or master level, in all academic subject areas, with Joint Degrees as priority since 2003.

3. Socrates/Erasmus Program – Curricula Development Projects

During the first Socrates/Erasmus phase, the CD projects had as classification criteria the level of students to whom the CD project was addressed to:

- CDA projects – for advanced level (master students),
- CDI projects – for initial level (undergraduate students),
- DISS projects (for a former CDA and CDI project dissemination).

Starting with the academic year 2001-2002, the CD projects were differently approached, taking into account the covering level of the study program:

- PROG projects (for an entire study program)
- MOD projects (for a partial study program/module)
- DISS projects (for a former PROG and MOD project dissemination).

Statistical reports are always welcome, providing analytical data focused on specific subjects of interest. A statistical consideration of the Socrates/Erasmus CD projects shows the evolution of the large number of partner institutions that put their efforts, experience, and expertise in order to develop joint new study programs. Statistical data, as well as an ex-post evaluation report of the Erasmus CD projects is available on the Socrates website (http://www.socleoyouth.be) for the period of 1997-2001 (Sahlin et al., 2005a). A less detailed presentation, but still containing an analytical overview of the CD projects was also subject of interest for our group (Draghici et al., 2005).

The evolution of the CD projects granted by EC during 2002-2006 is given in Table 1 (http://www.socleoyouth.be). There are presented, in an alphabetical order of the Erasmus country codes, the CD projects granted during the academic years of 2002-2006, for 26 eligible countries participating as coordinators.

TABLE 1. Evolution of all type of CD projects (PROG, MOD and DISS) granted during 2002-2006.

Erasmus country code	2002-2003	2003-2004	2005-2006			
			submitted	granted	% of submitted	% of total granted
AT	6	3	6	4	66,67	11,4
BE (I)	18	12	9	3	33,33	8,57
BG	3	1	0	0	0	0
CY	1	0	0	0	0	0
CZ	1	0	1	1	100	2,86
DE (IV)	10	6	9	3	33,33	8,57
DK	2	2	2	1	50	2,86
ES	8	2	3	2	66,67	5,71
FI	6	5	5	2	40	5,71
FR (II)	14	9	1	1	100	2,86
GR	4	2	1	1	100	2,86
HU	2	3	4	2	50	5,71
IE	1	0	0	0	0	0
IS	1	1	0	0	0	0
IT	10	3	6	0	0	0
LT	1	1	4	0	0	0
MT	1	1	2	1	50	2,86
NL	1	2	6	3	50	8,57
NO	0	1	1	0	0	0
PL	0	0	4	2	50	5,71
PT	4	4	4	3	75	8,57
RO (V)	9	4	4	3	75	8,57
SE	3	3	4	1	25	2,86
SI	2	0	1	1	100	2,86
SK	0	1	1	0	0	0
UK (III)	13	8	4	1	25	2,86
Total	121	74	82	35		100

Countries that coordinated no projects during this period were not included in the statistics (i.e. EE, LI, LU, LV,). Taking into account that data for the academic year of 2004-2005 were not available from the EC website, this table can only give an approximate evaluation.

Countries like Belgium (with 33 projects during the evaluated period and available data), France (with 24 projects), United Kingdom (with 22 projects), Germany (with 19 projects), and Romania (with 16 projects) are occupying the first five positions in the top of the list (I-V), followed by Austria, Finland, Italy, Spain and Portugal with more than ten projects. At the opposite side, countries like Estonia (EE), Lichtenstein (LI), Luxembourg (LU), and Latvia (LV) show no interest in the CD projects, as coordinating institutions. Romania, from all the East-European countries, occupies a strong position, situated among the five first countries. This position underlines the willingness of the Romanian high education institutions to adapt their curricula at high education level, according to the EU countries ones.

A drastically decrease in the number of projects granted by the EC during the selected evaluation period is registered. If for 2002-03 a total number of 121 CD projects were granted, their number was reduced for the academic year of 2005-06 to only 35 projects that represents a reduction of 71.07%.

There are several observations that should be underlined looking at the data available for 2005-2006. Comparing the number of submitted with the number of granted projects, a high decrease can be noticed, only 35 projects being granted from a total of 82 submitted (representing 42.68% success rate). To be also noticed the success rate of different coordination countries: CZ, FR, GR, and SI with 100% (but only one project submitted); PT and RO with 75% (3 projects granted of 4); AT and ES with 66.76% (4/6 and 2/3, respectively).

From this perspective, BE, DE and UK with good results in the cumulative period of 2002-2006 seemed to loose their leading positions for coordination CD projects during the last granted year, even with high interest in submitting for (9 submitted projects, 9 and 4, respectively), while PT, RO and AT constantly applied and obtained with high success rates EC financial support, via the Erasmus projects (please note that EC is considering every project year as an individual project, meaning that a project granted for three years counts as three projects).

Another relevant approach of the statistics should be in terms of Erasmus subject fields of interest for CD projects, as given in Table 2. (academic years of 2002-2006, except 2004-2005, with no data available).

To be noticed that there are several subject fields predominant in the granted CD projects: "Natural sciences" (with a total of 33 projects);

"Medical sciences" (with a total of 31 projects); and "Social sciences" (with a total of 29 projects).

TABLE 2. Evolution of the CD projects granted during 2002-2006, by Erasmus subject fields: PROG (P), MOD (M), DISS (D) and Total (T).

	Subject fields	2002/03				2003/04				2005/06				T 2002/06
		P	M	D	T	P	M	D	T	P	M	D	T	
1	Agricultural sciences	2	0	0	2	0	0	0	0	0	0	0	0	2
2	Architecture, urban and regional planning	4	1	0	5	0	1	1 (P)	2	0	0	0	0	7
3	Art and design	1	1	0	2	0	0	0	0	0	0	0	0	2
4	Business studies, management	3	3	1	7	4	1	0	5	3	1	0	4	16
5	Education, teacher training	8	3	1	12	7	2	0	9	1	0	0	1	22
6	Engineering, technology	3	3	1	7	2	0	2 (2x P)	4	0	1	0	1	12
7	Geography, geology	0	0	0	0	0	0	0	0	0	0	0	0	0
8	Humanities	4	2	0	6	4	1	1	6	3	0	0	3	15
9	Languages, philology	4	2	1	7	1	0	1 (M)	2	2	1	1	4	13
10	Law	0	1	0	1	1	0	0	1	1	0	0	1	3
11	Mathematics, information sciences	2	0	0	2	0	0	0	0	0	0	0	0	2
12	Medical sciences	12	1	1	14	9	2	2 (1x P)	13	4	0	0	4	31
13	Natural sciences	12	1	1	14	7	3	2 (1x P)	12	5	2	0	7	33
14	Social science	7	5	1	13	5	2	2 (2x P)	9	3	3	1	7	29
15	Communication and information sciences	1	0	2	3	4	0	0	4	0	0	0	0	7
16	Other fields	1	7	5	2	6	1	0	7	3	0	0	3	36

17	Total	7 7	3 0	1 4	12 1	5 0	1 3	11	7 4	2 5	8	2	3 5	23 0

Data available from the Database of the Socrates Technical Assistance Office (http://www.socleoyouth.be); allocation of the projects to different subject fields was appreciatively evaluated according to their titles, not according to the subject field declared by the coordinating institution (that was not available on the website)

This emphasis the higher interest for those subjects related to different aspects of life concern, and less related to technology development ("Engineering, technology" with only 12 projects). On the opposite side, with less interest for CD projects are subjects like "Geography, geology" (0 projects), "Mathematics, information sciences", "Art and design" and "Agricultural sciences" (with only 2 projects), or "Law" (3 projects).

Coming back to the most granted projects in terms of subject of interest, "Natural sciences" and "Medical sciences" are covering together 26.52% of the total CD projects. These are the subject areas including themes related to the environmental protection and public health concern. Assuming that less than half of the two groups of projects are strictly related to environmental protection and public health we arrive to an approximate 10% of the total CD projects that are in the announced subject of interest of this paper.

Table 2 is not only presenting the spread of the projects according to their subject field, but also the ratio between PROG, MOD and DISS projects. This ratio is approximated to 5.5/2/1 for the academic year 2002-2003, 4.5/1/1 for 2003-2004 and 12.5/4/1 for 2005-2006. Such an evaluation comes to the conclusion that most of the projects are developing entire study programs (PROG projects), less modular curricula development projects and a dramatically decreasing to the dissemination activity is registered. The last observation concerning the DISS projects is valid for both projects ratio and number of granted projects (14 at the beginning of the evaluation period and only 2 at the end if it), showing less interest of the coordinating institutions to disseminate their projects and related products and outputs. This could be due to the interest of the coordinating HEI only for the implementation of the curricula development projects results and outputs, and not to their dissemination too.

Considering the DISS projects, it is stated in the application eligibility criteria the option to disseminate a former project that have been developed in the immediate past year (one year before), or other former year as well. A survey of these projects granted for 2003-2004 show that from 11 DISS projects only 8 of them were developed during 2002-2003, while the rest of 3 were developed before the this year. Moreover, a great majority are dissemination activities of PROG projects (7 of 8) and only one for MOD (information given in brackets in Table 2).

4. Curricula Development Projects Based on Socrates Erasmus Projects in Transilvania University of Brasov

Since 1998 Romania joined the EC SOCRATES program, therefore, starting with 1997 the Romanian high education institutions were keen to apply for and participate to different projects, as coordinating or partner institutions. Involvement to this European program represents for Romania a substantial financial and logistic support for the reorganisation of the educational system and became a key tool for the national integration strategy in the European Union (http://www.soctares.ro).

The Ex Post Evaluation of the Curriculum Development Projects (Sahlin et al., 2005) demonstrates that Romanian participation in these projects provided the opportunity for universities to launch their internationalization process. This report was not only based on statistical data processing, but also relayed on consultation of the Erasmus CD Action representatives with groups from different countries participating to the CD projects as coordinators, like Romania is. This consultation provided valuable information and comments from the CD projects coordinators experience.

TABLE 3. The evolution of the Curricula Development projects during 1999-2006 in Romanian, as coordinating institution.

CD project	1999-00	2000-01	2001-02	2002-03	2003-04	*2004-05*	2005-06	Total
CDA	3	3						6
PROG			6 (1)	8 (2)	2 (1)	*1 (1)*		16 *(+1)*
MOD			1	1 (1)	1 (1)		3 (3)	6
DISS			1		1 (1)	*1 (1)*		2 *(+1)*
Total	3	3	8	9	4		3	30 *(+2)*

Presentations of the interest of Romanian HEI to develop new study programs and courses in the framework of Socrates Erasmus program were also done prior to this work (Draghici, et al., 2005). The evolution of the CD projects in Romania during 1999-2006 is given in Table 3. Data for 2004-2005 are only related to the Transilvania University of Brasov projects, as other data was not available, while numbers given in brackets relate to CD projects coordinated by Transilvania University of Brasov.

The spread of these projects on subject fields gives in overview of the interest of the Romanian Universities on developing new study programs and courses for the students' education (Figure 1). The figure only presents the percentage of the CD projects carried out in Romania during 2002-2004, for the next academic year data still not available, as for 2005-2006 all the three MOD projects are environmental management and protection oriented.

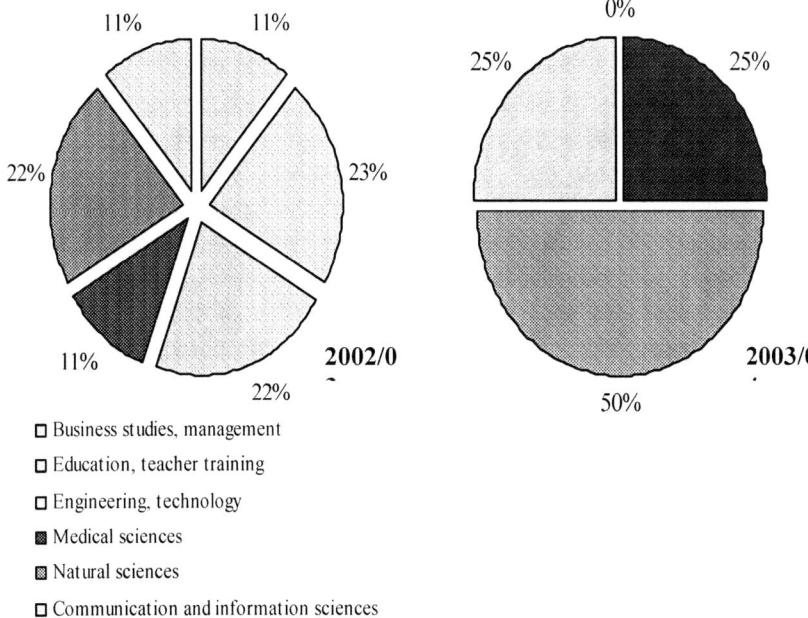

☐ Business studies, management
☐ Education, teacher training
☐ Engineering, technology
■ Medical sciences
▨ Natural sciences
☐ Communication and information sciences

FIGURE 1. Evolution of the CD projects in Romania during 2002-2004, by subject fields.

In this framework, Transilvania University of Brasov found out that CD projects are educational means for adapting and updating the knowledge at high education level. The developed CD projects under our university coordination are as follows:

5. PROG 1: Management of Chemical Investigation in Environmental Protection, 2001-2003 (undergraduate level),
6. DISS: Management of Chemical Investigation in Environmental Protection, 2003-2004,
7. PROG 4: e-Lab – Information and Communication Technologies in Applied Chemistry, 2002-2005 (master level),

8. MOD 3: Bioanalytical Methods – Linking Environmental Protection and Public Health, 2002-2004 (undergraduate level),
9. DISS: Bioanalytical Methods – Linking Environmental Protection and Public Health 2004-2005,
10. MOD 4: ECO-DESIGN – An Innovative Path towards Sustainable Development, 2005-2007 (undergraduate and master levels),
11. MOD 5: Forest Fires – Preventing and Monitoring, 2005-2007 (undergraduate level),
12. MOD 6: MASS – Management and Security Assessment for Sustainable Environment, 2005-2007 (undergraduate level).

The CD projects titles emphases the interest of our group for subjects fields like environmental protection and public health, as an obsessive target of our academic staff to contribute to the education of our students in this respect. This aim is according to the national and regional development policy and expects to cover educational gaps in this priority domain.

The main educational activities and the implementation level of the CD projects coordinated by Transilvania University of Brasov are given in Table 4. (the projects codes took into account the numbering system, given by EC, while the project year is given in brackets).

The first CD project, as coordinator, developed in Transilvania University of Brasov by the Chemistry Department (PROG 1) was a continuation of a CDI project started in 2000-2001, with Delft University of Technology (NL) as coordinating institution. During the project lifetime new syllabus and three new courses were developed and implemented at different study programs of our university, two European ones and 7 more from Romania. The project continued with the dissemination year (DISS of PROG 1), that gave the opportunity to launch the EnvEdu 2004, Trends in Environmental Education, International Conference (Draghici et al., 2004a; Draghici, 2004).

During 2002-2003 two more CD projects started: PROG 4 for master level and MOD 3 for undergraduate students, coordinated by Chemistry and Preclinical Medicine Department, respectively. A detailed description of the partnerships, inputs, activities and outcomes of these CD projects was previously presented (Draghici et al., 2004b). To be noticed that MOD 3 continued with the related DISS project, having as main activity the second edition of the EnvEdu 2005 (Badea et al., 2005), Trends in Environmental Education, International Conference. It is to underline that PROG 4 have also remarkable results, as a new master program, taught in English, was implemented in our university curricula, Applied Chemistry in Environment and Industry (ACEI), starting with the academic year 2004-2005.

For 2005-2007 three more MOD projects are developed under the coordination of different groups from Transilvania University of Brasov, demonstrating the constant increasing interest of the academic staff for curricula development process, with financial support from the EC Socrates Erasmus program.

TABLE 4. Evolution of the CD projects coordinated by Transilvania University of Brasov (2002-2006).

Year	Project code	Main educational activities	Implementation
2005/2006	MOD 4 (1)	syllabus, courses development	courses implementation
	MOD 5 (1)		
	MOD 6 (1)		
2004/2005	DISS of MOD 3	MOD 3 dissemination	EnvEdu 2005 conference
	PROG 4 (3)	courses development	ACEI – master program implementation
2003/2004	DISS of PROG 1	PROG 1 dissemination	EnvEdu 2004 conference
	MOD 3 (2)	courses development	courses implementation
	PROG 4 (2)	courses development	ACEI – master program submitted
2002/2003	PROG 1 (3)	courses development	courses implementation
	MOD 3 (1)	syllabus, courses development	courses implementation
	PROG 4 (1)	curricula development, syllabus, courses content	ACEI – master program preparation
2001/2002	PROG 1 (2)	courses development	courses implementation

Transilvania University of Brasov is not only participating to the **Socrates Erasmus Curricula Development** projects as coordinating institution, but also as partners in another PROG project (2005-2006). Partnership allows us to collaborate in **Socrates Erasmus Thematic Networks** (TN) projects too, one as coordinator (2003-2006) and too more as partners (2002–2004; 2004–2007). Moreover two **Socrates Comenius** projects are coordinated by different groups in our university (during 2001-2003 and 2005-2007), three **Socrates Minerva** partnership (all of them during 2002–2004), one **Socrates Grundtvig** (2003–2005) as partner and one as coordinator (2005-2007). All these Socrates projects are thus covering all the educational and training levels, on different subject fields that were found of interest in Brasov area educational needs. The projects developed in Transilvania University of Brasov in the last period (2000-2005) demonstrate the capability of the teaching staff to innovate the didactical processes their opening to European collaboration and experience exchange (Helerea et al., 2005).

In addition, there was strong link between the **Socrates Erasmus Organized Mobility** activity carried out by our university and the curriculum development projects. All our CD projects activities were combined with student or/and teaching staff mobility, as complementary inputs and feedback of the CD projects results.

5. Conclusions

As a general conclusion, EC and other initiatives or international organisms created the framework for educational and training programs, very close linked to the research ones, to promote international collaboration on subject fields of general interest for our communities. This offers the opportunity to institutions from different countries to participatory education and research projects, to agree and create cooperation networks, showing the common interest in the sustainable development of our society.

Among all these programs, attention was given to NATO Program for Security through Science, and more to the EC Socrates one, due to our participation at the NATO-ASI project and our experience in Socrates Curricula Development projects, respectively. Both of them are coming to promote and financially support international collaboration between different educational and research institutions in their efforts to find solutions for the most important subject of concern, human life and its security.

Romania takes a leading position in the assessment of the Socrates Erasmus CD projects, close to countries like Belgium, Germany, France, and United Kingdom. Moreover, with 12 CD projects as coordinator, of more than 32 projects developed in Romania, our university could be considered a driving force in this area. The experience gained during all these years of collaboration with European universities represents for us, and our students, inestimable goods.

All our Socrates Erasmus CD projects, as well as the Erasmus Thematic Networks, Comenius, Minerva and Grundtvig projects are springboard to international partnership and networks, that developed from other activities or generated other ones. As expected, the educational and training projects opened the doors for research projects too, in partnership with recognized Romanian and European academic institutions, and the reversal (research promoting education) is true as well. This mainly emphasis the most valuable good gained with these projects, meaning the trust in further collaboration, and the recognized abilities to innovate, create, manage, develop, implement and disseminate the projects' results and outputs.

All these complex activities carried out in Transilvania University of Brasov in the framework of the SOCRATES/ERASMUS CD projects show sustainability in the process of the curricula development at all educational levels (undergraduate and master programs) and at a very diverse subjects/area of our student's studies.

As a perspective, we intend to be involved in more educational activities in partnership with international institutions, combining different programs and projects, for a larger pallet of subjects of interest, and to promote networking, as a solution for a global approach of the themes.

ACKNOWLEDGEMENTS

Acknowledgements are addressed to the NATO-ASI project "Chemicals as Intentional and Accidental Global Environmental Threats" (ASI 981563) for offering the opportunity to participate to this event, and to the Socrates Erasmus Curricula Development projects carried out in Transilvania University of Brasov during 2000-2005, for the chance to develop new syllabus and courses based on international collaboration framework.

References

Badea, M., Sica, M., Draghici, C., 2005, EnvEdu 2005, Trends in Environmental Education Conference, *Book of Abstracts*, Transilvania University of Brasov.
Draghici, C., 2004, Conferences and European Projects, *Bull. Transilvania University of Brasov*; Series B2: Medicine, Chemistry, ISSN 1223-964X, (2004), 529–30.
Draghici, C., Perniu, D., Sica, M., 2004a, EnvEdu 2004, Trends in Environmental Education Conference, *Book of Abstracts*, Transilvania University of Brasov.
Draghici, C., Perniu, D., Coman, Gh., 2004b, Curricula Development by European Socrates Projects, *Ovidius University Annals of Chemistry*, 15(1), (2004), 119–124.
Draghici, C., Perniu, D., Coman, Gh., Badea, M., 2005, Curricula Development at High Education Level in the European Context, *Proceedings of CSE 2005*, International Conference on Sustainable Energy, CD-ROM, ISBN 973-635-539-X, O.5.4.
Helerea, E., Visa, I., Duta, A., Sofonea, L., Popescu, M., Draghici, C., Rogozea, L., Florea, D., Lache, S., 2005, Developing Projects as a Creative Way in Education, *Proceedings of CSE 2005*, International Conference on Sustainable Energy, CD-ROM, ISBN 973-635-539-X, O.5.2.
Sahlin, E., Gordon, J., Libert, I., Emin, L., Berg, A., Persson, M., 2005, Ex Post Evaluation of the Curriculum Development Projects Funded in the Framework of Socrates 1997–2001 report, (http://www.socleoyouth.be).
Pyörälä, P., 2005, Erasmus Curriculum Development Coordinators' Meeting, PowerPoint presentation, Brussels, 2005.

http://www.europa.eu.int
http://www.europa.eu.int/comm/education/programmes/socrates
http://www.nato.int/science
http://www.socleoyouth.be
http://www.socrates.ro

PLANNING AND EXECUTION OF A PILOT PHYTOREMEDIATION EXPERIMENT

BIANA SIMEONOVA[1] AND LUBOMIR SIMEONOV[2]*
[1]*Institute of Electronics, Bulgarian Academy of Science,*
72 Tzarigradsko Shossee, 1784 Sofia, Bulgaria
Central Solar-Terrestrial Laboratory,
Bulgarian Academy of Sciences
Acad. G.Bonchev Str. Bl.3, 1113 Sofia, Bulgaria

*To whom correspondence should be addressed: simeonov@bas.bg

Abstract. This paper deals with some specific problems of the procedure of planning and execution of a pilot phytoremediation experiment. Phytoremediation is environmentally-safe and -friendly technology, introduced to clean soils, contaminated anthropogenically with heavy metals, organics and radionuclides. The material is based on the experience from the first phytoremediation experiment in Bulgaria, carried in 1998 at the industrial site of Kremikovtzi Steel Works within a bilaterally initiated project with Phytotech, Inc. and using the genuine proprietary technology, developed by the company.

Keywords: phytoremediation; experiment planning; heavy metals; soil pollution

1. Introduction

1.1. TECHNOLOGY DESCRIPTION

The phytoremeiation technology uses specially selected and often genetically manipulated higher plants with considerable biomass production in processes of extraction of the pollutants with the root system of the plants and chemically forced translocation to the above-ground shoot – stem and leaves. Certain metal-accumulating plants have been discovered that contain unusually high concentrations of heavy metals in their tissue. At the present moment the number of identified metal-tolerant higher plants by different scientific groups exceeds several hundred. Hyperaccumulators of Ni and Zn, for example contain as much as 5% of these metals on a dry weight basis. Plants accumulating metals at a 5% (50 000 mg/kg) dry-weight concentration from a soil with a total metal concentration of 5000 mg/kg result in a ten-fold bioaccumulation factor. If the plant produces a significant amount of biomass while accumulating high concentrations, an important quantity of the metal can be remove from the soil via plant uptake. This process transfers the metal contaminant from an alumino-silicate soil matrix to a carbon-based plant matrix. The metal-rich plant material can be collected and removed from the site using established agricultural practices, without the loss of topsoil associated with traditional physical, chemical and mechanical remediation practices. The biomass can then be recycled to reclaim the metals that may have an economical importance; the latter procedure is even named "bio-mining" by several enthusiasts.

1.2. SYSTEM DESCRIPTION

Contaminated soils at industrial sites may have been exposed to numerous activities, i.e., compaction by heavy traffic, disposal of waste products, biuldings, and escavations, which disrupt the soil structure and alter the arable properties, presenting unique challenges for the application of phytoremediation in the field. Techniques to restore the land to a condition conductive to plant growth is the first step before phytoremediation can occur.

Field operations consist of an initial site visit and walkover to assess physical characteristics followed by an initial sampling and surwey. The use of portable x-ray fluorescence spectrometers can greatly aid in the identification of the magnitude and distribution of surface contamination. Another possibillity to perform the initial check is to use a portable laser time-of-flight analyser, which is capable to provide a multi-element and – isotope screening analysis of the site (sSmeonov et al., 1996; Simeonov and Managadze, 2006). Based on the goals and objectives for the site, the site is prepared for crpping using standard agronomic practices. Crop selection is based on site condition, the concentrations of contaminants and the geogrphical location. The crop is produced and metal accumulated through addition of appropriate soil amendments with attention to minimizing any potential downward migration of the contaminants through control of irrigation. When the crop has reached its optimum metal content (product of biomass and metal concentration) it can be harvested and disposed and the process repeated. Biomass disposal options are site dependent but focused on reducing the amount of material to be disposed of through various biomass treatment options. The process can then be repeated with successive crops until the desired regulatory goal is achieved.

1.3. ADVANTAGES AND LIMITATIONS

Phytoextraction has the potential to remediate many metal and radionuclide contaminated sites using a less invasive form of treatment than traditional methods such as escavation and disposal. There are four factors that influence or determine the ability of phytoextraction to effectively remediate a metal contaminated site: 1). Site arability and plant biomass yields; 2) metal solubility and availability for uptake; 3) the ability of the plant to accumulate metals in the harvestable plant tissues; and 4) regulatory criteria.

Phytoremediation is a technology, which has a history of nearly 15 years with the pioneered discoveries of Phytotech in identifying higher plants with the ability to produce substancial biomass on contaminat3ed soils while accumulatinf metals in their shoots. The application of soil and foliar amendments to enhance metal availability and uptake is also a key component of the technology. Heavy metals in soil may not be immediately available for plant uptake and often require slight adjustments to the soil chemistry to enhance removal rates. Phytoremediation is attractive as a remediation tool because it uses the natutal processes of mineral uptake by plant roots to remove toxic metals from the soil. With each successive crop an additional quantity of metal is removed resulting in a decrease in the soil concentration until the regulatory goal is met. Thus, the length of time (or number of crops) required is determined by the difference between the measured concentration and the regulatory goal. While phytoremediation is an effective method for most sites, not every site is conductive to phytoremediation due to excessively high contaminat concentrations or unsiutable conditions for plant growth.

2. Phytoremediation Project Design and Objectives

Every pilot experimental project is normally designed to demonstrate the technical, regulatory, and economic feasibility of the technology to be used. The Kremikovtzi 1998 summer trial experiment targeted presumably extraction of lead from two sites within the industrial territory of the Steel Works, situated some 15 km from the capital town of the country, Sofia. Accordingly the plant seed material was choosen to be Brassica juncea, selected espessialy for lead extraction. The seed material was generously provided by Phytotech Inc. The following tasks were carried out at Kremikovtzi Steel to meet these objectives:

- demonstrate and advance this technology in the field for two consecutive years;
- assess and document the potential for secondary risks (e.g. food chain transfers);

- evaluate the effectiveness and cost of selected methods to treat the harvested Pb-containing biomass.

The experiment could not be extended to its second year of application because Kremikovtzi Steel Works entered a cycle of privatization procedure and the access to the sites was to a greater extent limited despite of the positive results of the first experimental year.

3. Field Activities

The demonstration plot of the experimental site should be constructed to evaluate the effectiveness of phytoremediation. The field activities consist of site mobilization, plot layout an construction, and soil sampling combined with other agricultural practices designed secifically for phytoremediation and adapted for the conditions present at the site. The activities has to be conducted in accodance with the personal protective equipment, level of protection, action levels and other health and safety practices, i.e. hazard analysis; general safety reccomendations; evaluation of the mechanical, electrical, fire hazards, gas and power lines, heat stress, noise, chemical hazards.

3.1. SITE MOBILIZATION

Site mobilization activities should be managed by the project manager, responsible for the experiment. He has to ensure that the field personnel and subcontractors are familiar with the project protocols and obtain the necessary entry badges and approvals. An initial meeting with the field personnel and any personnel whose attendance is reqired, should be made to review the heath and safety guidelines. The personnel should be briefed on the proposed activities, schedules and any contingency plans for unexpected conditions.

3.2. PLOT LAYOUT AND CONSTRUCTION

The treatment plots of the two experimental sites at the Kremikovtzi Steel Works were situated between ajacent industrial biuldings and were with dimentions 30 x 20 and 40 x 10 m. Water supply for each site was ensured through locally-mounted hydrants. The irrigation of the experimental sites is important for normal plant growth and high biomass production and is especially critical in dry summer climates, like the Sofia region.

The plot location should be mowed and the clippings should be removed from the plot. The mowed area must include a border around and outside the treatment plots. A fence could be cobstructed around the treatment plots leaving a some distance on the inside of the fence and the plot.

3.3. EXPERIMENTAL DESIGN

Alongside to the treatment plots a smaller area is maintained as a control plot, at which no special amendments will be applied.

3.4. SOIL SAMPLING

Soil core samples are collected at several depth intervals to approximately 50 cm depth using for example a 25 x 25 cm sampling grid plan.

3.5. FERTILIZATION

The plot should be fertilized with nitrogen, phosphorous, potassium ans sulfur according to recommendations from the soil fertility analysis of the collected samples. Dolomite lime should be added as needed to ensure an adequate pH (>6) for plant growth. The plot is tilled to incorporate the fertlizers into the surface soil and create the seedbed. A light irrigation before tilling may be necessary to control dust. Additional plant nutrient applications are provided through the irrigation system as needed, according physiological development.

3.6. TILLAGE

Soil tillage operations are conducted to incorporate fertilizers and soil amendments as necessary. Tillage is also used to create the seedbed conductive to plant growth and encourage maximum root development and exploration of the soil profile. Tillage is usually limited to the depth of contamination. A tractor-mounted roto-tiller may be alternatively used. Surface debris and rocks that interfere with tilling operations should be removed prior to tilling.

3.7. SEEDING

After the seedbed preparation is completed, the plots are seeded with metal-accumulating plant cultivars. The selected crop, used in the pilot experiment of the Steel Works was a variety of *Brassica juncea* (Indian mustard), selected and provided by Phytotech Inc. The plots are seeded under recommendations for optimum row spacing and plant population.

3.8. PLANT GROWTH

Plant growth and development is monitored minimum three times a week best at the irrigation days. The inspector is obliged to report of any noted, suspected, or potential problems with respect to plant grows and development, irrigations, weeds, insects or other pests and plant deseases.

3.9. SOIL AMENDMENTS AND ADDITIVES

Metal uptake in plants can be enhansed by the application of soil and foliar amandments. Amendments consist of proprietary combinations of metal solubilizing agents, chelates and organic acids that increase plant availability and uptake of soil contaminant concentrations. The amendments are applied as a foliar or surface spray or through the irrigation system.

3.10. HARVEST AND PLANT SAMPLING

Plants are harvested one to two weeks after the final amendment application, depending on weather conditions and the physiological condition of the crop. The plant stems are cut at the soil surface. Representative plant samples are taken for laboratory analysis and treatability studies for evaluation of the phytoextraction.The remaining biomass is collected and removed from the site for further processing.

References

Simeonov L., C. Schmitt and K. Sheuermann, 1996 Scnelle semiquantitative Elementanalyse wässeriger Lösungen mit dem Laser-Massenanalysator LASMA, TerraTech, 6, pp. 29–31.
Simeonov L. and G. Managadze, 2006, Technological Transfer. Miniature Laser Mass Spectrometer for Express Analysis of Environmental Sample; In Simeonov, L. and Chirila, E. (eds) Chemicals as Intentional and Accidental Global Environmental Threats, pp 149–162, Springer, Brussels, Belgium.

DETERMINATION OF PROTEIN COMPLEXED PARALYTIC SHELLFISH POISONING. MECHANISM OF THE REMEDIATION (DETOXIFICATION) OF CHEMICALS (PESTICIDE, HEAVY METALS, OTHER TOXIC CHEMICAL COMPOUNDS)

Purification and Characterization of Paralytic Shellfish Poison Binding Protein from Acanthocardia Tuberculatum

NADIA TAKATI[1], DRISS MOUNTASSIF[2], HAMID TALEB[3] AND MOHAMED BLAGHEN[1]*
[1] *Laboratory of Microbiology, Biotechnology and Environment, Faculty of Sciences Aïn Chock, University Hassan II-Aïn Chock, Km 8 route d'El Jadida, B.P. 5366 Mâarif, Casablanca, Morocco.*
[2] *Laboratory of Biochemistry and Molecular Biology, Faculty of Sciences Aïn Chock, University Hassan II-Aïn Choc*
[3] *Laboratory of the Marine Biotoxins, National institute of Halieutic Research, Casablanca, Morocco*

*To whom correspondence should be addressed: blaghen@facsc-achok.ac.ma

Abstract. A paralytic shellfish poison binding protein (PSPBP) was purified with 16.6 fold from the foot of the Moroccan cockles *Acanthocardia tuberculatum*. Using the affinity chromatography, 2.5 mg of PSPBP showing homogeneity on SDS-PAGE was obtained from 93 mg of crude extract. The purified PSPBP exhibits a specific activity of 2.777 mU/mg proteins and having estimated molecular weight of 181 kDa. Observation of single band equivalent to 88 kDa on SDS-PAGE under reducing conditions suggested it to be a homodimer. The optimal temperature and pH for the purified PSPBP were 30 °C and 7.

Keywords: PSP; PSPBP; *Acanthocardia tuberculatum*; foot

1. Introduction

Paralytic shellfish poison (PSP) represents a significant public health and safety hazard concern worldwide and causes severe economic losses globally due to bans on harvesting of contaminated shellfish and the need for costly monitoring programmes. It is a serious illness, with predominantly neurological symptoms and, in severe cases, respiratory paralysis and death (Taylor, 1988; Kao, 1993). Paralytic shellfish toxins (PSTs) block the influx of sodium ions (Na+) through excitable membranes, interrupting signal transmission and causing paralysis.

Several species of dinoflagellate, such as *Alexandrium tamarense* (Prakash, 1967), *Pyrodinium bahamense var. compressum* (Harada et al., 1982) and *Gymnodinium catenatum* (Oshima et al., 1993) are known to transmit their toxins to shellfish.

Bivalve molluscs, the primary vectors of PSP in humans, show marked intre-species variation in their capacity to accumulate PSTs, and contain a mixture of several toxins, depending on the species of algae, geographic area and type of marine animal involved (Bricelj and Shumway, 1998).

Comparison between toxin profile of cockles and toxigenic algae (*Gymnodinium catenatum*) attributed the high toxicity of cockles to the biotransformation of C-toxins (with low specific toxicity) into decarbamoylsaxitoxin (dcSTX) with relatively higher specific toxicity (Taleb et al., 2001; Sagou et al., 2005). However, the uptake, elimination and the transfer of toxin in other organs follows a characteristic pattern in each bivalve species (Lassus et al., 1989; Bricelj and Shumway, 1998).

The cockle (*Acanthocardia tuberculatum*) is known to sequester PSP toxins for a long time in its tissues even when the potentially toxin producing microalgae are not present (Vale and Sampayo, 2002). Indeed, *Acanthocardia tuberculatum* sequester PSP toxins preferably in non-visceral organs (Foot, gill and mantle) (Sagou et al., 2005). However, the toxin accumulation and metablolism systems in *Acanthocardia tuberculatum* have not been well clarified. The studies of soluble toxin-binding proteins in different organs are of particular interest as these proteins are suspected to be implicated in these systems.

In Morocco, toxic algae blooms have occurred along the Atlantic and Mediterranean coasts. In November 1994, algal blooms caused a large number of intoxications (64) linked with shellfish consumption (Tagmouti-Talha et al., 1996). Twenty three persons were hospitalized and four died (Akalay, 1995). Acute neurologic illness was associated with paresthesia of the mouth and extremities, a sensation of floating, headache, cerebellar symptoms such as ataxia and vertigo, cranial nerve dyfonction and muscle paralysis. One child's illness rapidly progressed to respiratory paralysis and death. This situation caused the economic loss for fishermen and a significant effect on the local economy due to decreased of fishery revenue. *Acanthocardia tuberculatum* is mainly exploited in the canning industry in Morocco and Spain. A Spanish team has demonstrated that after a thermal treatment of cockles at 116 °C for at least 51 min, toxicity drops to undetectable levels by mouse bioassay (Burdaspal et al., 1998). On the basis of the later processing, an exceptional European legislation allows harvesting in Spain of cockles with PSP toxins levels less than 300 mg STXeq./100 g meat (OJEC, 1996).

Following the ulterior studies that attributed the hightoxicity of cockles to the biotransformation of C-toxins (with low specific toxicity) into decarbamoylsaxitoxin (dcSTX) with relatively higher specific toxicity (Taleb et al., 2001).

The present work focused on determining the causes of PSP toxins persistence in *Acanthocardia tuberculatum* especially in its non visceral organs. In bivalve molluscs, the viscera are invariably found to contain the highest toxicity levels immediately following the exposure to toxic algal blooms. Our aims are to isolate the protein that bind PSTs and named paralytic shellfish poison binding protein (PSPBP) from the complex environment and to clarify its physiological function in *Acanthocardia tuberculatum*.

Here, we present the purification and characterization of a PSPBP from the Morrocan cockle *Acanthocardia tuberculatum*.

2. Materials and Methods

2.1. MATERIAL

Cockle (*Acanthocardia tuberculatum*) specimens used in this study were collected from two stations (Kaâ Srass and M'diq) in Mediterranean coast of Morocco.

Cockles were washed and separated into digestive gland (hepatopancreas), foot, mantle, muscle, and gills. The cockle tissues were kept at -20°C until use.

2.2. EXTRACTION OF PARALYTIC SHELLFISH POISON (PSP)

Toxicity analysis for all organs was carried out by mouse bioassay according to AOAC method (1990). Tissues (100 g) collected from toxic cockles (Kaâ Srass) were blended with 100 ml of 0.1 M HCl and boiled for 5 min. The volume of mixture was brought to 200 ml with bidistilled water, stirred and centrifuged at 3000 x g for 10 min. The recuperated supernatant (water soluble extract) was analyzed to evaluate the toxicity levels by injection intraperitoneally of 1 ml of the supernatant in albinos mice (20 ± 2 g). The values of toxicity are expressed in terms of STX equivalents per 100 g of shellfish meat (µg STXeq/100 g meat).

2.3. CRUDE EXTRACT PREPARATION

All procedures were carried out at 4°C. Tissues extracts was prepared using both Ultra-turrax and a potter homogenizer from non toxic cockles (M'diq). Samples of tissues were quickly weighed and then homogenized 1/3 (w/v) in 50 mM potassium phosphate buffer pH 7.4 containing 1 mM EDTA, 1 mM DTT. The homogenates were then centrifuged for 15 min at 3000 x g at 4°C. The supernatant was collected, stored at -20°C until use.

2.4. PROTEIN ASSAY

Protein content was measured according to the Bradford procedure, using bovine serum albumin (BSA) as standard (Bradford, 1976).

2.5. PURIFICATION OF PSPBP

For the preparation of the affinity chromatography, we used as support an activated gel of agarose (Affigel 10, Bio-Rad) on which we attached the PSP (Figure 1) extracted according to the official method of the A.O.A.C. 5 ml of the affigel 10 in the isopropanol are treated beforehand by the ethylene diamine (50 µl) dissolved in the DMF (250 µl) during 30 min in room temperature. Then, the affigel was washed with 50 mM phosphate buffer, pH 7.4. Afterwards, we added glutaraldehyde for a final concentration of 5% and one leaves incubated during 30 min. Secondly, the affigel was washed with 50 mM tampon phosphate, pH 7.4. After, the PSP (20 µg STXeq) was attached on the column during 4 h at room temperature. The excess of the PSP is eliminated by successive washings by 50 mM phosphate buffer, pH 7.4. The formed complex was established by the addition of NaBH4 (0.1 M).Then, the crude extract of protein foot was added and one lets incubate all the night with 4°C under a weak agitation. The mixture is recovered and deposited in the column (1×10 cm), which was washed after with 50 mM phosphate buffer, pH 7.4. The protein adsorbed on the column was eluted by 1 M Tris-HCl, pH 10. Fractions of 1 ml were collected in tubes containing beforehand 100 µl 100 mM glycocolle, pH 2.5 for an immediate neutralization of the pH. After elution, the column was rebalanced with 50 mM tampon phosphate, pH 7.4 for a future re-use. Then, the fractions which fix the PSP were analyzed and gathered for mousse bioassay (AOAC, 1990).

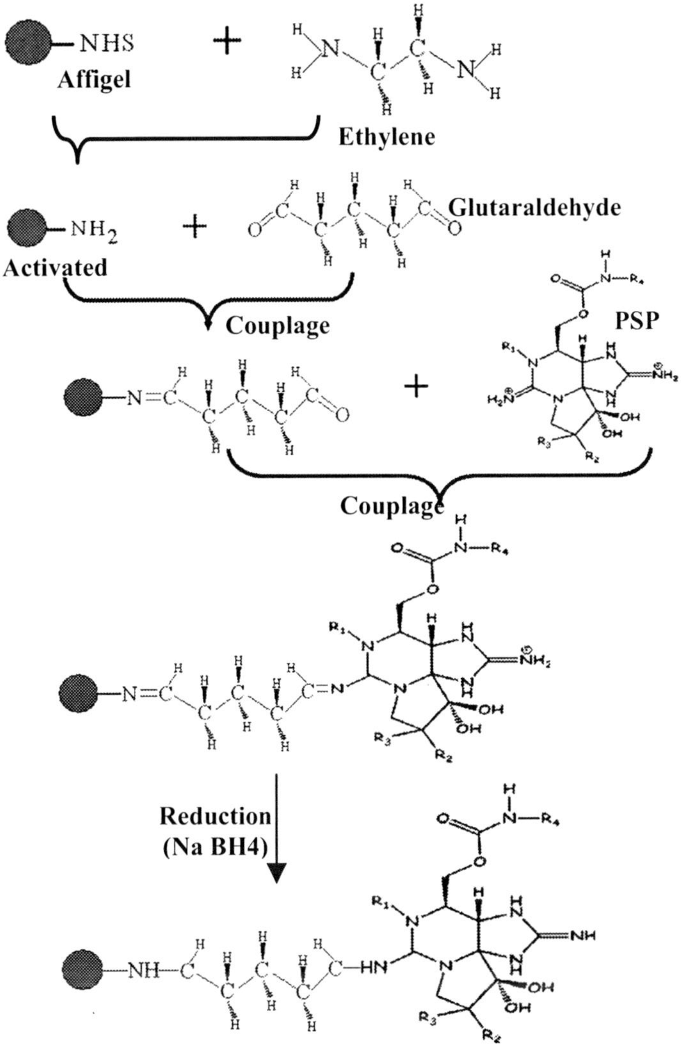

FIGURE 1. Protocol of the preparation of the affinity chromatography column. For experimental conditions, see Materials and Methods.

2.6. THIN LAYER CHROMATOGRAPHY ANALYSIS

Thin layer chromatography was carried out on dried silicagel plate (20 cm x 20 cm). 100 µg of PSPBP had been loaded before and after incubation with 0.75 µg STXeq at room temperature during 30 min. The migration was done with a medium containing pyridine/ethyl acetate/acetic acide/water

(75/25/15/20; v/v/v/v). After the end of the migration, the spots were revealed with iodine.

2.7. DENATURING POLYACRYLAMIDE GEL ELECTROPHORESIS

Sodium dodecyl sulfate polyacrylamide gel electrophoresis (SDS-PAGE) was performed as described by Laemmli (1970) on one-dimensional 12 % polyacrylamide slab gels containing 0.1 % SDS. Gels were run on a miniature vertical slab gel unit (Hoefer Scientific Instruments). Protein samples were prepared by boiling at 100°C for 5 min in 60 mM Tris-HCl, pH 6.8, 1% SDS, 10% glycerol, 1% mercaptoethanol and 0.01 % Bromophenol Blue.

After electrophoresis, gels were stained with Coomassie Brilliant Blue R-250 at 0.025 % (w/v) in methanol/acetic acid/water (4:1:5, v/v/v) for 30 min at room temperature (Diezel et al., 1972). Destained was done in methanol/acetic acid/water (4:1:5, v/v/v). The apparent subunit molecular weight was determined by measuring relative mobilities and comparing with the pre-stained SDS-PAGE molecular weight standards (Standard proteins, Sigma).

2.8. NATIVE MOLECULAR WEIGHT DETERMINATION

To determine the native molecular weight of purified PSPBP, non-denaturing polyacrylamide gel electrophoresis was carried out according to the method of Hedrick and Smith (1968). The separating gels (6, 8, 10 and 12 % polyacrylamide) were buffered with 1.5 M Tris-HCl (pH 8.8). The running buffer was composed of 25 mM Tris and 320 mM glycine (pH 8.6). All experiments were realized at 4°C. The electrophoresis running conditions, staining and destaining were as described for SDS-PAGE. The relative molecular weight of the native purified PSPBP was estimated using a commercial tetrameric urease (545 kDa), dimeric urease (272 kDa), dimeric BSA (132 kDa), monomeric BSA (66 kDa) and ovalbumin (45 kDa) (Sigma) as molecular weight markers. By constructing the Ferguson plot [(log (Rf × 100) versus the concentration of polyacrylamide gels (%)], the resulting slopes versus the standard native proteins of known molecular weight, permits to determine the molecular weight of purified PSPBP.

2.9. DETERMINATION OF OPTIMAL PH AND TEMPERATURE OF PURIFIED PSPBP ACTIVITY

The influence of the pH on the PSPBP activity was studied over a wide range of pH (from 4 to 10) using a mixture of different buffers that has different pKa (Tris, MES, HEPES, sodium phosphate and sodium acetate) adjusted to the same ionic strength than the standard reaction mixture. 0.75 μg STXeq was incubated with 100 μg of PSPBP at different pH during 30 min in room temperature. Controls containing PSP alone at different pH were also done. All the mixtures were after injected at swiss albinos mousse to evaluate the toxicity (AOAC, 1990).

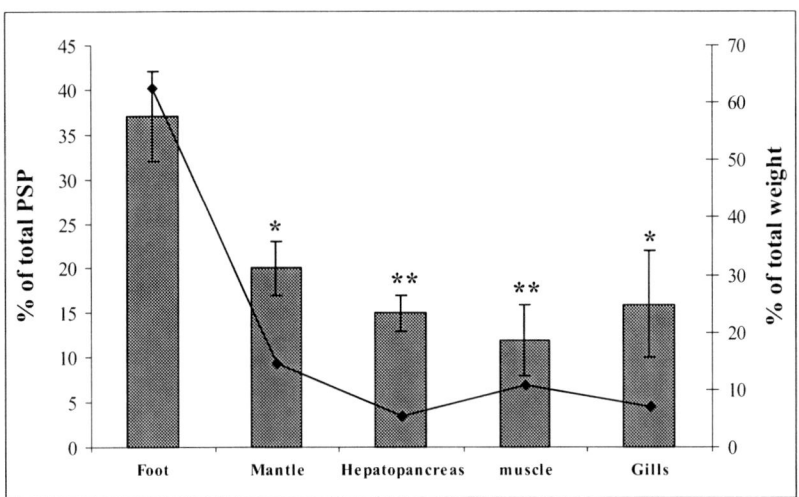

FIGURE 2. PSP and weight distributions in *Acanthocardia tuberculatum* tissues. PSP was extracted from all organs and the toxicity analysis was carried out by mouse bioassay. Weight percentage of all organs was also determined. Values are given as means ± standard deviations of three separate experiments. ** Very significantly different from the control value, $p<0.01$. * Significantly different from the control value, $p<0.05$ (Student t-test).

- The influence of the temperature on the PSPBP activity was carried out by incubating the assay mixtures at optimal pH 7.0 for 30 min in 15-60°C temperature range using a thermostated cuvette holder connected with a refrigerated bath circulator. 0.75 μg STXeq was incubated with 100 μg of PSPBP at different temperatures during 30 min at pH 7.0. Controls containing PSP alone at pH 7.0 and at different temperatures were also done. All the mixtures were after injected at swiss albinos mousse to evaluate the toxicity (AOAC, 1990).

2.10. STATISTICAL DATA ANALYSIS

In each assay, the experimental data represent the mean of three independent assays ± standard deviations. Means were compared using the

Student t-test. Differences were considered significant at the level p < 0.05 and very significant at the level p < 0.01.

3. Results

3.1. ANATOMICAL DISTRIBUTION OF PSP IN ACANTHOCARDIA TUBERCULATUM

The toxicity levels for all organs collected from toxic cockles *Acanthocardia tuberculatum* was carried out by mouse bioassay according to AOAC method (1990). The contribution of each organ to the total toxin body burden is function of both its absolute toxicity and relative weight contribution.

Figure 2 shows the percentage of toxicity and relative weight contribution in different organs of the cockles. There were substantial differences in the amounts of PSP toxins the various organs.

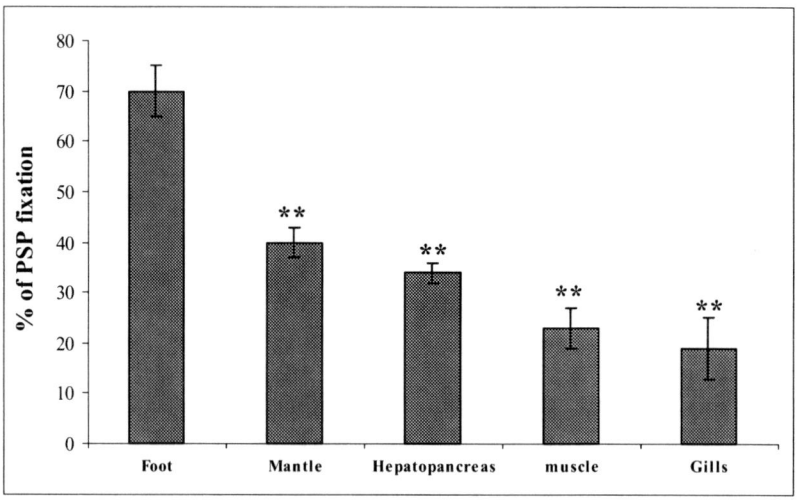

FIGURE 3. Anatomical distribution of PSPBP activity in *Acanthocardia tuberculatum*. Each tissue extract (1 mg) was added at 0.75 µg STXeq and incubated at room temperature and pH 7.0. After 30 min, the toxicity analysis was carried out by mouse bioassay. Values are given as means of three separate experiments. ** Very significantly different from the control value, p<0.01 (Student t-test).

The toxicity decreases as follows: foot> mantle > gills >digestive gland > muscle. Indeed, the foot accumulates the highest toxicity which represents 37 % of total toxicity and the majority of toxicity (85 %) is concentrated in the non-visceral bodies (Foot, mantle, adductor muscle and gills). Whereas, the hepatopancreas represents only 15 % of toxicity. This is can be explained by the fact that the foot represents the most significant contribution to the total body mass (62.36 %) (Figure 2).

3.2. ANATOMICAL DISTRIBUTION OF PSPBP ACTIVITY IN ACANTHOCARDIA TUBERCULATUM

Each tissue extract (1 mg) was added at 0.75 µg STXeq and incubated at room temperature and pH 7.0 during 30 min. Decrease of toxicity in this case can be explained by the ability of extract of each tissue for binding the PSP toxins. The ability of binding the PSP toxins shows anatomical distribution.

The results obtained in Figure 3 show that the foot binds 37 % of the toxicity, come thereafter the mantle (21.7 %), the hepatopancreas (18.6 %), the adductor muscle (12.4 %) and finally the gills (10.5 %). Thus, the foot presents a capacity of binding the PSP toxins of 66%, Indeed, it is able to bind the 2 thirds of initial toxicity. And thus we determined the foot to be a better source of PSPBP.

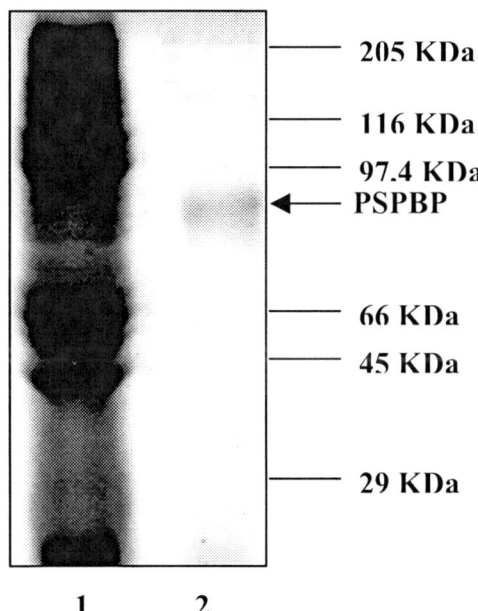

FIGURE 4. Sodium dodecyl sulfate polyacrylamide gel electrophoresis (SDS-PAGE) pattern showing the purification of the PSPBP purified from *Acanthocardia tuberculatum*. Lane 1 and 2 represent respectively crude extract and purified PSPBP fraction. For experimental conditions, see Materials and Methods.

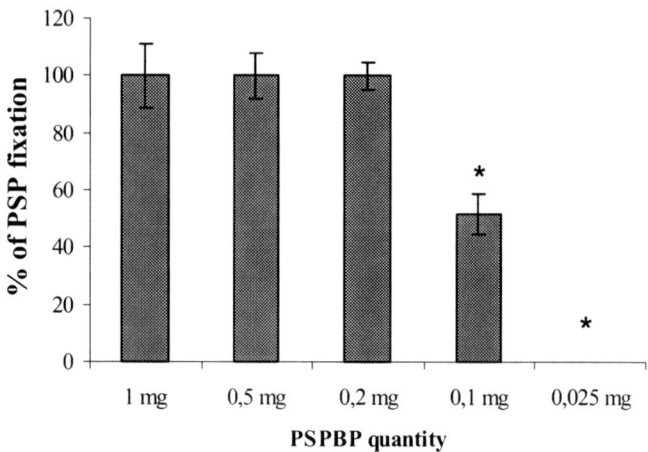

FIGURE 5. Purified PSPBP activity at different quantities. Different quantities of purified PSPBP (1 mg, 0.5 mg, 0.2 mg, 0.1 mg and 0.025 mg) were incubated separately with 0.75 μg STXeq and incubated at room temperature and pH 7.0. After 30 min, the toxicity analysis was carried out by mouse bioassay. Values are given as means of three separate experiments.
* Very significantly different from the control value, p<0.01.

3.3. PURIFICATION OF PSPBP FROM *ACANTHOCARDIA TUBERCULATUM*

A total amount of about 93 mg of protein, corresponding to approximately 15.48 mU of PSPBP, was obtained from foot crude extract of *Acanthocardia tuberculatum*. The purification of the PSPBP was performed by one step (affinity chromatography) (Figure 4).

TABLE 1. Summary of PSPBP purification from *Acanthocardia tuberculatum*.

Preparation	Total protein (mg)	Total activity (mU)	Specific activity (mU/mg)	Purification (fold)	Yield (%)
Crude extract	93	15.484	0.1665	1	100 %
Affinity chromatography	2,5	6.944	2.777	16.6	44.8 %

Table 1 summarizes a representative purification protocol. 2.5 mg of pure PSPBP was obtained. PSPBP was purified with 16.6 fold, a specific activity of 2.777 mU/mg proteins and a yield of 44.8 %. PSPBP test activity was performed using different quantity of purified protein incubated with

0.75 µg STXeq at room temperature and pH 7.0 during 30 min (Figure 5). The results obtained reveal that the fixation of the PSP by the purified PSPBP is higher and total until a quantity of 100 µg of PSPBP after which fixing decreases.

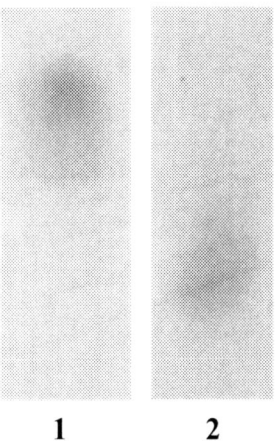

FIGURE 6. Chromatogram of the thin layer chromatography showing the PSPBP alone and the PSPBP incubated with PSP. 100 µg of PSPBP had been loaded before and after incubation with 0.75 µg STXeq at room temperature during 30 min. For experimental conditions, see Materials and Methods.

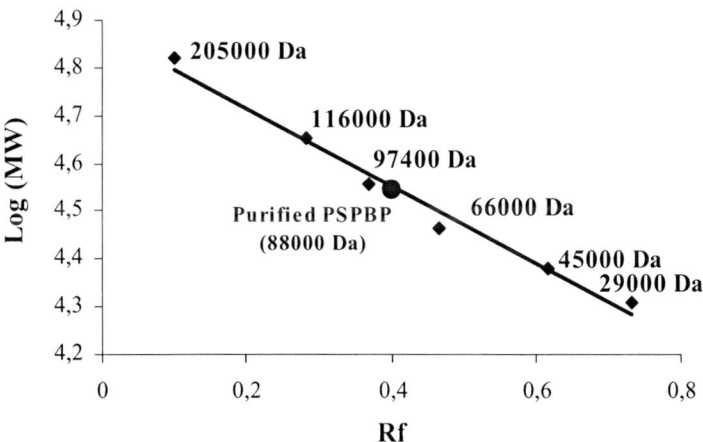

FIGURE 7. Determination of the molecular weight of the purified PSPBP from *Acanthocardia tuberculatum* foot by gel electrophoresis on denaturing conditions. Molecular weight marker proteins were commercial myosine (205 kDa), β-galactosidase (116 kDa), phosphorylase B (97.4 kDa), albumin (66 kDa), ovalbumin (45 kDa) and carbonic anhydrase (29 kDa). The plot represents the relative mobilities of proteins vs. Log (Molecular Weight).

3.4. THIN LAYER CHROMATOGRAPHY ANALYSIS

The revelation by iodine of the PBPSP alone and complex PSP-PBPSP made it possible to detect a delay of migration of the complex compared to the PBPSP alone. This delay is probably due to the binding of the PSP on PBPSP (Figure 6).

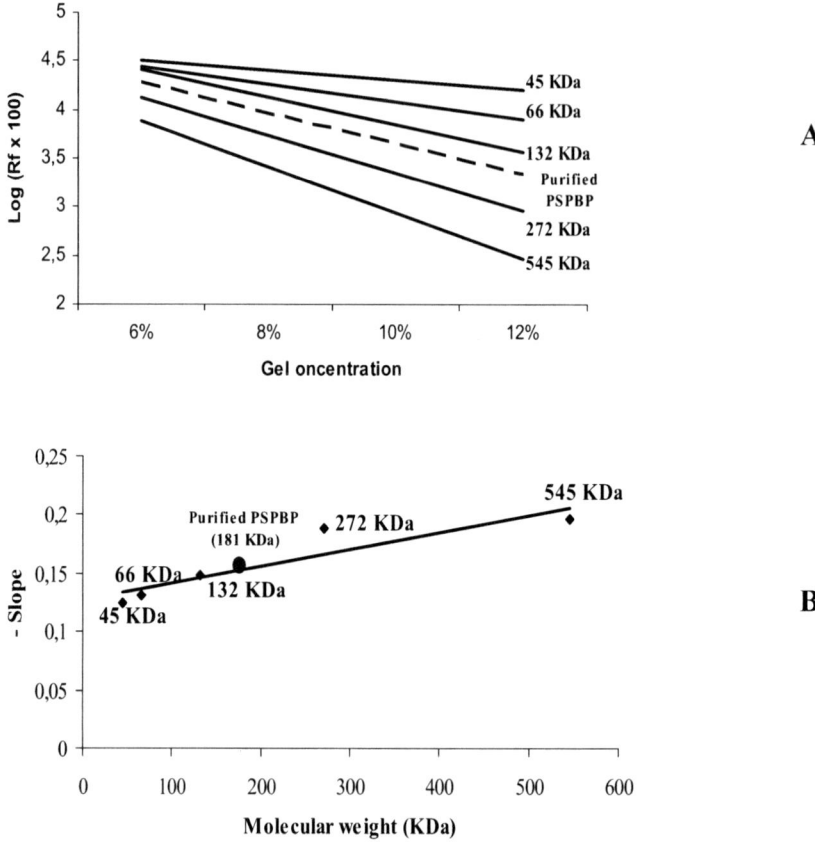

FIGURE 8. Determination of the native molecular weight of the purified PSPBP from *Acanthocardia tuberculatum* foot by native gel electrophoresis of various concentrations of polyacrylamide (6, 8, 10 and 12 %). Molecular weight marker proteins were commercial tetrameric urease (545 kDa), dimeric urease (272 kDa), dimeric BSA (132 kDa), monomeric BSA (66 kDa) and ovalbumin (45 kDa) (Sigma). (A) represents the relative mobilities of proteins plotted as Log ($R_f \times 100$) vs. gel concentration. A plot of the obtained slopes vs. molecular weight was linear and used to determine native PSPBP molecular weight (B).

3.5. MOLECULAR WEIGHT DETERMINATION OF PSPBP

The SDS-PAGE analysis (Figure 4) of the different fractions obtained during the purification procedure showed one protein band corresponding to the *Acanthocardia tuberculatum* PSPBP subunit whose molecular weight could be estimated to 88 kDa (Figure 7).

To determine the molecular weight of the native PSPBP, an electrophoresis in non denaturing system was performed using different separating gels (6, 8, 10 and 12 % polyacrylamide). From the Ferguson plot (Figure 8), a value of 181 kDa was estimated for the molecular weight of the native PSPBP. This result, compared with that obtained from SDS-PAGE, which shows a single band corresponding to 88 kDa protein (Figure 7), suggests that PSPBP purified from *Acanthocardia tuberculatum* should have an homodimeric structure.

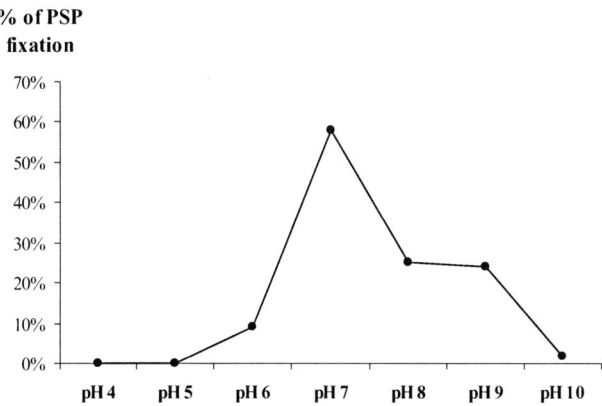

FIGURE 9. Purified PSPBP activity from *Acanthocardia tuberculatum* foot in 4.0-10.0 pH range using a mixture of different buffers. Values are given as means of three separate experiments. For experimental conditions, see Materials and Methods.

3.6. INFLUENCE OF PH AND TEMPERATURE ON THE PURIFIED PSPBP ACTIVITY

The pH activity profile of purified PSPBP was determined in a pH range from 4.0 to 10.0 using a mixture of different buffers (Figure 9). The protein had a typical bell-shaped profile covering a broad pH range. The maximum activity was occurred at about pH 7.0.

The influence of temperature on enzymatic activities was determined between 15°C and 60°C at pH 7.0 (Figure 10). The optimal temperature for PSPBP activity was 30°C and activity decrease significantly above 40°C.

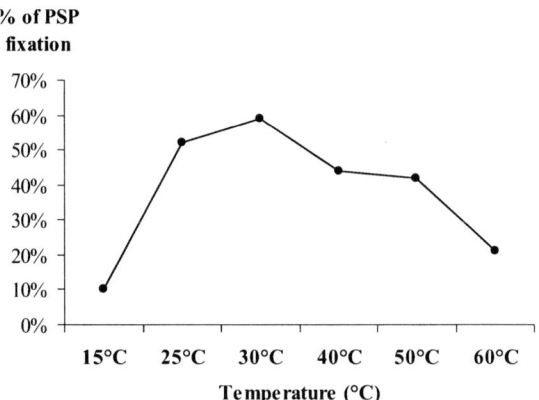

FIGURE 10. Purified PSPBP activity from *Acanthocardia tuberculatum* foot at temperature range from 15 to 60°C. Values are given as means of three separate experiments. For experimental conditions, see Materials and Methods.

4. Discussion

The present study shows a large variation in PSP toxins distribution and binding ability among different tissues of Moroccan cockles *Acanthocardia tuberculatum* and describes a rapid method for purification of new toxin binding protein named Paralytic sellfish poisons binding protein (PSPBP) from their foot.

A. tuberculatum showed a persistant contamination with high levels of toxicity, dcSTX and STX are the toxins that account for the most of this toxicity. Relative partitioning of PSP toxins among tissues is variable and the foot is most toxic organ followed by others organs (Taleb et al., 2001; Sagou et al., 2005). Similar studies were carried out in Spain (Berenguer et al., 1993) and Portugal (Vale and Sampayo, 2002) on the same species, revealed the conservation of PSP toxins mainly in the foot. Our work confirms also that the most toxicity is concentrates in foot (Figure 2) mainly in red part.

Indeed, *A.tuberculatum* is among slowly detoxifying bivalves of wich the relative toxin proportion increases gradually in non visceral tissues and can even surpass that of the digestive gland. The ulterior studies show that the highest PSP levels in bivalve species can be attributed to their nerves insensitive to PSP toxins. Consequently, bivalves didn't exhibit physiological and behavioral mechanisms for their detoxification (Bricelj et al., 1990). The mechanism of tissue-specific retention of PSP toxins in some species remains to be elucidated. Price and Lee (1971, 1972) suggested that STX was electrostatically and reversibly bound in the melanin fraction of pigmented tissues, but subsequent work does not validate this hypothesis (Beitler, 1992).

Incubation of extracted toxins by a hot acid AOAC method with tissue extracts from incontaminated cockles (*A. tuberculatum*) has show the decraese of total toxicity. This decrease can be explained by the PSP toxins binding ability of each crude extract. The foot presents the most PSTs binding activity (Figure 3). Thus, red foot is a good starting source for the purification of PSPBP.

According the pigmentation, foot shows two parts: red and yellow. *A.tuberculatum* concentrates most of the toxins as dcSTX and STX rapidly in foot especially in their red part (data not show). However, the toxin accumulation and prolonged retention systems in *A. tuberculatum* have not been well clarified. The studies of soluble toxin-binding proteins in differents organs are of particular interest as these proteins are expected to be implicated in these systems.

This is the first report about the isolation of a protein that binds PSP toxins in *A. tuberculatum*. Since it is difficult to purify PSPBP from foot of *A. tuberculatum* due to the variety of proteins, the large amounts of lipophilic substances in extracts, and after several tests. we describe in this paper a relatively simple and rapid method for the purification of PSPBP, a soluble PSP-binding protein form the foot of *A. tuberculatum* (Figure 1).

Method of purification (affinity column) is most effective than other methods (data not shown) since it gave pure protein with 44.8 % as yield and 17 as factor of purification and an important quantity of pure protein (2.5 mg of PSPBP from 93 mg of total proteins) (Table 1), in addition to the simplicity of the method which does not require much time. The purified PSPBP exhibits an activity of 2.777 mU/mg proteins and having estimated molecular weight of 181 kDa (Figure 8). Observation of single band equivalent to 88 kDa on SDS-PAGE under reducing conditions (Figures 4 and 7) suggested it to be a homodimer. The optimal temperature and pH for the purified PSPBP were respectively 30 °C and 7.0 (Figures 9 and 10).

The specificity of PSPBP to PSP extracted from *A. tuberculatum* was determined by thin layer chromatography analysis whish shows a delay of migration of the complex PSPBP-PSP compared to the PBPSP alone. This delay is probably due to the binding of the PSP on PSPBP (Figure 6).

Toxin profiles of *Gymnodinium catenatum* (toxinogen algae) and *A. tuberculatum*, (plankton feeder) differed significantly. *A. tuberculatum* contained mainly dcSTX (65%) and STX (8%), whereas *Gymnodinium catenatum* possessed C-toxins (77%). Supposing that all PSP toxins detected in *Acanthocardia tuberculatum* are derived from the PSP toxins contained in *Gymnodinium catenatum*, this difference may possibly be explained by both chemical and enzymatic conversion as suggested by Taleb et al. (2001). Thus, PSPBP may be binds especially to dcSTX and STX. This result was confirmed by thin layer chromatography analysis (Data not shown).

Other receptors for the saxitoxin were determined, by ulterior studies at other species, which bind this phycotoxin with different affinities and specificities and present very distinct molecular structures and physicochemical properties. On the one hand, saxiphilin, a soluble protein, that binds STX with high affinity ($K_d \sim 0.2$ nM) and specifity (Mahar et al., 1991). Saxiphilin was first discovered in frogs (Doyle et al., 1982; Moczydlowski et al., 1988), but similar soluble STX bindig activity has been observed in diverse species of arthropods, amphibians, reptiles, and fish (Llewellyn et al., 1997). Saxiphilin appears to be a monomeric (91 kDa) protein that is structurally related to the transferin family of proteins (Morabito and Moczydlowski, 1994a, 1995) but wich binds STX instead of Fe^{3+} (Li and Moczydlowski, 1991). Unlike the Na channel, saxiphilin does not possess any affinity for TTX making it a potentially valuble new means for specially detecting PSTs (Llewellyn et al., 1998; Negri and Llewellyn, 1997). Another receptor for the (saxitoxin) STX and tetrodotoxin (TTX) was purified from plasma of the puffer fish, *Fugu pardalis*, and named puffer fish saxitoxin and tetrodotoxin binding protein (PSTBP) (Yamashita et al., 2001). This glycoprotein possessed a binding capacity of 10.6 ± 0.97 nmol/mg protein and a Kd of 14.6 ± 0.33 nM for [^{3}H] saxitoxin in equilibrium binding assays. PSTBP seemed to consist of noncovalently linked dimers of a 104 kDa single subunit whose protein partwas estimed to be of 42 kDa. The predicted amino-acid sequences of PSTBP were not homologous to that of saxiphilin, or sodium channels, but their N-terminus sequences were homologous to that of the reported tetrodotoxin binding protein from plasma of *Fugu niphobles* (Matsui et al., 2000). Presumably, PSTBP is involved in accumulation and/or excretion of toxins in puffer fish.

In summury, prolonged retention (severs mounths to years) of PSP toxins as dcSTX and STX is a characteristic of *A. tuberculatum* that can be explained not only by the specific retention of dcSTX (Taleb et al., 2001) and differentiel accumulation of PSP toxins in non visceral organs (Sagou et al., 2005) but also by the presence of soluble toxin-binding proteins (PSPBP) in *A. tubercumatum* mainly in foot.

Future studies aimed at the sequencing the purified PSPBP and determining its tridimensionnel structure in order to understand its interactions with the PSP toxins. Cloning of the gene encoding for this protein, to see homologies with the already receptors of saxitoxin and saxitoxin derivatives, would be also of interest.

References

Akalay, D., 1995. Intoxications alimentaires par les coquillages (epidemic ponctuelle de novembre 1994, à propos de 6 cas). Thèse de Doctorat en Médecine, faculty of Medecine and Pharmacy of Casablanca, Morocco.

AOAC, 1990. Paralytic shellfish poison, biological method, final action. In: AOAC (ed.), Official Methods of Analysis, 15th Ed., Arlington, VA, method n° 959.08.

Beitler, M.k., 1992. Uptake retention and fate of PSP toxins in the butter clam (saxidomus giganteus). Ph. D. Thesis University of Washington, Seattle, W.A.

Berenguer J.A., Gonzalez L., Jimenez I., Legarda T.M., Olmedo J.B., Burdaspal P.A., 1993. "The effect of commercial processing on the paralytic shellfish poison (PSP) content of naturally contaminated Acanthocardia tuberculatum". L. Food Add. Contam., 10(2): 217–230.

Bradford, M., 1976. A rapid and sensitive method for the quantitation of microgram quantities of protein utilizing the principle of protein dye binding. Analytical Biochem. Ibid., 72: 248–254.

Bricelj V.M., Lee J.H., Cembella A.D., Anderson D.M., 1990. Uptake kinetics of paralytic shellfish toxins from the dinoflagellate Alexandrium fundyense in the mussel Mytilus edilus. Mar. Ecol. Prog. Ser., 63: 177–188.

Bricelj, V.M., Shumway, E., 1998. Paralytic shellfish toxins in bivalve molluscs: occurrence, transfer kinetics, and biotransformation. Rev. Fish. Sci., 6: 315–383.

Burdaspal, P.A., Bustos, J., Legarda, T.M., Olmedo, J.B., Vigo, M., Gonzalez, L., Berenguer, J.A., 1998. Commercial processing of Acanthocardia tuberculatum L. naturally-contaminated with PSP: evaluation after one year industrial experience. In: Reguera B., Blanco, J., Fernandez, M.L., Wyatt, T., (Eds.), Harmful Algae. Xunta de Galicia, IOC of UNESCO. Spain, pp. 241–244.

Diezel, W., Liebe, S., Kopperschlager, G., Hofmann, E., 1972. Association of proteins during polyacrylamide gel electrophoresis. Acta Biol Med Ger.; 28 (1):27–37.

Doyle, D.D., Wong, M., Tanaka, J., Barr, L., 1982. Saxitoxin binding sites in frog mycocardial cytosol. Science, 215: 1117–1119.

Harada, T., Oshima, Y., Kamiya, H., Yasumoto, T., 1982. Confirmation of paralytic shellfish toxins in the dinoflagellate Pyrodinium bahamense var. compressa and bivalves in Palau. Nippon Suisan Gakkaishi 48, 821–825.

Hedrick, J.L., Smith, A.J., 1968. Size and charge isomer separation and estimation of molecular weights of proteins by disc gel electrophoresis. Arch. Biochem. Biophys., 126: 155–164.

Kao, C.Y., 1993. Paralytic shellfish poisoning, pp. 75–86. In: Algal toxins in seafood and drinking water, IR Falconer (ed.). Academic Press, London and New York.

Laemmli, U.K., 1970. Cleavage of structural proteins during the assembly of the head of bacteriophage T4. Nature, 227: 4668–4673.

Lassus, P., Fremy, J.M., Ledoux, M., Bardouil, M., Bohec, M., 1989. Patterns of experimental contamination by Protogonyaulax tamarensis in some French commercial shellfish. Toxicon, 27 (12): 1313–21.

Li Y., Moczydlowski E., 1991. Purification and partial sequencing of saxiphilin, a saxitoxin-bindig protein from the bullfrog, reveals homology to transferrin. J. bio. Chem., 266: 15481–15487.

Llewellyn, L.E., 1997. Haemolymph protein in xanthid crabs: Its selective binding of saxitoxin and possiblerole in toxin bioaccumulation. Mar. Biol., 128: 599–606.

Llewellyn, L.E., Bell, P.M., Moczydlowski, E.G., 1997. Phylogenetic survey of soluble saxitoxin-binding activity in pursuit of the function and molecular evolution of saxiphilin, a relative of transferrin. Proc. R. Soc. Lond. B, 264: 891–902.

Llewellyn, L.E., Doyle, J., Negri, A.P., 1998. A high-through-put, microtiter plate assay for paralytic shellfish poison using the saxitoxin-specific receptor, saxiphilin. Anal. Biochem. 261: 51–56.

Mahar, J., Lukacs, G.L., Li, Y., Hall, S., Moczydlowski, E., 1991. Pharmacological and biochemical properties of saxiphilin, a soluble saxitoxin-binding protein from the bullfrog (Rana catesbeiana). Toxicon, 29: 53–71.

Matsui, T., Yamamori, K., Furukawa, K., Kono, M., 2000. Purification and some properties of a tetrodotoxin binding protein from the blood plasma of Kusafugu, Takifugu niphoblos. Toxicon, 38: 463–468.

Moczydlowski, E.M., Mahar, J., Ravindran, A., 1988. Multiple saxitoxin-binding sites in bullfrog muscle: tetrodotoxin-sensitive sodium channels and tetrodotoxin-insensitive sites of unknown function". Mol. Pharmacology, 33: 202–211.

Morabito, M.A., Llewellyn, L.E., Moczydlowski, E.G., 1995. Expression of saxiphilin in insect cells and localization of the saxitoxin-binding site to the C-terminal domain homologous to the C-lobe of transferrin. Biochemistry, 34: 13027–13033.

Morabito, M.A., Moczydlowski, E., 1994a. Molecular cloning of bullfrog saxiphilin: a unique relative of the transferrin family that binds saxitoxin. Proc. Natl. Acad. Sci. USA, 91: 2478–2482.

Negri, A., Llewellyn, L., 1998. Comparative analyses by HPLC and the sodium channel and saxiphillin 3H-Saxitoxin receptor assays for paralytic shellfish toxins in crustaceans and molluscs from tropical North West Australia. Toxicon, 36: 283–298.

Official Journal of the European Communities, 1996. n ; L 15, 20.1.96: 46–47.

Oshima, Y., Blackburn, S.I., Hallegraeff, G.M., 1993. Comparative study on paralytic shellfish toxin profiles of the dinoflagellate Gymnodinium catenatum from three different countries. Mar. Biol, 116: 471–476.

Prakash, A., 1967. Growth and toxicity of a marine dinoflagellate, Gonyaulax tamarensis. J. Fish Res. Bd. Canada, 24: 1589–1600.

Price, R.J., Lee, J.S., 1971. Interaction between paralytic shellfish poison and melanin obtainted from butter clam (Saxidomus giganteus) and syntetic melanin. J.Fish.Res. Bd. Canada, 28: 1789–1792.

Price, R.J., Lee, J.S., 1972. Paralytic shellfish poisoning in British Columbia. J.Fish.Res. Bd. Canada, 29: 1675–1658.

Sagou, R., Amanhir, R., Taleb, H., Vale, P., Blaghen, M., Loutfi, M., 2005. Comparative study on differential accumulation of PSP toxins between cockle (Acanthocardia tuberculatum) and sweet clam (Callista chione) ". Toxicon, 46: 612–618.

Tagmouti-Talha, F., Chafak, H., Fellat-Zarrouk, K., Talbi, M., Blaghen, M., Mikou, A., Guittet, E., 1996. Detection of toxins in bivalves on the Moroccan coasts. In: Yasumoto, T., Oshima, Y., Fukuyo, Y. (Eds.), Harmful and Toxic Algal Blooms. IOC of UNESCO, Paris, pp. 85–87.

Taleb, H., Vale, P., Jaime, E., Blaghen, M., 2001. Study of paralytic shellfish poisoning toxin profile in shellfish from the Mediterranean shore of Morocco". Toxicon, 39 (12): 1855–1861.

Taylor, S.L., 1988. Marine toxins of microbial origin. Food Technology 42 (3):94–8.

Vale, P., Sampayo, M.A.M., 2002. Evaluation of marine biotoxin's accumulation by Acanthocardia tuberculatum from Algarve, Portugal. Toxicon, 40 (5): 511–517.

Yotsu-Yamashita, M., Sugimoto, A., Terakana, T., Shoji, Y., Migazawa, T., Yasumoto, T., 2001. Purification, characterization and cDNA cloning of a novel soluble saxitoxin and tetrodotoxin binding protein from plasma of the puffer fish, Fugu pardalis. Eur. J. Biochzm., 268: 5937–5946.

EUROPEAN INSTITUTIONS FOR CONTROLLING CHEMICAL AIR POLLUTION: AN ANALYSIS OF CLRTAP-EUROPEAN UNION INTERPLAY

HEIKE SCHROEDER* AND DAYNA YOCUM
Donald Bren School of Environmental Science and Management
University of California, Santa Barbara
Santa Barbara, CA 93106, USA

*To whom correspondence should be addressed: schroeder@bren.ucsb.edu

Abstract. Chemical air pollution in Europe has been tackled by multiple institutions, in particular the Convention on Long Range Transboundary Air Pollution to Improve Air Quality (CLRTAP) and European Union air pollution policy. The two regimes differ in terms of membership and policymaking capacity which has, over the decades, resulted in significant interplay and the formation of linkages between them that has served to shape both institutions. Additionally, they have pushed each other toward increasingly stringent pollution targets, encouraged shared research and development of air pollution science, and have filled gaps in each other's policies.

Keywords: chemical air pollution policy; European institutions, EU Air Quality Framework Directive, Long Range Transboundary Air Pollution Convention (CLRTAP), UNECE, 1999 Gothenburg Protocol, NEC Directive, linkages, interplay.

1. Introduction

European air pollution policy has undergone significant developments and faced many challenges in its roughly 30 year history. To some extent, at least, environmental policy has been affected by interplay between horizontally and vertically linked institutions. This chapter seeks to investigate how this interplay has strengthened and/or weakened European environmental institutions for controlling air pollution, and to what extent they are institutionally integrated or ought to be. To demonstrate this interplay, this chapter will focus primarily on the United Nations Economic Commission for Europe's (UNECE) Convention on Long Range Transboundary Air Pollution to Improve Air Quality (CLRTAP) and its subsequent protocols, and the European Union's (EU) air pollution policy, particularly its Air Quality Framework Directive on ambient air and subsequent daughter directives, and National Emissions Ceilings Directive (NEC).

Both a local and transboundary problem, air pollution is caused by emissions of harmful pollutants which can travel hundreds of miles and give rise to environmental and health impacts far away from their source location. Sulfur dioxide, for example, forms into fine particulate soot, which causes breathing problems and particularly affects citizens with asthma, and it is also a major precursor of acid rain, which leads to acidified soils, lakes, and streams. Thus, effective regulations of one nation-state have direct impacts on neighboring nation-states, as air and water pollution travels across national borders.

Efforts to reduce transboundary air pollution have been undertaken in various multilateral and intragovernmental fora. The most comprehensive of these are the UNECE CLRTAP and two EU policies, the Air Quality Framework Directive and the National Emissions Ceilings Directive. By identifying linkages between these institutions, this chapter will examine how they have developed and impacted each other over the years. In doing so, this chapter will investigate past, present, and potential future trends in interplay between these agreements.

2. Theoretical Framework

Interplay among institutions is likely to occur in a setting where multiple institutions manage the same environmental problem. Institutions are understood as social artifacts that make up the 'rules of the game.' More specifically, they are "clusters of rights, rules and decision-making procedures that give rise to social practices, assign roles to participants in these practices and govern interactions among players of these roles." (Young 2002) Regimes – usually referred to as single-issue institutions or "sets of implicit or explicit principles, norms, rules, and decision-making procedures around which actors' expectations converge in a given area of international relations" (Krasner 1982) – do not operate in a vacuum, but rather are constantly affected by a multitude of other institutions. The resultant 'regime density' (Young 1996) – others call it 'treaty congestion' (Brown Weiss 1993) – will likely lead to interplay among these institutions governing the same public good – clean air in this case. Interplay may cause tension or conflict that can undermine the regimes' effectiveness with regard to addressing the problem at hand. It may also generate incentives, pressure, or competition for regimes to develop alongside each other, or for one regime to catch up with another in terms of adopting the higher standards and targets set by another regime, or in terms of adopting a new approach introduced by another regime.

Young et al. (2005) differentiate between two types of interplay, referring to 'vertical interplay' as interplay across levels of social organization and 'horizontal interplay' as interplay within the same level of social organization. Vertical interplay may occur between an international organization with a large, possibly universal, membership and a supranational regional entity with restricted regional membership, such as the EU. Horizontal interplay may occur between two directives within the EU or two international organizations. The issue of membership is important in this context; it may give rise to interplay if membership is different among these institutions. This chapter examines vertical interplay between UNECE and the EU, as they are both regional organizations with limited membership.

In addition, interplay is distinguished by characterizing it as either intended or unintended; while the former tends to be politically motivated (political interplay), the latter would likely result from an unintended linkage (functional interplay). Political interplay might arise when players seek to link (or de-link) institutions intentionally or engage in institutional bargaining in the interest of pursuing certain objectives or synergies from collaboration or linkage. Political interplay can thus arise from the consequences of institutional design. Functional interplay occurs when a problem addressed by multiple institutions is linked in biophysical or socioeconomic terms, meaning that it stems from a systemic interdependence among the related institutions. (Young 2002; Schroeder forthcoming) Interplay between CLRTAP and the EU comprises both intended and unintended features.

Interplay can be either 'reciprocal' or 'unidirectional.' Usually it occurs somewhere in between these polar categories, in the form of being, to some extent, asymmetrical. (Young 1996) This would result in one institution exerting somewhat more influence over the other than vice versa. Unidirectional linkages arise when the operation of one institution affects others significantly without triggering notable responses. Vertical interplay tends to trigger such unidirectional interaction. (Schroeder forthcoming) In the case examined in this chapter, interplay is reciprocal, but the level of reciprocity varies over time.

Regimes, such as the CLRTAP and EU framework directives, typically start out with a framework agreement that lays out the general principles, the ultimate objective, and the institutional architecture. Once member countries agree on the steps that must be taken to solve the problem at hand, protocols or daughter directives are negotiated and adopted. As a convention, CLRTAP allows for protocols that amend, refine, and update the initial principles and objectives. The EU Air Quality Framework Directive, on the other hand, incorporates previous directives on air pollution abatement to enhance inter-policy coordination and serves as an umbrella framework, under which daughter directives can be ratified as deemed necessary. This structure allows for the possibility of incorporating future scientific advances and for the passage of regulations on specific substances without waiting for decisions about more controversial topics, thus increasing flexibility.

The CLRTAP's so-called 'Convention-protocol approach' (Susskind 1993) is an example of institutional nesting. (Young 1996) It gives the regime the capacity to effectively adapt to changes that may occur such as the advancement of scientific knowledge or alternations in the biophysical properties of the environmental problem. (Schroeder forthcoming) The relationship between CLRTAP and the EU, alternatively, is one in which initially 'overlapping institutions' may be moving toward some form of 'institutional cluster.' Overlapping institutions are defined as individual regimes that were formed for different purposes and largely without reference to one another, but whose policy goals and regulations intersect or overlap. Such regimes would be addressing a common issue or problem with different policy objectives. This may lead them to cause substantial impacts on one another. Institutional clustering refers to a situation where several institutional arrangements are clustered into an institutional framework or package. This is most likely to be the result of political bargaining or increasing economies of scale in the operation of the regime, where the end result is likely to constitute a net benefit to all participants. (Young 1996; Schroeder forthcoming).

Overlapping institutions evidently rely on linkages, but such linkages do not necessarily function effectively. Linkages may be established that prove to be unproductive, or even disabling, for such overlapping institutions. Furthermore, missing linkages may lead to ineffective governance structures and processes. Establishing the right set of linkages tends to create momentum and opportunities for effective policy integration.

Recognizing that chemical air pollution is a multi-level and multi-faceted problem, and that today's world is characterized by high levels of institutional interdependence, the most effective governance approach is likely to be one characterized by integration and institutional linkage across and within levels of social organization. Without these linkages, separate decision-making processes may in practice easily lead to uncoordinated and fragmented decision-making. (Oberthür and Gehring 2001) This also poses the question of how much institutional complexity is desirable, and whether there exists a saturation point beyond which institutions and institutional linkages are in fact counterproductive and should be streamlined and simplified. (Ostrom 2005).

3. European Air Pollution Policy

In many ways, UNECE and the European Union have impacted each other, and their development of air pollution policy has occurred somewhat in parallel. While UNECE's CLRTAP had a greater influence initially on the EU than vice-versa, the CLRTAP is now impacted more strongly by EU policy. One determining factor here has been membership - or limits to membership. For example, in the early 1970s, Sweden discovered that its lakes were seriously acidic, which was caused mainly by air pollution from Britain. Since Sweden was not a member of the European Community, and Britain was still wavering on the question of membership (entering the EC in 1973), Sweden therefore did not present its case in the context of the EC institutions, but instead chose UNECE, which at the time had a membership comprising both North European and Soviet bloc countries. Sweden's push within UNECE towards adopting international air pollution policy served as the impetus to institutional action. Another such example is the wide-scale forest destruction in Germany in the 1980s resulting from acid rain ('Waldsterben'). Since the pollution originated in Eastern Europe, the EC was again not the proper forum, so UNECE was able to help mitigate the problem.

Also interesting to explore is the tendency of authorities to attempt to out-compete each other, intending to be perceived as the most relevant authority in tackling a given problem. To draw out the dynamics between the two institutions, this section will reference the historical landmarks in air pollution regulation from both governing bodies, and examine the history and development of air pollution policy in Europe more generally.

3.1. THE CONVENTION ON LONG RANGE TRANSBOUNDARY AIR POLLUTION (CLRTAP) AND ITS PROTOCOLS

The first major push toward combating transboundary air pollution in Europe on a broad regional basis was manifested in the 1979 Convention on Long-Range Transboundary Air Pollution (CLRTAP, in force in 1983), collaboratively composed by the member states of the United Nations Economic Commission for Europe (UNECE), in Geneva, Switzerland. CLRTAP was negotiated following concerns raised by the Swedish government at the United Nations Conference on the Human Environment, hosted in Stockholm in 1972, about transboundary air pollution that was causing acidification in Sweden's lakes and waterways. This claim was verified in subsequent years through the establishment of a monitoring program under the auspices of the Organization for Economic Cooperation and Development (OECD)[1]. The adoption of 'Principle 21' at the 1972 Stockholm conference led the way for substantive change in the future. It pointed out that states have an obligation to ensure that activities carried out in one country do not cause environmental damage in others, or the global commons. (Wettestad 1997) In 1979, 30 Western and Eastern European countries, the United States, Canada, and the European Community signed CLRTAP. This was a remarkable step given the context of world politics at the time; CLRTAP was one of only a small number of East-West cooperations at the time, and the first environmental treaty between East and West. (Lidskog and Sundqvist 2002).

Under this convention, member countries agreed to a non-binding commitment to aim to reduce transboundary air pollution. Focusing on the worst identified offender of air pollution, CLRTAP calls for an extensive international program for monitoring and evaluating air movement patterns of sulfur dioxide. This convention established a base for air pollution policy that mandated international cooperation in research, development, and publishing of emission levels, a trend the European Community has incorporated into its own subsequent policies.

The convention was succeeded by a protocol to support scientific data gathering (1984) and seven others designed to reduce emissions of specific hazardous air pollutants, including protocols to reduce sulfur dioxide emissions (1985), nitrogen oxide emissions (1988), volatile organic hydrocarbon emissions (1991), additional sulfur emissions (1994), heavy metals (1998), persistent organic pollutants (1998), and ground level ozone (1999). (Brachtl 2005) The 1985, 1994, and 1999 protocols are dealt with in greater detail below.

Spurred by more developed research findings and continued negative environmental and human health effects, UNECE member states adopted the first protocol in 1985 in Helsinki, Finland. This updated document acknowledges the effect of nitrogen oxides and other pollutants on air quality, but continues to focus on sulfur emissions. Great attention had been

---

[1] The OECD performed a study in 1977 on acidification of forest soils and forest productivity that concluded that "air quality in any European country is measurably affected by emissions from other European countries." (OECD 1977) This study, plus others, served as the background for CLRTAP's formation, and the beginning of the EMEP program, a co-operative program for monitoring and evaluation of the long-range transmissions of air pollutants in Europe.

brought to the damage of forests and water bodies by air pollution in the early part of this decade, and a request had been made by the attendees of the 'Multilateral Conference on the Causes and Prevention of Damage to Forests and Water by Air Pollution in Europe' (Munich, 24-27 June 1984) that a specific agreement on the reduction of annual national sulfur emissions or their transboundary fluxes be reached. Using 1980 national emission levels as a baseline, a 30 percent reduction of sulfur emissions by 1993 was agreed upon.

Building upon the Helsinki Protocol, UNECE strengthened its air pollution regime further with the 1994 Oslo Protocol. Utilizing research findings, the agreement outlines ways to additionally decrease sulfur emissions, among them increasing energy efficiency, encouraging the use of renewable energy, and reducing the sulfur content in fuels. Additionally, the Oslo Protocol names country-specific emission limit values to existing stationary combustion sources, and sets country-specific caps for national emissions. Setting the Oslo Protocol apart from the previous treaties, this treaty requires that each party match their national standards for the sulfur content of gas to a standard at least as stringent as those specified within two years, allowing for an extension of time in special cases. Following suit of its predecessors, the 1994 Oslo Protocol calls for continuing cooperation between nations involving technology exchange, shared results of research and development, and publishing data on sulfur levels gathered by monitoring. (Oslo Protocol 2004).

The most significant contribution to tackling the acid rain problem in Europe has been the management tool called 'critical loads.' This concept, carried over from the 1994 Oslo Protocol, played a strong role in defining emission ceilings in the 1999 Protocol on ground level ozone adopted in Gothenburg, Sweden. Critical loads is defined as "the highest load that will not cause chemical changes leading to long-term harmful effects on the most sensitive ecological ecosystems" in a designated area. The critical loads approach has a few key advantages over the more common flat-rate reduction approach as the latter was inevitably arbitrary from a scientific point of view. Also, the flat-rate approach ignored the fact that some countries had undertaken reductions prior to the negotiation of the protocol and that they overlooked that both sensitivity pollution and the cost of reducing emissions varied markedly across Europe. (Levy 1995) The thrust of the 1999 Protocol is towards abatement of acidification and eutrophication caused by air pollutants, and towards reduction of ground level ozone in Europe. The Protocol sets emission ceilings for 2010 for four pollutants, further diversifying the regulatory scope of CLRTAP: sulfur, NO_x, volatile organic compounds (VOCs) and ammonia. Additionally, the Protocol sets tight limit values for specific emission sources and requires best available technologies to be used.

CLRTAP and its protocols set a precedent for international environmental policy that the European Community adopted to a large extent. The following sections review the development of EU environmental policy and then highlight the linkages between the two institutions' policies and policy development.

3.2. EUROPEAN UNION AIR POLLUTION POLICY

The first European air pollution policies were implemented years before the European Community enshrined the environment into its treaties.[2] The first Community standards for spark-ignition engines for carbon monoxides and hydrocarbons were set in 1970 with EC Directive 70/220. Nitrogen oxides were added in 1978, and these standards were later tightened in 1987. In 1972, EC Directive 72/306 set limits on the opacity of emissions from diesel engine vehicles. Initial Community standards for the sulfur content in fuel were set in 1975 (EC Directive 75/716), and standards for lead and volatile organic compounds (VOC) emissions were added subsequently; these standards were tightened in 1996. Despite these steps, environmental policy prior to 1996 lacked a concerted effort at significantly reducing air pollution in Europe.

This first set of policies in the 1970s and 1980s recognized the destructive health and environmental effects of vehicle emissions and fuel additives, and served as a precursor to the more stringent and defining air quality policies introduced from 1984 onward. Framework Directive 84/360 began a series of strong air pollution regulations, setting emission standards of industrial plants. Various other air pollution regulations were passed in the late 1980s, notably EC Directive 88/77 and 88/436 which established standards for diesel engines, and EC Directive 88/609 setting both technology-based emission limits and overall percentage reductions for large combustion plants. A string of regulations followed, leading up to the establishment of the Air Quality Framework Directive of 1996, which had been instigated by the Fifth European Community Environment Programme (5th ECEP, 1992-2000).[3] The regulations included a Directive requiring public access to environmental information (1990), another establishing

---

[2] This happened with the Single European Act (adopted in 1986), into which an Environment Title was added, providing the Commission with the legal basis to act on environmental issues.

[3] The 5th ECEP's goals were first, to follow through with the Community's commitments set in the Single European Act of 1986 to develop and implement effective environmental policy, and second, to fulfill the Maastricht Treaty's objective of promoting sustainable growth while respecting the environment. (EUROPA website)

road vehicle emissions including VOCs (1991), another requiring ozone monitoring and setting standards (1992), a Directive setting limit values for sulfur content of gas oils (1993), and an Integrated Pollution Prevention and Control Directive establishing a multi-media permitting system and strengthening implementation mechanisms (1996). (Selin and VanDeveer 2003).

The European Union incorporated these former regulations into the 1996 Air Quality Framework Directive on ambient air and simultaneously initiated the development of an Acidification Strategy for combating acidification and eutrophication. Rather than establishing concrete air quality objectives, the Framework Directive lays out a framework with basic principles for ambient air quality monitoring and management, from which more specific 'daughter directives' can be adopted. Work begun on the Acidification Strategy, on the other hand, developed in a different direction; the final Directive has specific target values for emissions. The Acidification Strategy, which included a cost-effectiveness analysis of attainable environmental targets, was revised in the next few years and a focus on ground level ozone was added, constituting the National Emissions Ceilings (NEC) Directive, adopted in 2001. (AcidRain.org) As the name implies, the NEC Directive sets emissions ceilings, varying by country, for some of the worst air pollutants affecting environmental and human health.

Overall, the objectives of the Framework Directive are threefold: first, to support the preceding regulations (summarized above) by establishing quality objectives for ambient air, second, to determine consistent assessment methods, and third, to lay down guiding principles for information collection and dissemination. Due to its open, umbrella-like structure, the Framework Directive also allows for the introduction of new air quality standards for previously unregulated air pollutants. EU member states are required to transpose these regulations into national law, implement them, monitor pollutant levels, and devise an attainment program if limit values are exceeded. Member states must monitor emissions and make the following information available to the public: location where pollution is excessive, the nature of the pollution (i.e. chemical type), and the origin of pollution. There are four daughter directives so far under the Air Quality Framework Directive which specify limit values for pollutants: sulfur dioxide, nitrogen oxides, particulate matter, and lead (1999); benzene and carbon monoxide (2000); ground-level ozone (2010); and arsenic, cadmium, mercury, nickel, and polycyclic aromatic hydrocarbons in ambient air (2004). The daughter directives apply the principle of lowest possible exposure to set limits and establish common methods for assessing concentrations. (EUROPA website).

Aiming to limit acidification- and eutrophication-causing pollutants and ozone precursors, the NEC Directive is a step toward the EU's long-term objectives of not exceeding critical loads of harmful pollutants, and of effective protection of all people against recognized health risks from air pollution. Three interim environmental targets are identified: mitigation of acidification by reducing by at least 50 percent pollutant levels that exceed the critical load; reduction of health-related ozone exposure by a two-thirds decrease of ground-level ozone that exceeds the critical level for health and; a reduction of vegetation-related ozone exposure by a one-third decrease of ground-level ozone that exceeds the critical level. National emission ceilings for 2010 were also agreed upon for the ten new EU member countries as part of their accession treaties. (AcidRain.org).

The EU regulations have become progressively stronger through the years as research on environmental and health effects of air pollution continued. The precedent that CLRTAP set regarding the requirement for countries to share information about research findings is also strongly incorporated into EU air pollution policy. This is not the only parallel between these horizontally related institutions; in the next section, this paper explores the many linkages between UNECE's CLRTAP and EU air pollution policy.

4. Linkages between CLTRAP and EU Directives on Air Pollution

Before the EU adopted far-reaching policies to combat transboundary air pollution, its predecessor, the EC, and many, but not all, EC member countries were actively involved in the negotiations of CLRTAP during the 1970s. (Huber 2003) This partial overlap in membership has always triggered interplay and competition between the two institutions, and has often led to institutional development. Among the most synergistic is the horizontal interplay between the CLRTAP 1999 Gothenburg Protocol and the EU's 2001 National Emission Ceilings Directive (NEC). The following four elements of this synergistic interplay are particularly noteworthy:

- the influential nature of shared research and development in pollution science, i.e. the common dependence on International Institute for Applied Systems Analysis (IIASA) models (www.iiasa.ac.at);
- the shared use of norms and rules, and the consequent collaboration, i.e. the use of critical loads compiled by the Coordination Center for Effects (www.mnp.nl/cce);
- parallel strengthening of emission reduction rates, setting of similar target levels, and implementing a multi-pollutant/multi-effects approach; and

- the constant push towards stronger policies borne from the back-and-forth interplay.

This section will explore each of these and show how each commonality manifests within the policies and/or why it is important for future air pollution policy development.

Driven by both policies' common tenets requiring member states to share their scientific research and development with other member states, air pollution science has grown considerably since 1979. As this science becomes more accurate and scientists acquire more knowledge on the topic, policymakers have had to acknowledge the serious effects that VOCs, sulfur, NO_x, and other chemicals are having on our environment. This team effort to produce hard answers about the effects of air pollution may not have come about without participation from all member states or pressure from both policies that helped to push the science forward. In this politically motivated interplay, the institutions intentionally worked together to produce hard science, resulting in faster and more accurate research that can be used by both institutions. One hard example of CLRTAP's direct influence on EU policy is the shared use of the highly accurate IIASA modeling of atmospheric air pollution transport, first embraced by CLRTAP. The IIASA Regional Air Pollution Information and Simulation model (RAINS) is based on the concepts of critical loads and cost-effectiveness. RAINS computer models run scenarios that are cost-optimized for various rates of pollution reduction and follow the atmospheric path of pollutants, measuring their effects of acidification and eutrophication. (Selin and VanDeveer 2003, p 36; Wettestad 2002, p 34) Both the Gothenburg Protocol (and earlier CLRTAP protocols) and the NEC Directive make use of these models when setting policy. The double pressure and international coordination to develop more accurate science will continue to help in the future as far as enacting stronger and more effective air pollution policy is concerned, and ensuring that the measures in place are the right fit.

CLRTAP and EU air pollution policies also share norms and rules to determine boundaries and target levels. Perhaps the most engrained and significant linkage between CLRTAP and EU air policy is the common use of emission reduction goals and limits for transboundary pollutant transport. This is an example of political interplay, where institutions are linked intentionally. Links include using comparable reduction targets of the excess of atmospheric deposition over critical loads to determine desirable emission reductions, setting similar country-specific emissions ceilings, and using the same best available techniques (BAT) and emission limit value (ELV) standards on specific emissions sources. (Hettelingh et al. 2001) Serving as another example of intended political interplay, this data convergence produces benefits that primarily lie in the ease of converting data, comparing scientific data, and setting future targets, thereby increasing compatibility between the two policies. To design the daughter directives to the Air Quality Framework Directive on ambient air, for instance, experts from many organizations, including UNECE, were consulted and incorporated into a Steering Group. (EUROPA website) Furthermore, these norms and rules have changed and molded similarly over time. For example, emissions reduction requirements have increased through the years; both the Gothenburg Protocol and the NEC Directive set upper limits for pollutants responsible for acidification, eutrophication, and ground-level ozone pollution. (Selin and VanDeever 2003, p 32) Both policies also call for cuts in sulfur, NO_x, VOCs, and ammonia emissions by 2010, using 1990 baseline levels. Countries are assigned national ceilings in the Gothenburg Protocol, a concept first introduced in CLRTAP's 1994 Oslo Protocol, with rate reductions flat across the board. The NEC Directive also followed UNECE's lead on increasing regulatory scope – the multi-pollutant and multi-effects approach originated in the CLRTAP system, increasing the amount of substances regulated at the international level – and lowering national ceilings. Emission ceilings under the NEC, however, differ in each member state, employing policies tailored to the countries' circumstances. (Wettestad 2002).

The co-existence of the two institutions enables them to have constant influence on each other, having the effect of a push towards continually stronger regulations, thus serving as an example of reciprocal interplay. Megan Brachtl asserts that global environmental treaties like CLRTAP, in which treaty language is reworked and watered down so that it is universally accepted, can often be "ineffectual means of curing environmental ills." (Brachtl 2005) Regional legislation, however, can be made more specific and tight, more directly addressing the needs of the region. For instance, the EU can base the core of its environmental policy on CLRTAP, but can reject policy that is not strong enough, replacing it with stricter legislation. In this way, CLRTAP is acting as a 'springboard' from which the EU can leap forward. (Brachtl 2005) The EU Commission took advantage of this leverage when it refused to sign the CLRTAP 'multi-pollutant and multi-effects' protocol because it did not mandate a strict enough emission reduction. Commission representatives did not want to condone a protocol with weak emissions reduction targets. Despite this, all EU members of UNECE signed the protocol on the grounds that even small steps in global air pollution regulations would help, and a stricter protocol could be adopted later. EU leaders had aimed to pass the similar aforementioned NEC Directive before the signing of the 1999 Protocol in order to set a precedent of higher standards, in hopes of strengthening the CLRTAP agreement. (Selin and VanDeveer 2003, p 36) The NEC Directive, passed in 2001, did establish national ceilings for SO_2 that are, in fact, stronger than the CLRTAP agreement by 3 percentage points. The other regulated emissions – NO_x, VOCs, and ammonia – have set ceilings that are equal to or lower than the ceilings established by CLRTAP. Under the NEC Directive, sulfur emissions should decrease by 77 percent, NO_x by 51 percent, VOCs by 54 percent, and ammonia by 14 percent of 1990 levels (compared to the Gothenburg Protocol's expected reductions of sulfur by 63 percent, NO_x by 41 percent, VOCs by 40 percent, and ammonia by 17 percent). (UNECE website) With a more demanding policy, the EU Commission accomplished its goal of setting more stringent targets than CLTRAP. More importantly, the presence of legally binding national emission ceilings established in EU legislation gives stronger provisions for follow-up and control of member states' implementation and compliance with the emission ceilings.

5. Future Trends

If current standards and reduction cuts designated by CLRTAP and the EU air pollution policy are implemented, we can reasonably expect improved air quality in Europe. It is predicted that breaches in the World Health Organization's air quality guidelines for protecting human health can be expected to fall to about 20 days per year in the most affected parts of Europe, and that human health will improve due to the decrease in airborne particulate matter. However, acid deposition will continue to damage Europe's most sensitive forested and water-covered areas, ozone will remain a problem, as current regulation is not tight enough for significant improvements, and human health will still be affected due to remaining particulate matter and chemical pollutants. (Wettestad 2002, p 35).

The greatest potential for improving air quality in Europe lies in the more ambitious and institutionally more advanced EU – steps have already been taken towards an evaluation of air pollution policy within the Clean Air for Europe (CAFE) program. Initially formed with the objectives of reviewing current air quality policies and assessing progress towards attainment of the EU's long-term air quality goals, CAFE developed the analysis for the EU's Thematic Strategy on Air Pollution (TSAP), which was adopted by the Commission in September 2005. As a first milestone out of an ongoing five-year cyclical program, the TSAP suggests air pollution policy integration as a continuing goal for the EU. In 2002, the Community's Sixth Environmental Action Program (6th EAP) [4] set out the objective to attain "levels of air quality that do not give rise to significant negative impacts on and risks to human health and the environment." (Thematic Strategy on Air Pollution, p 1) In assessing current policy to

---

[4] The 6th Environmental Action Program sets environmental objectives for the next 10 years, focusing on four major areas of action: tackling climate change; nature and biodiversity; environment and health; and sustainable use of natural resources and management of waste. (EUROPA website)

evaluate to what extent the goals have been reached, the TSAP made various observations. In regards to the environment, it reported that reductions have had positive impacts, but still two thirds of the lakes and streams surveyed in Scandinavia, for instance, are at risk from acid deposition and some 55 percent of all EU ecosystems suffer from eutrophication. Human health effects due to particulate matter also earn mixed results: currently in the EU there is a loss in statistical life expectancy of over eight months due to particulate matter in air, which is equivalent to 3.6 million life years lost annually. Effective implementation of current policies will only reduce this figure to around 5.5 months (or two million life years lost). Estimates produced by the Gothenburg Protocol and EU policies indicate that there is a clear economic incentive built into both air pollution policies – the benefits of cleaner air outweigh the costs by a clear margin. However, there are still large barriers to tackle.

Some future goals for both CLRTAP and EU policy include the adoption of transboundary air pollution regulations that address particulate matter and a continued effort in scientific research (i.e. active measures such as technological improvements for stationary and mobile pollution sources and preventative measure such as improving knowledge of biological recovery of eutrophied and acidified waterways). In addition, the impacts of particulate matter on climate change and vice-versa need to be better understood.

As CLRTAP and EU policy characteristics converge, cases of interplay may further multiply. In harmonizing the monitoring system and approach to evaluating critical loads, the two institutions are developing a shared language. Aligned abatement strategies may be developed that further enhance communication and transferability of information between organizations. This same quality, though, may serve to hinder new policy development, impeding forward-moving development of creative programs designed to prevent and alleviate the effects of air pollution. This convergence of overlapping institutions, especially in terms of their harmonized monitoring system and approach to evaluating critical loads, may also benefit other countries and regions around the world, which have the opportunity to use this technology and approach to mitigate air pollution in their region.

6. Conclusion

Given the long list of cases of interplay among the various European institutions that have contributed to mitigating air pollution in Europe, one can conclude that this interplay has had a significant impact on the development of these institutions' approaches to tackling air pollution. In particular, differences and developments in membership and policymaking capacity have led these institutions to be at times more suitable and at other times less favorable a venue for countries to participate in. The limited number of countries that were EC members in the earlier years of transboundary air pollution management made it difficult for the EC to take a lead on this issue, while the EC/EU's greater policymaking strength and enforcement ability was, and continues to be, an advantage over CLRTAP. Because of this, greater policymaking strength was achieved by the 2001 EU NEC Directive than the 1999 Gothenburg Protocol. At the same time, the CLRTAP protocols expanded the somewhat less stringent regulations over a wider geographical scale, including Canada, the United States, and a number of Central and Eastern European countries. Despite its relatively weaker institutional architecture, CLRTAP was the institution that developed the most influential concepts and policymaking approaches, above all the concept of 'critical loads' and the 'multi-pollutants/multi-effects' approaches, which were subsequently adopted in the EU NEC Directive. It is conceivable that this trend toward convergence will continue in the near future and beyond, and that the two institutions will converge toward a clustered institutional architecture with potential for simplifying and streamlining its currently complex structures.

ACKNOWLEDGEMENTS

We thank David Huitema and Jean-Paul Hettelingh for substantive comments and the US National Science Foundation for funding this research as part of the project on the Institutional Dimensions of Global Environmental Change (IDGEC) under Grant Number BCS-0324981.

References

Primary sources

1972 United Nations Conference on the Human Environment in Stockholm.
1975 UNECE Convention on Long-Range Transboundary Air Pollution.
1985 Protocol to the UNECE 1979 Convention on Long-Range Transboundary Air Pollution, Helsinki, Finland.
1994 Protocol to the UNECE 1979 Convention on Long-Range Transboundary Air Pollution, Oslo, Norway.
AcidRain.org, 2006 The NGO Swedish Secretariat Website on Acid Rain. www.acidrain.org.
Air Quality Framework Directive 1996/62/EC.
Communication from the Commission to the Council and the European Parliament: Thematic Strategy on Air Pollution (COM (2005) 446)
Council Directive 1999/30/EC.
Council Directive 2000/69/EC.
Council Directive 2002/3/EC.
Council Directive 2004/107/EC.
EUROPA: Gateway to the European Union, http://europa.eu.int/index_en.htm.

International Institute for Applied Systems Analysis, 2006, http://www.iiasa.ac.at/~rains/doc/iiasa.htm.
OECD, 1977 The OECD programme on long range transport of air pollutants. Measurements and findings, Paris.
United Nations Economic Commissions for Europe, http://www.unece.org.

Secondary sources

Brachtl, M., 2005 Capitalizing on the Success of the LRTAP Regime to Address Global Transboundary Air Pollution, in L. Susskind and W. Moomaw, *Papers on International Environmental Negotiation*, 14, PON Books, http://www.pon.org/downloads/-ien14_3Brachtl.pdf.
Brown Weiss, E., 1993 International Environmental Issues and the Emergence of a New World Order, *Georgetown Law Journal* 81:3, p 675–710.
Krasner, S. D., 1982 Structural Causes and Regime Consequences: Regimes as Intervening Variables, *International Organization*, 36, 2, p 186.
Hettelingh, J.-P. et al., 2001 Multi-effect critical loads used in multi-pollutant reduction agreements in Europe. *Water, Air and Soil Pollution* 130, p 1133–1138.
Huber, M., 2003 Lessons Drawn from Burden Sharing Exercises. EC Acidification and Climate Change Policies, How Institutions Change, Leske + Budrich, Opladen, p 301-324.
Levy, M., 1995 International Co-operation to Combat Acid Rain, *Green Globe Yearbook 1995*, Oxford University Press, Oxford.
Lidskog, R. and G. Sundqvist, 2002 The Role of Science in Environmental Regimes: The Case of LRTAP, *European Journal of International Relations*, 8, 1, p 77–101.
Oberthür, S. and T. Gehring, 2001 Conceptualizing Interaction between International and EU Environmental Institutions, Ecologic, Berlin.
Ostrom, E., 2005 *Understanding Institutional Diversity*, Princeton University Press.
Schroeder, H., forthcoming "The IDGEC Approach to Analyzing Institutional Interplay between the Biosafety and Trade Regimes," in W.B. Chambers, J.A. Kim, and O.R. Young, *Institutional Interplay: The Case of Biosafety*, Tokyo, UNU Press.
Selin, H. and S.D. VanDeveer, 2003 Mapping Institutional Linkages in European Air Pollution Politics, *Global Environmental Politics*, 3, 3, p 14–46.
Susskind, L., 1993 *Environmental Diplomacy: Negotiating More Effective Global Agreements*, Oxford: Oxford University Press.
Wettestad, J., 2002 Clearing the Air: Europe Tackles Transboundary Pollution, *The Environment* 44:2, p 32–40.
Wettestad, J., 1997 Acid lessons? LRTAP implementation and effectiveness, *Global Environmental Change* 7, 3, p 235–249.
Wettestad, J. and A. Farmer, forthcoming The EU Air Quality Framework Directive: Shaped and Saved by Interaction? in S. Oberthür and T. Gehring, *Institutional Interaction*, MIT Press, Cambridge, MA, p 241–259.
Young, O., 2002 *The Institutional Dimensions of Environmental Change: Fit, Interplay, and Scale*, Cambridge, MA: MIT Press.
Young, O., 1996 Institutional Linkages in International Society: Polar Perspectives, *Global Governance* 2, 1, p 1–24.
Young, O. et al., 2005 (revised edition) *Institutional Dimensions of Global Environmental Change (IDGEC): Science Plan*, IHDP Report No. 16, Bonn.

DEVELOPMENT OF A PESTICIDES BIOSENSOR USING CARBON-BASED ELECTRODE SYSTEMS

ADINA ARVINTE*, LUCIAN ROTARIU AND CAMELIA BALA
*LaborQ - Laboratory of Quality Control and Process Monitoring,
Faculty of Chemistry, University of Bucharest,
4-12 Regina Elisabeta Blvd. 030018 Bucharest-1, Romania;*

*To whom correspondence should be addressed. adina_arvinte@yahoo.com

Abstract. The biosensors described in this work, for the monitoring of pesticides, are based on acetylcholinesterase immobilized on the surface of screen-printed electrodes. The principle of the biosensor is that the degree of inhibition of an enzyme sensor by a pesticide is dependent on the concentration of that pesticide. The DPV technique was used as a detection method and methyl-paraoxon as a reference pesticide for sensor calibration.

Keywords: biosensor, DPV; acetilcholinesterases; methyl-paraoxon.

1. Introduction

In the last years, the intensive agriculture has highlighted the need of procedures to study the presence of pollutants in crops and foodstuffs and other environmental matrices (Marco et al., 1996; Krämer, 1996). Pesticides and their metabolites have received particular attention during the last few years in environmental trace-organic analysis.

The current generation of pesticides, such as organophosphate and carbamates, are determined in several kinds of samples by using chromatographic techniques that are usually more expensive and time-consuming than the biological techniques, and need more sophisticated equipment. Electroanalytical sensors and biosensors provide an exciting and achievable opportunity to perform biomedical, environmental, food and industrial analyses away from a centralised laboratory due to their advantages such as high selectivity and specificity, rapid response, low cost of fabrication, possibility of miniaturization and easy to integrate in automatic devices.

Methods based on enzymatic recognizing techniques have been developed for a large variety of analytes, varying from natural components to contaminants in food and environmental samples. Also, biosensors able to analyze some natural components and additives in foodstuffs and crop samples have been developed and used to quality control (Maines et al., 1996; Campanella et al., 1996) and to industrial food processes monitoring (Deshpande et al., 1994).

Biosensor is an analytical tool that combines a biological recognition element (for example enzyme, cell, antibody and microbe) and a physical transducer, in intimate contact with each other. The role of the transducer is to convert the biochemical information into a physical signal related to the analyte concentration (Buerk, 1993). Analytical methods employing enzyme sensors based on the inhibition of cholinesterase with potentiometric or amperometric detection have been developed for the determination of organophosphate and carbamate insecticides (Skládal, 1996) – the most employed in agriculture due to their high activity, low bioaccumulation and relatively rapid degradation in the environment.

In living beings, these pesticides bind irreversibly to the active site of acetylcholinesterase (AChE) – enzyme involved in the transmission of nerve impulse. Electrochemical biosensors for measurement of these pesticides are based all on the inhibition of AChE and the inhibition degree is proportional to the pesticide concentration.

2. Cholinesterase-based Biosensors for the Analysis of Pesticides

With amperometric transducers, two types of pesticides biosensors can be constructed. One possibility is a two-enzyme system using AChE and choline oxidase (ChOD) and thus employing either an oxygen sensor (Campanella et al., 1991) or a hydrogen peroxide detector (Marty et al., 1992; Palleschi et al., 1992) as the internal transducer.

Acetylcholine is hydrolyzed to choline in the first enzymatic reaction catalysed by AChE. It is impossible to use only the first reaction to develop an electrochemical biosensor because no electroactive species is produced. Therefore the choline is oxidized to betaine in a second enzymatic reaction:

$$\text{Acetylcholine} + H_2O \xrightarrow{\text{ACh}E} \text{Choline} + \text{Acetic acid}$$
$$\text{Choline} + O_2 + H_2O \xrightarrow{\text{ChOD}} \text{Betaine} + H_2O_2$$

Use of two enzymes has some disadvantages regarding the optimization of the operational parameters of these two enzymatic reactions. The second alternative is by using only AChE enzyme, and in this case acethylthiocholine (ATCh) has been preferred as substrate. ATCh is hydrolyzed in the same proportion as the acetylcholine producing thiocholine and acetic acid. In a second step the thiocholine is oxidized at the surface of electrode applying a suitable potential (Fig.1).

Electrocatalytic oxidation of thiocholine is achieved at lower potentials using a modified graphite electrode featuring a redox mediator. However, TCNQ exhibits the most suitable characteristics to be used in cholinesterase biosensors.

Most biosensors based on AChE have the enzymes immobilized on the surface of the sensor. The inhibition reaction being irreversible, the membrane with immobilized enzyme has to be replaced after several measurements or the biosensor can be use for only one determination. Due to this fact, the researchers tried to realize pesticide biosensors with a renewable surface or disposable biosensors based on screen-printed electrodes (SPE). The screen-printing technology provides a simple, fast and inexpensive method for mass production of disposable biosensors for different biomolecules starting with glucose, lactate and finishing with environmental contaminants as pesticides (Kulys et al., 1991) and herbicides (Skladal, 1992).

3. Experimental

Materials and Methods

Acetylcholinesterase EC 3.1.1.7 type VI-S from Electric Eel and acetylthiocholine chloride were purchased from Sigma. Nafion and TCNQ were purchased from Fluka. The supporting electrolyte was an aqueous solution buffered at pH 8 with 0.1 M phosphate and 0.1 M KCl.

All voltammetric experiments were carried out using an electrochemical potentiostat BAS 100 B West Lafayette software. The three-electrode system consisted of the carbon paste electrode as the working electrode and Ag/AgCl as reference and auxiliary electrodes.

Enzyme Biosensor Construction

The biosensor described in this work, for the monitoring of pesticides, is based on acetylcholinesterase immobilized on the surface of screen-printed electrodes by adsorption and covered with a Nafion film. The incorporation of the TCNQ mediator in a carbon working electrode permits a dramatically decrease of the potential necessary for oxidation of thiocholine.

$TCNQ_{ox}$ immobilised on the surface of the graphite-working electrode is reduced by the thiocholine produced in the reaction of enzymatic hydrolysis of ACTh. TCNQ reoxidation is realized using DPV technique in the potential range of 0.35-0.7 V vs. AgCl reference electrode (fig. 1). Scan rate, pulse amplitude were optimised for DPV experiments in order to determine de AChE activity.

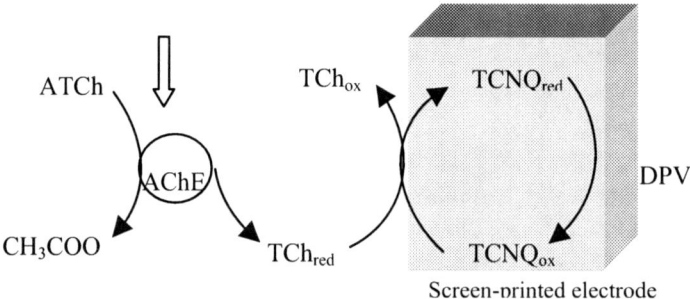

FIGURE 1. The sequence of reactions involved in operating way of a biosensor for pesticides based on the voltametric detection.

TCNQ (50 ng) mixed with Nafion was deposited on the surface of working electrode and allowed to dry for one night at room temperature. 10 mUI of enzyme solution in phosphate buffer were simply deposited onto the working electrode surface. After drying, the enzyme layer was covered by a second film of Nafion. For analytical measurements the enzyme electrode was placed horizontally and a drop of 150 microlitri covered completely the aria of the three electrodes.

Experimental Setup and Measurement

Methyl paraoxon was used as reference pesticide for inhibition tests. AChE activity was determined in the absence of pesticide and then after 10 min. incubation with methyl paraoxon standard solutions. The inhibition percentage was calculated using the equation (Skladal, 1992):

$$\% I = \frac{A_0 - A_1}{A_0} 100$$

where $I(\%)$ represent the inhibition percentage, A is the DPV peak area. The subscripts (0) and (1) correspond to measurements performed before and after incubation, respectively.

Optimisation of the assembly of cholinesterase sensors is commonly directed toward the improvement of substrate detection, i.e. to achieve maximum signal, prolonged shelf-life, stable and reliable response.

4. Results and Discussion

The electrochemical behavior of TCNQ-modified electrode was explored with cyclic voltametry (Fig. 2).

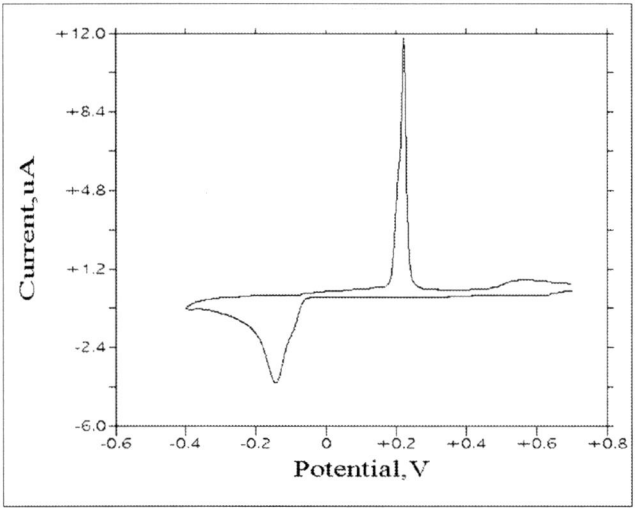

FIGURE 2. Cyclic voltammogram of the chemically modified SPE with TCNQ (phosphate buffer 0.1 M, pH=8, SR=50 mV/s).

$TCNQ_{ox}$ immobilized on the surface of the graphite working electrode is reduced by the thiocholine produced in the reaction of enzymatic hidrolysis of ACTh. Reoxidation of the reduced form of TCNQ is realized using DPV technique in the potential range of 0.35-0.7 V vs. Ag/AgCl reference electrode.

The influence of acetylthiocholine concentration on the sensor response is illustrated in Fig. 3. The peak area represents the analytical response of the sensor and it is proportional to the concentration of acetylthiocholine hydrolised. The peak occurs at a more positive potential owing to the higher pH of the solution. At concentrations above 0.5 mM ATCh, the main oxidation peak barely increases in magnitude.

For use in the inhibition experiments, the biosensor with 0.1 UI AChE was incubated with a sample of pesticide for 10 min at room temperature in a non-stirred solution and percentage of inhibition was calculated after the measurement of residual activity.

Methyl paraoxon was used as reference pesticides for sensor calibration. AChE activity, respectively the response of biosensor, was determined in the absence of pesticides and then after 10 min. incubation with methyl paraoxon.

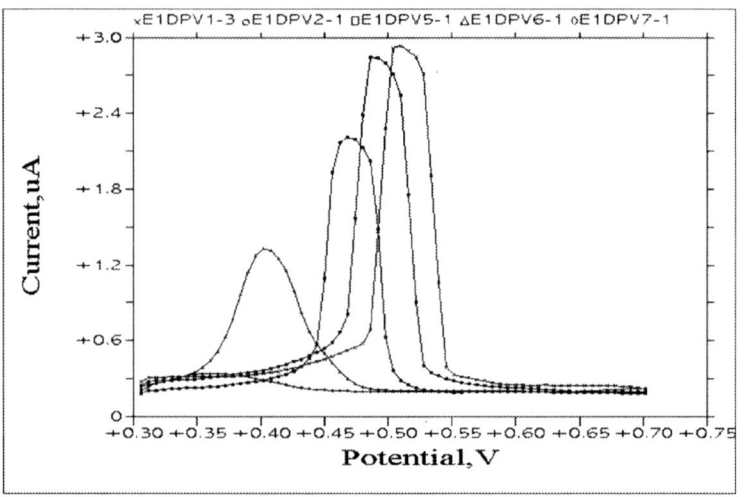

FIGURE 3. Differential pulse voltammograms for acetylthiocholine in 0.1 M phosphate buffer, pH 8 for 0.05 mM ATCh (1), 0.1 mM ATCh (2), 0.2 mM ATCh (3), 0.5 mM ATCh (4) and 1 mM ATCh (5).

Determination of methyl paraoxon can be realized in the range of $10^{-8} - 10^{-6}$ mol/L. The detection limit, defined as the concentration of pesticide that produce an inhibition percentage of 10% of the AChE activity correspond to a concentration of 5 10^{-9} mol/L methyl paraoxon.

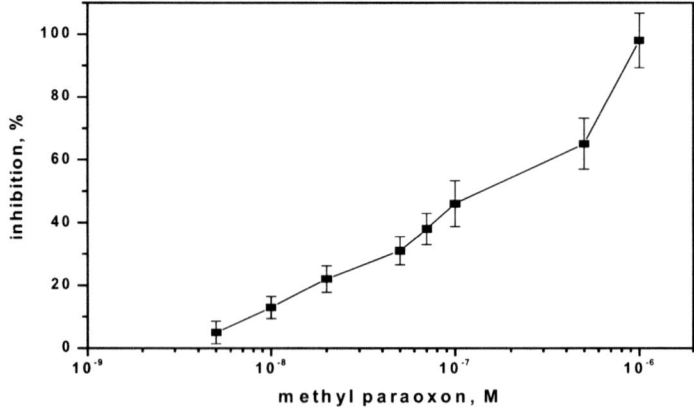

FIGURE 4. Calibration curve for methyl paraoxon.

5. Conclusions

A disposable pesticide biosensor based on voltammetric detection was realized in this study. Contrary to other sophisticated methods, adsorption is an easier, faster and cheaper method. It appears to be especially suitable for the manufacturing of SPE. Further improvement of detectability can be expected from further reduction of the amount of immobilized enzyme. Despite the high sensitivity and reproducibility of some ChE-based sensors, the development of methods employing biosensors to pesticide monitoring in food samples is still limited. Biosensors based on AChE inhibition represent a fast method that offers to different investigators an easy way to detect the presence of organophosphorus insecticide in various matrices, being use like a sensitive screening method for environmental monitoring and food quality control.

References

Bernabei, M., Cremisini, C., Mascini, M., Palleschi, G. *Anal. Lett.* **1991**, *24*, 1317.
Buerk, D.G. *Biosensors: Theory and Applications*, Technomic Publishing, Basel, Switzerland, **1993**.
Campanella, L., Achilli, M., Sammartino, M.P., Tomasseti, M. *Bioelectrochem. Bioenerg.* **1991**, *26*, 237.
Campanella, L., Tomassetti, M. *Food Technol. Biotechnol.* **1996**, *34*, 131.
Deshpande, S.S., Rocco, R.M. *Food Technol.* **1994**, 8, 146.
Krämer, P.M. *J. Assoc. Off. Anal. Chem.* **1996**, 79, 1245.
Kulys, J., D'Costa, E.J. *Biosensors Bioelectronics* **1991**, 6, 109.
Maines, A., Ashworth, D., Vadgama, P. *Food Technol. And Biotechnol.* **1996**, *34*, 31.
Marco, M.-P., Barceló, D. *Meas. Sci. Technol.* **1996**, 7, 1547.
Marty J.L, Sode K., Karube I., *Electranalysis*, **1992**, 4, 249.
Marty, J.-L., Sode, K., Karube, I. *Anal. Chim. Acta* **1989**, *228*, 49.
Navera, E.N., Sode, K., Tamiya, E., Karube, I. *Biosensors Bioelectronics* **1991**, 6, 751.
Palleschi, G., Bernabei M., Cremisini, C., Mascini, M.*Sensors Actuators,*B, **1992**, 7, 513.
Skládal, P. *Anal. Chim. Acta* **1991**, *252*, 11.
Skládal, P. *Food Technol. Biotechnol.* **1996**, *34*, 43.
Skladal, P. Kalab, T., *Anal. Chim. Acta*, 316, **1995**, 73.
Wollenberger, U., Setz, K., Scheller, F., Löffler, U., Göpel, W., Gruss, R. *Sensors Actuators B* **1991**, 4, 257.

STUDY ON THE CONTENT OF CHEMICALS IN LANDFILL LEACHATE

ANDREANA MAXIMOVA AND BOGDANA KOUMANOVA
University of Chemical Technology and Metallurgy
Department of Chemical Engineering
8, Kliment Ohridski, 1756 Sofia, Bulgaria

*To whom correspondence should be addressed. *bibolini@abv.bg*

Abstract. The nature of landfill leachate depends on the type of the wastes at the disposal site, the landfilling technique used, the waste degradability and their stage of degradation as well as of the climate. The aim of our investigation was to characterize the leachate from the landfill situated near by a town with 85,000 inhabitants. Samples were taken and then analyzed from three points: from the recultivated cell, from the cell under filling as well as from the mixed leachate of both cells. The common characteristics like pH, dissolved oxygen (DO), Chemical Oxygen Demand (COD), Biochemical Oxygen Demand (BOD), Total Organic Carbon (TOC), total nitrogen, organic nitrogen, [NH_4^+-N], [NO_2^--N], [NO_3^--N], total phosphorus, phosphates, suspended solids (SS), dissolved solids DS (total and organic fraction), electroconductivity, have been determined. The metals content has been studied using Emission Spectral Analysis and Atomic Absorption Spectroscopy. The values for DO, COD and BOD showed that the biodegradable components in the leachate are increasing and the anaerobic degradation in the waste body is in progress. During the period from March, 2003, till earlier 2004 the values for conductivity reached their maximum because of the increased quantity of the dissociated into water compounds. At the same time pH is decreased. The inorganic fraction in the DS is higher compared to that of organics. Dissolved organics in the leachate wasted from the operating cell are much more than that from the recultivated cell. The presence of B, Al, Si, Cu, Mn, Fe, Zn, Ti, Cr, P and Bi was detected to be negligible. It was established that the quantity of Na, K, Mg and Ca is significant. Ni, Cd, Hg, Pb, Sn, Sb were not found.

Keywords: solid waste; landfill leachate; organics; metals

1. Introduction

The nature of landfill leachate depends on the type of the wastes at the disposal site, the landfill technique used, the waste degradability and their stage of degradation as well as of the climate. The problem with the leachate purification is significant environmental problem. Infiltrated precipitation percolates through domestic or industrial wastes. In the course of this flow, moisture becomes heavily polluted with organic and inorganic matter. Therefore effective pollution control and waste management requires an understanding of the quantity and quality of landfill leachate.

The basic biological processes leading to the generation of methane from landfills have been well known for many years (Buswell and Mueller, 1952; Otieno, 1994; Reinhart, 1998).

Numerous landfill studies have suggested that the stabilization of waste proceeds in five sequential and distinct phases (Pohland and Harper, 1985; Augenstein, 1990; Farquhar, 1973; Clement and Thomas, 1995). The rate and characteristics of leachate produced and biogas generated from a landfill vary from one phase to another and reflect the microbially mediated processes taking place inside the landfill. The progress toward final stabilization of landfill solid waste is subject to the physical, chemical, and biological factors within the landfill environment, the age and characteristics of landfilled waste, the operational and management controls applied, as well as the site-specific external conditions.

Movement through the phases is reflected by significant changes in leachate and gas quality. Nonconservative constituents of leachate (primarily organic in nature) tend to decompose and stabilize with time, whereas conservative constituents will remain long after waste stabilization occurs. Conservative constituents include various heavy metals, ammonia, chloride, and sulfide.

Metals often are precipitated within the landfill and are infrequently found at high concentrations in leachate, with the exception of iron.

Cover material may influence water inflow from precipitation, and bottom and side liners, if present, will influence infiltration of ground water into the waste mass. Consequent varying levels of moisture content may strongly affect gas generation. Shredding, if carried out, is likely to facilitate generation through intermixing of refuse nutrients, and bacteria. These are only some of the cause-and-effect relationships existing between operational parameters and methane generation (Lee, 1998).

The variables, which influencing to the methane generation are mostly: waste management and processing (baling, shredding, material separation /removal), waste composition (organic/inorganic, proportion of yard/food/-paper/other organics), biological factors (moisture, nutrients, bacteria, pH, and temperature), design and land filling operation (day-to-day refuse handling and liquid addition), (Di Pula, 1991; Gomes et al., 2005; Lo I., 1996; O'Brien, 2005). It is evident that typical uncertainties in variables affecting to the methane generation will depends from:

- Waste placement/history/location/composition may by difficult to trace, especially for older landfills;
- Biological parameters: nutrients, temperature and pH: (difficult to measure; probability to vary);
- Collection efficiency;
- Moisture content: (difficult to measure or estimate, probability to vary; has an important effect on methane generation.

Gas from municipal waste sites is composed of methane (CH_4) and carbon dioxide (CO_2) as a lot of trace components. Especially the volatile chlorinated /fluorinated hydrocarbons in the range of mg/m^3 are originated from specific products like foam material and aerosol container. It was researched if these trace components should affect the biochemical degradation processes of waste.

Water percolating through a landfill forms a leachate, which will contain both organic and inorganic substances. These substances may be transported with the leachate out from the landfill and hence present an environmental risk.

Many of identified organic compounds found in leachates are volatile and identification of non-volatile substances is rare. There is a large discrepancy in the reported identification of organic compounds (Rees, 1980). The reason is that several factors influence the characteristics of leachates from landfills. The type of solid waste and the degree of transformation inside the landfill are important factors. An important factor influencing the analytical results is, besides the analytical method used, the sampling point. As organic substances may sorbs to dissolved organic matter, colloids or particles the amount of such substances present in the sample will influence the analytical result (Reinhart and Grosh, 1998; Schrab et al., 1993; Rees, 1980; Ribeiro et al., 2002).

The aim of this study is to analyze leachates samples from a municipal landfill situated near by town with 85 000 inhabitants on purpose to determine the quality of the leachate and based on it to conclude in which phase of decomposition is the waste body.

2. Experimental

The samples of leachate from the regional landfill for solid wastes before wastewater treatment plant were collected and analyzed for common physicochemical characteristics during three years period (2002-2004).

Samples of leachates from three sampling points of the landfill were taken also in June 2005 to estimate the actual transformation phase into the waste body (Fig. 1). The landfill was established in 2002 with a capacity about 146 500 m^3 per year. The first cell has been filled and recultivated. The second one is under filling. The leachates in every cell are collected into the draining system. Both are mixed before taking to the wastewater treatment plant.

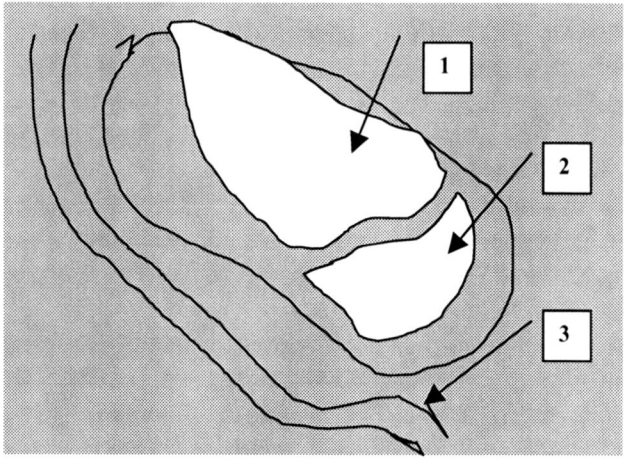

FIGURE 1. Sampling points for leachates (sample1 - from the recultivated cell; sample 2 – from the cell under filling; sample 3 - from the mixed leachate of both cells).

The common characteristics of the leachate samples were determined by the study of the following parameters:
- pH;
- Dissolved Oxygen (DO);
- Chemical Oxygen Demand (COD);
- Biochemical Oxygen Demand (BOD);
- Total Organic Carbon (TOC);
- Total Nitrogen;
- Organic Nitrogen;

- $[NH_4^+-N]$;
- $[NO_2^--N]$;
- $[NO_3^--N]$;
- Total Phosphorus;
- Phosphates;
- Suspended Solids (SS);
- Dissolved Solids DS (total and organic fraction);
- Electroconductivity.

The metals content has been studied using Emission Spectral Analysis, Atomic Absorption Spectroscopy (Perkin Elmer 5000) and UV-spectroscopy (Perkin Elmer 323).

3. Results and Discussion

The data for the parameters characterizing the quality of the leachates during three years period are summarized in Table 1. Typically, the BOD and COD concentrations will be low initially, and will increase as waste begins to solubilize. An eventual decline in BOD and COD concentrations is often observed as organic matter is being removed via washout and degradation.

TABLE 1. Characteristics of the leachate (2002 – 2004).

2002					
	Characteristics	February	May	August	November
1.	pH	7.4	7.42	7.57	8.01
2.	DO, $mgO_2\,dm^{-3}$	8.82	4.57	0	0.49
3.	Electroconductivity, Si	5610	755	853	1279
4.	BOD_5, $mgO_2\,dm^{-3}$	126.7	229	232	312
5.	COD, $mgO_2\,dm^{-3}$	200	610	1810	2302
6.	SS, $mg\,dm^{-3}$	73	270	261	318
7.	$NH_4^+ - N$, $mg\,dm^{-3}$	2.81	12.5	183.7	241.2
8.	$N_2^- - N$, $mg\,dm^{-3}$	2.015	0.157	0.067	0.113
9.	$N_3^- - N$, $mg\,dm^{-3}$	5.6	5.1	0.9	6.4
10.	Phosphates, $mg\,dm^{-3}$	0.46	0.21	1.1	1.98
11.	Total P, $mg\,dm^{-3}$	1.4	0.3	1.99	3.11

2003					
1.	pH	8.19	7.36	7.84	8.34
2.	DO, $mgO_2\,dm^{-3}$	1.57	2.13	2.14	0.84
3.	Electroconductivity, Si	746	10220	8570	10143
4.	BOD_5, $mgO_2\,dm^{-3}$	272	222	146.7	189.3
5.	COD, $mgO_2\,dm^{-3}$	1690	945	556	2945
6.	SS, $mg\,dm^{-3}$	267	477	243	300
7.	$NH_4^+ - N$, $mg\,dm^{-3}$	286.5	173.6	3293	7326
8.	$NO_2^- - N$, $mg\,dm^{-3}$	0.127	0.13	0.247	0.301
9.	$NO_3^- - N$, $mg\,dm^{-3}$	1.3	0.8	1.2	4.6
10.	Phosphates, $mg\,dm^{-3}$	-	-	-	-
11.	Total P, $mg\,dm^{-3}$	6.65	2.7	1.79	24.7
2004					
1.	pH	8.25	7.97	7.87	8.07
2.	DO, $mgO_2\,dm^{-3}$	5.31	2.44	2.32	-
3.	Electroconductivity, Si	1120	1368	7580	-
4.	BOD_5, $mgO_2\,dm^{-3}$	624	640	427	156
5.	COD, $mgO_2\,dm^{-3}$	895	1460	535	740
6.	SS, $mg\,dm^{-3}$	123	121	567	124
7.	$NH_4^+ - N$, $mg\,dm^{-3}$	405.1	395.3	149.7	137.7
8.	$N_2^- - N$, $mg\,dm^{-3}$	0.266	2.658	0.633	-
9.	$N_3^- - N$, $mg\,dm^{-3}$	4.7	6.88	7.57	-
10.	Phosphates, $mg\,dm^{-3}$	-	-	-	-
11.	Total P, $mg\,dm^{-3}$	20.1	38.3	2.7	-

Leachates generally comprise a mixture of compounds of varying biodegradability which need to be treated prior to disposal.

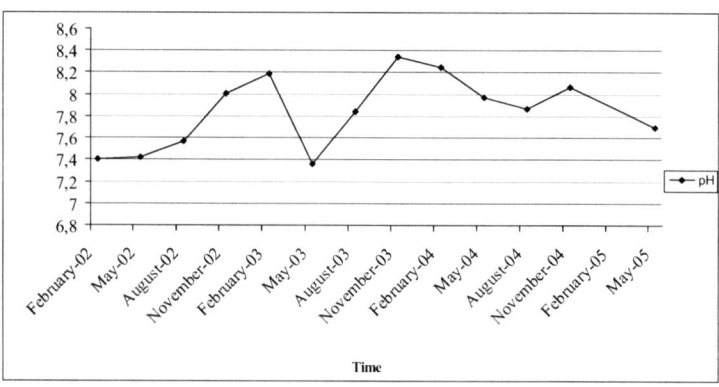

FIGURE 2. Values of pH.

The process of leachate formation in a landfill covers several phases, starting with initial decomposition to an acid forming period then to the methane formation phase. This biological process in which organic matter is catabolised to methane and carbon dioxide is anaerobic digestion.

The fluctuation of the values for pH in the landfill leachate during 2002-2004 is illustrated on Fig. 2.

During the first period studied the pH values are changed to the alkalinity. After May 2003 pH decreased and then started to increase again. The last registered value is to be neutral.

The fluctuation of the parameters DO, COD and BOD_5 is shown in Fig. 3.

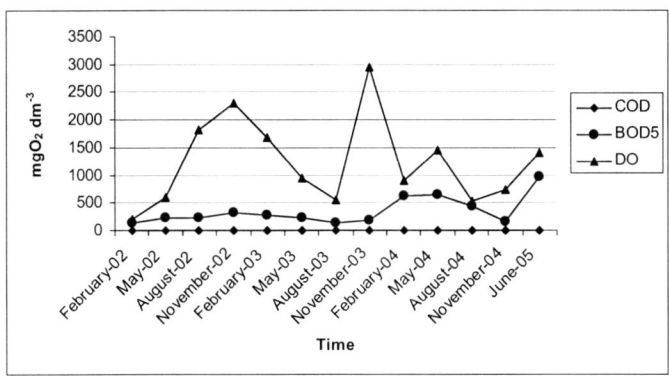

FIGURE 3. Values of DO, COD and BOD_5.

On Fig. 3 the unstable variation of the values for COD can be seen. The alteration of BOD_5 is low and the data for DO are responsible for anaerobic process.

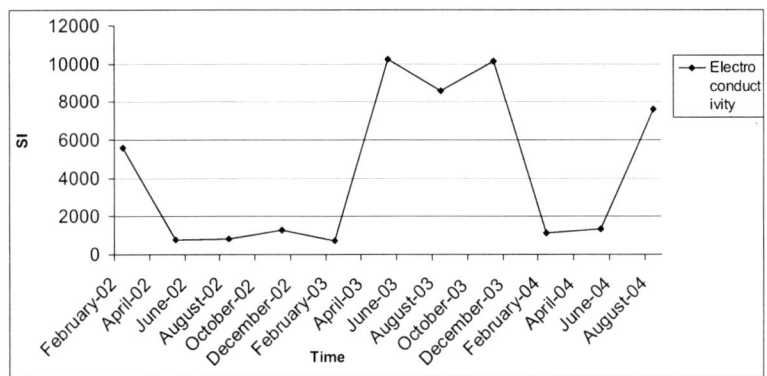

FIGURE 4. Data for electroconductivity.

The electroconductivity as a parameter for the determination of the ions presented into the aqueous samples is illustrated on Fig.4. Well formed maximum after March 2003 till the beginning of 2004 can be seen in the values of electroconductivity. The tendency of increasing after second half of 2004 is observed too.

Data for Total P, [NH_4^+-N], [NO_2^--N], [NO_3^--N] of the samples in three years period are shown on the Fig. 5.

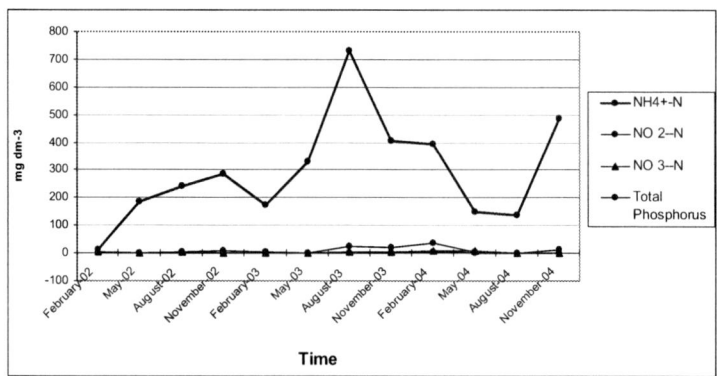

FIGURE 5. Data for Total P, [NH_4^+-N], [NO_2^--N], [NO_3^--N].

As can be seen on Fig. 5 the nitrogen mainly presents in ammonium form. The determined values for nitrites and nitrates are negligible. Fluctuations in the values for SS are given on Fig. 6.

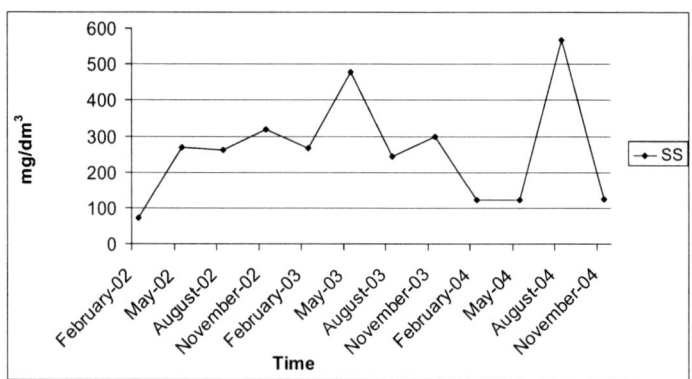

FIGURE 6. Data for SS.

The values are unstable and vary from 100 to 600 mg dm^{-3}. The chemical composition of the landfill leachate is a good indicator for the waste composition and age, the operating and climatic conditions.

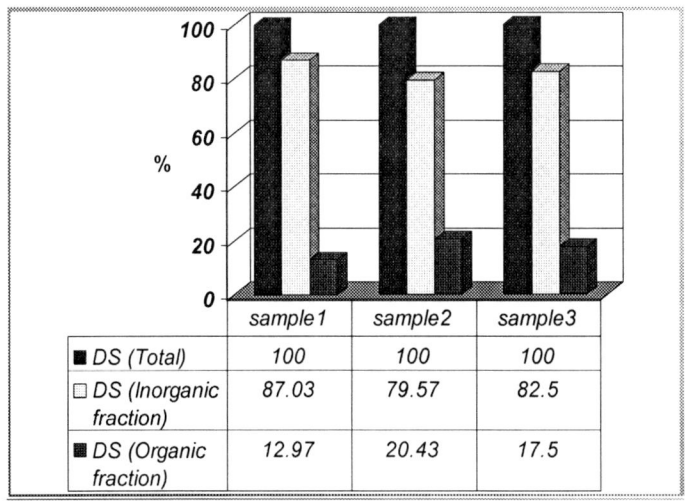

FIGURE 7. Dissolved solids (total, organic, inorganic).

For instance, low pH values (pH~6) and high chemical and biochemical oxygen demand (COD~20 000 and BOD$_5$~10 000) are typical for the acetic phase. Moderate to high pH values (pH~8) and relatively low values of chemical and biochemical oxygen demand (COD~3 000 and BOD5~200) are typical for the methanogenic phase (C. Gomes, 2005).

During 2005 samples were taken from three sampling points as shown in Fig. 1. Dissolved solids in the samples, including their organic and inorganic fractions are compared on Fig. 7.

Dissolved solids into the three samples are mainly inorganic in their origin independently of the sampling point. Slightly higher is the content of organics in the leachate from the new cell (sample 2).

The characteristics of the leachate samples (2005) are compared in Table 2. Heavy metals are metallic elements with relatively high atomic mass that are used in a variety of consumer products and industrial processes. At trace levels, many of these elements are necessary to support life. However, at elevated levels they become toxic and present significant health hazard. "RCRA heavy metals" are those metals and metalloids for which specific groundwater limits are established in the Resource Conservation and Recovery Act (RCRA), which was enacted by the US Congress in 1976 to address the management and disposal of municipal and industrial solid wastes. RCRA heavy metals, which include arsenic, barium, cadmium, chromium, lead, mercury, selenium, and silver, are on focus of the SWANA study (Jeremy O'Brien, 2005).

TABLE 2. Data for samples characteristics (2005).

Characteristic	Dimension	Sample 1	Sample 2	Sample 3
pH	-	7.2	7.8	7.7
Permanganate oxidation	$mgO_2\ dm^{-3}$	188	704	944
COD	$mgO_2\ dm^{-3}$	920	2400	1400
BOD_5	$mgO_2\ dm^{-3}$	270	1045	965
Total N	$mg\ dm^{-3}$	335	655	431
Organic N	$mg\ dm^{-3}$	18.0	70.0	51.0
NH_4^+- N	$mg\ dm^{-3}$	407	750	487
Total P	$mg\ dm^{-3}$	1.8	28.0	12.5
Total DS	$mg\ dm^{-3}$	7986,0	0808,0	8732,0
DS org. fraction	%	12.97	20.43	17,5
Total SS	$mg\ dm^{-3}$	15,0	121,0	68,0

The metals content has been studied using Emission Spectral Analysis for preliminary, semi quantitative information. Based on these data the samples were analyzed quantitatively by Atomic Absorption Spectroscopy and UV-spectroscopy (Al, Si, Ti, and P).

Emission Spectroscopy data showed that the amount of Na, K, Mg, and Ca is significant into the three samples. It was also found the presence of B, Al, Si, Cu, Mn, Fe and Zn.

In Sample1 and Sample 3 were detected some amounts of Ti, Cr, P and Bi. It was not found the presence of heavy metals Ni, Cd, Hg, Pb, Sn and Sb.

The content of the found elements is given in the Table 3.

Significant amounts of Na, K, Mg, and Ca were found both in the leachates from the old (sample 1) and new (sample 2) landfill's cells, as well as in their mixture (sample 3). The highest quantity was found in Sample 2. Relatively low are the quantities of Mn, Fe, Zn as well as of Cu.

TABLE 3. Quantitative data for detected elements.

Chemical element	Sample1 [µg/ml]	Sample 2 [µg/ml]	Sample 3 [µg/ml]
Na	1200.0	1450.0	1300.0
K	725.0	1550.0	1050.0
Mg	175.0	185.0	172.5
Ca	127.5	132.5	125.0
Cu	0.06	< 0.05	< 0.05
Cr	-	0.4	0.3
Mn	1.8	1.1	2.3
Fe	0.5	2.1	3.3
Zn	0.03	0.09	0.06
Bi	-	2.0	1.0
Al	< 0.05	-	0.13
Si	8.71	13.3	10.5
Ti	<0.05	< 0.05	-
P	-	-	9.00

4. Conclusions

- During the studied three years period the main characteristics COD and BOD_5 are increasing because of dissolving of organic compounds into the leachate as a result of biodegradation of wastes.
- The values of electroconductivity also are increasing because of dissolving of compounds which dissociate into the water.
- Inorganic fraction into dissolved solids is much more compared to organics. Dissolved organics in the leachate from the new cell are much more than those in the leachate from the recultivated cell.
- It was established the presence of B, Al, Si, Cu, Mn, Fe, Zn, Ti, Cr, P and Bi, but the content of Na, K, Mg and Ca is significant. There are not indications for heavy metals Ni, Cd, Hg, Pb, Sn and Sb.

References

Augenstein D.C. Greenhouse Effect Contributions of United States Landfill Methane GRCDA; 13th Annual Landfill Gas Symposium; Lincolshire; Illinois; 1990.
Buswell A.M., Mueller E.F. Mechanisms of Methane Fermentation.,Industrial and Engineering Chemistry 44 (1952) 550–552.

Clement B., Thomas O. Application of ultra-violet spectrophotometry and gel permeation chromatography to the characterization of landfill leachates Source: Environmental Technology 16 (4) 1995 367–377.

Di Pula M. "Third International Landfill Symposium; Biogas disposal and utilization, choice of material and quality control, landfill completion and aftercare, environmental monitoring" SARDINIA 91 14–18 October 1991.

Farquhar G.J. Rovers S.A. Gas Production from Landfill Decomposition; Water, Soil and Air Pollution 2 1973, 493.

Gomes C., Lopes M.L. et al. A Study of MSW properties of a Portuguese landfill, International Workshop "Hydro-Physico-Mechanics of Landfills", LIRIGM, Grenoble 1 University, France 21–22 March 2005.

Lo I.M.-C. Characteristics and treatment of leachates for domestic landfills; Environment International 22 (4) 1996 433–442.

Maw-Rong Lee, Yao-Chia Yeh, Wei-Shin Hsiang and Bao-Huey Hwang; Solid-phase microextraction and gas chromatography–mass spectrometry for determining chlorophenols from landfill leaches and soil Journal of Chromatography 806 (2) 15 May 1998 317–324.

O'Brien J. A summary of the SWANA Applied Research Foundation's findings June, 2005.

Otieno, F.A.O. Stabilization of solid waste through leachate recycling; Waste Management & Research 12 (1994) 93–100.

Pohland F.G., Harper S.R., Chang K.C., Dertien J.T., Chian E.S.K. Leachate generation and control at landfill disposal sites Water Pollution Resources Journal Canada;20(3) 1985 10–24.

Rees J.F. The fate of organic compounds in the landfill disposal of organic matter Journal of Chemical Technology and Biotechnology 30 1980, 361.

Reinhar D.R., Grosh C.J. Analysis of Florida MSW Landfill Leachate Quality; University of Central Florida Civil and Environmental Engineering Department; March 31 1997 Report #97–3.

Ribeiro A., Neves M.H., Almeida M.F., Alves A., Santos L. Direct determination of chlorophenols in landfill leachates by solid-phase micro-extraction–gas chromatography–mass spectrometry; Journal of Chromatography; 975(2); 1 November 2002 267–274.

Schrab G.E., Brown K.W., Donnelly K.C. Acute and genetic toxicity of municipal landfill leachate Water Air Soil Pollut. 69 (1-2) 1993 99–112.

ASPECTS OF EUTROPHICATION AS A CHEMICAL POLLUTION WITH IMPLICATIONS ON MARINE BIOTA AT THE ROMANIAN BLACK SEA SHORE

ALICE SBURLEA*, LAURA BOICENCO AND ADRIANA COCIASU
Department of Ecology and Environment Protection
National Institute for Marine Research and Development
Mamaia Blvd., nr. 300, RO - 900581, Constanta, Romania

*To whom correspondence should be addressed. alices@alpha.rmri.ro,

Abstract. Starting with 70ies, Romanian Black Sea shore is confronting with marine eutrophication, as a result of nutrients input from sources such as the Danube River and coastal anthropogenic activities. Due to economical decline in early 90ies, the effects of eutrophication (phytoplankton blooms, fauna mortalities) have had low occurrence.

Keywords: eutrophication; nutrients; impact

The intensification of the anthropogenic pressure induced strong increases of main nutrient stocks in the Romanian area of the Black Sea in the 70ies and 80ies. The sea water became an authentic "complex nutritive medium", the most important consequence consisting in the expansion of the frequency, amplitude and spatial extension of phytoplankton blooms. These fenomena induced serious perturbations in the ecosystem, dramatic changes of the environmental conditions, and substantial impact on benthic flora and fauna, which further endangered the exploitable stocks of living resources. Mass mortality of marine animals (Fig. 1a) caused by hypoxia and anoxia which accompanied and succeeded the microalgae blooms, and mass development of benthic flora (Fig. 1b) are still present phenomena at the Romanian shore, now with low occurrence do to economic decline starting in the early 90ies.

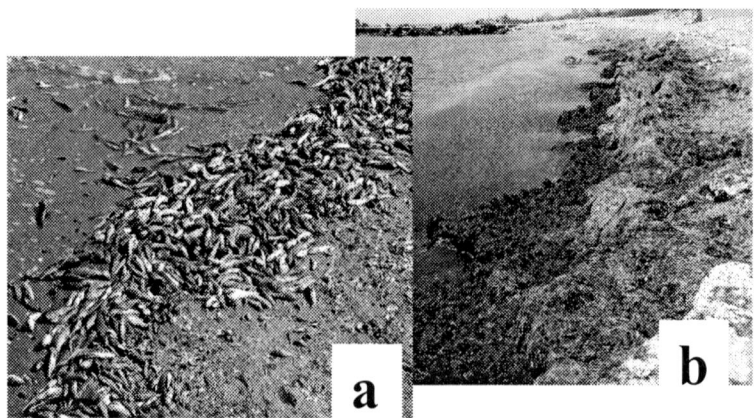

FIGURE 1. Eutrophication effects on Black Sea coastal biota: a) phytoplankton blooms caused mass fish mortalities (2001); b) macroalgae washed to shore after mass development due to anthropogenic eutrophication (2004).

Marine water nutrients concentration is responsible for phytoplankton structure variation. Therefore, the quantities of organic substance (Table 1) in 2001 and 2002 have triggered growth of non-diatom species, given that most of micro-algal consumers of organic substance are mixotrophic species of microalgae, which predominated in the respective time span.

TABLE 1. Long term mean values variation of most important nutrients in Constanta nearshore waters.

Year/Period	$P-PO_4$ (μM)	$N-NO_3$ (μM)	$N-NO_2$ (μM)	$Si-SiO_4$ (μM)	Organic substance ($mg\ O_2 \cdot l^{-1}$)
2000	**0.46**	**5.31**	**0.70**	**16.04**	**2.39**
2001	**0.66**	**5.03**	**0.85**	**16.13**	**2.66**
2002	**0.49**	**7.48**	**0.83**	**12.90**	**3.29**
1960-1970	*0.34*	*1.60*	*-*	*36.75*	*1.96*
1983-1990	*5.91*	*7.11*	*0.74*	*12.10*	*2.46*
1991-2000	*1.86*	*5.90*	*0.86*	*12.64*	*2.25*

In contrast, lower quantities of silicates recorded in 2002 influenced the decrease of diatoms share, which are known to be main consumers of silicates. Likewise, it should be pointed out that nitrogen compounds have reached higher concentrations in 2001 and 2002, contributing to an increase of phytoplankton abundance in general, and of cyanophytes in particular. The multi-annual dynamics of phytoplankton numerical density and biomass reveals a declining trend, starting the 80ies, which are years of intensive eutrophication (Bodeanu et al., 1998); as shown in Figure 3 the average values for 2003 of these two parameters are confirming this trend (Bodeanu et al., 2004).

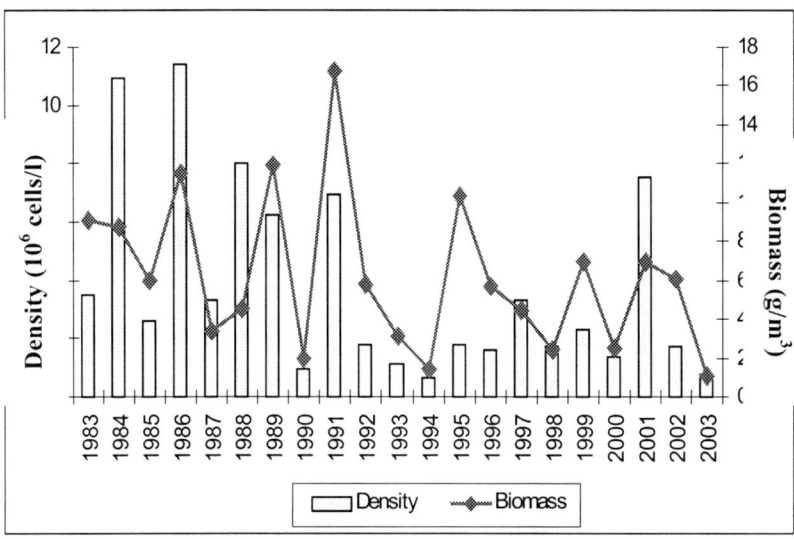

FIGURE 2. Multi-annual mean of phytoplankton quantities in Constanta near-shore marine waters between 1983 and 2003.

It should be pointed out that the numerical density of microalgae species involved in bloom episodes in the intensive eutrophication period showed a noticeable decrease in 2001, 2002 and 2003. Therefore, mass species (*Prorocentrum minimum, Eutreptia lanowii, Emiliania huxleyi, Skeletonema costatum, Cyclotella caspia*) developed with excessive rates in the 80ies, starting with 2000 had not intensive growth of population.

After last few decades of macroalgae and phanerogames decline (especially *Cystoseira, Phyllophora* and *Zostera*), a covering of hard substrata with red and green algae has been observed starting the 90ies. Thus, a higher macroalgae biodiversity in the Southern part of the Romanian coast has been noticed.

Fish stocks have a slow recovery, 26 species with high production in '70ies have lost their commercial value, only five of them being usualycaptured today (*Engraulis, Sprattus,* Gobiidae, *Merlangus*).

The Danube river was considered to be the main source of nutrients input into the Black Sea (Cociasu et al., 2004), associated with anthropogenic activities at the Romanian shore with high development in the '70ies (Navodari fertilizer plant, touristic resorts, untreated waste water, and harbour construction). The decline of economy at the Romanian shore in the early '90ies was the start for marine environment quality improving; therefore, lower phytoplankton densities and biomass were encountered in the last years (Fig. 3).

Eutrophication has long term effects on marine ecosystems, the existing nutritional base for microalgae being improved by nutrients trapped in sediments and slowly released in the bio-geo-chemical natural cycles. Thus, depending upon the conjuncture of the hydro-meteorological conditions, microalgal blooms are possible to reclaim in the next years, even if the nutrients input in the sea is maintained at existing level.

Conclusions

Eutrophication has damaging effects on marine biota, causing phytoplankton blooms with high fervency and magnitude, mass mortalities of benthic fauna due to hypoxia, and biodiversity loss.

The most intense period of eutrophication at the Romanian Black Sea shore were the '70ies and '80ies. The Danube river and coastal anthropogenic activities were the main sources of nutrients. Once the intensity of economic activities decreased (early '90ies), the marine environment is slowly evolving toward a normal status, some parameters of the marine biota (biomass and diversity) indicating the improving trend.

References

Bodeanu N., Moncheva S., Ruta G. and Popa L., 1998 – Long term evolution of the algal blooms in Romanian and Bulgarian Black Sea waters, Cercetari marine – Recherches marines, Constanta 31: 77–86.

Bodeanu N., Andrei C., Boicenco L., Popa L. and Sburlea A., 2004 - A new trend of the phytoplankton structure and dynamics in the Romanian marine waters, Cercetari marine – Recherches marines, Constanta 35: 77–86.

Cociasu A. and Popa L., 2004 – Significant changes in Danube nutrient loads and their impact on the Romanian Black Sea coastal waters, Cercetari marine – Recherches marines, Constanta 35: 25–37.

INVESTIGATION OF THE CONSTANTA SURFACE WATER'S POLLUTION SOURCES

ALINA COMAN*, ELISABETA CHIRILA AND IONELA POPOVICI CARAZEANU
Chemistry Department, Ovidius University of Constanta
124 Mamaia Blvd, 900527 Constanta, Romania

*To whom correspondence should be addressed: acoman@univ-ovidius.ro

Abstract. The aim of the paper is to investigate the pollution sources of Tabacarie Lake located in Constanta district Romania, by monitoring the quality parameters. Eight sampling sites were established around the lake and analyses were carried out twice on week, in July, 2005. Alkalinity, Chemical Oxygen Demand by potassium permanganate method (CODMn), Dissolved Oxygen (DO), Total and Calcium Hardness, Salinity, Sulfides, Cu, Cr and Fe were the monitored quality parameters. The monthly average of the quality parameters for all sampling sites ranged as follows: "p" alkalinity 0.91 – 1.51 meq/L; "m" alkalinity 4 – 5.15 meq/L; CODMn 8.42 – 15.09 mgO$_2$/L; DO 7.76 – 13.49 mg/L; calcium hardness 0.84 – 1.31 meq/L; total hardness 7.28 – 8.07 meq/L; salinity 0.39 – 0.54 g/L; sulfides 0.62 – 1.08 mg/L; Cu 11.41 – 18.97 mg/L; Fe 8.84 – 18.32 mg/L; Cr 0.96 – 4.9 mg/L. The values of quality-monitored parameters are variable in quasi-large ranges, depending on the position of the sampling sites. A direct correlation between the CODMn, Sulfides and the DO can be done.

Keywords: surface water; pollution sources; quality parameters

1. Introduction

Water is one of the most important environmental factors that contribute to the quality of the live.

Tabacarie lake is located in Constanta district, Romania, on the Black seaside coast, has a surface of about $9.5 \cdot 10^5$ m^2, an average water volume of about $1.7 \cdot 10^6$ m^3 (Breier, 1976) and is mainly used for agreement. The arrangement works concerning municipal sewage have decided the discharge of pluvial waters as a safety measure, only in the case of abundant rains or damages.

The aim of the paper is to investigate the pollution sources of Tabacarie Lake by monitoring the quality parameters.

2. Experimental

Eight sampling sites were established around the lake (as showed in the figure 1) and analyses were carried out twice on week, in July 2005.

Alkalinity, Chemical Oxygen Demand by potassium permanganate method (CODMn), Dissolved Oxygen (DO), Total and Calcium Hardness, Salinity, Sulfides, Cu, Cr and Fe were the monitored quality parameters, using standard titrimetric analytical methods according to Romanian regulations and UV-VIS spectrometry by standard addition method (Chirila et al., 2005).

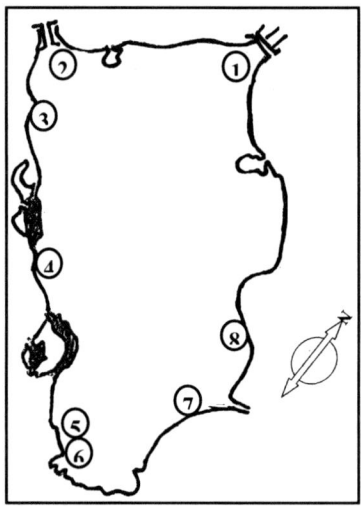

FIGURE 1. The position of sampling sites from Tabacarie lake.

3. Results and Discussion

The obtained results concerning the alkalinity, hardness and salinity of the studied pollution sources are presented in the Table 1.

The results for alkalinity and hardness are similar with other reported by Chirila et al. and Galatchi et al in 2005, but the salinity is a little over the mean reported values. That can be explained by the evaporation phenomenon that occurs in the summer time.

TABLE 1. Alkalinity, hardness and salinity of Tabacarie lake' pollution sources in July 2005.

Sampling site no.	Alkalinity, meq/L		Hardness, meq/L		Salinity, g/L
	"p"	"m"	Calcium	Total	
1	1.39	4.00	0.92	7.72	0.39
2	1.31	4.69	1.07	7.90	0.46
3	1.27	5.14	1.31	8.07	0.47
4	1.13	4.54	0.89	7.58	0.50
5	0.91	4.25	0.85	7.53	0.50
6	0.99	4.41	0.84	7.28	0.54
7	0.97	4.77	0.89	7.64	0.49
8	1.51	4.57	0.92	7.70	0.48

The distribution of the DO, CODMn and sulfides concentration in the analyzed samples during July 2005 is presented in the Figures 2, 3 and 4.

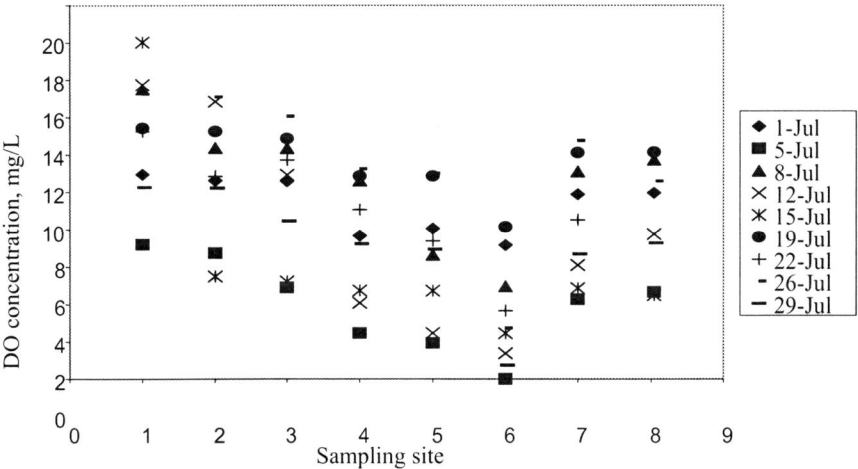

FIGURE 2. The distribution of dissolved oxygen (DO) concentration in Tabacarie Lake's pollution sources in July 2005.

The obtained values for DO vary in a wide range, ones of them being greater than the normal oxygen values in surface water. This fact can be explained by the interferences´ intervention. We can observe also that in the samples no 1 and 2, representing surface water discharge (no 2 from Siutghiol Lake to Tabacarie Lake and no 1 from Tabacarie Lake in the Black Sea), the DO is greater than in other samples.

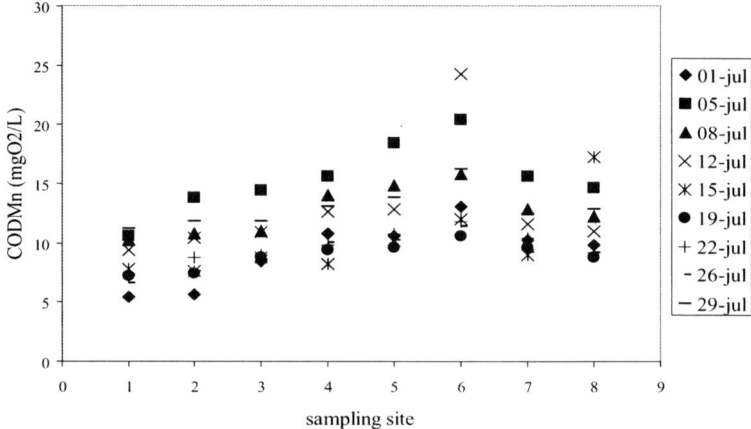

FIGURE 3. The distribution of CODMn concentration in Tabacarie lake's pollution sources in July 2005.

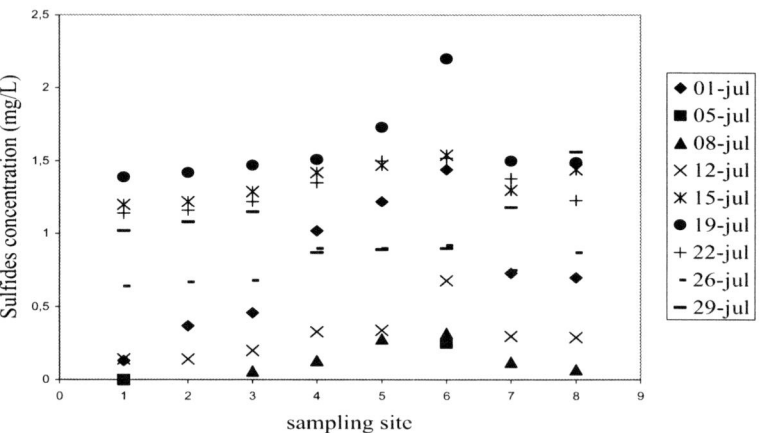

FIGURE 4. The distribution of sulfides concentration in Tabacarie Lake's pollution sources in July 2005.

The analysis of the COD and sulfides shows also a very wide variation of the values from one sample taking to another. The maximum values were recorded in the samples number 5 and 6 that are obviously the most polluted.

Comparing with the mean data reported by Chirila et al. in 2005, it can be noticed that the punctual values are different than the mean values, but in the same range of magnitude.

Dissolved oxygen can be consumed in the chemical, photochemical or biochemical oxidation processes of the dissolved organic compounds and sulfides (Chirila, 2000).

A direct correlation between DO, COD and sulfides can be done, and the figure 5 demonstrates it.

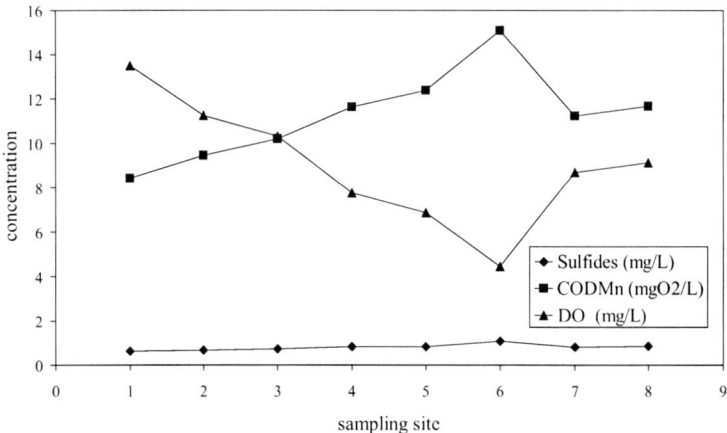

FIGURE 5. The distribution of dissolved oxygen (DO), chemical oxygen demand by potassium permanganate method (CODMn) and sulfides in Tabacarie lake's pollution sources in July 2005 (mean values).

When the sulfides and COD concentration increase, the DO decrease and this fact indicates the increasing of the pollution degree. The most polluted sources of Tabacarie lake are 4, 5, 6 and 7. The mean values of the total Fe, Cr and Cu concentrations determined by the standard addition method are presented in the Table 2.

Comparing with other values determined by the same technique or others (Chirila et al., 2002, Chirila et al., 2005) the reported metal concentrations in the analyzed samples are very high, over the limits imposed by Romanian regulations (in parentheses). This fact can be explained by the construction works for the extension of municipal wastewater treatment plant, situated on the Tabacarie Lake's northern part. Studying all the analyzed quality parameters it is obvious that urgent works of rehabilitation must be done, because the degree of pollution is over the accepted limits.

TABLE 2. Total Fe, Cu and Cr concentrations in Tabacarie Lake' pollution sources in July 2005 (mean values).

Sampling site no.	Fe, mg/L (1)	Cu, mg/L (0,05)	Cr, mg/L (0,5)
1	12.47	18.97	1.5
2	10.37	11.41	12.1
3	9.23	14.46	0.96
4	18.32	13.42	4.9
5	9.89	14.28	1.29
6	12.41	16.31	2.03
7	11.31	13.91	1.19
8	8.84	13.41	0.67

4. Conclusions

The monthly average of the quality parameters for all sampling sites ranged as follows: "p" alkalinity 0.91 – 1.51 meq/L; "m" alkalinity 4 – 5.15 meq/L; CODMn 8.42 – 15.09 mgO2/L; DO 7.76 – 13.49 mg/L; calcium hardness 0.84 – 1.31 meq/L; total hardness 7.28 – 8.07 meq/L; salinity 0.39 – 0.54 g/L; sulfides 0.62 – 1.08 mg/L being similar with other reported values.

Total metals concentrations are over the limits imposed by Romanian regulations (Cr 0.96 – 4.9 mg/L, Cu 11.41 – 18.97 mg/L and Fe 8.84 – 18.32 mg/L); urgent works of rehabilitation must be done.

The values of quality-monitored parameters are variable in quasi-large ranges, depending on the position of the sampling sites. A direct correlation between the CODMn, sulfides and the DO can be done.

References

Breier, E. 1976, Lakes from Romanian Black Seaside Coast, *Academy Press Bucharest*.
Chirila E., 2000, Study about analytical control in tertiary treatment stage of petrochemical effluents, *Rev.Chim.* (Bucharest), 51, 638–643.
Chirila E., Godeanu S., Godeanu M., Galatchi L.D. and Capota P., 2002, Analytical characterization of the Black Seacoast lakes, *Environmental Engineering and Management Journal*, 2, 205–212.
Chirila E., Dobrinas S., Carazeanu I. and Draghici C., 2005 Tabacarie lake water quality monitoring, *Environmental Engineering and Management Journal*, 4, 169–176.
Galatchi L.D., Dobrinas S. and Chirila E., 2005 Threats on the seaside lakes water and their acute and long-term consequences, in press, *proceedings ASI NATO*.

THE POTENTIAL USE OF *FESTUCA* CULTIVARS AND LIGNITE FOR PHYTOSTABILIZATION OF HEAVY METAL POLLUTED SOILS

JACEK KRZYŻAK[1], TYLER LANE[2] AND ANNA CZERWIŃSKA[1]*
[1]*Institute for Ecology of Industrial Areas
6 Kossutha Str., 40-844 Katowice, Poland,*
[2]*Harvard School of Public Health. 677 Huntington Ave., Boston, MA, 02115, USA*

*To whom correspondence should be addressed: anna_czerwinska@vp.pl

Abstract. Previous research has identified *Festuca* cultivars as good candidates for use as phytostabilizers in metal-contaminated soils. *Festuca* species combine good metal sorption properties in root tissues with high metal tolerance and proven cultivation methods. These advantages may allow *Festuca* use in remediation projects where funding is unavailable to support more expensive soil remediation methods. This project expands previous research by combining fescue cultivars with various soil additives in laboratory tests, and identifying the best cultivar/soil amendment combination for field applications. Following laboratory analysis, field test plots will be monitored over two growing seasons to assess biomass production and metal stabilization effectiveness.

Keywords: *Festuca*; phytostabilization; phytochemostabilization; heavy metals; soil remediation.

1. Introduction

Previous hydroponic and soil pot research investigations have identified a number of *Festuca rubra* and *Festuca arundinacea* cultivars suitable for binding and holding various heavy metals on or within their root tissue (Kucharski and Sas-Nowosielska, 2004a). *Festuca*'s ability to tolerate heavy metal-contaminated soils, combined with proven cultivation methods, renders it an excellent candidate for phytostabilization projects at metal-contaminated sites where alternative remediation methods may not be economically feasible.

For phytostabilization projects, the primary goal is the immobilization of soil contaminants to prevent or reduce introduction of pollutants into exposure pathways. Most phytoremediation research, however, has been conducted using metal hyperaccumulating dicots, with the goal of translocating metals from soils to aerial plant tissues through phytoextraction (Baker and Brooks, 1989). In the current study, plants have been used to stabilize metals in soil rather than extract them. Although the end goals of phytoremediation and phytostabilization are vastly different, many of the biogeochemical root processes for moving available metals from soil to root tissues are similar in both processes.

Plant roots stabilize bioavailable soil metal fractions through several mechanisms. Precipitation at the rhizosphere of the solubilized fraction of heavy metals in the soil is the primary mechanism of most metal adsorption to root surfaces (Blaylock et al., 1997). Adsorption also occurs through the binding of free metal cations by pectins on root cell walls and in pectin and other polysaccharide combinations from root-secreted mucilage (Waisel et al., 1996).

Studies have found many natural plant hyperaccumulators tend to have a higher density of metal transporters at the root-cell plasma membrane (Pence et al., 2000). The higher density of metal transporters allows these plants to readily take up metal cations from the soil solution. Once metals are accumulated, hyperaccumulating plant species usually exhibit a rapid translocation of accumulated metals from roots to shoots (Kramer, 2000). Translocated metals are then stored in vacuoles of the epidermal or mesophyllic cells of the stem to decrease toxicity to the plant (Mathys, 1977).

In highly contaminated soils these hyperaccumulating attributes are often fatal to the plant species, and therefore metal-tolerant phytostabilizers are a more viable remediation option. Further, unless a shoot treatment or isolation program is part of the phytoextraction process, the use of hyperaccumulators susceptible to herbivory should be avoided.

As a non-hyperaccumulating species, *Festuca* cultivars are much more suitable for phytostabilization processes. *Festuca* display several desirable phytostabilizing qualities, as they are highly metal-tolerant and the majority of cultivars appear to concentrate most accumulated metals in their root systems (Kucharski and Sas-Nowosielska, 2004a). In addition, *F. rubra* and *F. arundinacea* are indigenous in pasture lands and meadows throughout Europe, temperate Asia and through introduction in North America. *F. rubra* is generally cold-, salt- and drought-tolerant while *F. arundinacea* prefers calcareous, sandy soils and is also salt- and drought-resistant (Hubbard, 1984; Mossberg et al., 1992).

Although the mechanism is uncertain, like most monocots, root to shoot metal translocation does not readily occur in *Festuca* species. The lack of a translocation mechanism may be due to low concentrations of histidine in the xylem (Kramer et al., 1996), but this inhibition in *Festuca* requires further investigation. In many cases, *Festuca* species have also developed a symbiotic relationship with endophytic fungi, which appears to aid in reducing insect and some herbivore grazing. The low transport of metals into shoot tissues and symbiosis with endophytic fungi further reduce the possibility of metal introduction into the food chain, and increase the attractiveness of *Festuca* as a phytostabilizer.

2. Methods

Soil pot tests were used to identify the best *Festuca* cultivar and soil amendment combination for use in field trials. Ten cultivars of fescue seed were procured from the Swiss company DLF-Trifolium (Roskilde, Denmark). Table 1 outlines the cultivars identified as viable metal accumulators during previous hydroponic laboratory studies (Kucharski and Sas-Nowosielska, 2004a).

TABLE 1. Ten fescue test cultivars.

Festuca rubra		Festuca arundinacea	
1.	– cultivar Pernille	1.	– cultivar Cochise
2.	– cultivar Gondolin	2.	– cultivar Montserrat
3.	– cultivar Leonora	3.	– cultivar DP 50-9011
4.	– cultivar Napoli	4.	– cultivar Kora
5.	– cultivar Carina	5.	– cultivar Feline

The cultivar seeds were sown in pots filled with contaminated soil prepared from the Warynski zinc smelter site. The Warynski smelter is located near the town of Piekary Śląskie and 15 km northwest of the major population center of Katowice [pop. 345,000] (Central Statistical Office, 1999). The site is owned by the Orzel Bialy Mining & Metallurgical Works, S.A. and over 1.3 million people live within a 15 km radius of this site. Zinc and lead ore smelters operated at the site from 1927 until 1990. During this activity period, the smelters produced approximately 3,500,000 tons of mixed lead (Pb), cadmium (Cd) and zinc (Zn) waste, deposited in piles spread across a 60 hectare site. Although limited recyclable smelting activity continues at the site, the Piekary Slaskie municipal authorities are interested in redeveloping the land for alternative industrial purposes.

The Warynski soil was analyzed for compositional content and mixed to ensure all test samples received identical nutrient/contaminant exposure. The results of the soil parameter determination are displayed in Table 2.

TABLE 2. Warynski soil characterization.

Sand	Silt	Clay	OM[a] (mg/kg)	pH (H$_2$O)	PH (KCl)	EC (μg/cm)	CEC[b] (cmol/kg)
37.3%	56.3%	6.8%	8.52± 0.12	6.57± 0.07	6.71±0.03	154.0±11.0	6.67±0.24

[a] Soil organic matter, measured by content loss on ignition.
[b] Cation Exchange Capacity, measured according to ISO 13536.

A thorough analysis of Warynski soil samples was performed to assess the potential plant exposure to three major metal contaminants, Pb, Cd and Zn. Tables 3 and 4 describe the major metal contaminant component of the Warynski soil, displaying available fractions and soil composition differentiated by depth.

TABLE 3. Warynski soil metal concentrations.

	Pb (mg/kg)	% of Total	Zn (mg/kg)	% of Total	Cd (mg/kg)	% of Total
Total Metal Concentration[a]	9712 ± 562	100.0	11498 ± 417	100.0	537 ± 23	100.0
Potentially Available[b]	6533 ± 91	69.0	7673 ± 105	73.0	365 ± 2.89	68.0
Bioavailable[c]	5.23 ± 0.13	3.3	363 ± 7.47	0.06	41.78 ± 0.69	8.0

[a] Total soil element concentrations determined in *aqua regia*.
[b] Extraction with 0.43 N HNO$_3$.
[c] Extraction with 0.01 M CaCl$_2$.

TABLE 4. Warynski major contaminant soil characterization by depth.

Depth (cm)	EC (μg/cm)	Pb (mg/kg)	Zn (mg/kg)	Cd (mg/kg)
0-20	154 ± 11	8265 ± 1143	9673 ± 925	392 ± 45
20-40	125 ± 10	2890 ± 822	4854 ± 760	155 ± 23

Table 4 shows 65-70% of total metal contamination concentrating in the first 20 cm, indicating high accessibility for plant roots where phytostabilization would be most effective. Minor metal concentrations were analyzed (Table 5) to identify any additional elements that might be mobilized by soil amendment additions and act as potential confounders to plant toxicity.

TABLE 5. Warynski soil minor metal and semimetal concentrations.

	^{c}As (mg/kg)	^{c}Cu (mg/kg)	^{c}Cr (mg/kg)	^{c}Ni (mg/kg)	^{c}Hg (mg/kg)
Total Metal Concentration[a]	211.57±26.2	54.89±4.49	14.37±0.70	10.35±1.60	1.74±0.18
Bioavailable[b]	0.039±0.005	0.002±0.001	0.066±0.02	0.060±0.03	0.714± 0.04

[a] Total soil element concentrations determined in *aqua regia*.
[b] Extraction with 0.01 M $CaCl_2$.
[c] **Arsenic, copper, chromium, nickel, mercury**
Source: Kucharski et al. (2004a).

3. Experiment

A total of 120 test pots were used to examine the effects of three soil additive mixtures to root and shoot growth of the *Festuca* cultivars. Due to the extremely high metal content of the Warynski soils, additive mixtures were necessary to provide a nutrient substrate for plant development. The use of additives in phytoremediation projects may therefore be more properly termed phytochemostabilization.

Three replicates for each of the ten cultivars were established for each control and each of the three additive combinations. Thus, each cultivar was tested using 1) three control pots, 2) three pots with 2.5% Superphosphate (SP) additive [$Ca(H_2PO_4)_2$] , 3) three with 2.5% SP additive and 10% lignite, and 4) three with 20% lignite. Controls were potted in Warynski soil without any soil additives. Soil additive specifications were determined in previous tests (Kucharski et al., 2004b). Each pot contained 400 g of soil, including the addition of any soil additive combinations. Due to variance in seed sizes, *Festuca rubra* seeds were sown with 250 mg of seed per pot, while *Festuca arundinacea* seeds were sown with 500 mg seed per pot.

The test period ran for ten weeks in a growing room, controlling for light (11,000 Lumens), temperature (24°C), humidity (65-95%), water intake, drainage, and air circulation. Plant monitoring was performed weekly, examining height, coloration, soil saturation and general growth characteristics. At the end of the test period, all plant samples were removed from the pots and separated from the soil. Each sample was dried and evaluated for biomass ratio (root to shoot), root density, mass, metal binding and heavy metal uptake (analysis pending). Comparative analysis was performed to identify the species and cultivar with the greatest root mass and ability to stabilize heavy metal-contaminated soil.

4. Results and Discussion

Lacking any soil amendment additives, the control group displayed a stunted growth peak at three weeks followed by discoloring and decline by week eight. A lack of nutrients in the Warynski soil and high concentrations of metal toxicants combined to severely inhibit growth, possibly via decreased root elasticity (Lane and Martin, 1982) or synergistic toxicity (Kahle and Breckle, 1989). Further root analysis will be necessary to provide a definitive toxicant determination.

While the 2.5% SP additive provided additional soil nutrients, it also lowered the pH and thus further mobilized metals for binding. Growth was increased, especially in combination with the lignite additive, but at a cost of elevating the soluble metal content of the soil (Kucharski et al., 2004b). The 20% lignite soil additive provided all *Festuca* cultivars with the highest above-ground vegetative growth (see *Figure 1*). These results point to a reduction in the solubilized soil metal content, allowing for greater vegetative growth in both roots and shoots. Reducing soluble metal content by substituting lignite for SP fertilizers should further aid in soil stabilization remediation efforts.

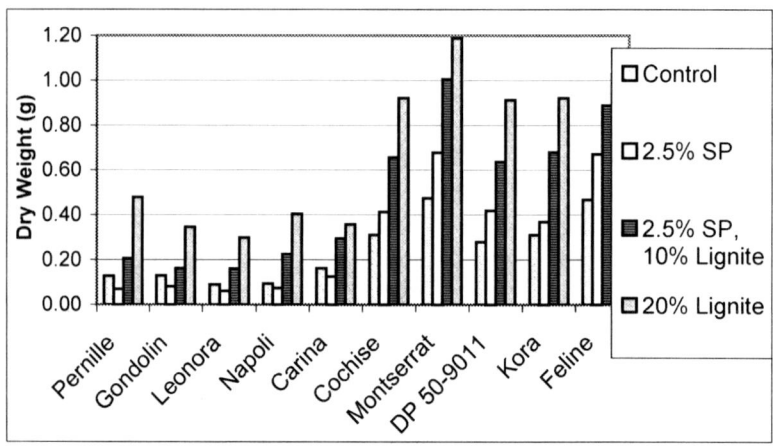

FIGURE 1. Above-Ground Tissue Comparison of *Festuca* Cultivars (at 10 weeks).

Atomic absorption spectroscopy (AAS) determination of solubilized soil metal content found the highest metal sorption for the roots of cultivars Montserrat and Feline (see Figure 2).

5. Conclusion

The soil pot test results are the first step in a total phytostabilization assessment using *Festuca* grasses. The next steps will evaluate the Montserrat and Feline cultivar and amendment combinations under field trial conditions.

The laboratory growth results indicated that the Montserrat and Feline cultivars held the greatest biomass potential for future field trials, due to high metal sorption at the root and low root to shoot metal transfer ratios (see Figure 3). The 20% lignite soil additive produced the best vegetative growth, but due to cost considerations at the field trial scale, the 2.5% SP and 10% lignite mixture may be a more cost effective additive. These field trials will be conducted over two growing seasons at two sites with lower levels of metal contamination, the Cooperative Farm (Bytom, Poland) and the Warynski smelter sites. These tests will assess shoot biomass production and metalstabilization effectiveness bioenergy analysis.

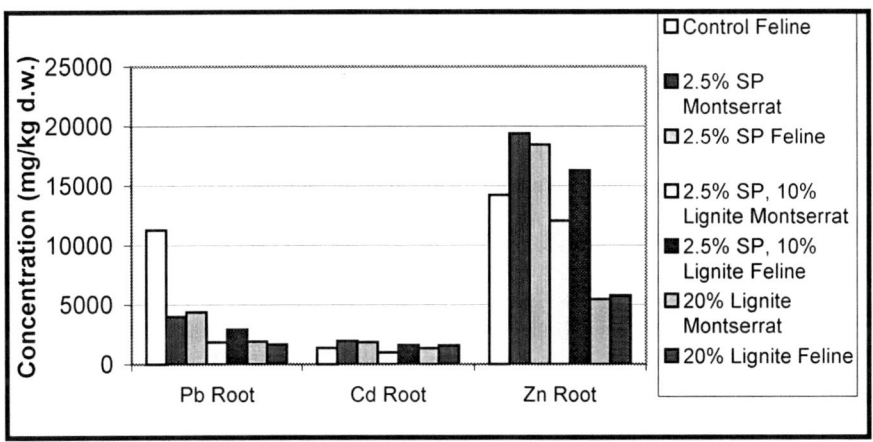

FIGURE 2. Selected *Festuca* Average Root Metal Concentrations.

The final goal will be the development of a harvestable *Festuca* crop yield for burning in local energy production facilities. *Festuca* species were chosen due to their low root to shoot transport of sequestered metals, thus avoiding secondary mobilization of the metals during biomass burning. As an additional precaution, the selected energy production facility will have sufficient scrubbers to remove any remaining mobilized metal particles from the outlet stack. Thus, the use of *Festuca* as a phytostabilizer may generate a sufficient bioenergy crop to render this exposure reduction method self-sustainable at sites where remediation would not otherwise be economical.

References

Baker, A., and Brooks, R., 1989, Terrestrial higher plants which hyperaccumulate metallic elements- a review of their distribution, ecology and phytochemistry, *Biorecovery*, **1**: 81–126.

Blaylock, M., Salt, D., Dushenkov, S., Zakharova, O., Gussman, C., Kapulnik, Y., Ensley, B., and Raskin, I. 1997, Enhanced accumulation of Pb in Indian mustard by soil-applied chelating agents, *Environmental Science And Technology*, **31**:860–865.

Central Statistical Office, 2000, *Basic Urban Statistics*, Central Statistical Office, Warsaw, Poland.

Hubbard, C., 1984, *Grasses: a guide to their structure, identification, uses and distribution in the British Isles*, 3rd ed., revised by Hubbard, JCE, Penguin Books, Middlesex, UK.

Jørgensen, R., Consequences of using genetically modified plants for phytoremediation, Risø National Laboratory, DK-4000 Roskilde, Denmark.

Kahle, H. and Breckle, S., 1989, single and combined effects of lead and cadmium on young beech trees (*Fagus sylvatica* L.) Proceedings of the 14th International Meeting, International Union of Forestry Research Organizations (IUFRO), Part 2, Interlaken, Switzerland, October 5, 1988, pp. 442–444.

Krämer, U., Pickering, I., Prince, R., Raskin, I., and Salt, D., 2000, Subcellular localization and speciation of nickel in hyperaccumulator and non-accumulator *Thlaspi* species. *Plant Physiol.*, **122(4)**:1343–1353.

Krämer, U., Cotter-Howells, J., Charnock, J., Baker, A., and Smith, J., 1996, Free histidine as a metal chelator in plants that accumulate nickel, *Nature*, **379**:635–638.

Kucharski, R., 2000, Warynski site project, Presentation (July, 2003), Institute for Ecology of Industrial Areas, Katowice, Poland.

Kucharski, R., and Sas-Nowosielska, A., 2004a, Metallophytes: An integrated approach towards removal by plants of toxic metals from polluted soils, Publication pending.

Kucharski, R., Sas-Nowosielska, A., Malkowski, E., Pogrzeba, M., and Krzyzak, J., 2004b, A decision support system to quantify the cost/benefit relationship of the use of vegetation in the management of heavy metal polluted soils and dredged sediments, Phytodec, Internal Final Report in cooperation with US Department of Energy. Unpublished.

Lane, S., and Martin, E., 1982, An ultrastructural examination of lead localization in germinating seeds of *Raphanus sativus*, *Zeitschrift Pflanzenphysiol.* **107**:33–40.

Mathys, W., 1977, The role of malate, oxalate, and mustard oil glucosides in the evolution of zinc-resistance in herbage plants, *Physiol. Plant.* **40**:130–136.

Mossberg, B., Stenberg, L., and Ericsson, S., 1992, *Den Nordiska Floran*, Walhstrom and Widstrand, Turnhuot, Belgium, 1992.

Nowak, W., 2003, Clean coal fluidized-bed technology in Poland, *Applied Energy*, **74(3)**: 405–413.

Pence, N., Larsen, P., Ebbs, S., Letham, D., Lasat, M., Garvin, D., Eide, D., Kochian, L., 2000, The molecular physiology of heavy metal transport in the Zn/Cd hyperaccumulator *Thlaspi caerulescens*, *Plant Biology*, **97(9)**:4956–4960.

Salt, D. and Kramer, U., 2000, Mechanisms of metal hyperaccumulation in plants,. *Phytoremediation of Toxic Metals*, John Wiley and Sons, Inc.

Tessier A., Campbell, P., and Bisson, M., 1979, Sequential extraction procedure for the speciation of particulate trace metals, *Analytical Chemistry*, **51**:844–851.

Hagemeyer, J., and Breckle, S., "Growth Under Trace Element Stress", Waisel, Y., Amram, E., and Kafikafi, U., Eds., 1996, *Plant Roots: the Hidden Half*, 2nd ed., Mercel Dekker, Inc. New York, pp. 415–428.

DEVELOPMENT OF ANALYTICAL METHOD FOR DETERMINATION OF PCBS IN SOIL BY GAS CHROMATOGRAPHY WITH ELECTRON-CAPTURE DETECTOR

ANNA DIMITROVA*, TEODOR STOICHEV, TOMISLAV RIZOV, ANASTASIYA KOLARSKA, NIKOLAY RIZOV AND ALEXANDAR SPASOV
National Center of Public Health Protection
15 Ivan Geshov Blvd, 1431 Sofia, Bulgaria

*To whom correspondence should be addressed: a.dimitrova@ncphp.government.bg

Abstract. The distribution of polychlorinated biphenyls (PCBs) and organochlorine pesticides (OCPs) in the environment has not been systematically studied in Bulgaria in spite of their negative effect on the human health. The aim of this study is to develop a cost-effective method for determination of low concentrations of PCBs and OCPs in soils. After extraction with hexane/acetone and column cleaning with silica, the analyses was performed by gas chromatography with electron-capture detector. The limit of detection is between 0.1 and 1 ng g^{-1}, the reproducibility at low environmental levels is about 15% RSD. The analytical recoveries for the individual compounds are between 65% and 100%. The method can be applied to study the sources of pollution, the migration and biogeochemistry of PCBs in the environment.

Keywords: PCBs; OCPs; soil; gas chromatography; electron-capture detector

1. Introduction

The polychlorinated biphenyls (PCBs) are synthetic organic chemicals with main sources in the environment being the processes of combustion and incineration (WHO, 1993). The organochlorine pesticides (OCPs), although banned internationally, can still be found in many environmental samples. Both classes of compounds are very persistent, tend to accumulate in the food chain and have negative effect on the human health (Cousins et al., 1998; Wade et al., 2002). For that reason many efforts have been made worldwide to develop analytical methods for their determination in different matrices. However up to now the behavior of PCBs and OCPs has not been studied systematically in soils from Bulgaria. The aim of this study is to develop a cost-effective method for simultaneous determination of low concentrations of PCBs and OCPs in soils.

2. Materials and Methods

Average soil sample was collected from five different points (method of diagonals), was dried, sieved through 2 mm sieve and homogenized. All reagents and solvents were of analytical reagent grade (Merck). The used analytical vessels were rinsed with hexane before use. A soil sample (10 g) was spiked with PCB 30 and PCB 204 (0.25 µg each). Another aliquot of the sample (10 g) was spiked with OCPs – γ-HCH (0.05 µg), ppDDE (0.2 µg) and pp-DDT (0.45 µg). The sample was covered with 10 g anhydrous sodium sulfate and extracted (18 hours) in Soxhlet apparatus with acetone/hexane (1/1).

After concentration, the extract was purified with a silica column using alternating layers of acidic silica (obtained by addition of 44 g concentrated sulfuric acid to 100 g silica) and basic silica (obtained by addition of 25 ml 1M sodium hydroxide to 100 g silica). The analytes were eluted with 50 ml hexane and concentrated.

The analysis was performed using a capillary column with intermediate polarity (SPB 608-Supelco, 30m x 0.25 mm i.d, 0.25 µm film thickness) with a Perkin-Elmer 8310 gas chromatograph equipped with electron capture detector. Isothermal oven conditions were applied (210°C), the temperature of the injector was 260 °C and of the detector – 400°C. The column was operated with a helium carrier gas (99.996%) with 190 kPa head column pressure. The detector makeup gas consisted of 5% methane and 95% argon. The runtime was set at 40 min.

3. Results and Discussion

In Table 1 are presented some analytical parameters of the proposed method. None of the analytes were detected in the blanks. The detection limits are in the range from 0.1 to 1.0 ng g^{-1}. These values are sufficiently low to allow the measurement of real PCBs and OCPs concentrations frequently found in soils (Grimalt et al., 2004). The repeatability was about 15% at low environmental concentration (three to six times the detection limits). The recovery for the PCBs spiked to the soil sample was between

65% and 75%. For the OCPs higher yields were obtained. On Figure 1. is presented a chromatogram for PCBs and OCPs in a soil sample.

The presented method is sensitive, cost-effective and it permits simultaneous determination of PCBs and OCPs in soils. The method can be applied to study the sources of pollution, the migration and the biogeochemistry of these persistent organochlorine compaunds in the environment.

TABLE 1. Analytical parameters for determination of PCBs and OCPs in soil sample.

Parameters	PCB 30	PCB 204	γ-HCH	p,p- DDE	p,p- DDT
Detection limit (ng g^{-1})	0.5	1.0	0.1	0.1	0.4
Determination limit (ng g^{-1})	1.0	2.0	0.2	0.2	0.8
Repeatability RSD (%), n=4	15	15	15	15	11
Recovery (%)	75	65	100	80	86

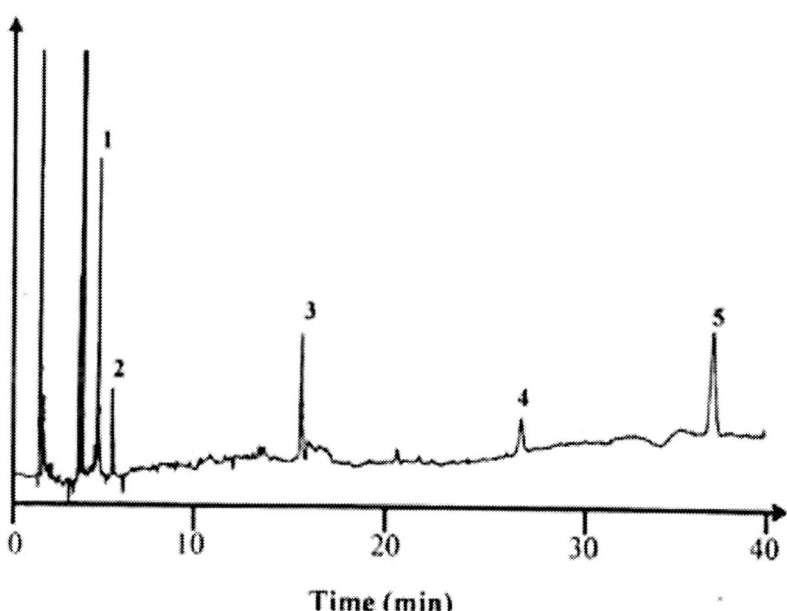

FIGURE 1. Chromatogram for PCBs and OCPs in a soil sample.
1 – PCB 30: 25 ng g^{-1} (spiked);
2 – γ-HCH: 1.5 ng g^{-1};
3 – p,p- DDE: 3.5 ng g^{-1};

4 – p,p- DDT: 1.9 ng g^{-1};
5 – PCB 204: 25 ng g^{-1} spiked)

References

Cousins I.T., Mclachlan M.S., Jones K.C. Environ. Sci. Technol. 1998; 32: 2734–40.

Grimalt J.O., van Drooge B.L., Ribes A., Vilanova R.M., Fernandez P., Appleby P., 2004 Chemosphere; 54: 1549–61.

Wade M.G., Parent S., Finnson K.W., Foster W., Younglai E., McMahon A., Cyr D.G., Hughes C., 2002, Toxicol. Sci., 67, 207–218.

WHO, EHC 140: Polychlorinated biphenyls and terphenyls (Second edition), Geneva, 1993.

METHOD FOR THE DETERMINATION OF SOME SELECTED STEROID HORMONES AND BISPHENOL A IN WATER AT LOW NG/L LEVEL BY ON-LINE SOLID-PHASE EXTRACTION COMBINED WITH LIQUID CHROMATOGRAPHY-TANDEM MASS SPECTROMETRY

[1]BORISLAV LAZAROV* AND [2]J.A. VAN LEERDAM
*[1]National Center of Public Health Protection
15, Akad. Ivan Geshov, 1431 Sofia, Bulgaria*
*[2]Kiwa Water Research
P.O. Box 1072, 3430 BB Nieuwegein, The Netherlands*

*To whom correspondence should be addressed: b.lazarov@ncphp.government.bg;

Abstract. A sensitive analytical method based on automatic on-line Solid-Phase Extraction (SPE) combined with Liquid Chromatography-Tandem Mass Spectrometry (LC-ESI-MS-MS) has been developed for the determination of six Endocrine-Disrupting Compounds (EDC's): 17-α-estradiol, 17-α-ethynylestradiol, 17-β-estradiol, bisphenol A, Estriol and Estrone in water samples. In order to enhance the sensitivity for these analytes, post column addition of different bases such as ammonia and 1,8-diazabicyclo(5,4,0)undec-7en (DBU) was evaluated. The post column addition of base is proposed here to raise effluent pH, helping in the ionization process of the compounds. The use of ammonium hydroxide diluted in methanol-acetonitrile mixture, proved to be the most efficient post column reagent for enhancing the MS signal. This strategy permitted direct determination of the six compounds at low ng/L levels. For the application to real water samples, an extraction and preconcentration step using SPE was carried out with cross-linked styrene-divinylbenzene polymer (PLPR-s) material. The recovery for all six compounds was satisfactory percentage (80-110%). The limits of quantification are between 1.0 and 2.0 ng/L. Application of the whole method, SPE-LC-ESI-MS-MS, to nature waters permitted low nanogram-per-liter determination of all six compounds.

Keywords: water analysis; post-column addition; solid-phase extractions; endocrine disruptors

1. Introduction

1.1. GENERAL INTRODUCTION

Many chemical substances display estrogenic activity and may be suspended of causing adverse effect in humans and/or environmental organisms. Those compounds have been linked to alterations in the endocrine system of animals and to adverse effects principally on the reproductive system.

Some environmental estrogens have natural origin. Estradiol, its main metabolites estriol and estrone, and their conjugates (basically sulfates and glucuronides) are naturally present in females and in much shorter extent in males. However, synthetic estrogens, such as the potent estrogen 17-α-ethynylestradiol, are extensively used in both, as contraceptives and for therapeutic purposes (management of the menopausal syndrome and in diverse cancers, mainly prostate and breast cancer) (Diaz-Cruz et al., 2003). Following excretion and incomplete removal in wastewater treatment plans, these compounds enter the at aquatic environment and reach concentrations normally in the nanogram and subnanogram per liter level. Estrogens are predominantly excreted in human urine as glucuronides and sulfates.

The presence in the environment of compounds with estrogenic properties has become a major subject of worldwide concern. Endocrine – disrupting compounds are environmental contaminants that interfere with the function of the endocrine system of wildlife and humans (Colborn et al., 1993). Among the wide range of substances with endocrine – disrupting properties, estrogens are of particular interest due to their high estrogenic potency. Natural steroid estrogens have been shown to exert estrogenic effects in fish at much lower concentrations in water (ng/l) (Purdom et al., 1994, Hansen et al., 1998), than industrial chemicals, such as nonylphenol, which are effective in the microgram per liter range (Routkege et al., 1998).

According to recent reports, municipal sewage effluents constitute a major source of estrogens in the aquatic media (Purdom et al., 1994; Larsson et al., 1999). Their presence in rivers downstream of sewage treatment works has been positively correlated with increased plasma vitellogenin levels in male and juvenile female fish (Purdom et al., 1994, Harries et al., 1997) and with distinctly high intersexuality rates (Jobling et al,1998, Petrovic et al., 2002) Moreover, it has been hypothesized that the statically derived decrease in sperm counts observed over the past decades, the increasing incidence of testicular cancer, and other disorders regarding male infertility may be caused by the intake of estrogens via food or drinking water (Sharpe et al., 1993).

1.2. METHODS REVIEW

These very low levels (ng/l) and the complexity of the environmental matrixes require the use of high sensitivity and selectivity methods for their analysis. For this purpose, gas chromatography/mass spectrometry (GC-MS) has been the technique most used (Larsson et al., 1999, Johnson et al., 2000). Biological assays, aimed at the determination of either target compounds (Rodriguez-Mozaz et al., 2004, Nilsen et al., 2004) or their estrogenic activity in standards or samples have also been developed (Scrimshaw et al., 2004, Korner et al., 1999).

Liquid chromatography – mass spectrometry (LC-MS) has gained in popularity in recent years. LC-MS, unlike GC-MS is not limited by such factors as non-volatility and high molecular weight and enables the determination a both conjugated and non-conjugated estrogens without the need for derivatization or hydrolysis Lopez de Alda et al., 2001). LC-MS analysis of estrogens has been carried out in most instances with an electro spray (ESI) interface operated in the negative ionization (NI). With this technique, the sensitivity achieved is considerably better than that obtained in the positive ion mode with either ESI or atmospheric pressure chemical ionization (Baronti et al., 2000, Labadine et al., 2005). The introduction of tandem mass spectrometry coupled with liquid chromatography (LC-MS-MS) has largely improved the performance of the technique by reducing the detection and quantification limits and enhancing analyte identification (Diaz-Cruz et al., 2003, Lagana et al., 2004; D'Ascenzo et al., 2003). Another benefit of LC-MS^n methods in the possibility of integrating sample preparation and enrichment online with the analysis. In the last year, a few attempts have been made in this line for the analysis of estrogens in water Ramsey et al., 1997; Labadien et al., 2005). One of them reported the determination of along with two other drugs in water by on-line supercritical fluid extraction coupled with ultraviolet-visible (UV-VIS) diode array LC-MS. The other two works represent previous attempts of the authors of develop suitable, fully automated methods for the analytical determination of environmental estrogens in water. However, because of detection technique used (UV, DAD or MS), none of them produced the necessary sensitivity and selectivity required for the reliable quantification of the most active estrogens estriol and 17-α-ethynylestradiol.

1.3. AIMS OF WORK

The experimental work is focused on develop automatic method based on solid phase extraction LC-ESI-MS-MS for determination of the most environmentally relevant endocrine disrupting chemicals in water samples on nanogram per liter levels, with a minimum amount of sample volume.

A set of estrogens was selected based on their abundance in the human body, there estrogenic potency, and the extent of their use in contraceptive pills. It included the two isomers of natural estrogen estradiol (α- and β-estradiol), its mine metabolites estriol (E3) and estrone (E1) and synthetic 17-α-ethynylestradiol (EE). Also part of this research is the determination of bisphenol-A, which have endocrine disruption effect.

The structure formulas of the compounds are shown on Fig. 1

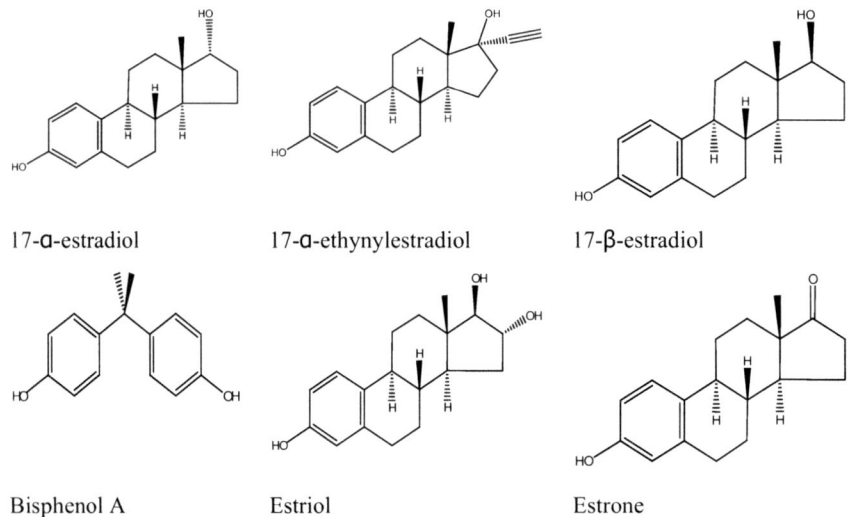

FIGURE 1. Chemical structures of the studied compounds.

2. Experimental Section

2.1. CHEMICALS

All the chemicals and solutions used were of analytical reagents grade. HPLC-grade water, acetonitrile and methanol were purchased from Merck (Germany) and were used to prepare the mobile phase and standard solutions. All pure standards of the components were purchased from

Sigma-Aldrich (Steinheim, Germany). Individual stock solutions of the analytes were initially prepared at 100 mg/l by dissolving 5 mg of each compound in 50 ml methanol. Standard mixtures of the compounds were prepared in methanol at different concentrations by appropriate dilution of the individual stock solutions. The standard mixtures were used as spiking solutions for preparation of the aqueous calibration standards and in the recovery study. Aqueous standard solutions did not contain more than 0.1% methanol.

2.2. SAMPLE PREPARATION

All water samples were filtrated through 0.45-µm syringe filters Spartan 30/0.45 RC (Dassel, Germany) to eliminate particular matter and other suspended solid matter immediately before the analyse. All the samples were stored at 4 °C in the dark. The extraction of the samples was always carried out within 24 h of collection to keep microbial degradation to a minimum.

2.3. INSTRUMENTAL CONDITIONS

2.3.1. *On-line Extraction*

The preconcentration columns were mounted on the injection valves of the auto-sampler Gilson 233 XL (Middleton, USA), and replaced the sample loop. The auto-sampler unit is equipped with two clamps with two cartridges and two high-pressure valves. This configuration permits the elution of the cartridge in one clamp while the following sample in a sequence is being loaded in another cartridge in the other clamp. The auto-sampler is controlled by Gilson 735 Sampler Software® version 5.1.

The dimensions of each cartridge were 10mm x 3mm i.d. Each cartridge was filled with PLPR-s (cross-linked styrene-divinylbenzene polymer, 300 Å (15-25 µm) particle size) provided by Polymer Laboratories Ltd. (Amherst, USA). This material is the most preferable one for solid phase extraction of estrogens in water (Lopez de Alda et al., 2002).

The parameter of the solid phase extraction procedure that has been chosen depends of the technical possibilities of the auto-sampler. The Gilson auto-sampler can only work which low backpressure and consequently the maximum flow rate in this configuration is about 1.5 ml/min.

The cartridges were previously conditioned with 4 ml acetonitrile and 4 ml water (flow rate 1 ml/min) in accordance with other works Lopez de Alda et al., 2002).

2.3.2. LC-MS-MS Analysis Conditions

LC-MS-MS analyses were carried out in a system consisting of a LC quaternary pump Finnigan SpectraSYSTEM® P4000 and Thermo Finnigan TSQ 7000 Triple-Quad triple-quadruple mass spectrometer all from Thermo Electron Corporation® (San Jose, USA) and binary LC pump Perkin Elmer (Fremont, USA) for post column addition.

Chromatographic separation was performed using a reversed phase Intersil® ODS-3 analytical column (100 × 4.0mm i.d, 3 μm particle diameter) preceded by guard column Phenomenex® C_{18} (ODS Octadecyl) (4 × 3.0 mm i.d). The LC mobile phase was consisting of (A) water and (B) acetonitrile/methanol (70:30). The LC gradient elution conditions are shown in Table 1.

TABLE 1. LC gradient elution conditions.

Time, min	Milli Q water (A), %	Acetonitrile/methanol 70/30 v/v (B),%
0	90	10
5	50	50
15	0	100
25	0	100
27	90	10
37	90	10
Temp. of the column: 20°C		Flow rate 1 ml/min

MS-MS detection was performed in the selected reaction monitoring (SRM) mode using ESI in the NI mode.

TABLE 2. MS-MS precursor/product-ion transitions for the compounds analyzed.

Compound	Q1 m/z Precursor ion [M-H]$^+$	Q3 m/z product ions	Time window (min)
Estriol	287	145	0 – 7,5
		171	
Estrone	269	**145**	7,5 – 16
		143	
α-Estradiol	271	**145**	
		143	
β-Estradiol	271	**145**	
		143	
d4-Estrone	273	147	
		145	

Bisphenol-A	227	**212**	
		133	
17-α-ethynylestradiol	295	269	
		145	

As it is shown in Table 2, two different SRM transitions were monitored per compound: the bold one, and most abundant, was used for quantification and the other one for purposes. To maximize sensitivity, data acquisition was performed under time scheduled conditions: Estriol was detected in the first time window, estrone, 17α-, 17β-estradiol, bisphenol A, d4-estrone and 17-α-etynylestradiol were detected in the second one (see Table 2). The optimization of the capillary temperature and collision energy was performed by measuring of the MS response of each transition of constant amount of each compound at different capillary temperatures and different collision energies. The optimization of capillary temperature was performed at 250, 300, 325, 350 and 400 °C, and the optimization of collision energy was performed at range 20 to 60 V.

To improuve sensitivity was used post column addition of base (Hernandes et al., 2001). Two different base solutions with different concentration were used for optimization of the conditions of post column addition – ammonium hydroxide (NH_4OH) [49] and 1,8-diazabicyclo (5,4,0)undec-7en (DBU) (Carabinas-Martínez et al., 2004) diluted in mixture of water/methanol (1:1 v/v). The conditions of the different experiments for optimization of this parameter are shown in Table 3.

TABLE 3. Conditions of the post column addition optimization.

Base	Molarity of base, mM	Concentration, %	Flowrate, μL/min	pH_{ESI}
NH₄OH	0	0	10	7
	0,2	0,05	10	9,5
	0,8	0,25	10	9,8
	8,5	2,5	10	10,7
	85	30	10	11,3
DBU	0	0	10	7
	1,3	0,02	10	12,3
	6,6	0,10	10	12,3
	20	0,33	10	12,6
	66	1,0	10	14,0

All other optimized MS parameters are shown in Table 4.

Instrument control, data acquisition, and evaluation were done means of Xcalibur 1.4 software of Thermo Electron Corporation® (San Jose, USA).

TABLE 4. Mass spectrometric conditions.

MS Conditions	
Ionization mode	Negative
Heated capillary temperature	380 °C
Spray voltage	4,2 kV
Sheath gas (N_2) pressure	70 PSI
Auxiliary gas (N_2) flow	30 L/min
Collision cell	2 mTorr Argon

The MS system was initially tuned with a MRFA/myoglobin solution by a flow injection pump and afterwards with a solution of estrone with concentration 5 mg/l to improve sensitivity.

2.3.3. *Identification and Quantification*

Identification of the target analytes was accomplished by comparing the retention time and the LC-MS-MS signals of the target compounds in the samples with those of standard analyzed under the same conditions. For positive identification the following criteria had to be met: (1) HPLC chromatographic retention time agreement with 2%; (2) relative abundance of the two selected precursor ion – product ion transitions with a margin of ±20%.

For quantification of the analytes, the external standard method was used, based on the peak obtained in the bold SRM transition (see Tabl.2). Six or eight points calibration curves were constructed from the online analysis of Milli Q water spiked with the standard mixture of the analytes at concentrations ranging between 0.5 and 30 ng/l in Milli Q water using a least- squares linear regression analysis.

2.4. RESULTS AND DISCUSSION: DEVELOPMENT AND OPTIMIZATION

2.4.1. *Optimization of Post Column Base Addition*

To increase the sensitivity of the MS detection of steroid compounds was tried with post column addition of base solutions. Two different base solutions with different concentration were used for optimization of the conditions of post column addition – ammonium hydroxide and 1,8-diazabicyclo(5,4,0)undec-7en (DBU) diluted in mixture of acetonitrile/methanol (1:1). The conditions of the different experiments for optimization of this parameter are shown in Table 3.

The optimization was performed without guard and analytical column but analyzing of 20 µL standard mixture of test compounds in methanol with a concentration 5 mg/L.

Addition of 85mM NH₄OH yielded the highest MS-response, indicating the best compromise between response (as area) and signal to noise ratio. As optimal conditions of post column addition of base were accepted: post column addition of 85mM NH₄OH solution in methanol/water 1:1 v/v with flow rate 10 µL/min, and were therefore used throughout the remaining experiments.

The intensity of MS-response of 20 µL standard solution of estrone with concentration 5mg/L in different molarity of NH₄OH addition is shown on Fig. 2.

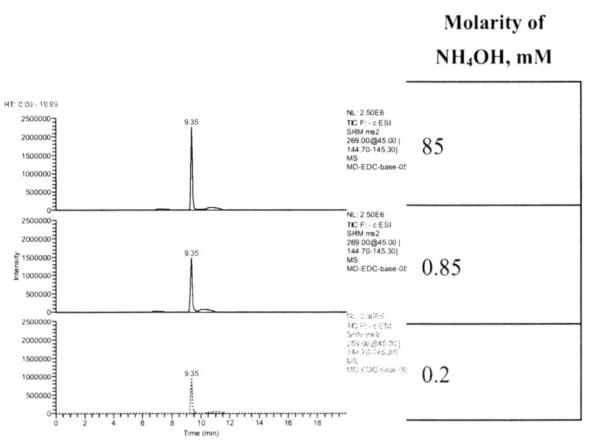

FIGURE 2. MS-response of 5mg/L Estrone in different molarity of base addition.

2.5. METHOD VALIDATION

The method validation was done according to requirements of European community legislation procedure (Eurachem Guide, 1998, 96/23/EC, 2002).

2.5.1. *Linearity*

Linearity was examined in the concentration range 0,5 – 30 ng/L. The method was found to be linear in that range. During the validation the correlation coefficient (R) was observed to be greater than 0,996 for each compound.

2.5.2. *Limit of Detection (LOD)*

Limit of detection is the lowest concentration of analyte in the sample that can be detected, but not necessarily quantified under the stated conditions of the test (Eurachem Guide, 1998).

According (Eurachem Guide, 1998) the limit of detection is 0.5 ng/l for all compounds in ultra pure water, and 1ng/l for all compounds in drinking and surface water samples.

Figure 3 show the TIC respond of 1.0 ng/l estrone in ultra pure, drinking and surface water.

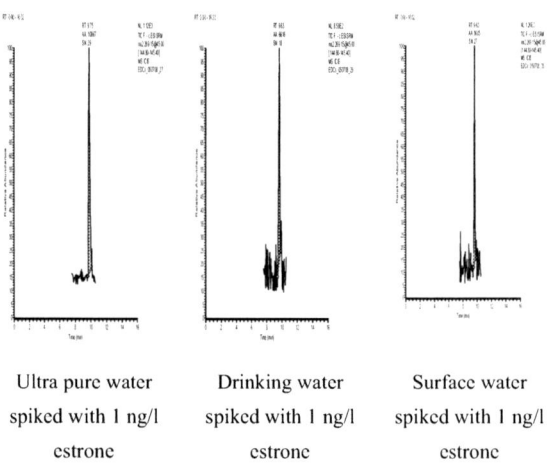

| Ultra pure water spiked with 1 ng/l estrone | Drinking water spiked with 1 ng/l estrone | Surface water spiked with 1 ng/l estrone |

FIGURE 3. TIC responds of 0,5 ng/l estrone addition in different water types.

2.5.3. *Limit of Quantification (LOQ)*

Limit of quantification (Limit of reporting) is the lowest concentration of an analyte that can be determined with acceptable precision (repeatability) and accuracy under the stated conditions of the test (Eurachem Guide, 1998).

According (Eurachem Guide, 1998) the limit of detection is 1 ng/l for all compounds in ultra pure water, and 2 ng/l for all compounds in drinking and surface water samples.

2.5.4. *Recovery From Spiked Water Samples*

Recovery was determined from drinking and surface water samples spiked at a concentration level around 5 ng/L (see Table 5).

TABLE 5. Recovery and coefficient of variation of each component in drinking and surface water.

Compound	n	Recovery in drinking water	CV%	Recovery in surface water	CV%
Estrone	4	110%	17%	122%	15%
Estriol	4	97%	14%	68%	11%
α-Estradiol	4	90%	8%	134%	20%
β-Estradiol	4	85%	16%	118%	24%
Bisphenol-A	4	109%	16%	110%	15%
17-α-ethynylestradiol	4	102%	9%	119%	12%

2.5.5. Analysis of Some Water Samples

As a part of the validation procedure, the method developed was applied to the analysis of the target analytes in various different real samples.

The present method was applied to the monitoring of the target components in a surface and drinking water from different rivers in The Netherlands.

Of all compounds investigated, only 17-α-ethynylestradiol and bisphenol A were detected in some surface water samples. 17-α-ethynylestradiol was found to be present in some river waters at concentration of 1.1 ng/L.

Bisphenol A, on other hand, was found to be present in the surface water samples at a concentration of 3.5 to 11.3 ng/L. Figure 4 shows that Bisphenol A was detected in surface water at concentration 7.1 ng/l and compared with a standard addition of 10 ng/l.

3. Conclusions

A fully automated method, based on online SPE-LC-MS/MS analysis, has been developed for the determination of some most relevant endocrine disrupting chemicals in water samples.

The method validation was done according to the requirements of European community legislation procedure (Eurachem Guide, 1998). A linear range with a limit of quantification between 1 ng/l and 2 ng/l was found, which made it possible to use the method for quantitative determinations. Recovery in drinking and surface water was acceptable (68 – 134%).

The results obtained during the validation of the method and the analysis of at least 50 drinking and surface water samples proved the suitability of the LC-MS-MS method for the determination of the most important endocrine disrupting chemicals.

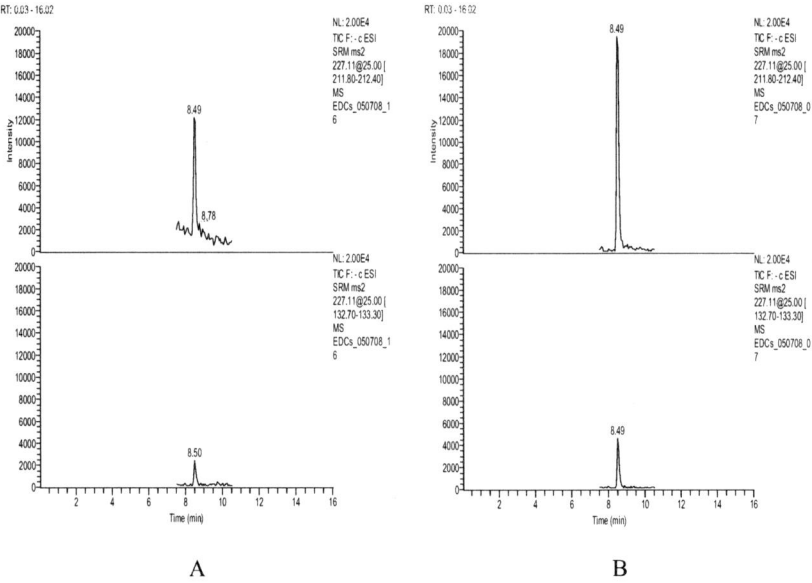

FIGURE 4. Shows that Bisphenol A was detected in surface water at concentration 7.1 ng/l (A) and compared with a standard addition of 10 ng/l (B).

References

96/23/EC. *Off. J. Eur. Communities Legis.* 2002, L221/8
Baronti, C., Curini, R., D'Ascenzo, G.C., Sumpter, J.P., Waldock, M., 2000, *Environ. Sci. Technol.*, *34,* 5059–5066.
Belfroid, A.C., Van der Horst, A., Vethaak, A.D., Schafer, A.J., Rijs, G.B.J., Wegener, J., Cofino, W.P., 1999, *Sci. Total Environ.*, *225,* 101–108.
Carabias-Martínez, R., Rodríguez-Gonzalo, E., Revilla-Ruiz, P., 2004, *J. Chromatogr, A.*, *1056,* 131–138.
Carcouet, M., Perdiz, D., Mouatassim-Souali, A., Tamisier-Karolak, S., Levi, Y., 2004, *Sci. Total. Environ.*, *324,* 55–66.
Céspedes, R., Petrovic, M., Raldúa, D., Saura, Ú., Piña, B., Lacorte, S., Viana, P., Barceló, D. 2004, *Anal. Bioanal. Chem.*, *378,* 697–708.
Colborn, T., Vom Saal, F.S., A.M., 1993, *Environ. Health Perspect. 101.*
Collie, I., Reder, S., Bucher, S., Gaugliz, G., 2002, *Biomol. Eng.*, *18,* 273–280.
D'Ascenzo, G., Di Corcia, A., Gentil, A., Mancini, R., Mastropasqua, R., Nazzari, M., Samperi, R., 2003, *Sci. Total. Environ.*, *302,* 199–209.
Desbrow, C., Routledge, E.J., Brighty, G.C., Sumpter, J.P., Waldock, M., 1998, *Environ. Sci. Technol.*, 1549–1558.
Diaz-Cruz, M.S., Lopez de. Alda, M.J., López, R., Barcelo, D., 2003, *J. Mass Spectrom.*, *38,* 91–93.

Eurachem Guide. *The finess for purpose of analytical methods – A laboratory guide to method validation and related topics.* 1998, 42, LGC UK.

Fine, D.D., Breidenbach, G.P., Price, T.L., Hutchins, S.R., 2003, *J. Chromatogr. A*, *1017*, 167–185.

Gomes, R.L., Scrimshaw, M.D., Lester, J.N. 2003, *TrAC, Trends Anal. Chem.* 22, 697–707.

Gutendorf, B., Westendorf, J., 2001, *Toxycology*, *166*, 79–89.

Hansen, P.D., Dizer, H., Hock, B., Marx, A., Sherry, J., McMaster, M., Blaise, C., 1997, *TrAC, Trends Anal. Chem.,17, 448–451*

Harries, J.E., Sheanan, D., Jobling, S., Matthiessen, P., Neall, P., Sumpter, J.P., Tylor, T., Zaman, N., 1997, *Envirron. Toxicol. Chem.* 16, 534–542.

Hernandes, F., Sancho, J.V., Pozo, O., Lara, A., Pitarch, E., 2001, *J. Chromatogr., A.*, *939*, 1–11.

Huang, C.H., Sedlak, D.L., 2001, *Environ. Sci. Technol.*, *20*, 133–139.

Ingrand, V., Herry, G., Beausse, J., de Roubin, M.R., 2003, *J. Chromatogr. A.*, *1020*, 95–100.

Isobe, T., Shiraishi, H., Yasuda, M., Shinoda, A., Suzuki, H., Morita, M., 2003, *J. Chromatogr., A.*, *984*, 195–202.

Jobling, S., Nolan, M., Tyler, C.R., Brighty, G., Sumpter, J.P., 1998, *Environ. Sci. Technol.*, *32*, 2498–2506

Jonson, A.C., Belfroid, A., Di Corcia, A., 2000, *Sci. Total Environ.*, *256*, 163–173

Korner, W., Hanf, V., Schuller, W., Kempter, C., Metzger, J., Hagenmaier, H., 1999, *Sci. Toatal. Environ.*, *225*, 33–48.

Labadie, P., Budzinski, H., 2005, *Anal. Bioanal. Chem.*, *381*, 1199–1205

Lagana, A., Bacaloni, A., De Leva, I. Faberi, A., Fago, G., Marino, A., 2004, *Anal. Chim. Acta.*, *501*, 79–88.

Larsson, D.G.J., Adolfson-Erici, M., Parkkonen, J., Petterson, M., Berg, A.H., Olsson, P.E., Förlin, L., 1999 *Aquat. Toxicol.*, *45*, 91–97.

Legler, J., Dennekamp, M., Vethaak, A.D., Brower, A., Koeman, J.H., van der Burg, B., murk, A.J., 2002, *Sci. Total Environ.*, *293*, 69–83.

Lopez de Alda, M.J., Barcelo, D. 2001, *Fresenius J. Anal. Chem.*, *371*, 437–447.

Lopez de Alda, M.J., Barcelo, 2000, D. *J. Chromatogr., A.*, *892*, 391–406.

Lopez de Alda, M.J., Barcelo, D. 2001, *J. Chromatogr., A.*, *911*, 203–210.

Lopez de Alda, M.J., Barcelo, D. 2001, *J. Chromatogr., A.*, *938*, 145–153.

Lopez de Alda, M.J., Diaz-Criz, S., Petrovic, M., Barcelo, D. 2003, *J. Chromatogr., A.*, *1000*, 503–526.

Lopez de Alda, M.J., Gil, A., Paz, E., Barcelo, D. 2002, *Analyst*, *127*, 3–8.

Nielen, M.W.F., Van Bennekom, E.O., Heskamp, H., Van Rhijn, J.A., Bovee, T.F.H., Hoogenboon, L.A.P. 2004, *Anal. Chem.*, *76*, 6600–6608.

Nilsen, N.M., Berg, K., Eidem, J.K., Kristiansen, S., Brion, F., Porcher, J., Goksøyr, A. 2004, *Anal. Bioanal. Chem.*, *378*, 621–633

Petrovic, M., Sole, M., Lopez de Alda, M.J., Barcelo, D. **2002**, *Environ. Toxicol. Chem.*, *21*, 2146–2156

Purdom, C.E., Hardiman, P.A., Bye, V.J., Eno, N.C., Tyler, C.R., Sumpter, J.P. **1994**, *Chem. Ecol.*, *8*, 275–285.

Quintana, J.B., Crpinteiro, J., Rodrigues, I., Lorenzo, R.A., Carro, A.M., Cela, R. **2004**, *J. Chromatogr., A*, *1024*, 177–185.

Ramsey, E.D., Minty, B., Rees, A.T. **1997**, *Anal. Commun.*, *34*, 261–264.

Rodriguez-Mozaz, S., Reder, S., Lopez de Alda, M., Gauglitz, G., Barceló, D. **2004**, *Biosens. Bioelectron.*, *19*, 633–640.

Routkege, E.J., Sheahan, D., Desbrow, C., Brighty, G.C., Waldock, M., Sumpter, J.P. **1998,** *Environ. Sci. Technol. 32, 1599–1565*

Scrimshaw, M.D., lester, J.N. **2004,** *Anal. Bioanal. Chem.*, 576–581

Seifert, M., Haindl, S., Hock, B. **1999,** *Anal. Chim. Acta, 386,* 191–199.

Sharpe, R.M., Skakkebaek, N.E. **1993,** *Lancet, 341,* 1392–1395

Snyder, S.A., Keith, T.L., Verbrugge, D.A., Snyder, E.M., Gross, T.S., Kannan, K., Giesy, J.P. **1999,** *Environ. Sci. Technol.*, *33,* 2814–2820.

Ternes, T.A., Stumpf, M., Muller, J., Haberer, K., Wilken, R.D., Servos, M. **2004,** *Sci. Total. Environ.*, *225,* 81–90.

Watabe, Y., Hosoya, K., Tanaka, N., Kondo, T., Morita, M., Kubo, T. **2004,** *Anal. Bioanal. Chem.*, *381,* 1193–1198

Zhang, H., Henion, J. **1999,** *Anal. Chem.*, *71,* 3955–3964.

EVALUATION OF BIOCIDE – FREE ANTIFOULING SYSTEMS

LUCICA BARBES
Chemistry Department, Ovidius University
124 Mamaia Blvd., 900527 Constanta, Romania
E-mail: lbarbes@univ-ovidius.ro

Abstract. The performance of biocide-free test coatings is directly related to the operational profile of ship: service speed, activity level and trading waters are decisive for the success of biocide-free antifouling systems. For antifouling coatings, a significant parameter into the environment is entry of active compounds into the aquatic system as they are leached from the paint film. The leaching rate is a critical parameter in an environmental risk assessment. The following issue describes a proposed method for calculating release rate of active compounds (ACs) from antifouling paints.

Keywords: natural products; active compounds; antifouling systems

1. Introduction

Biofouling is a major economic and management problem for the aquaculture industry which been managed in the past by the use of toxic chemicals (biocide) that have had serious environmental repercussions (Mc Cloy and De Nys, 2000).

The results of an intensive many-years project show that there are convincing alternatives to organotin and other biocidal paints. This means organotins considered to be the most toxic substances deliberately released into marine environmental can now be replaced with confidence. For example, marine organisms are affected by low environmental concentrations of tributyl tin (TBT), just few ng/L, due to accumulation (Omae, 2003). There are not rivals of TBT in the "imposex" of gastropod mollusks. The International Maritime Organization (IMO) resolution called for a global prohibition on the application of organotin compounds, which act as biocides in antifouling systems on ships by 2003, and a complete prohibition by 2008 (IMO, 2005).

The natural products may be expected to be used, as new environmentally friendly antifouling agents, having high anesthetic, repellant and settlement inhibition properties, without biocidal effects. Many kinds of natural compounds like as: terpens, steroids, nitrogen containing compounds, enzymes, vitamins and so one, are reported to have antifouling properties (Berlinck and Kossuga., 2005; Omae, 2003; Scott and Soderberg, 2003). It is known that antifouling coatings are classified by: conventional, a long-time paints and self-polishing paints (Terlizzi et al., 2001). In concordance by the new requirements economical aspects this report mentions a proposed one method for calculation the release rate of ACs from antifouling paints.

2. Methodology

The defense of the new potential solutions in biofouling control can involve physical, mechanical as well as chemical processes.

2.1. DEFINITIONS

X = amount of ACs released during first 14 days (µg/cm2)
Y = average leaching rate during the rest of the lifetime (µg/cm2 per day)

2.2. ASSUMPTIONS

- first 14 days linear drop in release rate
- rest of the lifetime - constant leaching rate
- $X/Y = 30$ – constant by experience

These assumptions are based on extensive experience within the paint industry of the measured release rates of copper and tin biocides from organotin copolymer and first generation TBT-free antifouling paints.

2.3. CALCULATION OF TOTAL AMOUNT OF ACS RELEASED DURING THE LIFETIME T

The value to be applied depends on the type of paint

$$X + 30\left(t - \frac{1}{2}\right)Y = \frac{L_a \mathrm{x} a \mathrm{x} W_a}{SVR \mathrm{x} SPG \mathrm{x} DFT} \mathrm{x} 100 \qquad (1)$$

$$\frac{X}{Y} = 30 \qquad (2)$$

t - specified lifetime (months) of the paint for the dry film thickness (DFT)

30 - 1 month; ½ - half a month

La - fraction of the active ingredient in the dry film released during the life time t, assumed to be 0.7

a - weight fraction of active ingredient in the products

Wa - concentration of AC in the wet paint in weight (%)

SVR - solid volume ratio in % (volume of dry paint versus volume of wet paint)

SPG - specific gravity of the wet paint (g/cm^3)

DFT - dry film thickness (in micron) specified for the lifetime t

3. Discussion

Policies on antifouling evaluation have mainly focused on strategies to regulate effectively the use of different antifoulants and to know the total amount of Acs released during the life time t of paints.

Equations (1) and (2) and data on the type of antifouling paints permit the calculation of assumed total release rate during the first 14 days and assumed average release rate of ACs during the rest of the lifetime. In order to establish the amount of ACs released during the lifetime t, the graphic method is possible to be used.

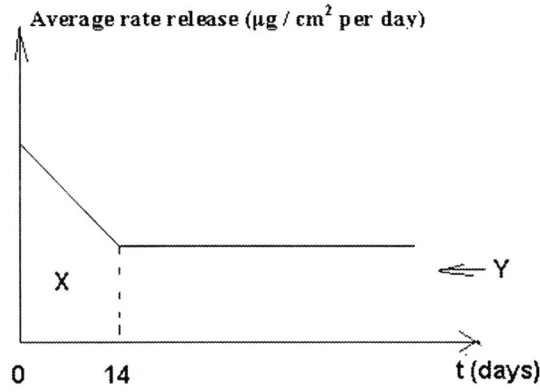

FIGURE 1. Graphic calculation of total amount of ACs released during the lifetime t.

4. Conclusions

Available relationships between biocide release rates and the true environmental inputs of biocides from antifouling paints is uncertain and release rate data from these methods cannot be reliably used for environmental risk assessment

The proposed calculation method to estimate active compounds (ACs) release rates from antifouling coatings can be used for the evaluation of biocide-free antifouling systems

References

Berlinck, G.S. and Kossuga, M.H., 2005, *Nat.Prod.Rep.*, 22, 516.
IMO. 2005. International Maritime Organization, Antifouling Systems, http://www.imo.org.
McCloy S and R. De Nys. 2000. Novel Technologies for the reduction of biofouling in shellfish aquaculture. In: Fisheries N (ed) Flat Oyster Workshop, Sydney, p 19–23.
Omae I., 2003, General Aspects of Tin-Free Antifouling Paints, *Chem. Rev.* 103, 3431–3448.
Omae I., 2003, Organotin antifouling paints and their alternatives, *Appl. Organometal. Chem.*, 17, 81–105.
Terlizzi, A., Fraschetti, S., Gianguzza, P., Faimali, M. and Boero, F., 2001, Environmental impact of antifouling technologies: state of the art and perspectives, *Aquatic Conserv. Mar. Freshw. Ecosyst.*, 11, 311–317.
Scott, T.L. and Soderberg, B.C.G., 2003, Palladium-catalyzed synthesis of 1,2-dihydro-4(3*H*)-carbazolones. Formal total synthesis of murrayaquinone A, *Tetrahedron*, 59, 6323.

REMOVAL OF DISSOLVED ORGANIC COMPOUNDS BY GRANULAR ACTIVATED CARBON

BERND SCHREIBER[2], DIANA WALD[1], VIKTOR SCHMALZ[1] AND ECKHARD WORCH[1]*
[1]*Institute of Water Chemistry, Dresden University of Technology, 01062 Dresden, Germany*
[2] *TU Bergakademie Freiberg, Germany*

*To whom correspondence should be addressed: eworch@rcs.urz.tu-dresden.de

Abstract. Granular activated carbon (GAC) adsorption is a widespread water treatment technology especially applied in waterworks for removing dissolved organic compounds (natural organic matter, micropollutants). Adsorption of organic compounds by GAC filters is influenced by a number of water characteristics as e.g. ionic strength, pH, water temperature and the initial concentration of organic compounds. The objective of our investigations is to find out if water temperature has to be considered for applying water purification technologies under various climatic conditions.

Keywords: granular activated carbon; adsorption; natural organic matter; micropollutants; atrazine; water treatment; water temperature

1. Introduction

Water treatment by granular activated carbon (GAC) filters is an effective technology for removing dissolved organic compounds. GAC filters are therefore applied for drinking water purification as well as decontamination of polluted water, as for example in pump and treat procedures. Unfortunately, removal capacity towards micropollutants is often reduced by natural organic matter (NOM) in water due to competitive interaction on GAC adsorption sites.

NOM in aquatic systems is a multicomponent organic mixture being composed of a number of compounds as e.g. humic acids, fulvic acids, carbohydrates, proteins and carboxylic acids (Frimmel et al., 2002). Basically NOM is not considered to be toxic when consuming drinking water. However, NOM has been found to act as precursors for toxic and mutagenic compounds as e.g. trichloromethane and chlorophenols originated by disinfection of the drinking water by chlorine. Especially in countries in which sewage water treatment is not common, surface water quality is partially very low in terms of microbial contamination. Thus, chlorine is commonly used when treating for drinking water purposes, leading to a widespread emergence of chlorinated organics. Disinfection byproducts have been found in groundwater, caused by corroded water distribution networks, and from surface water and cesspool water infiltration (Salameh et al., 2002). Additionally high NOM concentrations could enhance microbial growth in the drinking water distribution system. Thus, amongst micropollutants, reduction of NOM is also a goal for drinking water treatment.

Adsorption of NOM onto activated carbon has been found to be influenced by a number of physicochemical properties such as e.g. pH value, NOM initial concentration, type and molecular size distribution of NOM, ionic strength, and water temperature (Lee et al., 1981; Cornel et al., 1986; Fettig and Sontheimer, 1987; Summers and Roberts, 1988; Johannsen et al., 1991; Kilduff et al., 1996; Bjelopavlic et al., 1999). Besides physicochemical characteristics of the process water adsorption is also dependent on properties of the activated carbon as e.g. pore volume, pore size distribution (Lee et al., 1981; Bjelopavlic et al., 1999; Ebie et al., 2001) surface area accessible for adsorption and surface functional groups as e.g. carboxyl, hydroxyl and carbonyl groups (Cookson, 1980). Adsorption of organic micro-pollutants is also affected by water and activated carbon properties.

2. Experiments

Our work focuses onto the influence of water temperature on the adsorption of dissolved organic compounds on activated carbon. This study was conducted with subsidy of the German Ministry of Education and Research (BMBF) and is part of the joint research project "Adjusted water treatment technologies and water distribution under regional conditions". Compared to influences of e.g. pH and ionic strength temperature effects of NOM and micropollutant adsorption are still not satisfactorily investigated. In respect to different climatic conditions and global warming, temperature is a key parameter being of importance. To describe the influence of the water temperature onto adsorption laboratory batch and column experiments have been carried out using different surface water samples. Furthermore, a herbicide (atrazine, $C_8H_{14}ClN_5$) was used as a model micro-pollutant added to the surface water. Atrazine was (and in some countries is still) used to avoid weed growth. Caused by its very low tendency for soil adsorption atrazine possesses a high mobility in the environment and is often detected in groundwater (Akkan et al., 2003). As many water supplies of urban regions are situated in agriculturally used areas atrazine locally causes big problems relating to drinking raw water qualities.

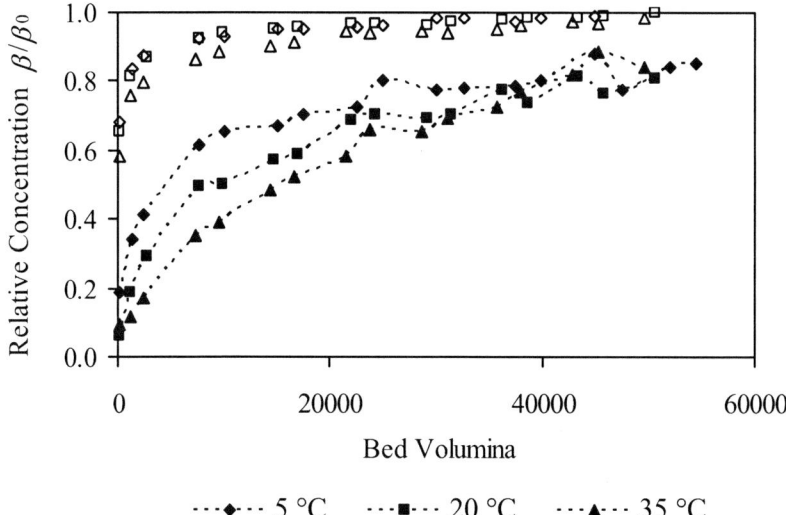

FIGURE 1. GAC adsorber breakthrough curves of atrazine (filled symbols with dashed lines) and NOM (open symbols) dissolved in lake water (initial concentrations: DOC = 12 mg/L, atrazine = 1 mg/L).

Our research indicates that the water temperature has an influence onto the removal of natural compounds and atrazine by granular activated carbon. NOM and atrazine removal was highest at 35°C and usually decreased with declining temperature (see Fig. 1). Mechanisms influenced by water temperature affecting adsorption onto GAC are supposed to be very diverse and are not always resulting in exothermic adsorption behavior. Some theoretical assumptions of NOM behavior are summarized by Schreiber et al. (2005) as for example changes in macromolecular structure of NOM (molecular self association) or changes in activated carbon surface hydrophobicity caused by changes in water temperature. An augmentation of atrazine adsorption as a consequence of increasing temperature is likely to be due to NOM-atrazine interaction. As temperature has got a distinct influence onto solute removal we therefore suggest that water temperature should always be considered in adsorption studies to be adjusted to realistic conditions.

References

Akkan, Z., Flaig, H. and Ballschmiter, K., 2003 *Pflanzenbehandlungs- und Schädlingsbekämpfungsmittel in der Umwelt - Emissionen, Immissionen und ihre human- und ökotoxikologische Bewertung*, Erich Schmidt Verlag, Berlin.

Bjelopavlic, M., Newcombe, G. and Hayes, R., 1999 Adsorption of NOM onto activated carbon: effect of surface charge, ionic strength, and pore volume distribution, J. *Colloid Interface Sci.* 210, 271–280.

Cookson Jr., J.T., 1980 *Physicochemical changes of substances by or within carbon adsorption beds*, in I.H. Suffet and M.J. McGuire (eds) *Activated carbon adsorption of organics from the aqueous phase*, Ann Arbor Sci., Michigan.

Cornel, P.K., Summers, R.S. and Roberts, P.V., 1986 Diffusion of humic acid in dilute aqueous solution, J. *Colloid Interface Sci.*, 110, 149–164.

Ebie, K., Li, F., Azuma, Y., Yuasa, A. and Hagishita, T., 2001 Pore distribution effect of activated carbon in adsorbing organic micropollutants from natural water, *Wat. Res.* 35, 167–179.

Fettig, J. and Sontheimer, H., 1987 Kinetics of adsorption on activated carbon: I. single-solute systems, J. *Environ. Eng.* 113, 764–779.

Frimmel, F.H., Abbt-Braun, G., Heumann, K.G., Hock, B., Lüdemann, H.-D. and Spiteller, M., 2002 *Refractory organic substances in the environment*, Wiley-VCH, Weinheim.

Johannsen, K., Assenmacher, M., Kleiser, M., Abbt-Braun, G., Sontheimer, H. and Frimmel, F. H., 1991 Einfluß der Molekülgröße auf die Adsorbierbarkeit von Huminstoffen. *Vom Wasser* 81, 185–196.

Kilduff, J.E., Karanfil, T., Chin, Y.-P. and Weber Jr., W.J., 1996 Adsorption of natural organic polyelectrolytes by activated carbon: a size-exclusion chromatography study, *Environ. Sci. Technol.* 30, 1336–1343.

Lee, M.C., Snoeyink, V.L. and Crittenden, J.C., 1981 Activated carbon adsorption of humic substances, J. *Am. Water Works Assoc.* 73, 440–446.

Salameh, E., Alawi, M., Batarseh, M. and Jiries, A., 2002 Determination of trihalomethanes and the ionic composition of groundwater at Amman City, Jordan, *Hydrogeol. J.* 10, 332–339.

Schreiber, B., Brinkmann, T., Schmalz, V. and Worch, E., 2005 Adsorption of dissolved organic matter onto activated carbon – the influence of temperature, absorption wavelength, and molecular size, *Wat. Res.* 39, 3449–3456.

Summers, R.S. and Roberts, P.V., 1988 Activated carbon adsorption of humic substances: I. heterodisperse mixtures and desorption, J. *Colloid Interface Sci.* 122, 367–381.

EXTERNAL AND INTERNAL EXPOSURE TO CARBON DISULFIDE AT THE WORKING PLACE

CHRISTINA KOPCHEVA*, TEODOR PANEV, TZVETA GEORGIEVA, VIDKA NIKOLOVA AND TODOR POPOV
Chemical Analyses, Spectrometry and Chromatography, Air Pollution Laboratory, National Center of Public Health Protection, 15 Ivan Geshov Blvd, 1431 Sofia, Bulgaria

*To whom correspondence should be addressed: kopcheva@yahoo.com

Abstract. Carbon disulfide is a well-known occupational hazard in viscose industry. This study was focused on assessment of the external and internal exposure to carbon disulfide of occupationally exposed workers with a view to make a comparative analysis of the results and determine the correlation ratio between amounts of carbon disulfide introduced in human organism and amounts that are kept there. These studies are a basis for further identification of methods to protect the workers against the hazardous effect of carbon disulfide as a chemical factor of the workplace.

Keywords: carbon disulfide; workplace air; external exposure assessment; internal exposure assessment; biomarkers

1. Introduction

Carbon disulfide is widely used in production of rayon, carbon tetrachloride, rubber chemicals and cellulose film, and is a by-product of widely used dithiocarbamate pesticides. Chronic low level and long term exposure to CS_2 can cause eye, ear, cardiovascular, nervous system and reproductive effects (Tan et al., 2001), (WHO: Criteria 10, 1979), (Kaloyanova, 1981). There are scientific reports that the long term exposure to low concentrations of CS_2 is related to endocrine disturbances as well (Lancranian I. et al., 1972); (Lyubomirova K. et al., 2006). Carbon disulfide is mostly used in viscose industry to yield sodium cellulose xanthate from alkali cellulose. (Tan et al., 2001).

2. Aims of the Study

To conduct a special targeted investigation of external and internal exposure implementing current methods and to assess the correlation between them.

3. Objects

The study involved workers from a plant for manufacturing man-made silk in Bulgaria. The study group involved 76 workers from three different shops of the plant producing man-made silk, exposed to different levels of carbon disulfide. The control group comprised 26 individuals from the same region without occupational exposure. The distribution by sex, age, total and specialized employment record length and smoking is presented in Table 1.

TABLE 1. Distribution by sex, age, total and specialized employment record length and smoking.

Indicator	Control group (n = 26)	I^{st} exposed group (n=26)	IInd exposed group (n=25)	IIIrd exposed group (n=25)
Sex				
Women (n, %)	11 (42.3 %)	16 (61.5 %)	13 (52.0 %)	
Men (n, %)	15 (57.7 %)	10 (38.5%)	12 (48.0 %)	25 (100%)
Age ($\overline{X}$, SD)	45.6 ± 9.2	40.8 ± 6.1	46.0 ± 6.7	47.4 ± 9.4
Total employment record length ($\overline{X}$, SD)	26.1 ± 8.3	19.4 ± 6.9	26.08 ± 7.4	27.4 ± 10.7
Specialized employment record length ($\overline{X}$, SD)	18.41± 0.44	13.23 ± 8.78	18.8 ± 9.41	17.8 ± 11.3
Smoking (n,%)				
yes	13 (52.0 %)	21 (80.8 %)	14 (56.0 %)	15 (60.0 %)
no	12 (48.0 %)	5 (19.2 %)	11 (44.0 %)	10 (40.0 %)

4. Materials and Methods

4.1. MONITORING OF WORKPLACE AIR

Strategy and methods for sampling at assessment of external exposure to carbon disulfide

The sampling strategy for this task covered the organization of average shift sampling for all operators from the plant. The personal exposure in the workers' breathing area was measured. The samples for quantitative analysis were collected on sorbent tubes filled with activated charcoal. A drying tube filled with silicagel covered with 20% solution of sodium sulfate was mounted before the sorbent tube in order to prevent the effect of high air humidity. The air samples were taken by low-flow personal pumps type Gilian 113 PS (USA) with sampling flow 50 cm^3/min.

Analytical methods for assessment of external exposure to carbon disulfide

The analysis of the collected average shift samples was made with a method validated by us, based on NIOSH method 1600/1994 (National Institute for Occupational Safety and Health). After elution of the activated charcoal with toluene the test was made with a gas chromatograph with mass selective detector "Perkin Elmer" and capillary column DB-5, 30 m long and thickness of the coating 0.25μm. Helium was used as carrier gas. Apparatus conditions: injector temperature 250°C, detector temperature 250°C, carrier gas pressure – 7 psig. The quantitative assessment of the samples was performed after absolute calibration with standard solutions of carbon disulfide in toluene. The limit of detection of the method is 0.01 mg/m^3 at 25dm^3 air sample.

4.2. BIOMARKERS OF EXPOSURE

Indicators for assessment of internal exposure

It is known that at exposure to carbon disulfide some 70-90% of the amount absorbed in the organism is biotransformed forming different metabolites. Part of them are slowly excreted with urine.

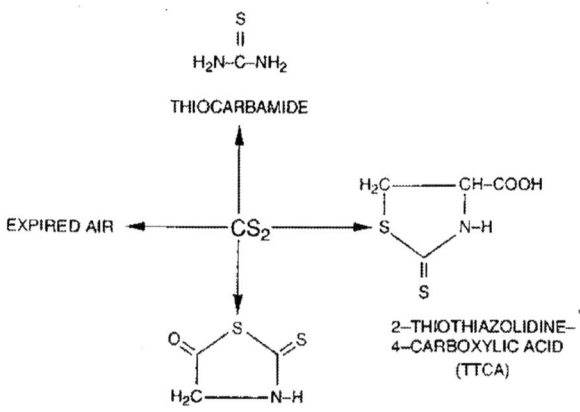

The assessment of internal exposure is made by testing biomarkers – metabolites of carbon disulfide in urine. The samples were collected in the end of the working shift. The results were re-calculated based on creatinine. The correction of the results vs. creatinine excretion is applied for reducing the confounding factors, respectively to eliminate the external and internal factors that are not associated with exposure to carbon disulfide.

The metabolites determined were:

1. Creatinine in urine.
2. Total metabolites of carbon disulfide through titrimetric iodine-azide test in urine (Jacubowski, 1965).
3. 2-thiothiazolidin-4-carboxylic acid (TTCA) in urine through HPLC with UV detector. (Van Door R. et al, 1981), (Rosier J. et al., 1982).

5. Results and Discussion

5.1. CHEMICAL ANALYSIS OF WORKPLACE AIR

The measured average shift concentrations in the "Spinning" shop showed that the level of workers' exposure to carbon disulfide there was 1.2 to 4.1 times the TLV for 8-hour exposure.

Workers engaged in transport of intermediate products have exposure levels to carbon disulfide from 2.2 to 3.1 times the 8-hour exposure TLV (TLV $10mg/m^3$).

The individuals responsible for the quality control of intermediate products revealed exposure levels of 2.5 to 3.8 times the 8-hour TLV.

The section supervisors had an exposure level exceeding 3.3 times the 8-hour TLV.

The measured average shift concentrations in the "Spinning" shop showed that the level of exposure to carbon disulfide in spinners, transport operators, quality control executives and section supervisors is from 3.11 to 40.82 mg/m^3.

In technological process operators from the II^{-nd} group a substantial share of carbon disulfide concentrations were within the 8-hour TLV, only two samples exceeded 1.3 times the TLV.

The operators of the acid station (II group) also provided samples within the limits with only one sample exceeding 1.7 times the TLV. The measured control average shift concentrations revealed that the level of exposure to carbon disulfide is below the LOD of the method (0,01 mg/m^3).

The results obtained for repair staff showed values from 0.09 to 9.15 mg/m^3, i.e. below the 8-hour TLV.

TABLE 2. Concentrations of CS_2 (mg/m^3) in workplace air (personal sampling) – GC/MS determination.

Groups	X	SD	Min	Max	P
Control	0.15	0.08	0.090	0.280	
Spinning (1)	21.38	10.91	0.650	37.850	**0.001 (0 vs 1)**
Improvement and drying (2)	3.96	4.95	0.060	17.080	**0.0001(1 vs 2)**
Repair (3)	1.99	2.90	0.001	9.150	0.003 (0 vs 3) 0.0001 (1 vs 3)

5.2. INDICES FOR ASSESSMENT OF INTERNAL EXPOSURE

TTCA in urine (mg/g creatinine)

The determined mean value 0.620 ± 0.600 mg/g creatinine (X ± SD) of TTCA for the control group consisting of individuals without occupational exposure to carbon disulfide is similar to data listed in publications on other studies(Table 3). The highest excretion of the metabolite, in the range 2.08 ± 0,96 mg/g creatinine was found in the urine of workers from the "Spinning" shop. The individual results follow the course of log-normal distribution in a broad concentration interval from 0.001 mg/g creatinine to 7.248 mg/g creatinine. Four workers (15.6%) had TTCA excretion exceeding the adopted biological limit of 4 mg/g creatinine (Ordinance №13/30.12.2003). The recorded average group value for the "Spinning" shop exceeds three times the value of the control group (p<0.001) and the value of the repair team (p<0.001) (Table 3).

Total metabolites of carbon disulfide in urine - molJ$_2$/mmol creatinine (iodine-azide test)

The calculated mean value of total metabolites of 3.77 ± 2.34 molJ$_2$/mmol creatinine in the control group is within the range of physiological fluctuations. Individual results above the biological limit - 5.7 mol J$_2$/mmol creatinine (Branch norm – Ministry of Health, №105, 1985) were found in 14% of the tested control persons. The highest mean values of iodine-azide test were found in workers from "Spinning" (I group), "Improvement and drying" and "Acid station" (II group) (Table 3). The determined average contents of total metabolites in the particular exposed individuals are statistically significantly higher than those of the control group as follows: "Spinning" shop - p< 0.001, "Improvement and drying" and "Acid station" - p< 0.05 and repair staff- p< 0.03.

TABLE 3. Biomarkers for assessment of internal exposure to carbon disulfide: 2-thiothiazolidin-4-carboxylic acid (TTCA) and iodine-azide test in urine collected at the end of the working shift.

	Groups	X	SD	Min	Max	p
TTCA/g creat	control	0.620	0.600	0.001	1.480	
	1	2.087	0..960	0.01	7.2484	0.001 (1 vs 0)
	2	0.689	0.500	0.001	1.895	0.001 (1 vs 2)
	3	0.720	0.630	0.103	2.93	0.001 (1 vs 3)
mol J$_2$/mmolc	control	3.770	2.340	0.560	6.370	
	1	5.080	2.530	2.300	13.540	0.001 (0 vs 1)
	2	5.140	2.030	2.600	10.760	0.050 (0 vs 2)
	3	4.090	1.890	1.160	7.840	0.030 (0 vs 3)

The comparative analysis of the data for external exposure and the biomonitoring results enable the study of the relationships between exposure level and deviations in applied biomarkers for prediction of the health risk at chronic exposure to carbon disulfide (Table 4).

The results of the repair staff at mean exposure to carbon disulfide about 2 mg/m^3 in the workplace air the excretion of total metabolites and TTCA in urine varied within the limits of physiological fluctuations. In the groups from "Improvement and drying" the level of total metabolites increased vs. that for the control group (p<0,05) while the excretion of TTCA varied within the normal range (p>0,05) at twofold increase of the exposure (Table 4).

TABLE 4. Correlation analysis between parameters of external and internal exposure.

Parameters	Correlation coefficient	P
CS_2 mg/m^3 vs TTCA	1 group – r = 0.414	**P < 0.05**
	2 group – r = 0.428	**P < 0.05**
Iodine-azide test vs CS_2 mg/m^3	3 group – r = 0.506	**P < 0.01**
Iodine-azide test vs TTCA	1 group – r = 0.478	**P < 0.05**

In the "Spinning" shop there was a recorded marked biological response to both biomarkers at carbon disulfide concentrations exceeding twice the TLV - 10 mg/m^3 for 8-hour exposure (Ordinance №13/30.12.2003). This group also recorded the highest mean values of TTCA and total metabolites compared to those of the control group (p< 0,001) and the highest individual values. In the "Spinning" shop the determined exposure to carbon disulfide exceeded twice the limit value (10 mg/m^3) – a prerequisite for triggering a biological response to the studied biomarkers for exposure assessment. This is also supported by the derived correlation ratio that outlines the relationship between the two indexes as moderate (r = 0.478; P < 0.05) and the correlation between exposure to carbon disulfide and TTCA urine excretion (r = 0.414; P < 0.05).

The derived significant correlation for repair workers (r = 0.506; P < 0.01) between the concentrations of carbon disulfide and total metabolites excretion can be explained with the intermittent character of the exposure related to job specifics.

6. Conclusions

The recorded values of carbon disulfide up to 4.1 times exceeding the TLV in two shops are due to technological processes.

The results obtained categorically confirm the adequate choice of biomarkers for assessment of internal exposure. The determination of specific metabolites of carbon disulfide in urine can be widely applied, as the methods are specific and non-invasive. As mentioned above the only source of TTCA appearance is the inhaled carbon disulfide. It should be underlined that the high correlation between external and internal exposure proves the credibility of the used sampling strategy and implemented GC methods for determination of carbon disulfide concentrations in workplace air. On the other hand the existing correlation excludes the possibility for "false" positive results.

The obtained results for the assessment of external and internal exposure are the major arguments at health risk assessment for the exposed workers.

References

Iodine-azide test. Branch normal №105. Collection of methods for obligatory laboratory tests for hygienic-epidemiological control in the field of work hygiene – industrial toxicology, Ministry of Health, 170–176, Sofia, (1987) (in Bulgarian).

Jakubowski, M., Piotrowski, J.: Badania nad ocena stonia ehpozyci na dwusiarczek wegla. II J. Medycyna Pracy, XVI, 2, 86–95 (1965).

Kaloyanova F. Industrial toxicology. General part. Publishing House "Meditsina i fizkultura", Sofia (1981) (in Bulgarian).

Lancranian I., Sukmansky M., Stanuca L., Antonescu C, Popescu HI. Study of the thyroid function in chronic carbon disulfide poisoning. MedLav, 63(3):123–5 (1972).

Lyubomirova K., Georgieva Tz., Panev T., Popov T., Mihailova A., Nikolova V.: Follow – up of the adverse health effects of workers exposed to carbon disulfide. International Journal of Occupational Medicine and Environmental Health. (in press) (2006).

NIOSH Pocket Guide to Chemical Hazards: www.skcinc.com.

Ordinance № 13/30.12.2003 for workers protection against risks related to occupational exposure to chemical agents.

Rosier J. Veulemans, H., Maschelein, R. Vanhoome, M., Van Petegham, C.: Experimental human exposure to carbon disulfide (1987).

Tan, X., Wang, F., Bi, Y., He, J., Su, Y., Braeckman, L., De Bacquer, D., Vanhoorne, M.: Carbon disulfide exposure assessment in a Chinese viscose filament plant. Int. J. Hyg. Environ. Health 203, 465–471 (2001).

Van Door R., Delbressine, L.P., Leijdekkers, C.M., Vertin, P.G., Henderson, P.Th.: Identification and determination of TTCA in urine of workers exposed to carbon disulfide. Arch. Toxicol. 47, 51–58 (1981).

WHO: World Health Organization. Task group on environmental health criteria for carbon disulfide. Environmental Health Criteria 10. Carbon disulfide, Geneva (1979).

ACCUMULATION OF COPPER, CADMIUM, IRON, MAGNESIUM AND ZINC IN THREE DEVELOPMENT STAGES OF RED PEPPER

*SIMONA DOBRINAS, SEMAGHIUL BIRGHILA AND MARIUS BELC
Chemistry Department, Ovidius University
124 Mamaia Blvd, 900527 Constanta, Romania

*To whom correspondence should be addressed: sdobrinas@univ-ovidius.ro

Abstract. Accumulation of copper, cadmium, iron, magnesium and zinc was studied in *Capsicum Solanaceae* plant (red pepper). A population of *Capsicum Solanaceae* from one urban site (Constanta city, Romania) was used to assess whether the five co-accumulated metals have a similar or different distribution in the plant. Metals accumulation in red pepper was quantified by flame atomic absorbtion spectrometry (FAAS). Analyses were performed after the chemical mineralization of the samples with nitrogen acid and hydrogen peroxide in a Digesthal device. Mn as well as Fe and Zn are preferentially accumulated in the roots, which exhibit higher concentrations than the other organs of *Capsicum Solanaceae* plant and Cu in plant's leaves. The accumulation of studied metals in red pepper varied according to the organ and the development stage of the plant.

Keywords: red pepper; Cu; Cd; Fe; Mn; Zn; FAAS

1. Introduction

Metal pollution is one of today's most serious problems. Requests for the determination of copper, cadmium, iron, manganese and zinc in plant materials are increasing mainly in view of their relevance in plant nutrition studies.

Copper is an essential element for plant growth and important in various biochemical process, but toxic in higher concentrations, when it interferes with numerous physiological processes. While manganese, iron and zinc are essential microelements that are indispensable for normal plant growth at low concentration and are toxic only at high concentration, cadmium has no vital function in plants developing under "natural" conditions.

The aim of this study was to analyze quantitatively the distribution of five investigated metals in green pepper plant collected from an urban site using flame atomic absorbtion spectrometry.

2. Experimental

The organs of red pepper plant (Capsicum Solananceae population grown in a orchard from Constantza area) were investigated. In order to determine Cu, Cd, Fe, Mn and Zn concentrations, the samples were washed with deionised water, dried and homogenised and then submitted to digestion in a Digesdhal device (a Birghila et al., 2004). After the complete digestion the samples solution was filtered, made up to 50 mL with deionized water and the investigated metals were determined by FAAS. The instrumental analysis and characteristics of metal calibration were presented in a previous paper (Matei et al., 2005).

3. Results and Discussion

Data on green pepper plants collected showed that while roots accumulated high concentrations of Fe, Mn and Zn, leaves accumulated lower concentrations (Table 1). Cu as well as Cd is preferentially accumulated in the leaves, which exhibit higher concentrations than the other organs of plant. Plants absorb cadmium through their roots and leave and this affects the plant metabolism and development. The highest concentrations of cadmium in plants are always reported for the leaves (Ivanova et al., 2001). Concentrations of Cu, Cd and Zn found in green pepper leaves in all development' stages are higher than those encountered by Silva (1998) in spinach and apple leaves and in tomato leaves (Birghila et al., 2004). Quantitative data obtained showed that the concentrations of five studied metals in the plant vary with the development stage (copper higher concentrations were observed in red pepper of 5 cm and lower concentrations in red pepper of 30 cm).

Iron was found in higher quantities than the other studied metals in the all three development stages of plant. This result means that in time, this element is suffering some changes, being transferred from soil and water sources.

Based on the obtained results, it can be stated that these values are situated within the limits imposed by the last regulations of the specialized international commissions such as EC's Scientific Committee for Food.

TABLE 1. Cd, Cu, Fe, Mn and Zn distribution in green pepper plant in three development stages.

Pepper plant		Cd	Cu	Fe	Mn	Zn
Pepper plant of 5 cm	roots	0.52	4.32	138.44	88.55	21.12
	stem	0.79	4.12	134.44	55.24	9.45
	leaves	0.98	4.77	92.90	32.57	5.87
Pepper plant of 20 cm	roots	0.22	3.23	176.71	81.87	30.69
	stem	0.25	3.12	24.01	6.54	22.69
	leaves	0.12	3.33	57.14	28.34	2.57
Pepper plant of 30 cm	roots	0.22	2.97	125.51	86.04	30.69
	stem	0.25	3.01	99.69	44.14	18.34
	leaves	0.99	3.12	33.76	52.11	26.56
	flower	0.21	0.73	58.98	5.24	1.80
	fruit	0.01	0.47	4.75	0.19	3.52

References

[a] Birghila S., Dobrinas S., Belc M., 2004, Determination of copper by standard addition method in leafy vegetables from apiaceae family, *Ovidius University Annals of Chemistry*, 15, 19–21.
[b] Birghila S., Dobrinas S., Matei M., Magearu V., Popescu V., Soceanu A., 2004, Distribution of Cd, Zn and ascorbic acid in different stages of tomato (Lycopersicum esculentum solanaceae) plant growing, *Rev. Chim. (Bucharest)*, 55, 683–685.
Ivanova J., Korhammer S., Djingova R., Heidenreich H., Markert B., 2001, Determination of lanthanoids and some heavy and toxic elements in plant certified reference materials by inductively coupled plasma mass spectrometry, *Spectrochim. Acta B*, 56, 3–12.
Matei N., Dobrinas S., Birghila S., Rasanu N., Belc M., 2005, Accumulation of Cd, Cu, Fe, Mn, Zn and ascorbic acid in different stages of prunus persica plant growing, *Proceeding of International Conference "Agricultural and Food Sciences, processes and Technologies"*, Sibiu, Romania, 2, 1-7, ISBN:93-739-093-8.
Silva, F.V., Nogueira A.R.A., Souza G.B., Zagatto E.A.G., 1998, A polyvalent flow injection system for mutielemental spectrophotometric anaysis of plant materials, *Anal. Chim. Acta*, 370, 39–46.

DETERMINATION OF ORGANOCHLORINE AND POLYCYCLIC AROMATIC HYDROCARBONS PESTICIDES IN HONEY FROM DIFFERENT REGIONS IN ROMANIA

SIMONA DOBRINAS[1]*, SEMAGHIUL BIRGHILA[1] AND VALENTINA COATU[2]
[1] *Chemistry Department, Ovidius Universit*
124 Mamaia Blvd., 900527 Constanta, Romania
[2] *National Institute of Marine Researches and Development Mamaia 300, ROMANIA*

*To whom correspondence should be addressed: sdobrinas@univ-ovidius.ro

Abstract. In this study honey collected from local markets or made by private peasants from eighteen regions of Romania during years 2002-2004 were analyzed for organochlorine pesticides (OCPs). An analytical procedure based on liquid-liquid extraction with n-hexane followed by gas chromatography with electron capture detection (GC-ECD) has been developed. For clean-up and preconcentration purposes was used a usual sorbent material (florisil). Limits of detection were from 0.02 and 0.05 μg /Kg. The pesticide endrin was the most frequently detected in 61% of the samples, followed by dieldrin in 16% of the samples. DDT and their metabolites were detected in 11% of the samples. Results indicate that honey consumers should not be concerned about the amounts of organic pollutants found in Romanian honey.

кeywords: GC-ECD; OCPs; honey

1. Introduction

Nowadays there is an increasing concern over the potential hazardous effects that pesticides may have on human health. Also, honey got an increasing importance as essential natural resources in promoting healthy food.

Honey is an easily digestible foodstuff containing a range of nutritiously important complementary elements. The bees collect for their own developmental needs the nectar, honeydew, resinous substances (to produce propolis), pollen and water from the environment, which unfortunately is often exposed to various contaminants than can be found in the human consumed foods (Antonescu and Mateescu, 2001).

Organochlorine pesticides (OCPs) have been restricted or banned in agriculture since 1978 in Europe because of their persistence and bioaccumulation in the environment. However, these pesticides are still frequently found in soil (Doong and Liao, 2001), from which they continue to cycle through the environment. Different studies demonstrated the bioaccumulation of organochlorine from contaminated soil to different plants (Colume et al., 2001) and to organisms (Porrini et al., 2003).

The purpose of this work was to develop a liquid-liquid extraction method for the analysis of nine OCPs in honey samples from various botanical origins followed by GC-ECD and evaluate the level of contamination with OCPs.

2. Experimental

Honey samples used in the research represent several Romanian honey sorts and these samples were taken from local market or made by private peasants, during 2002 –2004 years. These samples were stored in their original containers at room temperature in a dark place. Aliquots of ca.1 g of studied samples were made up to 100 ml with deionized water and then where transferred into the glass vessels for the determination of OCPs.

An analytical procedure based on liquid-liquid extraction with n-hexane followed by gas chromatography with electron capture detection (GC-ECD) has been developed. The preconcentration step was presented in a previous paper (Dobrinas et al., 2004).

3. Results and Discussion

Once DDT is released into the environment, it begins to degrade and can be found in two other forms, DDE and DDD. DDE is DDT's main metabolite and also the most persistent one. DDT and its metabolites were detected only in 2 samples (11%) at concentrations of 0.003 and 0.014 µg/g from linden honey (Dambovita district) and from locust honey (Constantza district) collected in 2002. Also, HCB was found in these 2 samples at 0.024 and 0.057 µg/g. However, Romanian honey was less contaminated than Portuguese and Spanish honey (range of DDT was from 0.020 to 0.658 mg/Kg and of HCB was from 0.12 to 3.24 mg/Kg) (Blasco et al., 2003).

Endrin was the most frequently detected in 61% of samples at concentration ranging from 0.062 to 1.620 µg/g and just one sample was contaminated with dieldrin at 0.025 µg/g. A total of 4 samples (linden and locust honey collected in 2002) were contaminated by aldrin at levels from 0.001 to 0.003 µg/g. Heptachlor was detected only in 2 samples (linden and locust honey collected during 2003-2004) at 0.021 and 0.061 µg/g. Two honey samples (linden and locust) contained lindane at concentrations of 0.01 and 0.225µg/g.

From the obtained data result that the most frequently detected OCPs were in linden and locust honey and in sunflower, mountain flower, fir and conifer honey was detected only one or two pesticides. Organochlorines were not detected in honey obtained from various flowers in two regions of Romania.

The results obtained could be expected, because the OCPs have been extensively used and are still present in the environment due to their high persistence. The harmful effects of OCPs (lipophilic substances and consequently, soluble and stable in the beewax) are observed directly on bees or in time, as a result of a small but continuous intake of pesticides (Porrini et al., 2003).

ACKNOWLEDGEMENTS

This work was realized by financial support of CNCSIS Grant no.137/2005.

References

Antonescu C., Mateescu C., 2001, Environmental pollution and its effects on honey quality, *Roum. Biotechnol. Lett.*, 6, 371–379.

Blasco C., Fernandez M., Pena A., Lino C., Silveira M.I., Font G., Pico Y., 2003, Assesment of pesticide residues in honey samples from Portugal and Spain, *J. Agric. Food Chem.*, 51, 8132–8138.

Colume A., Cardenas S., Gallego M., Valcarcel M., 2001, Semiautomatic multiresidue gas chromatographic method for the screening of vegetables for 25 organochlorine and pyrethroid pesticides, *Anal. Chim. Acta*, 436, 153–162.

Dobrinas S., Birghila S., Matei N., Coatu V., 2004, Determination of PAHs and organochlorine pesticides in tomato and green pepper, *Rev. Chim. (Bucharest)*, 55 (12), 942–944.

Doong R.A., Liao P.L., 2001, Determination of organochlorine pesticides and their metabolites in soil samples using headspace solid-phase microextraction, *J. Chromatogr. A*, 918, 177–188.

Porrini C., Sabatini A.G., Girotti S., Ghini S., Medrzycki P., Grillenzoni F., Bortolotti L., Gattavecchia E., Celli G., 2003, Honey bees and bee products as monitors of the environmental contamination, *Apiacta*, 38, 63–70.

EVALUATION OF THE WOMEN FERTILITY HEALTH AND THE ENVIRONMENT. NICKEL EXPOSURE IN INDUSTRY

ELENA MAKAROVA-ZEMLYANSKAYA*, LUDMILA
DUEVA AND MARINA FESENKO
Research Institute of Occupational Health
Russian Academy of Medical Science
31 Prospect Budennogo, 105275 Moscow, Russia

*To whom correspondence should be addressed: helen456@mail.ru

Abstract. Nickel and its compounds :are widely used in various manufactures such as machine-building plants, galvanic industry, metallurgy and others. The basic chemical factors in the galvanic industry are aerosols of sulfate and chloride of nickel, which concentration in air of the working zone did not exceed the maximum permissible level Influence of solutions on skin is limited by the use of individual protective devices. The analysis of condition of reproductive health of working women exposed to nickel has not revealed any statistical connection between morbidity and exposure to this metal. Prevailing diseases were a climacteric syndrome, hysteroptosis and colpoptosis. Activation humoral immunity and the sensibility to nickel were found in workers exposed to nickel

Keywords: environment; nickel; reproductive health; immunity.

1. Material and Methods

According to the literature data (Antoshina, 2005; Dujeva and Sivochalova, 2002) the compounds of nickel possess toxic effect, influencing blood, respiratory, digestive, endocrine, reproductive, and immune systems. Nickel is carcinogen and mutagen (Vos et al., 1996; Toxicological Profile for nickel 1997), We analyzed the results of the clinical gynecologic examination of working women exposure to nickel.

Conditions of work, pollution of air of workplace of galvanic industry in the machine-building plant were studied.

Immune system at 76 workers exposure to nickel is investigated. The program of the immunological investigation included: concentration of the basic classes of antibodies (A, M, G and E), circulating immune complexes in the blood and specific markers of influence of the nickel on workers - antihapten antibodies to nickel. We analyzed of results of research according to standards of immunological parameter (Methodical recommendations of Institute of Immunology RAMS). Results of immunology studies analyzed depending on the seniority of work in galvanic industry and states of health.

Results were analyzed by Student's t-test, and changes were considered significant at $p<0,05$(*), $p<0,01$(**).

2. Results and Discussion

The basic chemical factors in the galvanic industry are aerosols of sulfate and chloride of nickel which concentration in air of the working zone did not exceed the maximum permissible level. Influence of solutions on skin is limited to use of individual protective devices.

The age the workers was 30-50 years (62% of persons) and the seniority of work in galvanic industry till 5 years (68,4% of persons), 5-10 years (11,8%) and over 10 years (19,7%).

Chronic gynecologic pathology was discovered at 78,4% of the working women exposed on nickel and 89,8% of working women not exposed to this metal (Table 1). Climacteric syndrome and menopause were the prevailing diseases among the working women, who either were or were not exposed to nickel. On the second place among all working women were hysteroptosis and colpoptosis (23,5% and 27,2% accordingly). The third place among working women was cervical erosion (17,6% and 22,8% accordingly).

The analysis of immunological disturbances of the workers showed that activation of all humoral immunity parameters had 82% of workers. We analyzed the relationship between condition of immune system of the workers exposed to nickel and seniority in galvanic industry. The most considerable and frequent changes of parameters of immunity are revealed at workers with low seniority (till 5 years). There are high titer of antibodies to nickel and high level of circulating immune complexes in blood. Combined changes of parameters humoral immunity and sensibility to nickel are noted at 39% of workers, that statistically significant distinguish from group of workers with the seniority over 10 years (p <0, 01). In the group of workers with the seniority of 5-10 years the disturbance of humoral immunity is revealed at some smaller number of the surveyed persons that is connected with development of adaptation to negative factors of industry. At the same time, in group of workers with the seniority over 10 years the increase of frequency of disturbance of humoral immunity and isolated reactions of sensibility to nickel is observed.

TABLE 1. Reproductive health of working women depending on exposure to nickel.

Reproductive pathology	Exposure to nickel n=102		Not exposure to nickel n=206	
	Absolute	P±m, %	Absolute	P±m, %
Chronic reproductive pathology	80	78,4±4,1	185	89,8±2,1
Colpitis	5	4,9±2,1	16	7,8±1,9
Cervicitis	16	15,7±3,6	30	14,6±2,5
Cervical erosion	18	17,6±3,8	47	22,8±2,9
Salpingitis	10	9,8±2,9	15	7,3±1,8
Hysteromyoma	13	12,7±3,3	43	20,9±2,8
Endometriosis	4	3,9±1,9	8	3,9±1,4
Hysteroptosis and colpoptosis	24	23,5±4,1	56	27,2±3,1
Climacteric syndrome, menopause	35	34,3±4,7	73	35,4±3,3

Thus immunological tolerance to nickel interrupts among workers with seniority over 10 years. We observe the hyperergic status of immunity that determines unsatisfactory condition of the worker's health.

TABLE 2. Humoral immunity and sensibility to nickel according to the pathology in %.

Condition of the imumune system	Rare resp. infections	Often diseased resp. Infections	Inflammation	Allergic diseases
Disturbance of humoral immunity (HI)	56,3	90*	72,4	47,6
Antibodies to nickel	37,5	50	41,4	52,4
Disturbance of HI and antibodies to nickel	25	50	31	23,8
Antibodies to nickel without disturbance of HI	12,5	0	10,3	28,6**
Increase Ig E	18,7	60*	24,1	23,8

The analysis immunological status (Table 2) depending on character of pathology were shown the highest frequency of changes humoral immunity (90%) among persons with often respiratory infections ($p<0,05$). Immunological status at workers of often diseased respiratory infections group showed overstrains of immune system which stimulated pathological processes, primarily acute respiratory diseases. Immunological changes were less frequent in the group of workers with chronic inflammations than it was among often-diseasing workers and that speaks about a chronization of pathology. Insignificant increase concentration Ig E among workers in this group connected with mainly not allergic pathogenesis chronic inflammations. The frequency of changes of parameters humoral immunity among the workers with allergic diseases are less and frequency of antihapten antibodies to nickel are highest than in others groups. In this group sensibility to nickel have mainly isolated character (28,6%, $p <0,01$). While atopic allergic status (on level Ig E) among these workers feebly marked in comparison with often diseased workers. These results evidence that the group with allergic diseases form workers mainly with skin diseases (delayed-type hypersensitivity) which development under exposure to nickel. At the same time diseases in group of often ill workers are Ig E-dependent allergic reactions of immediate hypersensitivity.

3. Conclusions

1. The analysis of reproductive health of women exposure to nickel has not revealed any statistical connection between morbidity and exposure. Prevailing diseases were a climacteric syndrome, hysteroptosis and colpoptosis.

2. Activation humoral immunity and the sensibility to nickel were found in workers exposed to nickel.

References

Antoshina L.I. Medical and ecological problems of workers. – 2005. - № 1. – P. 58–63.
Dujeva L.A., Sivochalova O.V. // In: Sixth Intern. Nickel Conference. - Murmansk, Kola Peninsula, Russia. 2002. - P. 99.
Toxicological profile for nickel (Update) U.S. Department of health and human services Public Health Service Agency for Toxic Substances and Disease Registry. – 1997. P. 262.
Vos J.G, M.Younes, E. Smith Ed. Allergic Hypersensitivities Induced by Chemicals. Recommendation for Prevention.- WHO: Geneva, Switzland, 1996.

STUDY ON THE INFLUENCE OF VEGETATION ON THE QUALITY OF AQUATIC SYSTEMS

SILVIYA LAVROVA AND BOGDANA KOUMANOVA
University of Chemical Technology and Metallurgy
Department of Chemical Engineering,
8 Kliment Ohridsky Str.,1756 Sofia, Bulgaria

*To whom correspondence should be addressed: silviya@lavrova.com

Abstract. The aim of this study is to show the influence of plants growing in aquatic systems on their water characteristics. There are many various compounds in the aquatic system, which undergo different transformations under surrounding environment. Several processes occur in the water basin: settling down, resuspension, evaporation, uptake by microorganisms, and accumulation in the structure of the aquatic plants. Under certain conditions some of these compounds may return to the water after the plants death. Three aquatic systems with different level of pollution were studied. The first aquatic system is located in comparatively clean area, the second one is in industrial area, and the third one is polluted with oily products. The common reed (***Phragmites australis***) is the main vegetation in these reservoirs. The water samples have been examined by physicochemical analysis to determine their characteristics: Chemical Oxygen Demand (COD), Biochemical Oxygen Demand (BOD_5), Total Nitrogen (TN), Total Suspended Solids (TSS), Total Dissolved solids (TDS), pH, etc. The highest value of BOD_5 was measured for the third sample, which is polluted with oily products. It means that the water contains variety of organic compounds due to the pollution and the decay processes. Emission Spectroscopy was used for preliminary (semi-quantitative) information of the elements presented in the water samples. Atomic Absorption Spectroscopy and UV-spectroscopy were used to determine the significant metals Fe, Mn, Mg, Ca and Si. It was established that the concentration of metals is relatively low. The plant samples preliminary cut and dry were extracted using two solvents. The extracts were analyzed consequently by Thin Layer Chromatography, Gas Chromatography and Mass Spectrometry. The results show similar hydrocarbons content but difference in their quantity. It was confirmed that the plants growing in oily polluted water accumulate to some extent the hydrocarbons.

Keywords: aquatic systems; common reed; hydrocarbons; metals.

1. Introduction

A wetland is composed of water, substrate, plants, plant litter, invertebrates (mostly insect larvae and worms), and microorganisms (Halverson, 2004). Processes controlling contaminant retention in the aquatic system sediment may be abiotic (physical and chemical) or biotic (microbial and botanical) and are often interrelated (USDA, 1995; ITRC, 2003), (Fig. 1).

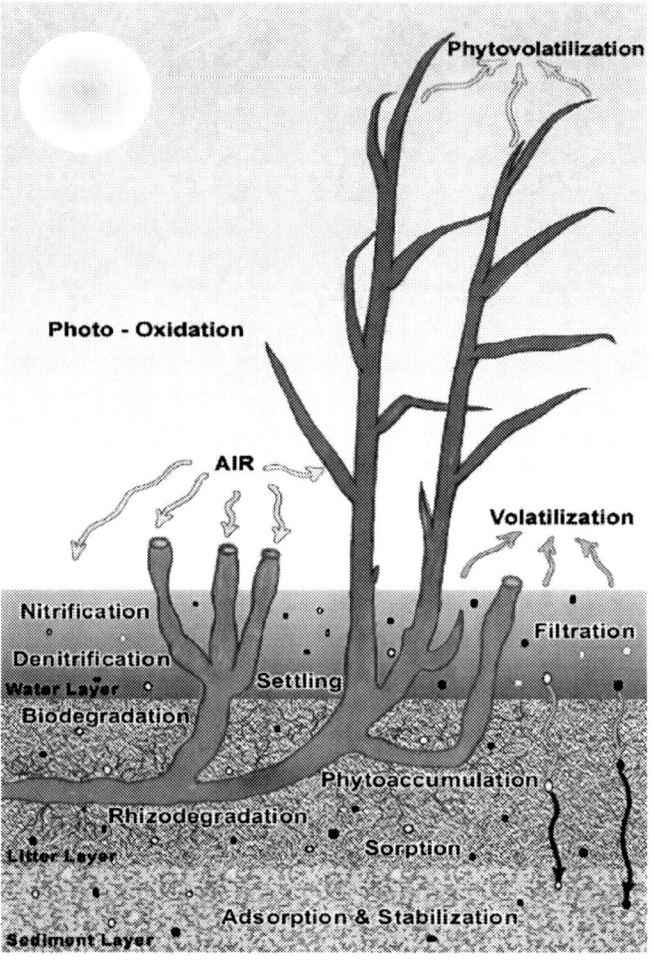

FIGURE 1. Abiotic and biotic mechanisms treating compounds in aquatic systems.

The primary abiotic processes that are responsible for removing water contaminants in the wetland are:
- settling and sedimentation - which achieve efficient removal of particulate matter and suspended solids;
- sorption - including adsorption and absorption, the process occurring on the surfaces of plants, substrate, sediment, and litter that results in short-term retention or long-term immobilization of contaminants;
- chemical oxidation/reduction - conversion of metals, through contact of the water with the substrate and litter to an insoluble solid form that settles out. This is an effective method for immobilizing toxic metals in the wetland;
- photodegradation/oxidation – degradation/oxidation of compounds in the presence of sunlight;
- volatilization - which occurs when compounds with significant vapor pressures partition to the gaseous state (USDA, 1995; ITRC, 2003).
- Biotic processes like biodegradation and plant uptake are also responsible for contaminant removal, in addition to the abiotic processes. Some microbial and botanical processes taking place in a wetland are:
- aerobic/anaerobic biodegradation - metabolic processes of microorganisms, which play a significant role in removing organic compounds in wetlands;
- phytoaccumulation - uptake and accumulation of inorganic elements in plants;
- phytostabilization - the ability to sequester inorganic compounds in plant roots;
- phytodegradation - plant-produced enzymes break down the organic and inorganic contaminants that enter into the plant during transpiration;
- rhizodegradation - plants provide exudates that enhance microbial degradation of organic compounds;
- phytovolatilization/evapotranspiration - uptake and transpiration of volatile compounds through the leaves (ITRC, 2003).

Combinations of biological and no biological processes enable effective treatment of a wide range of contaminants (ITRC, 2003).

Vegetation plays an important role in wetlands. Plants provide a substrate for microorganisms, which are the most important processors of water contaminants (Rovira, 1965). They also provide microorganisms with carbon. Stands of vegetation reduce current velocity, allowing solids to settle out of the water column. Plants assimilate nutrients, but as the plants are dying, some nutrients are released back into the water, (Vymazal, 1998). A portion of the nutrients is retained in the undecomposed fraction of the plant litter and accumulates in the soils. Plants translocate oxygen from the atmosphere to the underground organs through a lacunar system of intercellular air-spaces or through aerenchyma, (Armstrong, 1992; Brix, 1993), and provide aerobic conditions within the reduced soil. Oxygen produced in the photosynthesis may also be used. The vegetation has additional site-specific values by providing habitat for wildlife and making wastewater treatment systems aesthetically pleasing (Reddy et al., 1989). However, the widespread species are robust species of emergent plants, described as macrophytes, such as the common reed, cattail and bulrush (Brix, 2003). These are the dominating life form in wetlands and marshes, growing within a water table range from 50 cm below the soil surface to a water depth of 150 cm or more. In general they produce aerial stems and leaves and an extensive root and rhizome-system. The plants are morphologically adapted to growing in a water-logged or submersed substrate by virtue of large internal air spaces for transportation of oxygen to roots and rhizomes. This life form comprise species like *Phragmites australis* (Common Reed), *Glyceria* spp. (Mannagrasses), *Eleocharis* spp. (Spikerushes), *Typha* spp. (Cattails), *Scirpus* spp. (Bulrushes), *Iris* spp. (Blue and Yellow Flags) and *Zizania aquatica* (Wild Rice) (Brix, 2003; Marks et al., 1994).

The aim of the present study was to compare the characteristics of three water systems situated in different areas and the influence of the plants growing in them.

2. Experimental

Three aquatic systems with different level of pollution were studied. The first aquatic system is located in a comparatively clean area, the second one is in an industrial area, and the third one is polluted with oily products. The common reed (***Phragmites australis***) is the main vegetation in these aquatic systems.

Water and plant samples were taken from each of these three aquatic systems.

The water samples have been examined according to Standard Methods for the Examination of Water and Wastewater (18th edition, 1992) to determine their characteristics: pH, Permanganate oxidation (PO), Chemical Oxygen Demand (COD), Biochemical Oxygen Demand (BOD$_5$), Total Nitrogen (TN), Organic Nitrogen (ON), Ammonium Nitrogen (NH$_4^+$-N), Total Phosphorus (TP), Total Suspended Solids (TSS), Total Dissolved

Solids (TDS), Organic Dissolved Solids (ODS). For elemental content determination of the water samples Emission Spectroscopy was made and for the quantitative determination of significant metals - Atomic Absorption Spectroscopy ("Perkin Elmer-323") and UV-spectroscopy ("Perkin Elmer-5000"). Based on the preliminary experiments (Koumanova et al., 1983) the following optimum conditions for extraction of the hydrocarbons from polluted with oily products water samples were established: an adsorption on natural sorbent, followed by an eluation with methylene chloride and additional extraction with methylene chloride at pH 2. The concentrated extract was analyzed with Gas Chromatography and Mass Spectrometry for identification and quantitative determination of the extracted hydrocarbons.

It was of interest to examine how the vegetation which grows in polluted wetlands contributes to their additional purification. For that purpose plant samples, after preliminary preparation, were extracted in an Soxlet's apparatus using a mixture of methanol/chloroform. After that the extracts were let through a column filled with aluminum oxide and by Thin Layer Chromatography were separated to three fractions: sterols, terpenic and aliphatic alcohols and hydrocarbons. The hydrocarbon fractions were analyzed by Gas Chromatography and Mass Spectrometry.

The analyses of hydrocarbons content in water and plant samples were carried out by:

- Gas Chromatograph "Perkin Elmer 2B" with glass capillary column OV-17 SCOT and temperature programming in interval 70-260°C and velocity 5°C/min;
- Mass Spectrometer, GC-MS LKB 2091 at the referred above conditions.

3. Results and Discussion

Figure 2 to 5 illustrate the comparison between characteristics of the water samples from comparatively clean area and industrial area.

The pH values both in the water from the clean and the industrial areas are in the neutral zone. The values of PO, COD and BOD_5 of the water sample from the industrial area are higher than those of the water sample from comparatively clean area (Fig. 2). Obviously this is a result of the contamination, bigger density of the vegetation and the smaller area where the decay products are with higher concentration.

The values of nitrogen in the industrial area are less than those for the water sample from comparatively clean area (Fig. 3). Total nitrogen (TN) gives an idea of the organic and inorganic nitric compounds. As a whole the nitrogen plays important role for the synthesis of chlorophyll in the plants leaves and it is "element of growth", but there is an important detail - the plants cannot assimilate it directly. The nitrogen must be in the form of NH_4^+, NO_3^- or NH_3. This precondition gives us a reason to clime that the dense occupied industrial area with *Phragmites australis* has low content of nitrogen, because plants assimilate it. The nitrogen and the phosphorus are considered to be two of three basic nutrient elements in the soil.

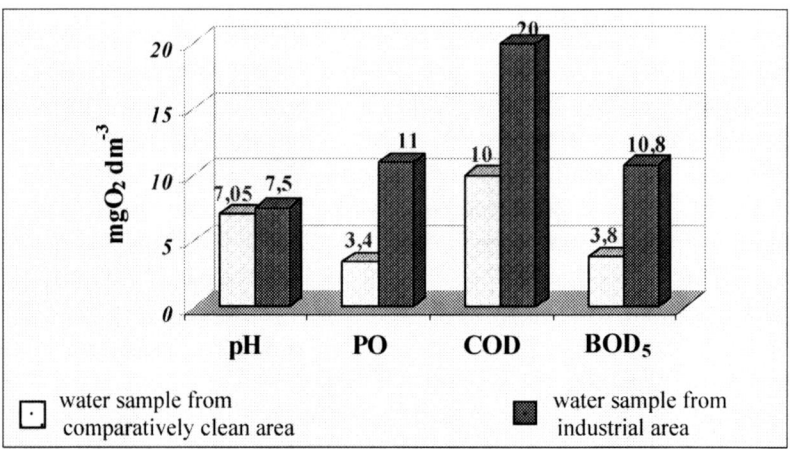

FIGURE 2. Values of pH, PO, COD and BOD_5 for water samples from comparatively clean area and industrial area.

The phosphorus joins in the structure of the plant stem and after the end of life-cycle of vegetation their leaves fall down in the water body, which is precondition for decay processes, in which result phosphoric compounds return into the water (Van der Putten, 1997).

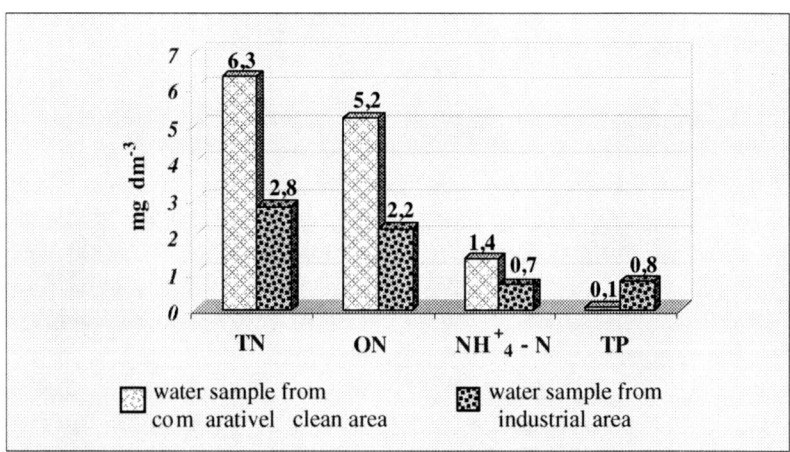

FIGURE 3. Values of TN, ON, NH_4^+- N and TP for water samples from comparatively clean area and industrial area.

Figure 4 shows that the values of TSS are low in the both wetlands. The values of total TDS are similar for the both wetlands – 292.0 mg dm^{-3} in the comparatively clean area and 334.0 mg dm^{-3} in the industrial area. TDS gives an idea of the dissolved organic and inorganic compounds. About 40 % in the both wetlands are organic (Fig. 5) and 60 % are inorganic. The organic part is mainly due to the growing vegetation and the decay processes, while the inorganic part is due to the mineral solids from the soil, sorption of oxides from the atmosphere or from leakage of the ground waters.

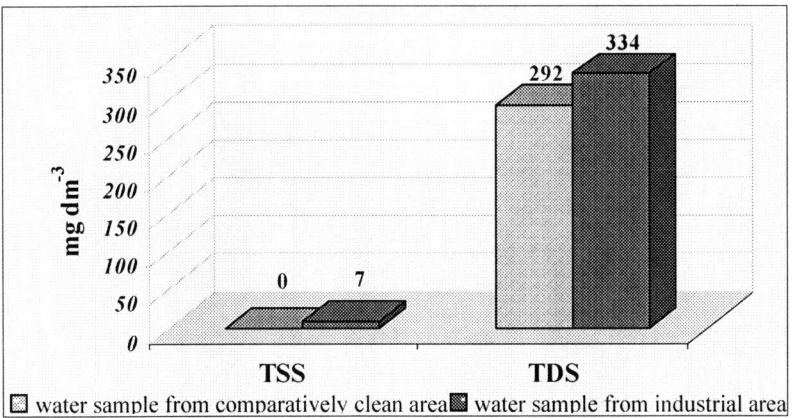

FIGURE 4. Values of TSS and TDS for water samples from comparatively clean area and industrial area.

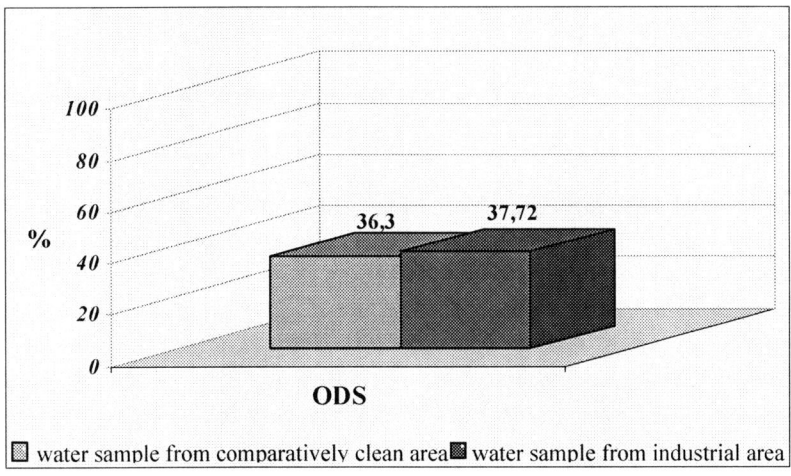

FIGURE 5. Values of ODS for water samples from comparatively clean area and industrial area.

Emission Spectroscopy was carried out for obtaining of preliminary information for metals content. This semi-quantitative analysis showed that the water from comparatively clean area contains only Mg, Ca, Si, and Fe, while the water from industrial area contains Mn besides Mg, Ca, Si, and Fe. The quantitative determination of Mg, Ca, Mn, and Fe was made by Atomic Absorption Spectroscopy and the quantitative determination of Si was made by UV-spectroscopy. Fig.6 illustrates the metal content of the water samples from the comparatively clean area and the industrial area. It shows that the metal content in the water from the industrial area is higher in contrast with this in the water from the comparatively clean area. The quantities of Mg and Ca are vastly great than those of Si, Fe, and Mn. As a whole metal content in the aquatic systems is low and there is no presence of heavy metals.

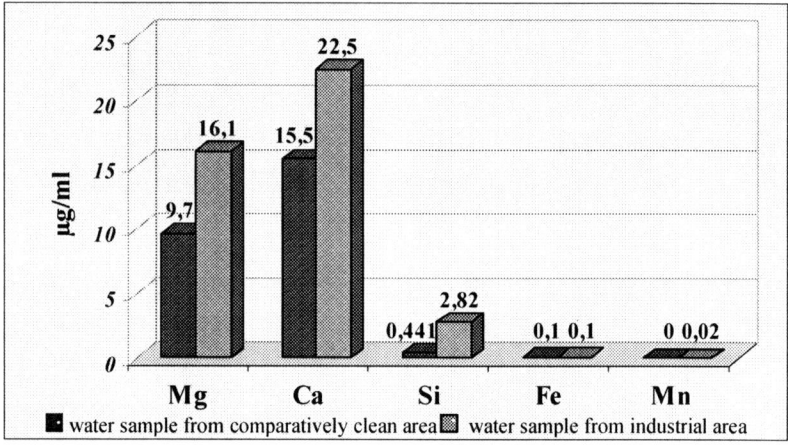

FIGURE 6. Metals content in water samples from comparatively clean area and industrial area.

The water sample which was taken from polluted with oily products area was analyzed by Gas Chromatography and Mass Spectrometry for identification and quantitative determination of their composition (Table 1). It was established presence of hydrocarbons ($C_{14} - C_{34}$), phenols, esters and nonidentified compounds. The hydrocarbon fraction is dominated especially the hydrocarbons with abundant number of carbon atoms ($C_{21} - C_{28}$). Figure 7 illustrates gas chromatogram of hydrocarbons fraction.

In order to specify the influence of the vegetation on the quality of aquatic systems, i.e. to what extent the vegetation which grows in polluted wetlands contribute to their additionally purification, the plant samples were consequently analyzed.

TABLE 1. Gas chromatography data and Mass Spectral identification of organic content in water sample polluted with oily products.

Compound	Content, %
Hydrocarbons	
$C_{13}-C_{20}$	9.1
$C_{21}-C_{28}$	35.0
$C_{29}-C_{32}$	3.4
Phenols	11.5
Esters	16.4
Nonidentified	24.5

Preliminary cut and dried plants from comparatively clean area and oily contaminated area were extracted in Soxlet's apparatus using mixture of methanol/chloroform.

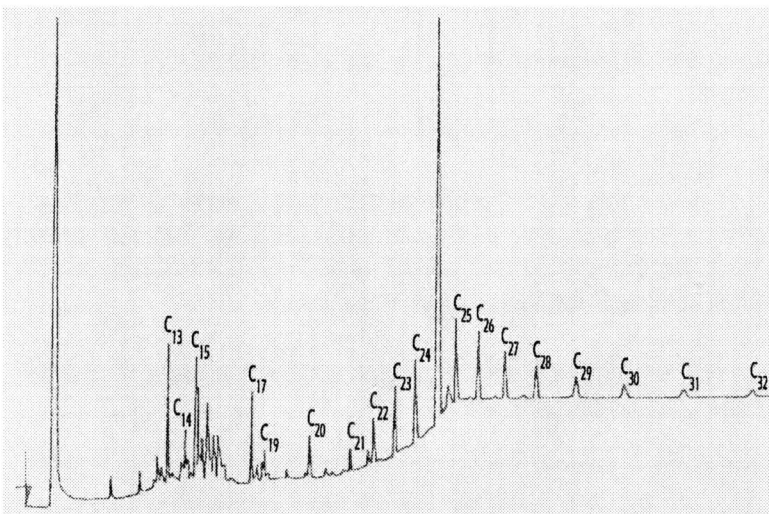

FIGURE 7. Gas chromatogram of the organic extract obtained from the water sample polluted with oily products.

After additional treatment the extracts were analyzed by Thin Layer Chromatography as a result of which they were separated to three fractions: sterols, terpenic and aliphatic alcohols and hydrocarbons. After that the hydrocarbon fractions were analyzed using Gas Chromatography and Mass

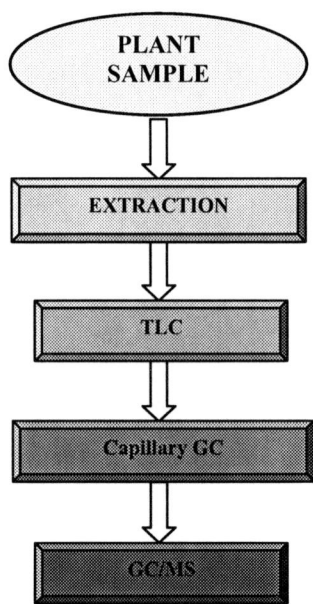

FIGURE 8. Extraction, concentration and analysis of plant samples.

Spectrometry for identification and quantitative determination of their composition (Fig. 8).

The results from Gas Chromatography and Mass Spectrometry are presented in Table 2. The data comparison shows that the percentage content of hydrocarbons from C_{14} to C_{20} and from C_{21} to C_{28} is vastly great in the plant sample from comparatively clean area.

TABLE 2. Gas chromatography data of the plant samples from comparatively clean and polluted with oily products aquatic systems.

Compound	Content, %	
	Plant sample	
	Comparatively clean area	Polluted with oily products area
$C_{14} - C_{20}$	37.4	14.1
$C_{21} - C_{28}$	33.5	15.6
$C_{29} - C_{34}$	18.0	47.6

There is considerable difference in content of hydrocarbons from C_{29} to C_{34}. In the plant sample from comparatively clean area it is quite low, whereas in the plant sample from polluted with oily products area it reaches 47.6%. Figures 9 and 10 illustrates gas chromatograms of hydrocarbons obtained from both plant samples from comparatively clean and from oily contaminated area.

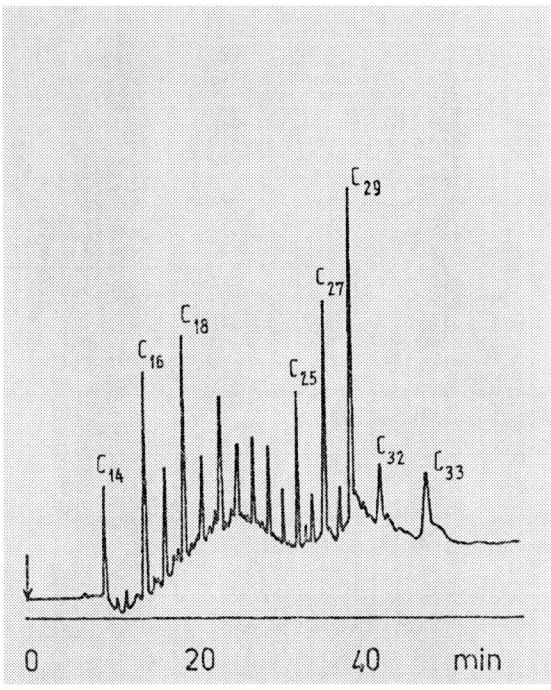

FIGURE 9. Gas chromatogram of hydrocarbons obtained from the plant sample from comparatively clean aquatic system.

It is obvious from the comparison of both gas chromatograms that the picks corresponding to hydrocarbons from C_{29} to C_{34}, obtained from the plants growing in the oily polluted aquatic system are considerably higher than those of the hydrocarbons from C_{29} to C_{34}, obtained from plants growing in comparatively clean aquatic system. This verifies the ability of vegetation growing in polluted areas to accumulate the hydrocarbons, especially these with abundant number of carbon atoms. Based on the results obtained from analyses of water and plants from polluted with oily products aquatic system (Table 3), we can confirm that the *Phragmites australis* contribute to the purification of the waters because accumulates the hydrocarbons as longer-chain hydrocarbons ($C_{29} - C_{34}$) in its tissue.

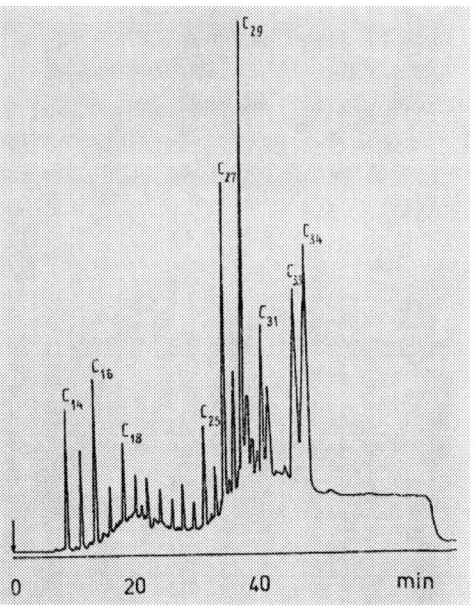

FIGURE 10. Gas chromatogram of hydrocarbons obtained from the plant sample from aquatic system polluted with oily products.

TABLE 3. Comparison of the established hydrocarbon contents of the analyzed water and plant samples from polluted with oily products aquatic system.

Compound	Content, %	
	Oily contaminated aquatic system	
	Water	Plant
$C_{14} - C_{20}$	9.1	14.1
$C_{21} - C_{28}$	35.0	15.6
$C_{29} - C_{34}$	3.4	47.6

4. Conclusions

The quality of three aquatic systems situated in different areas and the plants growing in them have been studied. It was established that the metals content in aquatic systems is low and there is no presence of heavy metals.

The content of the organic compounds is lowest in the relatively clean aquatic system because the pollution extent is insignificant, as well as the vegetation density.

The content of the organic compounds in the sample from water body after wastewater treatment plant of refinery is highest because of oily contamination.

The analyses of plant samples showed similar hydrocarbons content but different in their quantity.

The comparison between established hydrocarbon contents of the water and plant from a wetland polluted with oily products showed that the *Common reed* accumulates long-chain hydrocarbons ($C_{29} - C_{34}$).

It was confirmed that the plants contribute to the purification of wetland waters which are polluted with organic substances.

References

Armstrong, J., Armstrong, W. and Beckett, P.M., 1992. *Phragmites australis*: Venturi- and humidity-induced pressure flows enhance rhizome aeration and rhizosphere oxidation. New Phytologist 120: 107–207.

Brix, H., 1993. Macrophyte-mediated oxygen transfer in wetlands: Transport mechanisms and rates. In: Moshiri, G.A. (Ed.), Constructed Wetland for Water Quality Improvement, Boca Raton, London, Tokyo, pp. 391–398.

Brix., H., 2003. Plants used in constructed wetlands and their functions. 1st International seminar on the use of aquatic macrophytes for wastewater treatment in constructed wetlands.

Halverson, N., 2004. Review of Constructed Subsurface Flow vs. Surface Flow Wetlands. WSRC-TR-2004–00509. Westinghouse Savannah River Company, Savannah River Site, Aiken, SC 29808.

Interstate Technology & Regulatory Council (ITRC), 2003. Technical and Regulatory Guidance Document for Constructed Treatment Wetlands. The Interstate Technology & Regulatory Council Wetlands Team.

Koumanova, B., Popangelova, M., Angelova, L., Dimitrov, D., 1983. Hydrocarbons content in aquatic plants growing in polluted water body. Periodica Polytechnica, vol.27, No. 4., Budapest, Hungary.

Marks, M., Lapin, B., and Randall, J., 1994. Phragmites australis (P.communis): threats, management, and monitoring. Natural Areas Journal, 14: pp. 285–294.

Reddy K.R. and Smith, W.H., 1989. Aquatic plants for water treatment and resource recovery. pp. 27 - 48. Magnolia Publishing, Orlando, Florida. ISBN: 0-941463-00-1.

Rovira, A.D., 1965. Interactions between plant roots and soil microorganisms. Ann. Rev. Microbiol., 19, pp. 241–266.

United States Department of Agriculture (USDA), 1995. Handbook of Constructed Wetlands. A Guide to Creating Wetlands for: Agricultural Wastewater, Domestic Wastewater, Coal Mine Drainage, Storm water in the Mid-Atlantic Region. Volume 1: General Considerations. USDA-Natural Resources Conservation Service/US EPA-Region III/Pennsylvania Department of Natural Resources. Washington, D.C.

Van der Putten, W., 1997. Die-back of Phragmites australis in European wetlands: An overview of the European research program on reed die - back and progression (1993 - 1994), Aquatic Botany, 59: pp. 263–275.

Vymazal, J., Brix, H., Cooper, P.F., Green, M.B., and Haberl, R., 1998. Constructed Wetlands for Wastewater Treatment in Europe. Backhuys Publishers, Leiden, The Netherlands.

HEAVY METALS CONCENTRATIONS IN AQUATIC ENVIRONMENT AND LIVING ORGANISMS IN THE DANUBE DELTA, ROMANIA

MIHAELA-IULIANA TUDOR, MARIAN TUDOR, CRISTINA DAVID, LILIANA TEODOROF, DANA TUDOR AND ORHAN IBRAM
Department of Limnology
Danube Delta National Institute for Research and Development
165 Babadag Str, Tulcea, Romania

*To whom correspondence should be addressed: mtudor@indd.tim.ro

Abstract. A major role in terms of danger to the aquatic ecosystem is played by heavy metals, which stand out by their persistence and accumulability. In this study an overview on recent monitoring (2003-2005) of some heavy metals (Cu, Zn, Cd, Pb) analyzed in fish species, plants and benthic macro invertebrates from Danube Delta Biosphere Reserve.In the system of contamination classes elaborated by (Wachs B. 2000), for the grading of the situation of ecosystems in term of heavy metal content, of all the various compartments, fish constitute the only critical part. The fish organs - in particular the muscles – show a very wide spread of concentration. Fish are very sensitive to cadmium and lead poisoning, but certain species, for example roach, perch and gibbel carp in aquatic ecosystem from Danube Delta was found with high concentration of cadmium and lead. The accumulation of contaminants in aquatic organisms can prove harmful to a number of fish species during sensitive stages within the lifecycle. Concentrations in aquatic plants was higher special in *Ceratophyllum demersum* Cd (1.89 mg/kg wet substance) and Pb (23,2 mg/kg w.s.).Variations of heavy metal concentrations from different sampling stations are discussed. Results indicate that the levels of lead and cadmium were highest concentration

Keywords: heavy metals; fish species; aquatic vegetation; macroinvertebrates; bioconcentration

1. Introduction

In the paper there are analyzed the heavy metals concentrations in aquatic environment and living organisms in the Danube Delta, during the last three years.

The effects of heavy metals on freshwater ecosystems heave been investigated by examining a number of organisms including fish, invertebrates, and aquatic macrophytes. Fish have been widely studied because of their economic and regression value.

Monitoring programs and research for metals in the environmental samples have become widely established because of concerns over accumulation and toxic effects, particularly in aquatic organisms and to humans consuming these organisms.

Factors known to influence metal concentrations and accumulation in aquatic organisms include metal bioavailability, season of sampling, hydrodynamics of the environmental, size, sex, changes in tissue composition and reproductive cycle (Boyden 1981).

Element concentrations (Cd, Pb) in fish muscle at the same location (lake) differ between different species and individuals due to species-specific capacity to regulate of accumulate trace metals (Otchere 2003). Different individuals in the same community at the same trophic level could accumulate heavy metals differently due to differences in habitat/niche's physical and chemical properties.

2. Methods

Previous investigations concerning freshwater contamination, aquatic macrophytes, inveretebrates and fish.

In the field there were 9 sampling points represented by fish collecting points Figure 1, so the fish samples were representative for and area of the Danube Delta Biosphere Reserve.

There are three categories of areas in the Danube Delta Biosphere Reserve: first are the core areas – strictly protected areas, were no humans activities are allowed, except research. The second category are the buffer areas, were are allowed traditional activities and the third category are the economic areas were economic activities are allowed, including commercial fishing.

In each sampling point were collected 6 adults individuals for each of sixteen fish species, monitored. For all the species were determined the concentrations of four heavy metals: zinc, copper, cadmium and lead.

Fish samples were chosen from different ages and weights to be analyzed along with samples of the aquatic plant, macroinvertebrate and lake water.

'Metal concentrations in fish were compared using guidelines established by the Romanian Council of Ministers of the Environment (Ordin No. 356/2001): Maximum limits for metals accumulated in fish tissues (fillet):

- Lead (Pb) 0.2mg/kg
- Cadmium (Cd) 0.05 mg/kg

Zinc and Cupper is not includes in this guidelines.

Mean concentrations were calculated (Figures 4, 5.) and compared to consumption values established for the protection of human health.

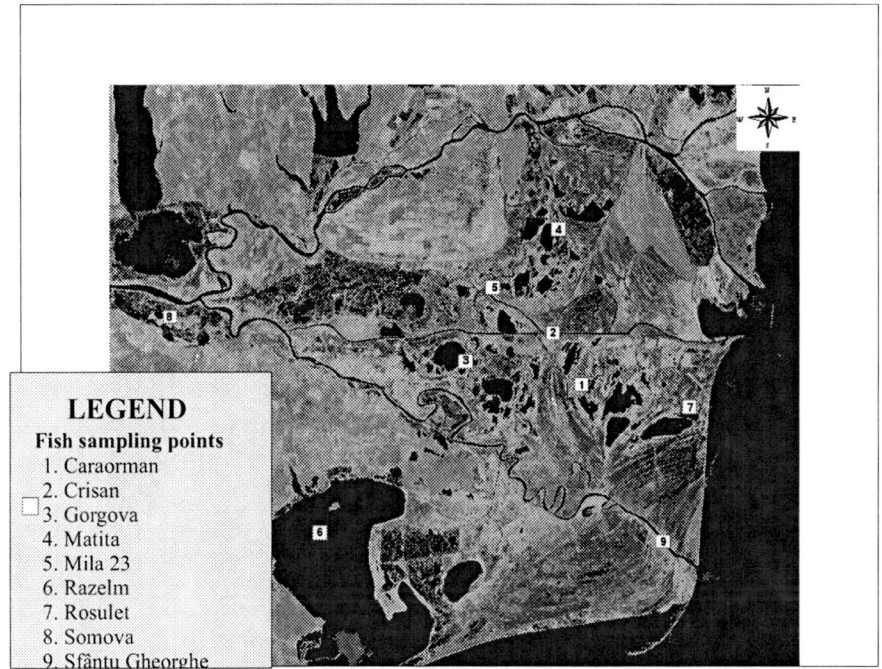

FIGURE 1. Sampling points collection in Danube Delta Biosphere Reserve.

3. Discussion and Conclusion

The result of this activity is the main purpose of the research due to the potential risk for human health represented by the heavy metals accumulated in fish. The concentrations of all 4 metals increased in macrophytes and fish.

The highest levels of Cd in muscle tissue were found in fish *Carassius auratus gibelio* (0.19mg/kg-0.06 mg/kg), *Perca fluviatilis* (0.20 mg/kg-0.08 mg/kg) and were high the upper limit recommended for human consumption (0.05 mg/kg) (Figure 2).

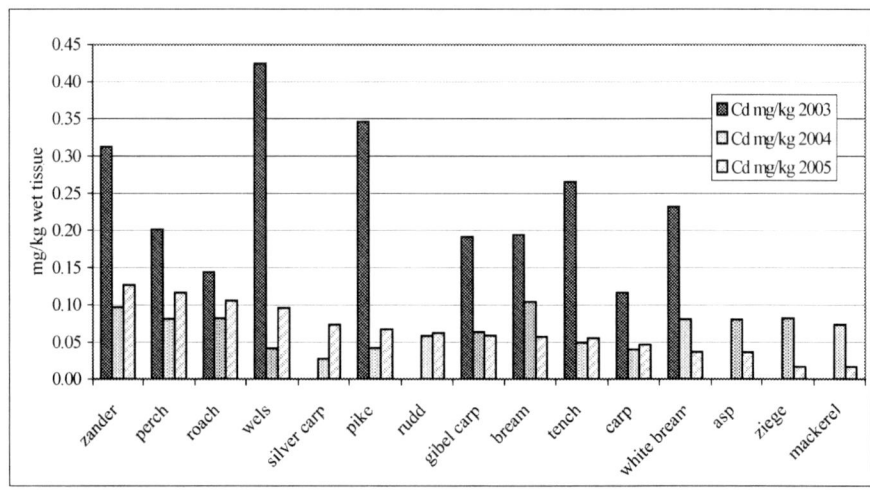

FIGURE 2. Content of cadmium in fish muscle (Danube Delta) – 2003-2005.

The highest lead concentrations were determined in pike (2,4 mg/kg w.t.), gibbel carp (2.6 mg/kg w.t.), perch (3,9 mg/kg w.t.) and roach (1.3 mg/kg w.t.), (Figure 3).

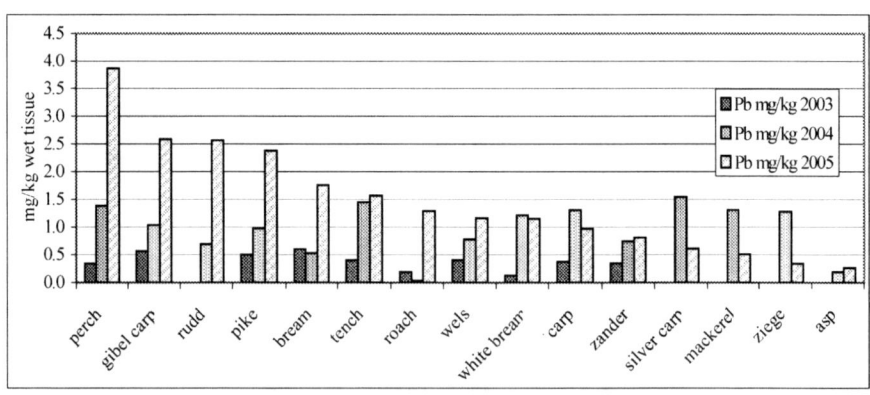

FIGURE 3. Content of lead in fish muscle (Danube Delta) – 2003-2005.

Levels of cadmium acumulation in the muscles of perch, zander, pike, caught in lakes 2003 was very high.

TABLE 1. A comparison of the bioconcentration factors for fish with those for the faunal species fish feed on.

bioconcentration factors (BCF)			
Heavy metal	Fish muscle (Wachs B. 2000) // RBDD* (constricted range)	zoobenthos(Wachs B. 2000) // RBDD (constricted range)	leech (Wachs B. 2000) // RBDD (constricted range)
Cd	50 -100 (-200) // 18-283	1,000-50,000 // 1,121-4,387	1,400 // 0,336
Cu	40 -100 (-200) // 19-381	800-17,000 // 2,810-2,934	1,700 // 2,330
Pb	30 –100 // 12-90	500-6,000 // 5,058-8,608	1,400 // 0,474
Zn	150- 400 (-700) // 56-155	700-16,000 // 9,662-27,202	6,000 // 72,187

(RBDD*-Constricted range of heavy metals analysed in Danube Delta Biosfere Reserve)

In the ecosistem of flowing waters suspended matter and the periphyton mass exhibit the highest of heavy metal concetration. A mean level of acumulation can be observed in the bentic fauna and in the aquatic macrophites, whereas the smallest concentration factors are found in fish muscle(Wachs B. 2000).

Based on the results of this study cadmium and lead concentration in whole fish in 2003 to 2005 are high.

The concentration factors listed in Table 1 serve to indicate the topological sequence of specific element enrichment in the muscles of fish: Zn>Cu>Pb>Cd = sequence of the heavy metal accumulation in the fish (muscles) in the Danube Delta lakes.

Sence fish in their natural environment do not accumulate the persistent contaminating group of heavy metals on the parth of biomagnification, they therefore exibit relatively low bioconcentration factors based on wet weight.

Biomagnification is a phenomenon which is to date not fully understood. It is imperative to understand biomagnification if the protection of organism of all trophic levels is to be considered. An understanding of the metal content of food source of higher trophic levels, and the environment in which they live, must be assessed (Timmermans 1992; Goodyear 1999).

Concentrations of heavy metals (Cd, Pb, Zn and Cu) were measured in three species of aquatic macrophytes collected from the lakes (Fortuna, Nebunu, Razelm, Sinoe, Somova) – *Ceratophyllum sp.*, *Myriophyllum sp.*, *Potamogeton sp.*, and aquatic macro-invertebrates –Chironomidae, Gammarus sp.

Analyses of water, and plant samples indicated that the lakes (Figures 4, 5) were polluted with Pb, Cd, and partly with Cu and Zn.

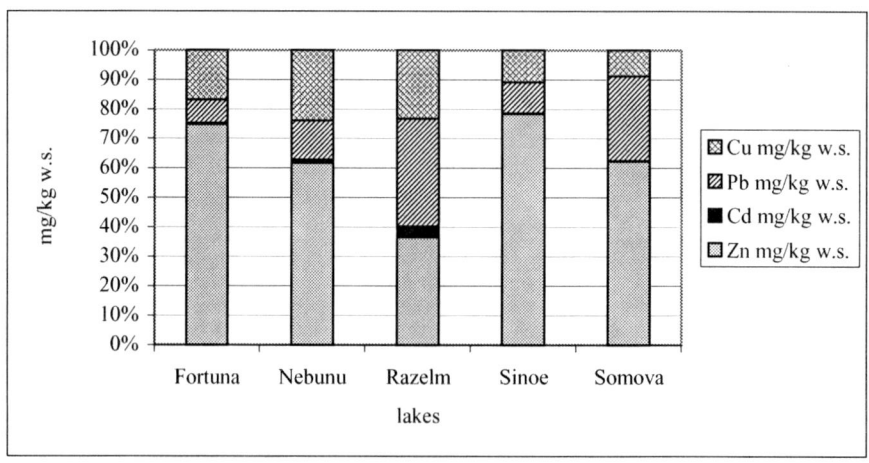

FIGURE 4. Metal concentration in aquatic macrophytes.

In macroinvertes and aquatic macrophytes, levels for toxic metals, such as lead and cadmium is more than in water. A generalised concentration gradient of metal concentrations in aquatic plants and macroinvertebrates follows thus: Cd<Pb<Cu<Zn.

As a result, relationships between environmental metal concentrations and those in organisms are not well understood.

In conclusion, the metals Cu, Zn in lakes water and fish are within permitted levels, except Pb and Cd which are more higher than maximum limits admitted established by the Romanian Council of Ministers of the Environment (Ordin No. 356/2001).

According to national guidelines, the high concentration of lead and cadmium in fish was find in fish species like: zander, perch and gibel carp.

In general, however, piscivorous fish show lower contents than non predatory fish; depending on the species of adult fish selected and compared.

The mean values reported for each species may be useful as indicators of metal concentrations in tissue of fish from Danube Delta lakes and be considered possible danger for human health.

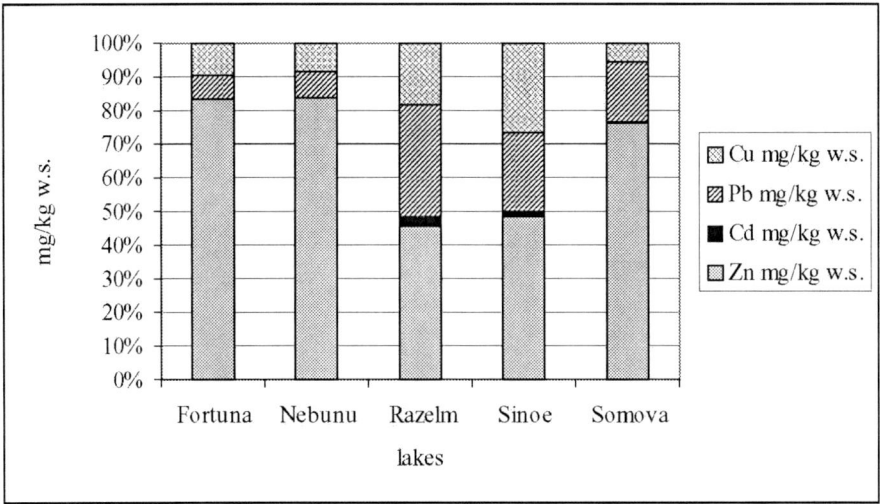

FIGURE 5. Concentration in aquatic macroinvertebrates.

Further investigation of more species, for more metals, and over greater concentration ranges are needed to develop the understanding of bioacumulation in fish in Danube Delta lakes.

ACKNOWLEDGEMENTS

The authors would like to thank to all the technical staff involved in collecting and preparing of the samples and to the commercial companies which allowed this research activities in their fishing areas.

References

Boyden, C.R., Phillips D.J.H., (1981). "Seasonal variation and inherent variability of trace elements in oysters and their implications for indicator studies." *Mar. Ecol. Prog.* **5**: 29–40.

Goodyear, K.L., McNeill S., (1999). "Bioaccumulation of heavy metals by aquatic macro-invertebrates of different feeding guilds: a review." *Science of the Total Environment* **229**: 1–19.

Maletin S., D.N. (1992). "Heavy metal content in fish from Backwater Tisza (Biser island)." *Tiscia* **26**: 25–28.

Otchere, F.A. (2003). "Heavy metals concentrations and burden in the bivalves (*Anadara* (*Senilia*) *senilis, Crassostea tulipa* and *Perna perna*) from lagoons in Ghana: Model to describe mechanism of accumulation/excretion." *African Journal of Biotechnology* **Vol. 2(9)**(1684-5315): 280–287.

Timmermans, K.R., Spijkerman E., Tonkes M., (1992). "Cadmium and zinc uptake by two species of aquatic invertebrate predators from dietary and aqueous sources." *Can J. Fish Aquatic Science* **49**: 655–662.

Wachs B. (2000). "Heavy metal content in Danubian fish." *Arch. Hydrobiol. Suppl.* **115/4**: 533–556.

DETERMINATION OF CYANIDES FROM DISTILLED ALCOHOLIC DRINKS

SEMAGHIUL BIRGHILA AND NALIANA LUPASCU*
Department of Chemistry, Ovidius Universiy, of Constanta
124 Mamaia Blvd., 900527, Constanta, Romania

*To whom correspondence should be addressed: nlupascu@univ-ovidius.ro

Abstract. Cyanides and hydrocyanic acid are one of the priority pollutants being most toxic. Therefore, it is very important to monitor cyanide concentration with specific and sensitive analytical methods. Some analytical methods for cyanide determination were presented in the last years: spectrophotometer, potentiometer with silver cyanide electrode and titrimetric method. The paper presents original results concerning the spectrometric method and argentometric titration utilization for the measurement of cyanide concentration in some distilled alcoholic drinks from plums, grapes and apricots. The obtained cyanide concentrations vary between 0.0162 and 0.0970 mg/100mL, being under the imposed limits.

Keywords: cyanides; hydrocyanic acid; titrimetric; spectrometric; alcoholic drinks.

1. Introduction

Cyanide is produced naturally in the environment by various bacteria, algae, fungi and numerous species of plants including beans (coffee, chickpeas and lima), fruits (seeds and pits of apple, cherry, pear, apricot, peach and plum), almond and cashew nuts, vegetables of the cabbage family, grains (alfalfa, and sorghum), roots (cassava, potato, radish and turnip), white clover and young bamboo shoots. Incomplete combustion during forest fires is believed to be a major environmental source of cyanide, and incomplete combustion of articles containing nylon produces cyanide through depolymerization.

Cyanide in aqueous matrices is usually measured by colorimetric, titrimetric, electrochemical methods or by headspace gas chromatography with a nitrogen-specific detector, after pretreatment to produce hydrogen cyanide and absorption in sodium hydroxide solution (Agrawal et al., 1991; Chinaka et al., 1998).

Fruits that have a pit, such as cherries or apricots, often contain small quantities of hydrogen cyanide in the pit (Okafor et al., 2005).

Hydrogen cyanide can be produced by hydrolytic reaction catalysed by one or more enzymes from the plants containing cyanogenic glycosides.

Apple and pear seeds and the inner stony pit of apricots and peaches contain a naturally occurring substance, amygdalin, is converted to glucose, benzaldehyde, and hydrogen cyanide.

Water + amygdalin → 2 glucose + cyanide + benzaldehyde

The purpose of this study was to determine the hydrogen cyanide in plum brandy disttiled using the titrimetric and colorimetric methods.

2. Experimental

The samples used in the research represent several natural plum brandy "țuica", vodka and palinca (Transylvanian apricot brand).

In the titrimetric method hydrogen cyanide is produced by hydrolytic reaction with ammonium hydroxid; the solution is titrated with standard silver nitrate in the presence of iodine potassium solution.

The excess of silver nitrate solution reaction with iodine potassium and a opalescence color of iodine silver indicate the end point of the titration.

$$NH_4CN + AgNO_3 \longrightarrow NH_4Ag(CN)_2 + NH_4NO_3$$

$$KI + AgNO_3 \longrightarrow AgI + HNO_3$$

In the colorimetric measurement the cyanide is converted to cyanogen chloride, by reaction with chloramine-T at a pH less than 8 without hydrolyzing to the cyanate. After the reaction is complete, color is formed on the addition of pyridine – pyrazolone reagent and the absorbance is read at 620 nm.

3. Results and Disscutions

The obtained results are presented in Table 1

TABLE 1. The hydrogen cyanide concentration in disstiled drinks.

Nr. Crt	Samples	Colorimetric method HCN (mg /100mL)	Titrometric method HCN (mg /100mL)
1	Plum brandy	0.0694	0,0772
2	Apple brandy	0.0121	0,0162
3	Grapes brandy	0.0734	0,0756
4	Apricot brandy	0.0929	0,0972
5	Vodka with apricot arome (Prodvinalco)	0.0599	0,0648
6	Transylvanians apricot brandy.("palinca" Rieni)	0.0611	0,0648
*			max. 0,2

* literature date

Fruits that have a pit, such as cherries or apricots contain a naturally occurring substance, amygdalin, is converted to glucose, benzaldehyde and hydrogen cyanide.

The paper presents original results concerning the spectrometric method and argentometric titration utilization for the measurement of cyanide concentration in some distilled alchoolyc drinks from plums, grapes and apricots.

It can be observed that the concentration of hydrogen cyanide in these samples is smaller than literature date for these samples.

References

Chinaka S., Takayama N., Michigami Y., Ueda K., (1998) Simultaneous determination of cyanide and thiocyanate in blood by ion chromatography with fluorescence and ultraviolet detection. *Journal of Chromatography B*, 713:353–359.

Agrawal V., Cherian L., Gupta V.K., (1991) Extraction spectrophotometric method for the determination of hydrogen cyanide in environmental samples using 4-aminosalicylic acid. *International Journal of Environmental Analytical Chemistry*, 45:235–244.

Okafor P.N., Nwogbo E., (2005), Determination of nitrate, nitrite, N-nitrosamines cyanide and ascorbic acid contents of fruit juices marjeted in Nigeria, *African Journal of Biotechnology, 4(10),* 1105–1108.

THE QUALITY OF THE DRINKING WATER – THE MAIN FACTOR INTO THE SOCIO-HYGIENIC MONITORING

GRIGORE FRIPTULEAC[1], SERGIU CEBANU[1]* AND VLADIMIR BERNIC[2]
[1]*State Medical and Pharmaceutical University*
165, Stefan cel Mare Blvd., 2048 Chisinau, Moldova
[2]*National Scientific and Practical Center of Preventive Medicine*
67 Gh. Asachi Str., 2028 Chisinau, Moldova

*To whom correspondence should be addressed: sergiucebanu@yahoo.com

Abstract. This study was performed to evaluate the drinking water quality used for drinking purposes by the population of Chisinau. There were present the comparative data of chemical composition of drinking water from aqueducts, the wells from Chisinau and suburbs as well. All quality indices from aqueduct in comparison with the water from rural sectors and town wells correspond to hygienic standards. High mineralization and high concentration of some chemical compounds are common for water from suburb and Chisinau wells.

Keywords: drinking water; chemical indices; bacteriological indices

1. Introduction

A serious problem in Republic of Moldova constitutes the drinking water supply, as the water sources are not uniformly distributed on the territory, and the quality of the water, mostly, does not correspond to the existing national standards. The expectation concerning the national economy development as well as the people's health in our country, in a large measure depends on the aquatic resource deficiency which permanently increases (Environment performances review, 1998).

The total country water resources available for various purposes are estimated at 7,2 km^3/year, including internal surface water resources (1,3 km^3), the groundwater (1,1 km^3) and the share of Moldova from transboundary rivers, Dniester river (2,5 km^3) and the Prut river (0,8 km^3) (Ropot, 2002).

2. Aim

The hygienic evaluation of the drinking water quality used for drinking purposes by the population of Chisinau.

3. Material and Methods

In the condition of sanitary–chemical laboratory and the bacteriologic one of National Scientific and Practical Center of Preventive Medicine and Municipal Center of Preventive Medicine Chisinau after the standardized methods there was accomplished the sample analysis of water from aqueducts, the well from Chisinau and suburbs as well. There were determined the following indicators: turbidity, ammonia, nitrite, nitrate, total dissolved solid, concentration of fluoride, hardness, content of calcium, magnesium, K+Na, chloride, sulfate, carbohydrate, iron, NTG, coliform bacterium.

4. Results

The majority of the population from Chisinau town use drinking water from aqueduct. In principle all quality indices from aqueduct are diminished and more favorable in comparison with the water from rural sectors and town wells and correspond to hygienic standards, depending on the majority-studied indices. A little part from population of Chisinau town use for drinking the water from wells, as seen from the existing facts in Table 1 don't differ from the drinking water from suburb's wells.

Especially such a water mineralization index as that of total dissolved solid corresponds to the hygienic normative being 600,2 mg/dm^3. The highest values are characteristic for the towns wells and village wells constitute respectively 1624,16 mg/dm^3 and 1544,8 mg/dm^3.

From total samples number from suburb villages 29 % exceed the hygienic normative of 1500 mg/dm^3 of content of total dissolved solid (Friptuleac, 2005). There is no correspondence to the medium value of sulfate in the wells from villages.

The same rule happens concerning such indices as chemical composition of water as the presence of magnesium (town-153,03 mg/dm^3, villages-150,7 mg/dm^3, aqueduct-62,3 mg/dm^3), chloride (respectively 161,1 mg/dm^3 and 138,8 mg/dm^3, aqueduct-68,9 mg/dm^3). As well for the wells from the town it is characteristic a more increased concentration of nitrates and nitrites, which constitute respectively 0,07 and 95,8 mg/dm^3, but in rural sector 0,05 and 93,4 mg/dm^3, aqueduct 0,03 and 10,32 mg/dm^3.

TABLE 1. The comparative data of chemical composition of drinking water from mun.Chisinau.

Indices	Town		Villages
	Aqueduct	wells	wells
	M±m	M±m	M±m
Turbidity	0,0056±0,0018	0,25±0,09	0,055±0,016
Ammonia, mg/dm^3	0,078±0,056	0,004±0,02	0,096±0,043
Nitrite, mg/dm^3	0,034±0,0089	0,066 ±0,01	0,047 ±0,016
Nitrate, mg/dm^3	10,32±0,82	95,79±11,64	93,04±9,95
Total dissolved solid, mg/dm^3	600,25±29,83	1624,16±100,24	1544,83±77,24
Fluoride, mg/dm^3	0,15±0,02	0,14±0,01	0,18±0,016
Hardness, mmol/dm^3	5,0±0,32	16,61±0,67	17,74±0,87
Calcium, mg/dm^3	50,21±4,94	85,37±5,64	104,33±5,98
Magnesium, mg/dm^3	62,25±6,74	153,03±6,85	150,74±7,26
Carbohydrate, mg/dm^3	322,68±22,55	466,81±16,4	532,08±20,04
Sulfate, mg/dm^3	229,08±24,89	698,23±64,05	703,89±47,48
Chloride, mg/dm^3	68,89±7,32	161,08±9,63	138,77±12,38
$\sum$ (Na+K), mg/dm^3	114,12±13,36	255,29±19,21	237,56±20,02
Iron, mg/dm^3	0,07±0,02	0,134±0,016	0,07±0,008

Another rule was observed concerning such indices as sulfates, carbohydrates and calcium.

The water samples taken from the wells examined which do not correspond to the existing rules to the microbiologic measures in the last years have increased essentially from 13,6 % in 2001 to 18,2 % in 2003. In the majority cases the observing are concerned to such microbiologic indices as coli index and NTG.

5. Conclusions

1. There is a need for a permanent monitoring of such indices as total dissolved solid, total hardness, nitrate, coli index, NTG.

2. The results of the investigation have the necessity of taking prophylactic measures in order to improve the system of water supply, which can be realized with common state actions and actions of everybody.

References

Environmental performances review. Republic of Moldova (1998). Economic Commission for Europe Committee on Environmental Policy. UN, New York and Geneva; pp.67–72.

Friptuleac Gr. (2005). Problemele de sănătate a populației în relație cu calitatea apei potabile. (The population health in relation with quality of drinking water). Mediul ambiant, nr.19; pp 23–25 (in Romanian).

Ropot, V. (2002) Resursele de apa, cantitatea, calitatea, utilizarea si protectia lor. (Resources of water, their quantity, utilization and protection). Materialele Seminarului din 20-23 noiembrie 2001. Chisinau, Editura "Universul"; pp. 16–31 (in Romanian).

DETECTION OF HEAVY METALS AND ORGANIC POLLUTANTS FROM BLACK SEA MARINE ORGANISMS

GABRIELA STANCIU AND *SIMONA LUPŞOR
*Chemistry Department, Ovidius University,
124 Mamaia Blvd., 900527 Constanta, Romania*

*To whom correspondence should be addressed: sgutaga@univ-ovidius.ro

Abstract. This work presents aspects regarding heavy metals (Cd, Pb, Cu, Zn, Cr, Mn, Ni and Hg) and organic pollutants in marine organisms collected from different zones of the Romanian Black Sea Coast. The heavy metals were analyzed using atomic absorption spectrometry. For the organic pollutants analysis was used a FISONS gas chromatograph.

Keywords: heavy metals; organic pollutants; marine organisms

1. Introduction

The sea holds vital natural resources and great commercial potential, both of which demand consideration for the marine environment.

Heavy metals, even in small quantities, are very toxic for life. The gravity of toxic effect depends on nature, metal concentration, body resistance and presence of other contaminants. So, the analytical determination of the heavy metals is a very important task for environment studies (SHAWI A.W.Al., DAHL R. and PABLOS ESPADA M. C. et al.).

Because of the large scale dilution of contaminants in the aquatic matrices, concentrations of many organic pollutants are below the detection limits of standard analytical and sampling methods. Thus, gas chromatography with specific detection methods such as electron capture detector and HPLC has been frequently used for analysis of pesticides and polycyclic aromatic hydrocarbons in water and biological samples.

This work presents aspects regarding the pollution with heavy metals (Cd, Pb, Cu Zn, Cr, Mn, Ni and Hg) and pesticides of mussels (*Mytilus galloprovincialis*) and small fishes (*Gobius platyrostris, Engraulis encramicholus*) collected from the Romanian Black Sea Coast.

2. Materials and Methods

Mussel's samples (adult exemplars) and small fishes were collected on the Romanian Black Sea Coast, during May 2005.

Pesticide residues were extracted from samples (total lipid extract from mussels tissue and small fishes tissue) with ether of oil and acetone and then were purified on fluorisil column with a layer of anhydrous Na_2SO_4. A total of 10 g fluorisil or aluminium oxide was packed in a glass column with ether petroleum. Pesticides were eluted from the column with ethyl ether / ether petroleum in the 20 mL fraction. The fraction was concentrated in KUDERNA-DANISH apparatus for concentrating to about 1 mL.

For the pesticides analysis was used a FISONS gas chromatograph equipped with an electron capture detection (ECD) and a capillary chromatograph column filled with mixture of silicone oils (QF-1, OV-11, XE-60) on chromosorb WHP.

Conditions: a 1µL aliquot of the extract was injected; column temperature 200 ^0C; injector temperature 210 ^0C; detector temperature 250 ^0C; carrier gas: nitrogen at a flow rate of 4 mL/min.

For to analyze the heavy metals concentrations, the fresh tissue samples were washed, hashed, dried at 105 ^0C and mineralized by wet digestion method (HNO_3 - H_2SO_4). About 0.5g of dried tissue sample was predigested in 2 mL 65% HNO_3 for 24 hours at room temperature, then 2 mL of 98% H_2SO_4 were added and the mixture was digested in a VELP DK6 heating digester. After cooling, solution was made up to 25 mL of deionizer water. All used reagents were of analytical reagent grade (Merck). The resultant solutions were analyzed with an atomic absorption spectrophotometer GBC-AVANTA (air / acetylene flame) for the analysis determine the heavy metals concentration: Cd ($\lambda = 228.8$ nm), Cr ($\lambda = 357.9$ nm), Mn ($\lambda = 279.5$ nm), Ni ($\lambda = 232$ nm), Cu ($\lambda = 324.7$ nm), Zn ($\lambda = 213.9$ nm) and Pb ($\lambda = 217$ nm). Hg ($\lambda = 253.7$ nm) and As ($\lambda = 193.7$ nm) were determined with hydrates generator. Two replicate analysis were done for each solution.

3. Results and Discussions

In Table 1 are presented the concentrations of pesticides in mussel tissue.

The data from Table 1 show an increase concentration of β HCH over admitted limit [12]. The others are below maximal limits, but their presence indicates a contamination of marine environmental with such organic pollutants.

The admitted limits for heavy metals on marine organisms are (*Ordinul MAP 352/2001,* MO 812/2001): 50 µg/g for Zn, 6 µg/g for Cu, 1µg/g for total Cr, 0.5µg/g for Mn, 1µg/g for Ni, 1µg/g for Pb, 0.1 µg/g for Cd and 0.5µg/g for Hg.

TABLE 1. Concentrations of organochloride pesticides in mussel tissue.

Pollutant	Concentration ppm	Concentration ppm	Maximal admitted limit	Retention time
α HCH	0.1591	0.2362	0.2	1.18
γ HCH	0.1243	0.2541	2	1.46
β HCH	0.4820	0.5400	0.1	1.85
δ HCH	0.1227	0.0131	-	2
op DDE	0.0759	0.0204	-	3.10
op DDD	0.0007	0.1241	-	4.69
pp DDD	0.0023	0.0602		6.19
pp DDT	0.0010	0.2041	1	7.34

HCH - hexachlor cyclohexane; DDE - dichlor-diphenyl-dichlor-ethene; DDD - dichlor-diphenyl-dichlor-ethan; DDT - dichlor-diphenyl-trichlor-ethan.

The heavy metals concentrations in the samples analyzed are presented in Fig. 1.

We remark from figures above the high concentrations of some heavy metals like Pb (1.12 µg/g) and Hg (0.95 µg/g) in mussels tissue and Pb (1.23 µg/g), Hg (0.80 µg/g) in small fishes tissue, which are over admitted limits. The other heavy metals are below maximal admitted limits (*Ordinul MAP 352/2001,* MO 812/2001; *Ordinul MAP 975/1998,* MO 268/1998).

μg/g in dry tissue samples

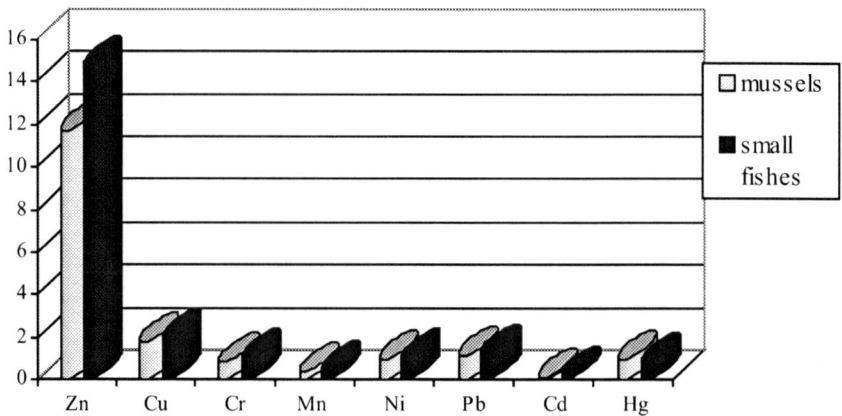

FIGURE 1. Concentration of heavy metals in mussel tissue and small fish's tissue.

4. Conclusions

Marine organisms retain small particles from sea water, so the presence of some pollutants in mussel and small fish's tissues indicate a contamination of marine environmental.

The concentration values of some pesticides and heavy metals are not so high, but their presence indicates a high degree of pollution and permits the identification of the principal contamination sources. We have detected a high concentration of β HCH, Pb and Hg over maximal admitted limits which represent an alarm signal for human and environmental health.

References

Ordinul MAP 352/2001, MO 812/2001;
Ordinul MAP 975/1998, MO 268/1998;
Pablos Espada M.C., Garrido Frenich A., Martinez Vidal J.L., Analytical Let., 34 (4), 602 (2001);
Shawi A.W.A., Dahl R., Anal. Chem. Acta, 391, 1, 35, (1999).

METHYLMERCURY IN SURFACE SEDIMENTS IN COASTAL ENVIRONMENTS. COMPARISON BETWEEN A DYNAMIC MACROTIDAL ESTUARY AND A MICROTIDAL LAGOON

TEODOR STOICHEV[1,2]*, DAVID AMOUROUX[1], OLIVIER F.X. DONARD[1] AND CHRISTO DAIEV[3]
[1]*Laboratoire de Chimie Analytique Bio-Inorganique et Environnement, UMR 5034, CNRS UPP, Avenue P. Angot 2, 64053 Pau, France*
[2]*Laboratory of Food Chemistry National Center of Public Health Protection, 15 Ivan Geshov Blvd, 1431 Sofia, Bulgaria*
[3]*Department of Analytical Chemistry, Sofia StateUniversity 1 James Bouchier Blvd, 1164 Sofia, Bulgaria*

*To whom correspondence should be addressed: tstoichevbg@yahoo.com

Abstract. The biogeochemical cycle of mercury is studied for coastal environments, particularly for a macrotidal dynamic estuary influenced by industrial and urban activities in its downstream part (Adour, France) and for a microtidal lagoon (Varna, Bulgaria) formerly polluted with mercury due to industrial activities. In both ecosystems the methylation in the sediments is connected directly to the bacterial activity and bioavailability of inorganic mercury. Possible higher methylation in the lagoon of Varna due to generally lower salinity is proposed. The results show the importance of the specific characteristics of each ecosystem for the fate of the mercury species.

Keywords: mercury; methylmercury; speciation; sediments; estuaries; lagoons

1. Introduction

The surface sediments are considered as the most probable place in which one part of the inorganic ionic mercury (Hg^{2+}) is converted to monomethylmercury ($MeHg^+$) subsequently bioaccumulated in the aquatic food chain (Stein et al., 1996). The ecological risk induced by this extremely toxic species should be especially high in coastal areas, since many wild animal species feed there. The study of the coastal environments is rather complicated due to the complex gradients of major physicochemical conditions, which gives to each ecosystem a unique character.

In this study we compare methylmercury concentrations in surface sediments for two contrasting coastal environments: a dynamic macrotidal estuary (Adour River, Southwest France) and a microtidal lagoon (Varna lake, Bulgarian Black Sea Coast).

2. Material and Methods

2.1. STUDY AREAS

The Adour estuary is situated in SW France (Atlantic Ocean). The tidal amplitude ranges between 2 and 5 m height. Depending on the tides and the water discharge the surface sediments may be influenced from denser ocean water masses with high salinity. Two sampling campaigns for surface sediments were undertaken on October 2000 and June 2001. More about the Adour estuary can be found in the literature (Stoichev et al., 2004a).

The lagoon of Varna is the deepest and biggest (by volume) lake on the Bulgarian Black Sea coast. The tidal amplitude is negligible. It is characterized by relatively low salinity, 11-12 psu on average, which decreases with the distance away from the sea. The water quality is seriously affected and the lagoon is euthrophicated. Surface sediments were taken on two occasions (November 1999 and October 2000). Additional information on the Varna lagoon can be found elsewhere (Stoichev et al., 2004b).

2.2. ANALYTICAL METHODS

The sampling and the analytical vessels were cleaned in 0.1% detergent RBS50 biocide solution and later in acid baths. After the sampling, the surface sediments were immediately frozen and in the laboratory they were freeze-dried and moulded. Hg^{2+} and $MeHg^+$ were determined after 6M HNO_3 extraction in open-focused microwave field. An on line ethylation and analysis by cryogenic trapping, gas chromatography and atomic fluorescence spectrometry was used for simultaneous mercury species determination in the sediments. The procedure for the speciation of mercury in sediment samples was described in details elsewhere (Tseng et al., 1998; Stoichev et al., 2004c). Another aliquot of the freeze-dried sample was used for organic carbon and total sulphur measurements in induction furnace with infrared detection (Abril et al., 1999).

3. Results and Discussion

The co-dependence of MeHg$^+$ concentration on both organic carbon and total sulphur is presented as a 3D plot on Figure 1 for surface sediments from the Adour estuary and the lagoon of Varna. For each studied ecosystem and irrespectively of the sampling campaign there is a well-defined maximum in the MeHg$^+$ concentrations at about 0.3% total sulphur and 3% organic carbon. At higher sulphur levels speciation of Hg^{2+} shifts towards negatively charged complexes, less available for uptake by methylating sulphate-reducing bacteria, thus decreasing the MeHg$^+$ formation (Benoit et al., 2001), which explains the observed relationship. Therefore, for both studied systems the most probable areas for mercury methylation in the surface sediments may be determined only by considering the organic carbon and total sulphur levels.

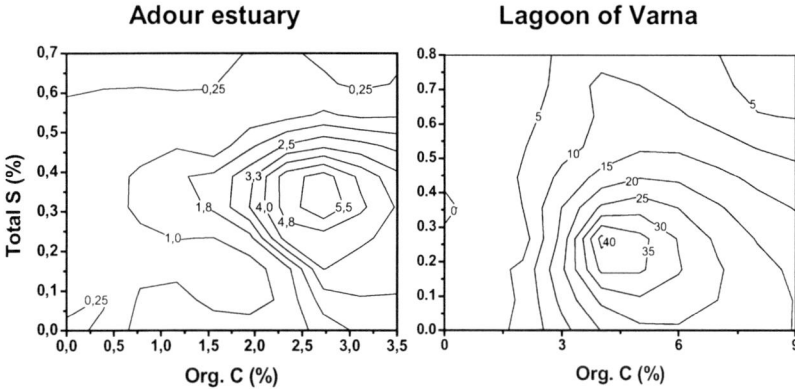

FIGURE 1. 3D contour plot of methylmercury concentrations (pmol g^{-1}) in surface sediments of the Adour estuary (October 2000 and June 2001, $n=48$) and the lagoon of Varna (November 1999 and October 2000, $n=32$) as a function of organic carbon and total sulphur content.

In spite of the similar behaviour of MeHg$^+$ there is, however, important difference between the studied coastal areas. The sediments from Varna have one order of magnitude higher concentration of MeHg$^+$ (see Figure 1), which is not related to any difference of the concentration of Hg^{2+}. As additional confirmation let us consider the percentage of MeHg$^+$ with respect to the total mercury for each studied ecosystem. For the Adour estuarine sediments the average %MeHg$^+$ is between 0.1% and 0.6% for June 2001 and October 2000, respectively. For the sediments from the lagoon of Varna much higher average %MeHg$^+$ is observed (3.3% for November 1999 and 1.8% for October 2000). Therefore, it is quite probable that specific geochemical conditions favor the methylation reactions in the lagoon of Varna. It is usually admitted that coastal lagoons with longer water residence time affect considerably the accumulation and transformation of the pollutants in the water column. However, it should not influence to a large extend the production of MeHg$^+$ in the sediments since steady state conditions for methylation can be developed within a few days (Compeau and Bartha, 1984). A more significant factor should be the intermediate salinity occurring in the lagoon. Together with the higher organic carbon concentration, the intermediate salinity in the water column should regulate sulphide concentration of the sediments under sulphate-reduction microbial activity. Therefore, the geochemical conditions developed in the surface sediments from the lagoon of Varna could be more favorable for the *in-situ* methylation of Hg^{2+} (Stein et al., 1996). Thus, the methylation in the sediments depends strongly on the specific characteristics of each ecosystem and should be studied separately for each case.

References

Abril, G., Etcheber, H., Le Hir, P., Bassoullet, P., Boutier, B., Frankignoulle, M. 1999 *Limnol. Oceanogr.* 44, 1304–1315.

Benoit, J.M., Gilmour, C.C., Mason, R.P. 2001 *Appl. Environ. Microbiol.* 67, 51–58.

Compeau, G., Bartha, R. 1984 *Appl. Environ. Microbiol.* 48, 1203–1207.

Tseng C.M., De Diego A., Pinaly H., Amouroux D., Donard O.F.X. 1998 *J. Anal. At. Spectrom.* 13, 755–764.

Stein E.D., Cohen Y., Winer A.M. 1996 *Crit. Rev. Environ. Sci. Technol.* 26, 1–43.

Stoichev T., Amouroux D., Wasserman J., Point D., De Diego A., Bareille G., Donard O.F.X. 2004a *Est. Coast. Shelf Sci.* 59, 511–521.

Stoichev T., Dushkin C., Daiev C., Amouroux D., Donard O.F.X. 2004b, *Annuaire Fac. Chimie, Université de Sofia*, 96, 83–90.

Stoichev T., Rodriguez Martin-Doimeadios R.C., Tessier E., Amouroux D., Donard O.F.X. 2004c, *Talanta*, 62, 433–438.

A QUICK AND EASY RETRIEVAL OF INFORMATION ON PESTICIDE TOXICITY TO HUMANS AND ENVIRONMENT: PESTIDOC

FRANCESCA VELLERE*, TERESA MAMMONE, ROMILDE BASLA AND CLAUDIO COLOSIO
*International Centre for Pesticides and Health Risk Prevention (ICPS), Azienda Ospedaliera L. Sacco- Polo Universitario,
74 Via G.B.Grassi, 20157 Milan, Italy*

*To whom correspondence should be addressed: ukvelle@tin.it

Abstract. The extensive use of pesticides in agriculture entails risks for human health and the environment, hence risk assessment and management of pesticide exposure is a key activity in any country, particularly in the developing world. Easy real-time access and retrieval of pesticide related information is essential in any preventive activity, and in any emergency situations. A documentation center on pesticides, "PESTIDOC", has been created and developed by ICPS, aimed at collecting and recording selected technical and scientific documentation on all the different fields of interest related to pesticides: human health, environmental health, and regulatory issues.

Keywords: pesticides; database; human and environmental health; risk assessment and risk management

1. Introduction

Pesticides are chemicals specifically manufactured to be toxic to target organisms, deliberately introduced in the environment in large amounts. The extensive use of pesticides in agriculture can entail risks for human health, for environment and for non-target organisms. Therefore, the problem of pesticide environmental contamination is a cause of growing concern both in the scientific community and among the large public.

Under the pressure of EU public bodies and consumers associations, many initiatives have been developed to increase the overall awareness about pesticides and their risk by Intergovernmental Organizations (World Health Organization -WHO, International Labour Organisation -ILO, United Nations Environment Programme -UNEP, US EPA Pesticide Factsheet), by Ministry of Health, Ministry of Agricultural Policies. Several toxicological databases are available online (EXTOXNET, IARC, PANpesticide), environmental databases (FADINAP), and legislation databases (European Union Law - EU-lex). However, this copiousness of knowledge is not always easily accessible because of its spreading among different sources and the variety and complexity of the scientific disciplines involved. This situation makes often difficult to find the specific information required and to interpret the available data.

2. The PESTIDOC Project

The International Centre for Pesticides and Health Risk Prevention (ICPS) promoted and developed the data bank PESTIDOC, in order to collect and record a selected technical and scientific documentation on pesticides, and to make a quickly available user-friendly information for different kinds of users (both specialized and non-specialized users at local, national and international level). The database has been developed through a multidisciplinary approach, by unifying the information spread in different data repositories and by collecting data on human health, environmental health, and regulatory issues.

PESTIDOC can be a useful tool for the healthcare scientists, researchers and public health managers for planning an investigation on toxic effects of pesticides and for planning survey campaigns. PESTIDOC can be useful also for technicians and agronomists because it collects all the national Plant Protection Products (PPPs) registered in the national market, including their labels. So, it is possible easily evaluate the human and environmental risks linked to the use of a specific PPP. The database can be also easily consulted by farmers and citizens because it provides the first aid procedures and the principles of healthcare, with information on "what to do" in case of accidental poisoning, contamination, ingestion.

3. The PESTIDOC Structure

PESTIDOC is a relational and user-friendly database able to provide complete and update information on all the different fields of interest related to pesticides.

The data have been collected and analysed using a Microsoft Access database; the web pages have been developed in ASP (Active Server Pages) system. The database has been created to make easily available the information for the different kinds of users. In order to achieve such a result, it has been devised a file-user interface "user-friendly", guiding the users by means of easy-access menus.

Data are stored and organized in the following main archives: Active Ingredients, Plant Protection Products (PPPs) authorized in Italy, Pesticide producers, Maximum Residue Limits (MRLs) in foodstuff, scientific literature. From each of these main archives, independent queries can be done to find complete information on the following topics: chemical-physical properties, European and Italian regulatory status, uses, toxicological classification, ecotoxicological characteristics, environmental fate; MRLs in food commodities, protocols for health surveillance and for monitoring of exposure to pesticides, list of PPPs authorized in Italy and labels, pesticide producers.

The active ingredients stored in the database are 1130, of which roughly 500 are authorized in Europe and in Italy. For each active ingredient one web page has been developed, containing all the relevant information.

The relational database allows the user to find information on the various issues on pesticides and commercial products starting from several sections, in order to get data needed for the human and environmental risk assessment.

4. Conclusions

PESTIDOC has been designed and created as a quick and easy retrieval of documentation on pesticides. The relational database allows the user to find information on the various issues on pesticides and commercial products, to get data needed for the human and environmental risk assessment. So, the database could provide operational guidance and problem solving in case of response and recovery from environmental, human contamination episodes. Furthermore the flexibility of the software allows the implementation of country-specific information and knowledge, such as pesticide commercial names, national MRLs, producers and so on. So the information currently spread in different data repositories could be unified under one ontology, managed by proper content management workflows and published for user-friendly query purposes.

The PESTIDOC will be soon available at the ICPS web site, at the following URL: www.icps.it.

References

ChemFinder (http://chemfinder.cambridgesoft.com/reference/chemfinder.asp)
Compendium of Pesticide Common Names (http://www.hclrss.demon.co.uk)
Environmental Protection Agency (US-EPA) (http://cfpub.epa.gov/oppref/rereg/status.cfm?-show=rereg)
EXTOXNET - The EXtension TOXicology NETwork (http://extoxnet.orst.edu/ghindex.html)
Gazzetta Ufficiale della Repubblica Italiana (http://www.gazzettaufficiale.it/)
IARC Monographs Programme on the Evaluation of Carcinogenic Risks to Humans (2001) (http://www.iarc.fr)
International Programme on Chemical Safety (IPCS) (http://www.inchem.org)
Italian Ministry Decree August 27th 2004. Limiti massimi di residui di sostanze attive dei prodotti fitosanitari tollerati nei prodotti destinati all'alimentazione.
Italian Ministry of Health database (http://www.ministerosalute.it/alimenti/sicurezza/fitosanitari/ricerca.jsp)
Maroni M., Colosio C., Fait A., Ferioli A., 2000, Biological monitoring of pesticide exposure: a review, Toxicology, Vol. 143, Issue n.1.
Muccinelli M., Prontuario dei fitofarmaci (VIII, X editions), Edagricole.
Official Journal of the European Union (http://europa.eu.int/eur-lex/)
Pesticide Action Network (http://www.pesticideinfo.org/Index.html)
PubMed (National Library of Medicine, Bethesda, MD, USA)
The WHO recommended classification of Pesticides by hazard and Guidelines to classification 2000-2002.
Tomlin C.D.S., The Pesticide Manual (XI, X, IX, VIII, X, XI editions), British Crop Protection Council.
Tordoir W.F., Maroni M., He F., 1994, Health Surveillance of pesticide workers. A manual for occupational health professionals, Elsevier, Vol. 91.
UNEP (United Nations Environment Programme) (http://www.unep.org)

STUDY OF BIOFILM FORMATION ON DIFFERENT PIPE MATERIALS IN A MODEL OF DRINKING WATER DISTRIBUTION SYSTEM AND ITS IMPACT ON MICROBIOLOGICAL WATER QUALITY

ZVEZDIMIRA TSVETANOVA
Water Resources Quality Problems Department
Institute of Water Problems
Bulgarian Academy of Sciences,
G. Bonchev Str, Bl. 1, 1113 Sofia, Bulgaria
E-mail: zvezdimira@yahoo.com

Abstract. The biofilm formation in drinking water distribution systems depends on many factors and may cause a number of technological and hygienic problems. In this study, the influence of pipe material and flow velocity on the biofilm growth dynamics and its impact on microbiological water quality in a model of drinking water distribution system were assessed.

Keywords: biofilm; drinking water distribution system; microbiological water quality

1. Introduction

The biofilm formation is a usual phenomenon closely related to the type of material because of its surface characteristics, such as roughness, surface energy, "biological affinity", etc. (Flemming, 1991). Water flow may influence on biofilm growth with its chemical and hydraulic characteristics.

Testing the potential of pipe materials to promote biofilm growth is needed because it may affect the taste, odor or turbidity of drinking water and may cause non compliance with microbiological water quality parameters and a risk to consumers' health. Because EU Drinking Water Directive (98/83/EC) controls the water quality at the consumers' tap instead of the point of supply, it is important to be assessed if the pipe material in drinking water supply systems release substances that may enhance or inhibit attached microbial growth.

It was determined that plastic materials PVC and polyethylene (PE) promoted biofilm growth better than mild steel (Schwartz et al., 2000). Similar density of fixed bacterial biomass on PVC and polyethylene was proved (Niquette et al., 2000), but the other study found that pipe material influenced biofilm activity far less than chlorine with biofilm growth ranked in the order glass < cement < PE < PVC (Hallam et al., 2001).

The *goal* of this study was to assess if the studied materials have influence on biofilm formation and impact of biofilm on microbiological water quality.

2. Material and Methods

Biofilm study was carried out in the drinking water supply system DWSS"Yovkovtsy", that supply with water the region of Veliko Tarnovo, and in the laboratory model of water distribution system. Some experiments were performed: 1) to determine microbiological composition of biofilm samples scraped from mild steel or reinforced concrete's main pipe line and concrete tank of DWSS"Yovkovtsy"; 2) to study biofilm formation process on test pipe from PVC, PE, stainless steel and carbon steel in a laboratory model of drinking water distribution system under flow velocity 0,006 cm/s. 3) to study dynamics of biofilm formation process on polypropylene, the pipe material used during the last years in Bulgaria, in a model water distribution system under flow velocities 0.3 m/s, 0.5 m/s, 0.7 m/s and 1 m/s.

Biofilm samples and microbial contamination of drinking water flowing through the model in result of biofilm formation were analyses with standard methods (APHA, 1992). Number of heterotrophic bacteria (R2A/22°C/7d) coliforms and bacteria from different physiological groups, including corrosion related bacteria were determined.

3. Results and Discussion

Biofilm samples collected from carbon steel and concrete pipes of the DWSS "Yovkovtsy" had different appearance, content of water, organic and corrosion products, while ones from concrete pipe and water tank's surfaces were the same. Biofilm community mainly consisted of heterotrophic bacteria and some fungi, actinomyces and autotrophic bacteria. Corrosion related bacteria were determined in some of the biofilm samples.

Comparison between heterotrophic bacteria count in biofilm samples from carbon steel and concrete pipes didn't show significant difference (P=95%) in bacterial density under residual chlorine content in water higher than 0.3 mg/dm^3. Below that value, the significant difference in bacterial density of the biofilms on surfaces from contrite, galvanized steel, mild steel and stainless steel was determined (Fig. 1).

Because of complex effect of pipe material, water quality and flow velocity on the biofilm in the studied water distribution system, the biofilm formation process was studied in a model system. It was determined the highest bacterial density of the biofilm developed on carbon steel and the lowest ones of the biofilms on plastic pipes (Fig. 2). The results showed strong influence of the pipe material on biofilm density during initial phases of the process compared with mature biofilm.

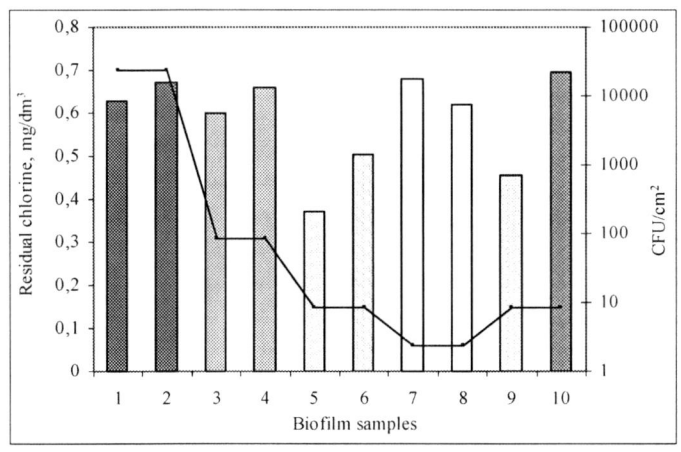

FIGURE 1. Bacterial density of biofilms developed on main pipe lines from carbon steel (1,2) or reinforced concrete (3,4) and concrete tank (5,6) in drinking water distribution system "Yovkovtsy", domestic installation from galvanized steel (7,8) and test surfaces from stainless (9) or carbon steel (10).

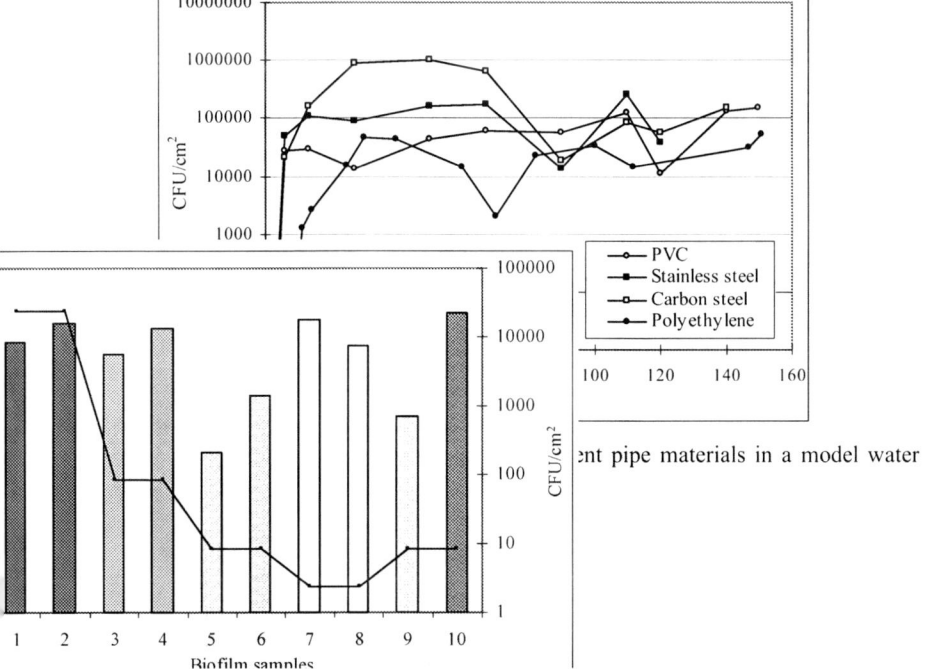

Results support the view that surface characteristics of materials had great importance for biofilm formation, because there was no evidence for enhancement of biofilm growth by organic substances released from plastic pipes. Contribution of substances from the plastic materials to biofilm growth has to be determined in waters with very low concentration of organic matter.

The data at Table 1. demonstrate that the biofilms formed on the studied materials can deteriorate microbiological water quality. It was determined that the microbial contamination of water depends on the bacterial density of the biofilms.

TABLE 1. Microbial contamination of drinking water in a model water distribution systems from studied pipe materials in result from biofilm impact.

№	Age of biofilm, d	Heterothrophic bacteria count in inlet water, $N \pm \Delta N$, CFU/cm^3	Heterothrophic bacteria count ($N \pm \Delta N$, CFU/cm^3) in outlet water from		
			PVC pipe	carbon steel pipe	stainless steel pipe
1	6	1440 ± 249	1090 ± 176	3960 ± 1220	1820 ± 710
2	13	390 ± 62	700 ± 224	3050 ± 263	1290 ± 125
3	27	780 ± 380	800 ± 286	380 ± 162	2200 ± 131
4	50	1060 ± 158	840 ± 143	1080 ± 208	3940 ± 550
5	67	1130 ± 163	1480 ± 62	1400 ± 300	1040 ± 149

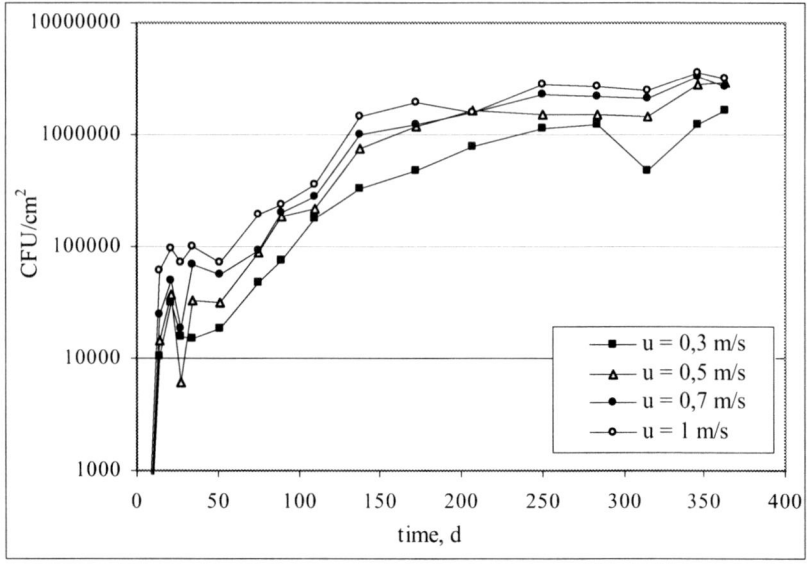

FIGURE 3. Dynamics of biofilm formation on polypropylene in a model drinking water supply system under flow velocities 0.3 m/s; 0.5 m/s; 0.7 m/s and 1 m/s.

The data presented on Fig. 3 showed fast bacterial colonization of the polypropylene pipe surfaces under high flow velocities when residual chlorine in water tended to zero. It was found out that initial bacterial attachment and biofilm density were direct depending on flow velocity in a model.

Comparing number of bacteria in the drinking water samples from inlet and outlet of the model system (Fig. 4) it was found higher microbial contamination of water under lower flow velocities 0.3 m/s or 0.5 m/s during initial 140 days of biofilm formation process.

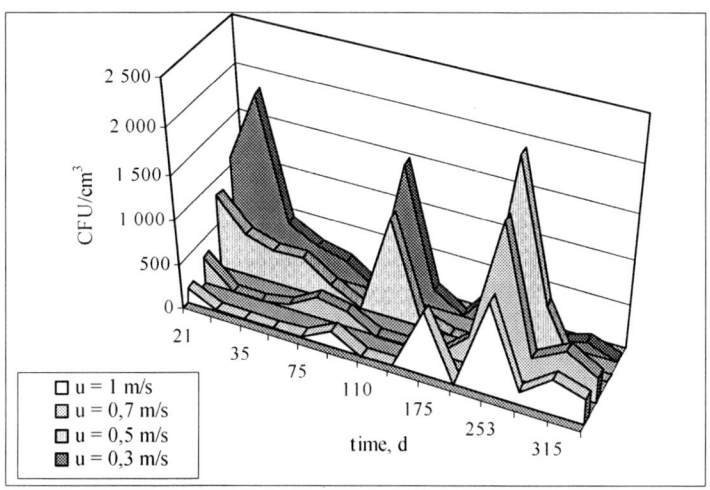

FIGURE 4. Microbiological contamination of drinking water in a model distribution system under flow velocities 0.3m/s; 0.5 m/s; 0.7 m/s and 1 m/s.

It can be supposed that biofilms grew under higher flow velocities were more rigid and strongly attached because of shear stress and as a result bacteria detachment was reduced. Increasing impact of mature biofilms under higher flow velocities 0.7 m/s or 1 m/s found 175 day after the beginning of the process can be related to the critical value of biofilm thickness reached to and shear stress effect on the biofilms' surface layer.

4. Conclusions

1. It was determined that microbial density of the biofilms on the equipment of drinking water supply system "Yovkovtsy" didn't depend on material type under high chlorine residual content in water.

2. Pipe material and flow velocities had influence on the biofilm growth, especially during initial phases of the process.
3. Biofilms formed on different materials and under different flow velocities in the model water distribution system made an impact on microbiological quality of drinking water depending on their bacterial density.

References

APHA. Standard Methods for the Examination of Water and Waste Water. 18th edition, 1992.

Flemming, H.-C., Biofilme und Wassertechnologie, I:Entstehung, Aufbau, Zusammensetzung und Eigenschaf ten von Biofilmen, Gas Wasser Fach., 1991, 132, №4, pp. 197–207.

Hallam, N.B., J.R. West, C.F. Foster, J. Simms, The Potential for Biofilm Growth in Water Distribution Systems, Wat. Res., 2001, Vol. 35, № 17, pp. 4063–4071.

Niquette, P., P. Servais, R. Savoir, Impacts of Pipe Materials on Densities of Fixed Bacterial Biomass in a Drinking Water Distribution Systems of Brussels, Wat.Res., 2000, Vol. 34, №6, pp. 1952–1956.

Schwartz, Th., S.Hoffmann, U. Obst, Formation and Bacterial Composition of Young, Natural Biofilms Obtained from Public Bank-Filtered Drinking Water System, Wat.Res., 2000, Vol. 32, № 5, pp. 1495–1502.

MODELING THE TRANSPORT AND FATE OF CONTAMINANTS IN THE ENVIRONMENT: SOIL, WATER AND AIR

MARIA DE LURDES DINIS* AND ANTÓNIO FIÚZA
*Geo-Environment and Resources Research Center (CIGAR),
Engineering Faculty, University of Porto,
Rua Dr. Roberto Frias, 4465-024, Porto, Portugal,*

*To whom correspondence should be addressed: mldinis@fe.up.pt

Abstract. The environmental effects originated by uranium mining activities result mainly from the wastes generated by the ore processing. Large quantities of radioactive wastes are generated in this extractive process requiring a safe management. Besides the radioactivity these wastes may also hold different amounts of chemicals used in the extraction process, toxic pollutants associated with the mineralization and precipitates provoked by pH or Eh alterations.

The main concern of waste management and long term stabilization is to confine the residues in order to reduce the dispersion of contaminants to concentrations that not exceed the trigger values considered to be safe: there is thus a need to ensure that the environmental and health risk from these materials are reduced to an acceptable level. However, the confinement will always represent a potential source of environmental contamination to the air, soil, superficial water and groundwater, due to the contaminants release and transport in the environment, which may occur by natural erosion agents like rainfall or wind.

Keywords: waste, disposal, release, dispersion, risk, environment.

1. Introduction

Uranium mining activities generates large volume of wastes composed by overburden, waste rock and tailings. Generally, these materials are deposited in waste rocks piles or dumps. Uranium mill tailings are the solid residues resulting from the leaching of the ore.

The Environmental effects resulting from the uranium mining activities are mainly derived from the wastes generated by the ore processing. Wastes constituents of concern include radionuclides (uranium, radium, radon and thorium), arsenic, copper, selenium, vanadium, molybdenum, heavy metals and dissolved solids. The radionuclides in the tailings are more mobile and chemically reactive than in the original ore and may enter into the environment becoming a contamination source to the soil, air, superficial water and groundwater.

When modeling contaminant release and transport mechanisms for each environmental compartment, the major output chosen is the contaminant concentration in each exposition point selected (for instance, breathing air zone, superficial soil and well water). This will allow the assessment of doses, if an additional exposure scenery is created. The objective is to quantify the potential exposure levels of the hazard at the receptor location answering to these main questions: what is the contaminant concentration in the exposition point? How much of a contaminant do people inhale or ingest during a specific period of time? Is the situation acceptable?

This work proposes a generic exposure model that incorporates simultaneously an atmospheric and a hydrologic transport model. In the atmospheric transport a two-dimensional model is used for calculating the flux diffusion from a radioactive waste disposal, having as result the hazard concentration at a defined distance from the soil (the breathing or mixing height) which will be the starting point for the dispersion each can be considered either simultaneously in each wind octant direction or considering only a prevailing wind direction. For the hydrologic transport a two-direction model is proposed for simulating the contaminants release form the waste disposal and its migration process through the soil to the groundwater. The final result is the contaminant concentration in the groundwater as function of the elapsed time, at a defined distance from the waste disposal, generally the location where the exposition point is considered being represented by an hypothetical well.

2. Methods and Results

2.1. ATMOSPHERIC TRANSPORT MODEL

For the gaseous contaminants the release mechanism from the soil is generally based in the principles of diffusion across a porous medium. Basic diffusion equations are used for estimating the theoretical values of the gaseous flux from the waste material. The generic diffusion equation can be represented by a one-dimensional steady-state equation:

$$D\frac{\partial^2 C}{\partial x^2} - \lambda C + \frac{R\rho\lambda E}{\varepsilon} = 0 \qquad (1)$$

In this equation D ($m^2.s^{-1}$) represents the radon diffusivity, λ the radioactive decay constant (s^{-1}), C ($Bq.m^{-3}$) the radon concentration in the pore space, R ($Bq.kg^{-1}$) the radium concentration in the material, ρ ($kg.m^{-3}$) the bulk density of the dry material, E (dimensionless) the radon emanation power coefficient for the pore spaces, ε (dimensionless) the total porosity and θ (dimensionless) the moisture. The solution of the diffusion equation for an homogeneous medium represents the flux release from the waste material to the surface, J_t ($Bq.m^{-2}.s^{-1}$). For a system without cover we obtain (Rogers, 1984):

$$J_t = R\rho E \sqrt{\lambda D_t} \tanh\left(\sqrt{\frac{\lambda}{D_t}} x_t\right) \quad (2)$$

The contaminant concentration released is estimated by a box model formulation which has implicit a mass balance formulation. The box volume (V) is defined by its length (L), width (W) and the mixing height (h). As a consequence of a steady state assumption, we have that the pollutant concentration (C) is constant in time, the mass flow rate entering (ϕA) into the box is equal to the flow rate leaving the box (uSC):

$$V \frac{dC}{dt} = \phi A - uSC \quad (3)$$

The atmospheric dispersion is modeled by a modified Gaussian plume equation which estimates the average dispersion of the contaminants released from the source in each wind direction. The Gaussian model of a plume dispersion accounts for the gaseous contaminant transport from the source area to a downwind receptor and is represented by the equation of Pasquill as modified by Gifford (Chacki, 2000):

$$C = \frac{Q}{2\pi\mu\sigma_y\sigma_z} e^{\left[-1/2\left(\frac{y}{\sigma_y}\right)^2\right]} \left\{ e^{\left[-\frac{1}{2}\left(\frac{z-H}{\sigma_z}\right)^2\right]} + e^{\left[-\frac{1}{2}\left(\frac{z+H}{\sigma_z}\right)^2\right]} \right\} \quad (4)$$

This equation represents a Gaussian distribution, where C ($Bq.m^{-3}$) represents the radionuclide concentration, Q ($Bq.s^{-1}$) the source strength, and H (m) the corrected source released height. Dispersion parameters, σ_y (m) and σ_z (m), are the standard deviations of the plume concentration in the horizontal and vertical directions respectively. The atmospheric transport is done at wind-speed (height-independent), u ($m.s^{-1}$), to a sampling position located at surface elevation, z (m), and transverse horizontal distance, y (m), from the plume centre.

2.2. HYDROLOGIC TRANSPORT MODEL

A leaching model based on a sorption-desorption process is used for describing the contaminant release from the waste disposal. The leachate concentration, C_L (Bq.m^{-3}), is determined by the a distribution or a partition coefficient, K_d (cm^3.g^{-1}) which describes the relative transport speed of the contaminant to the water existing in the pores; soil properties such as bulk density, ρ (g.cm^{-3}), and water content, θ, affect the extent of contamination, described by the contaminated zone thickness, x_t (m), area, A (m^2) and the amount of contaminant in the source, I_t (Bq), (EPA 1996; Hung 2000):

$$C_L = I_t / [A(x_t \theta + x_t \rho K_d)] \quad (5)$$

The transport for the dissolved contaminants is considered to occur either in the vertical direction through the unsaturated zone until an aquifer is reached either in the horizontal direction, through the saturated zone flowing to an hypothetical well, where the contaminants become accessible to humans or other forms of life. The vertical flow is considered to be one-dimensional. It is assumed that there is retardation during the vertical transport that is estimated assuming that the adsorption-desorption process can be represented by a linear isotherm, which means that there is a linear relationship between the radionuclides concentration in the solid and liquid phases.

Movement and fate of radionuclides in groundwater follow the transport components represented by the basic diffusion/dispersion–advection equation. The following expression describes the basic equation for the advective and dispersive transport with radioactive decay and retardation for the radionuclide transport in the groundwater:

$$D\frac{\partial^2 C}{\partial x^2} - V\frac{\partial C}{\partial x} - R\frac{\partial C}{\partial t} - \lambda RC = 0 \quad (6)$$

In this equation, D represents the molecular diffusion coefficient (L^2.T^{-1}), C represents the solute concentration (M.L^{-3}), V represents the interstitial velocity (L.T^{-1}), R represents the retardation factor due the sorption phenomena. The component dispersive and diffusive is represented by $\partial^2 C/\partial x^2$, the advection is represented by V $\partial C/\partial x$ and the concentration gradient is represented by $\partial C/\partial x$.

In the analytical solution the term for the contaminant concentration, C, was replaced by the rate of radionuclide transport, Q'(t), (Hung 1986):

$$Q'(t) = Q_0 \left(t - \frac{RL}{V} \right) e^{[-(\lambda_d R L)/V]} \quad (7)$$

The rate of radionuclide transport to the well point is represented by $Q'(t)$ (Bq.yr^{-1}), the rate of radionuclides transport at $x = 0$ is represented by Q_0, (Bq.yr^{-1}), t (yr) is the elapsed time and L (m) the distance from the disposal site to the well point.

2.3. A CASE STUDY

The model was applied to a specific case of contaminants from a uranium waste disposal. The contaminants of main concern were considered to be radon in air and radium in groundwater. The exposition point for the atmospheric model is a receptor located about 2 km from the source and for the hydrological model the exposition point is a hypothetical well located about 100 meters from the source.

The final output for the atmospheric model is the radon concentration at a defined distance from the source, in each wind direction and in the dominant wind direction, where is considered to be located the receptor. For the hydrological model the final outputs are the radium concentration in the well water and the corresponding cumulative rate of radium transported to the well after the time considered. Local meteorological data, namely wind velocity and frequency, was used for simulating the dispersion in each octant direction. These data was obtained from a local automatic meteorological station (INAG 2004). The dominant wind direction is NW. The unknown parameters were estimated from available data.

The contaminated site is composed by $1,6 \times 10^6$ ton of two different kind of wastes with a total area of approximately 75000 m^2. The total waste volume is about $1,5 \times 10^3$ m^3. It was assumed in the simulation that there is no covering system although there is some natural vegetation partially covering the site.

The air breathing tallness, or mixing height, was defined as 1,0 m. The radon concentrations were calculated in each sector at this height, taking into account the average wind speed and respective frequency of blowing. The concentration at the breathing height in the dominant wind direction refers only to the boundary side that limits the respective sector. The dispersion results can be seen in the figures below.

Local hydro-geological conditions were considered for each zone where the radionuclides transport occurs, namely for the contaminated zone, for the unsaturated zone and for the saturated zone, using different densities, porosities, hydraulic conductivities, radionuclide distribution coefficients and thicknesses. The well is located at the down-gradient edge of the contamination source.

Meteorological data, precipitation and evaporation from the same automatic meteorological station were used for estimate the infiltrating water rate into the contaminated zone. Figures 4 and 5 show the results obtained for radium in groundwater.

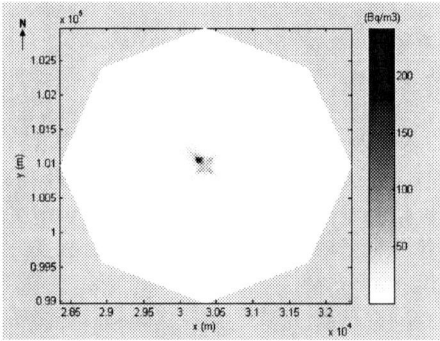

FIGURE 1. Radon dispersion in each wind direction, Bq.m^{-3}.

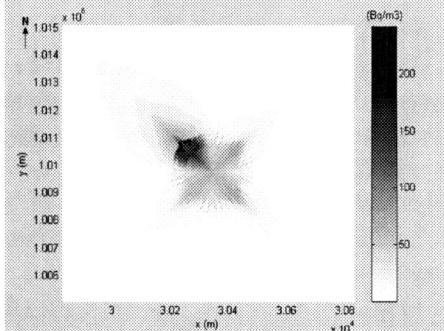

FIGURE 2. Radon dispersion in each wind direction, Bq.m^{-3}.

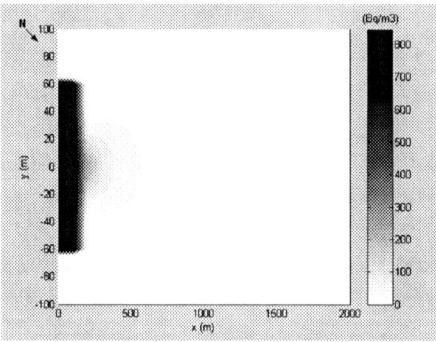

FIGURE 3. Radon dispersion, dominant wind direction, Bq.m^{-3}.

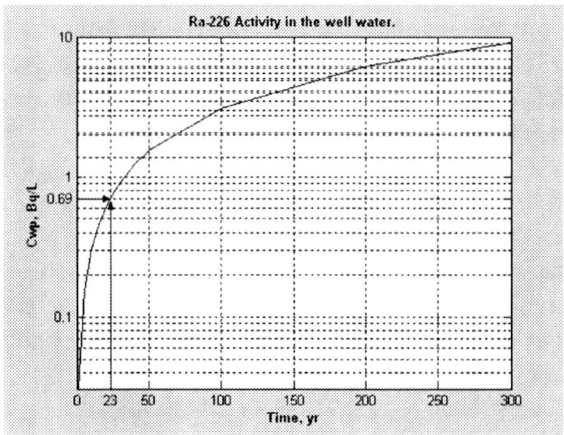

FIGURE 4. Radium activity concentration in the well water, Bq/L.

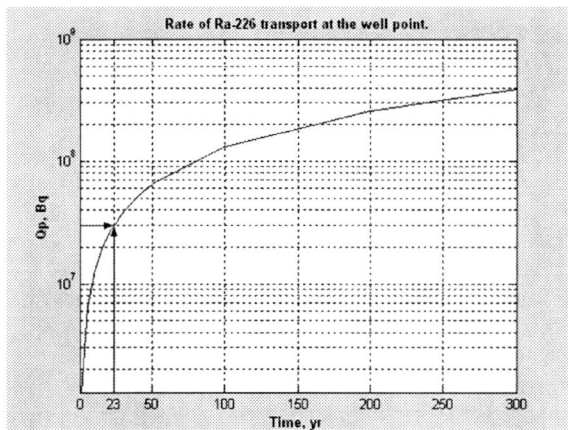

FIGURE 5. Cumulative rate of radium release, Bq.

3. Conclusions

Contaminants may travel through the atmosphere, soil, groundwater and superficial waters affecting the organisms that inhabit these media. The exposure modeling quantifies the impacts of contaminants as they travel through more than one of these compartments.

The atmospheric model describes the contaminant transport from the source area to a downwind receptor and the corresponding simulation quantifies the contaminant dispersion concentration in each wind direction.

The hydrologic model quantifies the movement of subsurface water and provides inputs to contaminant transport models. Its usage as a simulation tool allows previewing the contaminant behaviour in the groundwater al well as a quantitative assessment for the concentration of the contamination at a particular exposition point.

The model limitations are mostly related with the application of an analytical solution which is limited by its specific form of boundary conditions. The boundary conditions are stipulated in order to solve the unknowns in the problem domain. This will generate some error which magnitude will depend on the conformity between the local conditions and those stipulated in the model. The forecast accuracy of models are often compromised since some key parameters are imperfectly known and may have to be estimated from literature references in the absence of actual site specific measurements.

Many soil specific parameters show a great variability both in space and in time. Also these parameters will vary over the year due to climatic changing. This will generate some difficulties in obtain the most appropriate data needed to characterize the contaminated site.

References

Chacki S. and Parks B., (2000), Update's User's Guide for Cap88-PC, Version 2.0. U. S. Environmental Protection Agency, EPA 402-R-00-004.

EPA (1996), Documenting Ground-Water Modeling at Sites with Radioactive Substances. U.S. Environmental Protection Agency, 9355.0-60, EPA 402-R-96-003, PB96-963302, Washington, D.C.

Hung C.Y. (1986), An Optimum Groundwater Transport Model for Application to the Assessment of Health Effects Due to Land Disposal of Radioactive Wastes. Proceedings of Nuclear and Chemical Wastes Management, Vol. 6, pp. 41–50.

Hung C.Y. (2000), User's Guide for PRESTO-EPA-CPG/POP Operation System. U.S. Environmental Protection Agency, Version 4.2, EPA 402-R-00-007, Washington, D.C.

INAG (2004), http://snirh.inag.pt, Water Resources Database, data extracted from the Mesquitela automatic monitoring station.

Rogers V.C. and Nielson K. (1984), Radon Attenuation Handbook for Uranium Mill Tailings Cover Design. U.S. Nuclear Regulatory Commission, NUREG/CR-3533.

EQUILIBRIUM STUDIES OF Pb(II), Zn(II) AND Cd(II) IONS ONTO GRANULAR ACTIVATED CARBON AND NATURAL ZEOLITE

MIRKO MARINKOVSKI*, LILJANA MARKOVSKA AND VERA MESHKO
Faculty of Technology and Metallurgy
Sts. Cyril & Methodiu University
16 Ruger Boskovic Str., Skopje, Republic of Macedonia

*To whom correspondence should be addressed: mirko@ian.tmf.ukim.edu.mk.

Abstract. Many toxic heavy metals have been discharged into the environment as industrial wastes, causing serious soil and water pollution. Zn(II),Cd(II), Pb(II) are especially common metals that tend to accumulate in organisms, causing numerous diseases and disorders. The adsorption of Pb(II), Zn(II) and Cd(II) from single (non-competitive) aqueous systems onto granular activated carbon and natural zeolite has been studied as single equilibrium isotherms. The effectiveness of each adsorbent was measured in terms of its adsorption capacity towards individual constituents of the effluent. All the adsorption isotherms display a non-linear dependence on the equilibrium concentration. The equilibrium of a solution between liquid and solid phases is described by Langmuir, Freundlich, Langmuir-Freundlich and Redlich-Peterson model. The parameters in the adsorption isotherms were estimated from the experimental equilibrium data using non-linear regression software. From the Langmuir isotherms, maximum adsorption capacities of the NZ and GAC towards Pb(II), Zn(II) and Cd(II) were determined.

Keywords: equilibrium; adsorption; natural zeolite; activated carbon; heavy metals

1. Introduction

Heavy metals such as lead, cadmium and zinc are very toxic elements and their discharge into receiving water causes detrimental effects on human health and environment. Elevated levels of zinc may come from a variety of sources, such as effluents from manufacturing of batteries, pharmaceutical and agricultural chemicals, (Kandah, 2001). These toxic metals can cause accumulative poisoning, cancer, brain damages, when they are found above the tolerance levels. According to some surveys from the public health services of different countries, significant numbers of people have been exposed to the hazards of excess metals in the municipal water supplies. Therefore, different methods have been used to remove heavy metals from aqueous solutions such as chemical precipitation, membrane processes, ion exchange, adsorption, etc (Ricordel S. et al., 2001).

In advanced countries, removal of heavy metals in wastewater is normally achieved by advanced technologies such as precipitation-filtration, ion exchange and membrane separation (Katsumata et al., 2003). However, in development countries, these treatments cannot be applied because of technical level and insufficient funds. Therefore, the simple and economical removal method which can utilize in developing countries has to be established. Although the treatment cost for precipitation- filtration method is comparatively cheap, the treatment procedure is complicated. Adsorption has been found to be superior to other techniques for water re-use in terms of initial cost, simplicity of design, easy of operation and insensitivity to toxic substances. Activated carbon is the most popular adsorbent and has been used with great success, but is expensive (Mohan and Singh, 2002). The need of low cost replacement for activated carbon initialized a number of studies (Bailey et al., 1999). Heavy metals removal potential of some unconventional low cost as chitosan in prawn shell (Chu, 2002), red mud and fly ashes (Apak et al., 1998), sheep manure waste (Kandah, 2001), montmorillonite (Barbier et al., 2000), bagasse pith (Mohan et al., 2002), apple residue (Lee et al., 1998), peat (Allen et al., 1992), bentonite (Pradas et al., 1994), chelating resin (Lin et al., 2000), zeolite NaX (Barros et al., 2004), etc. The Langmuir (Apak et al., 1998; Chu, 2002; Mohan et al., 2002), Freundlich (Mohan et al., 2002, Lin et al.), Langmuir-Freundlich (Meshko et al., 2004) and Redlich-Peterson (Meshko et al., 2004) isotherms are usually applied to describe equilibrium of adsorption.

The chemical and physical characteristics of the sorbent materials as well as those of the heavy metals vary widely and it is difficult to recommend specific low-cost materials for specific pollutants. There is a necessity of experimental work with different materials and different heavy metals in order to understand the variations in the sorption phenomena. This paper presents the results of comparative equilibrium studies performed using activated carbon and natural zeolite. Three heavy metals, Zn(II), Cd(II) and Pb(II), from aqueous solutions as adsorbats were used.

2. Experimental

2.1. MATERIALS AND METHODS

Adsorbate characteristics. The adsorbates are hydrate metals (Zn, Cd, Pb) salt with determinate concentration of metal ion in solution. The adsorbates are supplied from "Topilnica-Veles"- metallurgical industry in Macedonia. The initial concentration for Zn and Pb are from 0-250 mg/dm^3, and for the Cd are from 900-1400 mg/dm^3, because the real wastewater from the metallurgical industry is polluted in this range of concentration with these heavy metals.

Adsorbents characteristics. Granular Activated Carbon (GAC) and Natural Zeolite (NZ) were used as adsorbents in the investigated systems. The GAC was supplied by "Miloje Zakic", Krusevac, Serbia&Montenegro. It was repeatedly washed with deionized water to remove any leachable impurities and adherent powder and then dried to constant weight at 110°C for 24 h. Then, it sieving to determinate granulation (particles in the size range 1-3 mm).

The NZ was supplied by "Nemetali", Vranjska Banja, Serbia&Montenegro. The mineralogical composition of the natural zeolite is 90% clinoptiolite and the rest is mordenite and haylandrite. The chemical analysis of the natural zeolite shows that the oxides of silicon, aluminum, calcium and iron are the main constituents while other oxides are present in trace amounts (Table 1). Prior to the experiments, the NZ was dried at 300°C for 48 h in order to remove any traces of moisture or other constituents. The properties of the adsorbents used are listed in Table 2.

For the determination of adsorption isotherms a constant mass of adsorbent (3g granular activated carbon and natural zeolite for Zinc and 5 g granular activated carbon and natural zeolite for Cd) and 0.2 dm^3 of solution with known initial concentration were mixed at 500 rpm at a constant temperature of 25°C. The equilibriums were established for 2-11 days for each heavy metal-adsorbent system. Aqueous samples were taken for heavy metals concentration measurements using AERL 3520 Atomic Absorption Spectrophotometer.

3. Results and Discussion

One of the parameters which determine the possibility of using a particular material as an adsorbent for the removal of heavy metals is its adsorption capacity. For the purpose the adsorption isotherms for the systems Zn(II), Cd(II) and Pb(II) onto granular activated carbon and natural zeolite were determined at 25°C. Amounts of metal taken up by the adsorbent were determined by the following mass balance equation:

$$q^* = (V/m)(C_0 - C^*) \tag{1}$$

TABLE 1. Chemical analysis of natural zeolite.

Constituents	Natural zeolite % by weight
SiO_2	64.88
Al_2O_3	12.99
Fe_2O_3	2.00
TiO_2	0.37
CaO	3.26
MgO	1.07
Na_2O	0.95
K_2O	0.89
Ignition loss	13.59

TABLE 2. Properties of the adsorbents.

Property	GAC	Natural zeolite
Particle diameter, mm	1-3	1-3
Particle density, g/cm^3	1.87	2.12
Bulk density, g/cm^3	1.0-1.2	1.43
Surface area, cm^2/g	900-1200	20-40

where q^* and C^* are respectively the adsorbent phase metal concentration and solution phase metal concentration at equilibrium; C_0, the initial metal concentration; V, the solution volume, and m is the mass of adsorbent. A number of equations exist which enable the equilibrium data to be correlated, and the four most frequently used are the Langmuir (Eq. 2), Freundlich (Eq. 3), Langmuir-Freundlich (Eq. 4) and Ridlich-Peterson (Eq. 5). The parameters in these equations are very useful for predicting adsorption capacities and also for incorporating into mass transfer relationship in the design of contacting equipment.

$$q^* = \frac{q_m K_L C^*}{1 + K_L C^*} \tag{2}$$

$$q^* = k_f C^{*1/n} \tag{3}$$

$$q^* = \frac{q_m k_C C^{*(1/n')}}{1 + k_C C^{*(1/n')}} \quad (4)$$

$$q^* = \frac{K_{RP} \cdot C^*}{1 + A \cdot C^{*\beta}} \quad (5)$$

The parameters in the adsorption isotherms were estimated from the experimental equilibrium data using MATLAB Curve Fitting Toolbox. The comparison of experimental and estimated data by Langmuir, Freundlich, Redlich-Peterson and combined Langmuir-Freundlich models for the investigated systems are presented in Figures 1 to 3 for six investigated systems.

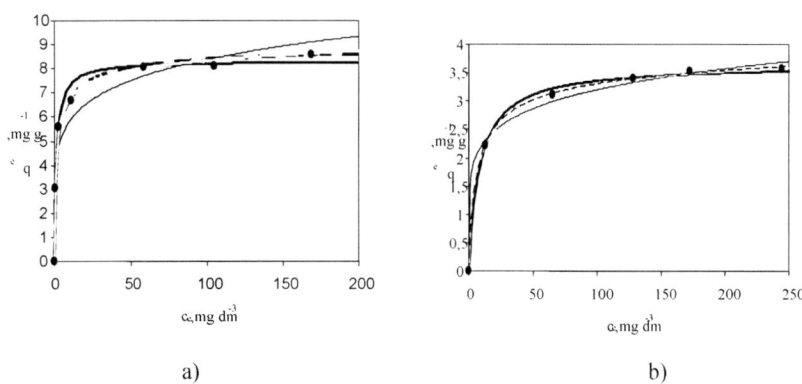

FIGURE 1. Comparison of experimental and estimated data by different isotherms for the systems a) Zn(II)-GAC b) Zn(II) –NZ at 25°C

● Experimental data;   ▬▬▬ Langmuir;   ▬▬▬ Freundlich;
   Langmuir-Freundlich;   ----- Ridlich-Peterson

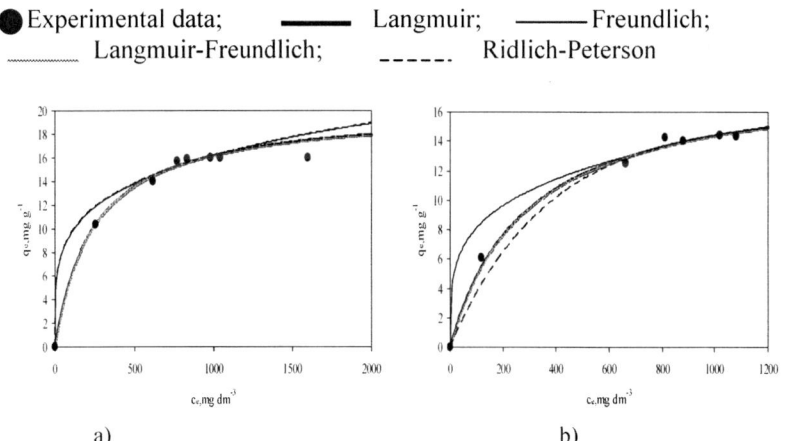

FIGURE 2. Comparison of experimental and estimated data by different isotherms for the systems a) Cd(II) – GAC and b) Cd(II)-NZ at 25°C

● Experimental data;   ▬▬▬ Langmuir;   ▬▬▬ Freundlich;
   Langmuir-Freundlich;   ----- Ridlich-Peterson

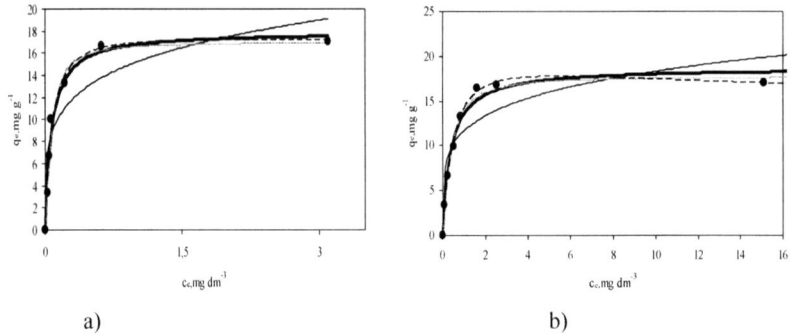

FIGURE 3. Comparison of experimental and estimated data by different isotherms for the systems a) Pb(II) – GAC and b) Pb(II)-NZ at 25°C

● Experimental data; ▬▬▬ Langmuir; ▬▬▬ Freundlich;
▬▬▬ Langmuir-Freundlich; ----- Ridlich-Peterson

According to the obtained data for the model parameters it is obvious that a general conclusion as to which proposed equilibrium model is favorable can not be deduced. The values for the investigated systems heavy metals-adsorbents are given in Tables 3, 4, 5 and 6.

For all investigated systems, the correlation coefficients were estimated. The values of the correlation coefficients were very high, near to 1. The best fit gives combined Langmuir-Freundlich isotherm for the most of equilibrium isotherms, but it contents three parameters. The lowest correlation coefficient has Freundlich isotherm. The investigator decides which of the proposed models will be chosen for further mathematical modeling of the adsorption process that is in accordance with the subsequent application in kinetic and dynamic studies.

Table 3. Langmuir parameters for the systems heavy metals-adsorbents at 25°C.

Langmuir parameters		Langmuir parameters	
Zn(II)-GAC		Zn(II)-natural zeolite	
q_m, mg g^{-1}	K_L, dm^3 mg^{-1}	q_m, mg g^{-1}	K_L, dm^3 mg^{-1}
8.325	0.6895	3.6444	0.1169
Cd(II)-GAC		Cd(II)-natural zeolite	
q_m, mg g^{-1}	K_L, dm^3 mg^{-1}	q_m, mg g^{-1}	K_L, dm^3 mg^{-1}
20.139	0.00409	18.438	0.003515
Pb(II)-GAC		Pb(II)-natural zeolite	
q_m, mg g^{-1}	K_L, dm^3 mg^{-1}	q_m, mg g^{-1}	K_L, dm^3 mg^{-1}
17.93	14.63	18.73	2.635

TABLE 4. Freundlich parameters for the systems heavy metals-adsorbents at 25°C.

Freundlich parameters		Freundlich parameters	
Zn(II)-GAC		Zn(II)-natural zeolite	
k_f, mg g^{-1}(dm^{3}mg^{-1})$^{1/n}$	n, no units	k_f, mg g^{-1}(dm^{3}mg^{-1})$^{1/n}$	n, no units
4.038	6.3172	1.538	6.2814
Cd(II)-GAC		Cd(II)-natural zeolite	
k_f, mg g^{-1}(dm^{3}mg^{-1})$^{1/n}$	n, no units	k_f, mg g^{-1}(dm^{3}mg^{-1})$^{1/n}$	n, no units
3.359	4.3937	2.602	0.2471
Pb(II)-GAC		Pb(II)-natural zeolite	
k_f, mg g^{-1}(dm^{3}mg^{-1})$^{1/n}$	n, no units	k_f, mg g^{-1}(dm^{3}mg^{-1})$^{1/n}$	n, no units
15.18	4.902	11.71	5.128

TABLE 5. Langmuir-Freundlich parameters for the systems heavy metals-adsorbents at 25°C.

Langmuir-Freundlich parameters			Langmuir-Freundlich parameters		
Zn(II)-GAC			Zn(II)-natural zeolite		
q_m, mg g^{-1}	k_c, (dm^3/mg)$^{1/n'}$	n', no units	q_m, mg g^{-1}	k_c, (dm^3/mg)$^{1/n'}$	n', no units
9.035	0.6241	1.5366	4.07708	0.2465	1.618
Cd(II)-GAC			Cd(II)-natural zeolite		
q_m, mg g^{-1}	k_c, (dm^3/mg)$^{1/n'}$	n', no units	q_m, mg g^{-1}	k_c, (dm^3/mg)$^{1/n'}$	n', no units
20.06	0.003988	0.994	18.315	0.003276	0.9862
Pb(II)-GAC			Pb(II)-natural zeolite		
q_m, mg g^{-1}	k_c, (dm^3/mg)$^{1/n'}$	n', no units	q_m, mg g^{-1}	k_c, (dm^3/mg)$^{1/n'}$	n', no units
17.073	37.38	0.777	17.845	3.745	0.787

TABLE 6. Ridlich-Peterson parameters for the systems heavy metals-adsorbents at 25°C.

Ridlich-Peterson parameters			Ridlich-Peterson parameters		
Zn(II)-GAC			Zn(II)-natural zeolite		
K_{RP}, dm^{3}mg^{-1}	A, (dm^3/mg)$^\beta$	β, no units	K_{RP}, dm^{3}mg^{-1}	A, (dm^3/mg)$^\beta$	β, no units
8.079	1.266	0.9398	0.6334	0.2442	0.9363
Cd(II)-GAC			Cd(II)-natural zeolite		
K_{RP}, dm^{3}mg^{-1}	A, (dm^3/mg)$^\beta$	β, no units	K_{RP}, dm^{3}mg^{-1}	A, (dm^3/mg)$^\beta$	β, no units
0.04286	9.311x10^{-5}	0.7724	0.04058	0.0004178	1.213
Pb(II)-GAC			Pb(II)-natural zeolite		
K_{RP}, dm^{3}mg^{-1}	A, (dm^3/mg)$^\beta$	β, no units	K_{RP}, dm^{3}mg^{-1}	A, (dm^3/mg)$^\beta$	β, no units
237.5	13.05	1.034	36.44	1.575	1.102

On the other hand, the Langmuir model is convenient for determination of the equilibrium capacity of every adsorbent for a particular adsorbate (Table 3).

According to the Langmuir model (Eq.2) the adsorption capacity q_m for Cd is 2.5 times grater than for Zn and adsorption capacity q_m for Pb is 2 times grater than Zn when granular activated carbon is used. When natural zeolite is used as adsorbent, the adsorption capacity q_m for Zn is 5 times lower than Cd and Pb. So, q_m varied in the order Cd (II)> Pb(II) >Zn(II) for GAC, and Pb(II)$\cong$Cd(II)>Zn(II) for the natural zeolite as adsorbent. Ricordel et al (2001) and Tsoi and Zhao (2004) reported a similar relationship when different adsorbents were used. This can be explained on the basis of their ionic radii, hydration energy, ionic mobility and diffusion coefficient. The explanations of different authors were given on the basis of the surface covered by the adsorbed metal ions or on the basis of metal surface complexation constants and thermodynamic parameters values.

Comparing values of q_m for the same adsorbate, but different adsorbent, it can be seen that in both cases for Zn and Cd, granular activated carbon is better adsorbent than natural zeolite, and for Pb natural zeolite is better adsorbent than granular activated carbon. In case of Zn, granular activated carbon has 2.5 times grater capacity from natural zeolite; in case of Cd, GAC has a little better capacity from natural zeolite, and in case of Pb, natural zeolite has a little grater capacity than GAC. These preliminary results show that natural zeolite is an effective adsorbent for removal of metal ions from aqueous solutions.

4. Conclusion

Certain general points may be deduced from the experimental and theoretical analysis on predicting adsorption equilibrium of heavy metals onto granular activated carbon and natural zeolite

- The experimental equilibrium data were fitted by Langmuir, Freundlich, Langmuir-Freundlich and Ridlich-Peterson models. The equilibrium studies have shown that all proposed equilibrium models indicate a good correlation between the theoretical and experimental data for the whole heavy metals concentration range.
- For all investigated systems, except in case Zn(II)-GAC, the Langmuir-Freundlich isotherm has shown the best agreement with experimental equilibrium data.
- Comparing the values of q_m (maximal capacity of adsorbent) for adsorption of Pb(II),Zn(II) and Cd(II) onto granular activated carbon

and natural zeolite, it can be concluded that the order of adsorption capacity of granular activated carbon is Cd(II)>Pb(II)>Zn(II) and for natural zeolite is Pb(II) ≅ Cd(II)>Zn(II).

- These preliminary results show that natural zeolite is an effective adsorbent for removal of metal ions from aqueous solutions. It would be useful for the economic treatment of wastewater.

5. Nomenclature

A parameter in the Redlich-Peterson equation, $(dm^3\ mg^{-1})^\beta$
C^* equilibrium liquid-phase concentration, $mg\ dm^{-3}$
C_o initial liquid-phase concentration, $mg\ dm^{-3}$
K_L parameter in the Langmuir equation, $dm^3 mg^{-1}$
k_C parameter in the Langmuir-Freundlich equation, $(dm^3\ mg^{-1})^{1/m}$
K_{RP} parameter in the Redlich-Peterson equation, $dm^3 mg^{-1}$
k_f parameter in the Freundlich equation, $mg\ g^{-1}(dm^3\ mg^{-1})^{1/n}$
m mass of adsorbent, g
n parameter in the Freundlich equation
n' parameter in the Langmuir-Freundlich equation
q^* equilibrium solid-phase concentration, $mg\ g^{-1}$
q_m parameter in the Langmuir equation, $mg\ g^{-1}$
V volume of solution, m^3
β parameter in the Redlich-Peterson equation

References

Allen S., Brown P., McKay G. and Flynn O., (1992) An evaluation of single resistance transfer models in the sorption of metal ions by peat, J. Chem. Tech. Biotechnol., 54, 271–276.

Apak R., Tutem E., Hugul M. and Hizal J., (1998), Heavy metal cation retention by unconventional sorbents (red muds and fly ashes), Wat. Res., 32, 430–440.

Bailey S.E., Olin T.J., Bricka R.M. and Adrian, (1999), A review of potentially low-cost sorbentsfor heavy metals, Water Research, 33, 11, 2469–2479.

Barbier F., Duc G., Petit-Ramel M., (2000), Adsorption of lead and cadmium ions from aqueous solution to the montmorillonite/water interface, Ohysicochemical and engineering Aspects, 166, 153–159.

Barros M.A.S.D., Silva E.A., Arroyo P.A., Tavares C.R.G., Schneider R.M., Suszek M. and Sousa-Aguiar E.F., (2004) Removal of Cr (III) in the fixed bed column and batch reactors using as adsorbent zeolite NaX, Chemical Engineering Science, 59, 5959–5966.

Chu K.H., (2002), Removal of copper from aqueous solution by chitosan prawn shell: adsorption equilibrium and kinetics, Journal of Hazardous Materials B90, 77–95.

Kandah M. (2001), Zinc adsorption from aqueous solutions using disposal sheep manure waste (SMW), Chemical Engineering Journal, 84, 543–549.

Katsumata H., Kaneco S., Inomata K., Itoh K., Funasaka K., Masuyama K., Suzuki T. and Ohta K., (2003), Removal of heavy metals in rinsing wastewater from plating factory by adsorption with economical viable materials, Journal of Environmental Management, 69, 187–191.

Lee S.H., Jung C.H., Chung H., Lee M. Y. and Yang J., (1998), Removal of heavy metals from aqueous solution by apple residues, Process Biochemistry, 33, 2, 205–211.

Lin S.H., Lai S.L. and Leu H.G., (2000) Removal of heavy metals from aqueous solution by chelating resin in a multistage adsorption process, Journal of Hazardous Materials B76, 139–153.

Meshko V., Markovska L., and Marinkovski M., (2004) Modeling of adsorption kinetics of zinc from aqueous solution by granular activated carbon and natural zeolite, XVIIIth Congress of Chemists and Technologists of Macedonia, Ohrid, Macedonia.

Mohan D., Singh K.P., (2002), Single- and multi-component adsorption of cadmium and zinc using activated carbon derived from bagasse-an agricultural waste, Water Research, 36, 2304–2318.

Pradas E.G., Sanchez M.V., Cruz C., Viciana M.S. and Perez M.F., (1994), Adsorption of cadmium and zinc from aqueous solution on natural and activated bentonite, J. Chem. Tech. Biotechnol., 59, 289–295.

Ricordel S., Taha S., Cisse I. and Dorange G., (2001), Heavy metals removal by adsorption onto peanut husks carbon: characterization, kinetic study and modeling, Separation and Purification Technology, 24, 389–401.

Tsoi L.Lv.G., and Zhao, H.S. (2004), Uptake equilibria and mechanisms of heavy metals ions on microporous titanosilicate ETS-10, Ind. Eng. Chem. Res. 43, 7900–7906.

NITRATE POLLUTION OF THE LESNOVSKA RIVER CAUSED BY FILTRATION OF CHEMICALS BY AGRICULTURAL AREAS

SVETLANA BOZHINOVA* AND GRIGOR VELKOVSKI
Institute of Water Problem
Bulgarian Academy of Sciences
G. Bonchev Str, Bl. 1, 1113 Sofia, Bulgaria

*To whom correspondence should be addressed: swetlana_bojinova@abv.bg

Abstract. The nitrate pollution caused by filtration of chemicals from agricultural areas is a subject to a number of studies during the last decade. The main purpose of this study was to specify the river flow (Q, m³/s), nitrate concentrations and conductivity in a part of the Lesnovska River Basin, situated near the Sofia City, the capital of Bulgaria. Incoming flow (Q, m³/s) by filtration in to the river and evaluation of the contaminated with nitrate water quantity were defined by means of balance methods, based on the results from the precise performed measuring. The main tasks of the study were verification of the mathematical model of the nitrate pollution formation in the river terraces groundwater and to define the contribution of this contamination in to the river. Thus, the affects of fertilizing on agriculture areas and on the river water quality were determinate on the basis of information about the quantity of the nitrate pollution coming from the groundwater and by filtration (Q, m³/s).

Keywords: surface and groundwater pollution; nitrate contamination from agriculture; fertilizing impact

1. Introduction

From the hydrology and hydrogeology is known that river flows appear to be successors of the surface and groundwater from the adjoining territories. It is obvious that river flow is formed under the influence from the surface and groundwater flow and bears the pollution caused of them. The understanding of groundwater pollution processes due to fertilizer nutritious circle, transporting mechanism of this substance from the soil layer through the sub ground and down until the ground waters, processes of there transformation are significant for assessment and forecast about state of ground water and river flow.

2. Object of the Investigation

Assessment of the stage of pollution of the water into the river flow of Lesnovska River in the part with in the bridge Dolni Bogrov and Chepinci is the aim of the investigation. The basic problems which we solve are:
- Estimation of the water flow in specific cross – sections.
- Estimation of the incoming water of filtration and its pollution

3. Methodology

In the part of the Lesnovska river where are made the investigation there are twelve cross – sections. There location is chosen according some conditions which is:
- negotiability of the place, where will be the cross-section
- if there are some artificial or natural disturbance of the river flow (duckweed, rapids)
- Incoming of water flows as feeders and channels.

It is very important to be mentioned that the velocity of the water flow is measuring at the same place from the same depth of the vertical section of the studied cross-section, where the water for chemical analyses is taken from.

To determine the concentrations of NO_3 are taken water samples from certain vertical lines from everyone cross-section. There are five such vertical lines. Water is taken into plastic bottles with capacity 500mg. With results from the chemical analysis is prepared profiles of the twelve cross section, which represent lines of point with equal value of concentration of nitrate Fig. 1.

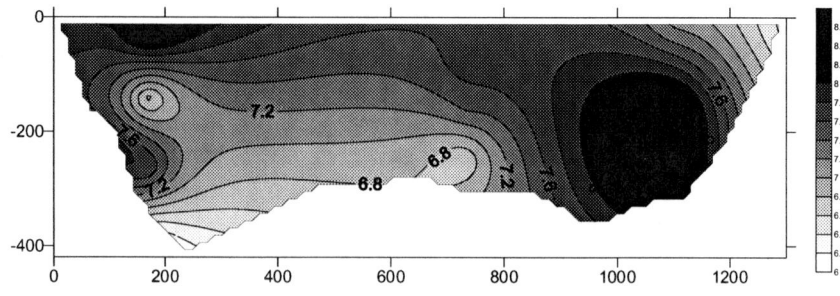

FIGURE 1. Line with equal concentration of NO_3 – izoline.

4. Estimation of the Incoming Filtrated Water Quantity into the River Flow and Quantity of Pollutant NO_3

To calculate the incoming filtrate water quantity and pollutant coming to the river flow from filtration is necessary to be known water flow and quantity of pollutant (in case NO_3) in examined cross-sections. Calculation of water quantity incoming by filtration and pollutant transported by it is performed by balance method Table 1. and Fig. 2 (a.b.c.d).

5. Summary

- Water quantity for certain profiles from the water flow and concentration of nitrates in typical points from the river profile were defined. The quantity of the flowing pollutant in chosen profiles.
- Filtrate water quantity from the bottom of the river, quantity and concentration of NO3 pollutant were defined.
- It is made a balance of water flow and amount of pollutant transported by water in the studied section. Balance is based on already determined pollution of the main flow (stream) and feeders including and filtration flow.
- It was found out that for selected area of the river flow there is no full mix of nitrate pollutant up to receipt of homogeneous solution. The flow consists of stream with high and low content of pollutant.

In the study is shown methodology about evaluation of real pollution in the river flow and average pollutant concentration.

Results in Table 2 shows that groundwater incoming into the river flow from terrace under village Chelopechene brings significant part of pollution of river flow with nitrate. Studying the contributions of each feeder that takes part in formation of water flow in the river and respectively its pollution became clear that water from filtration coming from excavation is low contaminated $C=2.1mg/l$. Consequently water is formed in the hydrologic cycle which means that this is fresh rain water.

6. Conclusion

Investigation that has been performed in the field along the Lesnovska River is basic for determination of the water flow of the certain cross-section in the part within Dolni Bogrov and Chelopechene.

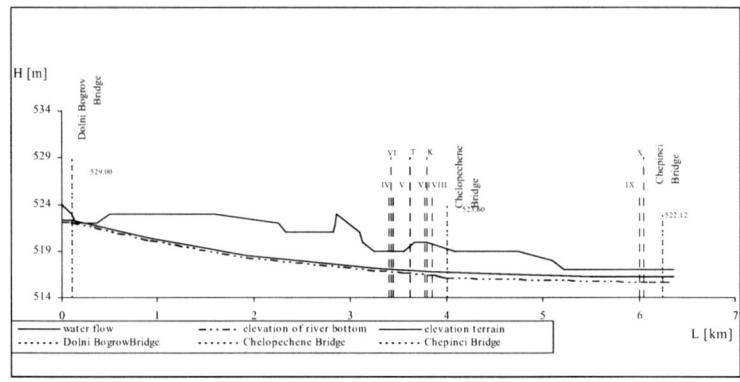

FIGURE 2a. Longitudinal profile.

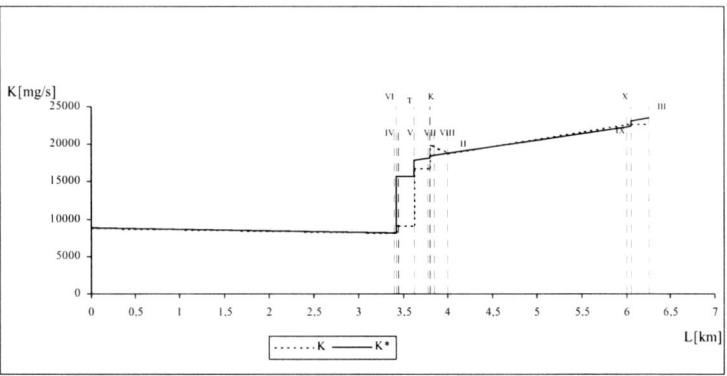

FIGURE 2b. Water flow.

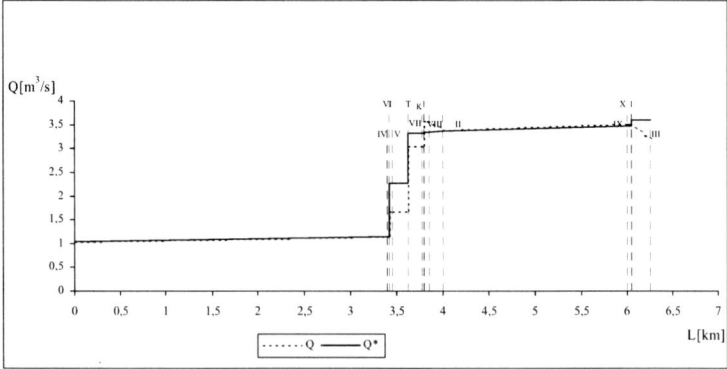

FIGURE 2c. Quantity of pollutant NO3 in chosen profiles.

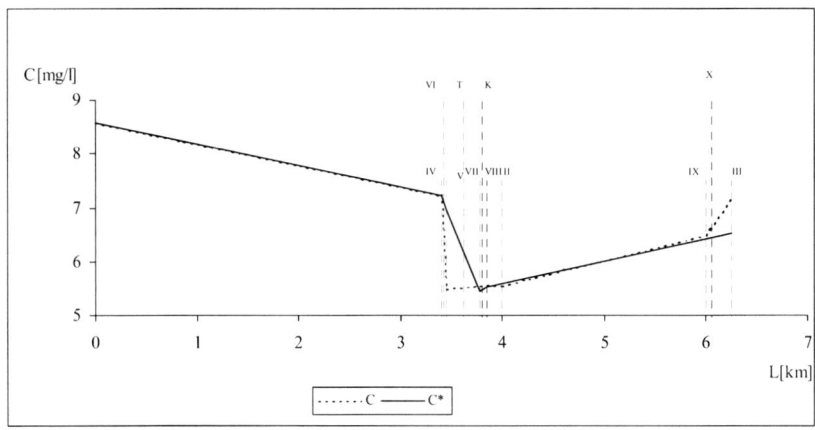

FIGURE 2d. Concentration of NO$_3$ in chosen profiles.

TABLE 2. Water flow and pollutant quantity NO$_3$ in them.

Q		m³/s	%	K		mg/s	%	C		mg/l
QI		1.0388	28.89	KI		8911.14 -650.92	37.95 -2.77	CI		8.5783
Qi	Q$_C$	1.111	30.90	Ki	K$_C$	7453.12	31.74	Ki	C$_C$	6.7079
	Q$_T$	1.06	29.48		K$_T$	2226.0	9.48		C$_T$	2.1
	Q$_K$	0.018	0.5		K$_K$	249.47	1.06		C$_K$	13.558
	Q$_X$	0.106	2.95		K$_X$	734.9	3.13		C$_X$	6.9396
	ΣQi	2.2954	63.83		ΣKi	10663.49	45.41			-
Qf	I-T	0.1119	3.11	Kf	I-T	-	-	Cf	I-T	-
	T-III	0.1498	4.17		T-III	4557.32	19.41		T-III	-
	ΣQf	**0.2617**	7.28		ΣKf	4557.32	19.41		30.42	
Qe= QI+ΣQi+ΣQf		3.5959	100	Ke= KI+ΣKi+ΣKf		23481.03	100	Ce=Ke/ Qe		6.531
Qex		3.5955	100	Kex		23481.03	100	Cex		6.531

Using this method for estimation of filtrate water flow and pollution are got results which is base of visualization of pollution of water in the river.

Reference

Ivanov I.M. 1979: "Hydrological reference book", Sofia.

PESTICIDE APPLICATIONS AND SUSTAINABLE AGRICULTURAL DEVELOPMENT IN ARMENIA

VARDAN SARGSYAN AND ARMAN SARGSYAN
Yerevan State Institute of Economy
128 Nalbandyan Str. 375025 Yerevan, Armenia

*To whom correspondence should be addressed: sarm@freenet.am

Abstract. Problems connected with sustainable agricultural development have no clear and universal solution. Solutions are to be developed locally, building on mutually agreed principles rather than recipes. Armenian agricultural sector has quite strategic importance for the country since USSR period. One of the generated problems since mentioned period is high volume of used chemical substances, mainly pesticides. From that point in this paper we touched to the problem of finding environmentally acceptable development paths of agriculture within "economical-legal-social-political-environmental" system.

Keywords: agriculture; pesticides; sustainable development; economical principles

1. Introduction

As one of the former USSR Republics, Republic of Armenia was economically developed region with poly branched industry, intensive agriculture, and vast system of organizations with industrial and social infrastructure. The Republic was distinguished as fruits, vegetables, grape, tobacco, fruit and vegetable tins producer. Industry was dynamically developed till 1988. After the destroying earthquake in 1988 the significant part of industrial potential was lost and due to the USSR disintegration, the most considerable reduction of production happened in 1992. Now agriculture is the major branch of national economy. Privatization of land has been implemented in the Republic (first among former Soviet Union countries).

The adjective "sustainable" derives from the following concern: can our current patterns of economic activity be continued over long periods without disastrous consequences for the environment, or for us? The sustainability is development that meets the needs of present without compromising the ability of future generations to meet their own needs. Environmental externalities from agricultural activities, both past and present, are pervasive. Since Neolithic times humans have been converting forests, wetlands into crop and grazing lands. These activities have shaped the rural landscape and the hydrology and ecology of agriculturally developed regions. In order to save the nature and the health of people we can resign from nuclear technologies or reduce some productions, close factories. The same can't be said about agriculture, where chemical substances mainly in shape of pesticides are nowadays widely used, which are dangerous for nature and people's health. At present, agriculture can't be imagined without chemistry, as "pure" agriculture isn't capable to meet the demands of the mankind, and it will result in famine. The usage of chemistry raises the effectiveness of production and brings great profit on the one hand, and on the other hand, causes pollution of nature and diseases which may be observed as an economic damage.

The environmental problems connected with pesticides are very actual in Armenia. First of all Armenia is one of those countries, where pesticides were used in inadmissible high quantities. If in the USSR in general 0.5 kg. of pesticides was used per ha, in Armenia it reached 10-15 kg/ha, and in some regions - up to 35 kg/ha. The 70-80% of the pesticides in Armenia is used in Ararat Valley where the density of population is very high. In Armenia about 60 pesticides, which fall under 7 chemical groups, are used. Medical investigations showed that in Ararat Valley frequency of some diseases is very high (Ambartsumyan, 1985; World Bank report, 1999). This fact first of all is connected with the usage of pesticides. It is evident, that a great damage was given to the nature too. Above related shows that in Armenia environmental control in agriculture was never realized. Economic aspects of this problem are not investigated here completely.

Though nowadays the scientists offer non-chemical ways of agricultural production, nevertheless, none of those methods can substitute chemicals completely. The only way out seems to be the adopting of a "favorable" strategy of chemistry usage in agriculture with the minimum possible ecological damage. Here, reconciliation of environmental protection and economic development is necessary.

2. Agriculture in Armenia-one of the Main Paths of Development

At the very beginning of transition from planned to market economy in Armenia, the agricultural sector was regarded as one of the most important sectors of economy taking into account its infrastructure on the general socio-economic condition of the Republic and the problem of food security. Agriculture is the main "working" branch in Armenia. During the first years of independence, reforms began in agriculture aiming to transform this sector into a stable and a productive one, based on private property and market relations. Land privatization in Armenia was initiated first among Newly Independent States. Rural households cultivate or use most of the available land resource through a combination of private ownership and leasing of state lands. Obviously, privatization and land ownership has resulted in new and unfamiliar legal and social conditions relating to land use. It is not clear how this will affect biodiversity. Difficult economic conditions and lack of resources for farming have acted to reduce the effectiveness of agriculture, which is another potential problem to environment. The absence of effective regulations for the use of private lands could result in even greater environmental impacts in the future.

Taking into account above-mentioned the problem of how to model and evaluate alternative and environmentally acceptable development paths of agriculture within "economical-legal-social-political-environmental" system is of grate importance.

Agriculture in Armenia contributes about one-third to Gross Domestic Product and accounts for 42 percent of employment. Crop production accounts for roughly 60 percent of agricultural output and livestock production about 40 percent. Armenia's agricultural lands amounts to 1391 thousand ha (including 483 thousand ha of arable land), or 45% of land resources, and practically there are no reserves for extension. The agricultural reform for new independent Armenia first of all was based on laws adopted by the Parliament of the Republic of Armenia: the laws "On property" (1990), "On rural and rural collective farms"(1991), "Code of land" (1991), "On enterprises and entrepreneurial activity" (1992), regulatory acts and many governmental resolutions.

Land privatization was initiated in 1991. The privatized land accounts for 72.2% of the agricultural land of the republic, 62.5% of arable land, 74.8% of plantations and 57.3% of grassland. Since March 1991, as a result of land privatization and some other agrarian reforms in the Republic, farms have been broken into smaller entities. 321 thousand rural households and 273 agricultural collective farms were created as of January 1, 1997. Land may be legally sold, mortgaged, leased and subleased but officially registered land transactions are few as the land market is only in the period of formation. As in other NIS countries the establishment of real working land market is a strong need for development.

3. The Problem- Balancing Agriculture and Environment

Environmental constraints of agriculture are the core of the problem. As it has been already mentioned, agriculture remains the largest sector in Armenia, and almost half of the total land area is devoted to agricultural land. As such agriculture is a key sector for natural resource use and has caused much damage to biodiversity. Main impacts from agriculture in Armenia include Pesticide use, soil pollution, and pollution of water resources. From the other side there are also problems with land degradation (soil erosion and increased salinity) and reduction in productivity, over-grazing (affecting vegetation composition of pastures).

Land pollution is mainly caused by farmers' uncontrolled use of pesticides. Uninformed and incorrect pesticide management and use were key environmental problems during Soviet period and they continue to require attention in Armenia today. The recent researches have proved a high concentration of pesticides in the soil and water. Problem includes excessive use of pesticides, improper timing of applications, poor enforcement of existing regulations, and failure to introduce and integrate non-chemical alternatives for pest control. Agriculture is the main source of pollution of water resources in Armenia.

In our days the concordance between economic development and rational use of natural resources is of the great importance. Without that the ecological disaster is inevitable. The problem stands more firmly for the ex-communist countries under conditions of transitional economy. The economic analysis of environmental problems needs fresh approach. This problem stands greatly also in the sphere of economical education, which must be greatly connected with environmental topics. As distinct from industry, where great attention is given to the environmental problems, in agriculture this problem is comparatively new. Among the least elucidated ecological problems is the rational usage of chemicals and, especially pesticides in agriculture.

The result in agriculture is possible to get only on condition, that effective chemicals against harmful insects and diseases are used. We can say that this problem is of great importance in the whole world. It was calculated, that harvest losses, which can be amended with the help of chemical means, in different countries of the world can be expressed by millions of dollars. If in the USA they stop using pesticides, the quantities of meat and milk would be reduced for 25%, vegetables and fruits for 50% (A.I.Stenberg). According to materials, published by UNO, annual world losses of corn crop, caused by harmful insects, reach 33 million tons, which would suffice for 150 million people. The investigations show that without systematic struggle with diseases and pests it is possible to gather not more than 37% of normal crop potatoes at best, 22% of cabbages, 10% of apples, 95% of peaches etc. According to exact data world crop losses are evaluated as 74500 million dollars a year, or 34% of possible crop yield. Naturally such kinds of losses are considered as a great economical damage. That is why it is impossible to organize agriculture without use of pesticides.

Now let's illuminate the second, more important, side of this problem.

The usage of chemistry in agriculture creates numerous environmental problems. A lot of investigations have proved the harmful effects of pesticides on the biosphere and the health of people. According to the data of UNESCO pesticides belong to the class of 10 main polluters of nature. Unlike other chemicals, pesticides very often are introduced into nature in inadmissible quantities, which results in circulation. It brings to pesticide accumulation in the soil, air, water and human food with grave consequences. This problem was deeply investigated by scientists (especially in USA). Particularly J.L. Schnoor (1981) has evaluated circulation of pesticides (particularly dildrin) in one of the water basins of Iowa by using a mathematical model. It was found out that about 50% of the pesticide is taken out by the current, 40% falls at the bottom and 10% accumulates in the tissues of fish. As a result, pesticides accumulate in human organism too, and cause a number of diseases. The problem was especially acute in Armenia, ex-USSR republics of Middle Asia and Moldavia, where the indices of pesticide usage are rather high. The medical investigations have confirmed that in the regions where pesticides were intensively used, morbidity is comparatively high.

4. Analysis and Modeling

The analysis of pesticide impact on people's health and effectiveness of agriculture and modeling of process in order to find out the ecologically optimal quantities of pesticides has been done.

We have targeted our research on following directions:

-The study of dynamics of pesticide usage and the human diseases in the regions of intensive usage of pesticides (regions of Ararat Valley). Statistic data have been processed. In the above mentioned district the dynamics of pesticide application has been investigated with the using the method of integral evaluation of pesticides, which takes into account not only quantitative, but also qualitative toxicological characteristics of each pesticide, because the toxicity of the pesticide, and not its quantity is more important. Pesticides were valued in special units with the help of the above-mentioned method. For regions of Ararat Valley and some foothills,

with the use of statistic methods, the dynamics of morbidity has been investigated and the increase of morbidity in regions of intensified applications of pesticides has been found out.

- Valuation of ecological damage caused by pesticide usage. A method of valuation of damage from morbidity of people of different age groups has been worked out and the damage from 23 diseases of adults and 15 diseases of children has been evaluated. The analysis shows that the great part of this damage is the loss of national income. That is why the chief damage is caused by morbidity of working people, which in different regions fluctuates from 88% to 95.9% of whole damage. Different investigators give different definitions of economical damage from pollution of environment. Many of former Soviet economists, who dealt with this problem didn't think it right to give a monetary valuation of this damage under conditions of socialism. They thought it improper to count how much pure air and pure water cost. That is why our economists have no much experience in this question. In the USA the first evaluations of economic damage from air pollution were done before First World War (8.5 million dollars per year). In 1978 the economic damage from the pollution of atmosphere in the USA was 25000 millions of dollars. Investigations showed that absolute damage from pollution is greater than expenses for protection of environment.

- Mathematical modeling of the process. A two staged integrated system of models was constructed (Sargsyan, 1993; Sargsyan, 1989; Sargsyan, 2001).

First stage - pesticide effect on human health and agricultural production (regression analysis). Two non-linear empiric dependencies "Pesticide-ecological damage" and "Pesticide-economical effect" have been revealed. Here, the pesticide effect on environment, human health and farm income was processed. In order to find out these ties 17 non-linear types of dependencies have been probed. The analysis was done to estimate the optimal strategy for use of DNOK (Dinitroortokresol) in the agricultural districts of the Ararat Valley in Armenia. DNOK is a high toxicity pesticide in the Nitrofenol group. It circulates very fast. In Armenia, in the Ararat Valley fruit agriculture is mostly developed. Farmers are using this pesticide on fruit trees in early spring as an insecticide and fungicide. In order to test the models statistical data for 5 districts of the Ararat Valley have been processed. Within the districts there is the same type of agricultural production and use of pesticides. In order to get more realistic results the years when the agriculture was not productive because of meteorological factors were not used in the model. This was done to avoid as much as possible the external influences. The following regression equation produced the best fit.

$$Y = \left(\frac{33.17}{1+1.004e^{-0.223X}}\right)^2$$

where:
Y - farm income from the apple trees where DNOK was used ($/ha).
X - usage of pesticide DNOK (kg/ha).
R=0.729
Criterion of Fisher - F=36.8,>F_{tab}=4.08.

The next analysis was done in order to evaluate the level of ecological damage from usage of this pesticide. The data used was taken from the economical damage from morbidity of people. Those forms of diseases which can be developed from DNOK were chosen. For this model a time lag of two years was used, because the pesticides effects on people are not immediate. DNOK enters the human organism with food water and air and accumulates up to dangerous amounts for health over time.

The following equation was estimated:

$$Y_1 = 7.23255 X^2 - 12.0206 X + 269.215$$

where:
Y_1 - the summary evaluated damage from morbidity (those forms of diseases, which can be caused by DNOK) for one hectare of soil, where DNOK was used ($/ha).
X - usage of pesticide DNOK (kg/ha).
R=0.756, F=58.5>Ftabl=3.23

Second stage - optimization. In case of having incomplete information about the system "pesticide-environment" determinated mathematical methods of analysis are of little use. That is why, in the block of optimization of the model system, a dynamic stochastic model based on Bellman's method of dynamic programming, has been used. Markov process was taken as a mathematical model of the system (Hovard, 1964). The main goal of the optimization model is to find out the optimal value of X taking into account the ecological negative influence of pesticide. In ecological systems we often meet situations, where great role was given to casual influences, because in ecological systems is impossible to plan processes and relations. This shows the stochastic character of the problem. That is why the main model was a dynamic stochastic optimization model.

Let's describe the optimization problem: The usage of chemicals in agriculture is accompanied with economical efficiency ($Y=F(X)$) from increasing yield. Simultaneously the usage of chemical gives ecological-economical damage ($Y1=D(X)$). The additive model ($Y-Y1$) characterizes the "actual" income[3]. The problem is to find out the optimal strategy of pesticide usage, when ($Y-Y1$) will be maximized.

5. Conclusion

According to the above mentioned, the results of implemented research will have its practical significance for the transitional economy of Armenia. Despite the fact that Armenia just now has begun to build it's economy and a lot of problems exist, our state gives much attention to the questions of environment, and such kind of investigations are useful.

The results obtained, show the necessity of carrying out further investigations of environmental problems from agricultural chemicals in Armenia. In order to make this work more practical, further investigations are aimed to make the model more realistic.

References

Ambartsumyan H.S., Approaches to the Integral Evaluation of Pesticides in Agriculture, Hygiene of Usage, Toxicology of Pesticides and Polymer Materials, vol. 15, Kiev, 1985.

Armenia, National Program of Protection of Environment, Main Report, (1999), The World Bank.

Hovard R.A., Dynamic Programming and Markov Processes, Moscow, 1964.

Sargsyan, V., Ecologic-Economical System of Models of Determination of Optimal Quantities of Pesticides. Economic Independence of Republic of Armenia (collection of articles), Erevan, Armenia: Erevan Institute of National Economy, 1993.

Sargsyan, V., Some Questions of Ecologic-Economical modeling. Materials of the Third Republican Conference of post-graduates of Armenia, Erevan, Armenia, 1989.

Sargsyan, V., Risk Assessment for Agricultural Pollutants (Armenia): Modeling and Optimal Control: "Assessment and Management of Environmental Risks," Kluewer, Amsterdam 2001.

---

[3] By actual income we mean the income, which considered the "environmental" charges

EVALUATION OF BIOFILMS OCCURING IN DRINKING WATER DISTRIBUTION SYSTEM OF BALATONFÜRED

ZSUZSA LUDMÁNY, MATYAS BORSÁNYI AND
MARTA VARGHA
*Department of Water Hygiene,
Fodor Jozsef National Public Health Centre,
National Institute of Environmental Health,
Budapest, Gyali ut. 2-6, H-1097, Hungary*

*To whom correspondence should be addressed: ludmanyzs:@okk.antsz.hu

Abstract. The paper deals with the problems of water quality in one of the biggest regional Hungarian water distribution system of Balatonfüred. The aim of this study was to survey the water-network of Balatonfüred with evaluating of water and biofilm quality and its interaction. Two main types of drinking water sources could be separated. According to various water quality (karstic, surface and mixed water) and various construction materials of pipelines-life conditions of microorganisms are very diverse. Results of chemical quality of water shows a wide variety on sampling sites and on date of sampling.

Keywords: biofilms; water distribution system; water quality; secondary contamination

1. Introduction

The Government Ordinance 201/2001 on the quality requirements of drinking water and on the course inspection (based on the Council Directive 98/83/EC) came into force in Hungary in the autumn of 2001. These regulations require waterworks to guarantee the water quality on taps that are normally used for human consumption. Its quality (among other factors) depends on the type of raw water, on construction materials in contact with drinking water and on secondary contamination. Biofilm growth within a water distribution system could lead to operational problems such as pipe corrosion, water quality deterioration and other undesirable impacts.

One of the biggest regional drinking water suppliers in Hungary is the Transdanubian Regional Waterworks Co. (DRV Rt.). DRV Rt. had a research and development agreement with the National Centre of Public Health for several years. The project included a survey of the water-network of Balatonfüred.

2. Sampling Strategy

Samples were collected from eight fire hydrants of drinking water distribution net on three separate occasions, in May, July and September, 2004 (see Results and discussion). Before filling bottles we have opened fire hydrants to maximum rate of flow for two minutes to flush off biofilms from the walls of pipes and to aviod stagnant water sampling.

3. Parameters and Methods

3.1. WATER QUALITY

In order to characterize water quality, the following parameters were measured from the water samples in accordance with Hungarian Standards: total organic carbon (TOC), chemical oxygen demand, concentrations of ammonium, nitrite and nitrate ion, iron, manganese, pH, conductivity, m-alkality, turbidity, and heterotroph colony count.

3.2. BIOFILM

For biofilm characterization the following parameters were obtained: detection of nitrifying, sulphate-reducing and *Sphingomonas* bacteria by PCR (Polymerase Chain Reaction) and community restriction fingerprint by ARDRA (Amplified rDNA Restriction Analyses).

3.3. DNA ISOLATION

5 liters of sample were filtered on 0.45 mm pore size membrane filter. It was resuspended with water obtained from the samples. DNA was isolated from biomass with Invitrogen PureLink™ Genomic DNA Purification Kit.

3.3.1. *PCR (Polymerase Chain Reaction)*

PCR amplification began with a 5-min predenaturation at 94°C; this was followed by 35 to 40 cycles of 94°C for 60 s, primer-pair's degree for 60 s, and 72°C for 2 min. The final cycle was extended at 72°C for 7 min.

3.3.2. *ARDRA (Amplified rDNA Restriction Analyses)*

Diversity of biofilm microbial communities was assessed by restriction analyses of PCR-amplified rDNA (16S-8V/16S-1387R, lenght: 1352 bp). PCR products were restriction digested with *Hinf*I (G↓ANTC) and *Hae*III (GG↓CC) enzymes (Boehringer-Mannheim) in accordance with the manufacturer's instructions.

4. Results and Discussions

4.1. LIFE CONDITIONS OF MICROORGANISMS

Results of chemical quality of water shows a wide variety on sampling sites and on date of sampling. The main reason of it is the source of drinking water.

Balatonfüred is located on the north side of Lake Balaton thus the major source raw water is surface water. It is disinfected with chlorine-dioxide. Particularly in the summertime, the water supplier supplements drinking water with untreated karstic water because of higher consumption. The mixing occurs in the pipework, but near the feedstock features of one or the other type of water are dominant.

Karstic water has higher hydrogen-carbonate (calculated from m-alkality) and nitrate ion concentration, and lower organic matter content (TOC, chemical oxygen demand) than surface water. There is a slightly difference within pH and conductivity, surface water's values are larger. Amounts of iron, manganese and turbidity are variable in the network, since these depend on the quality of pipes and on the distance from drinking water source rather than the quality of raw water. Values of ammonium and nitrite ions' concentrations set below detection limits.

Based on the chemical quality of drinking water, we could separate sampling points that were supplied with **surface water** (1 sample), **karstic water** (3 samples), or **mixed water** (2) and points that were supplied **alternately with surface or karstic water** depending on pressure characteristics (2).

TABLE 1. Microbial communities.

NAME	CHARACTERISATION	REPRESENTATIVES	APPLIED PRIMERS	REFERENCE
NIRTIFYING BACTERIA 1. Ammonium-oxidizers $NH_4^+ \rightarrow NO_2^-$	- aerob, chemolitoautotroph - ubiquitous: soil, freshwater and marine environments, biological wastewater treatment plants…	*Nitrosomonas spp.* $\rightarrow$ *Nitrosococcus spp.* $\rightarrow$ *Nitrosospira spp.* …	16S-Nitroso4E Target region: 638-657 with EUB338	Amann et al, 1990 Amann et al, 1990
2. Nitrite-oxidizers $NO_2^- \rightarrow NO_3^-$	- their metabolisms may be responsible for significant pollution of water supplies with NO_2^- and NO_3^-	*Nitrobacter spp.* $\rightarrow$ *Nitrospira spp.* …	16S-Nit3 Target region: 1035-1048 with EUB338	Hiorns et al, 1996 Amann et al, 1990
SULPHATE REDUCING BACTERIA $SO_4^{2-} \rightarrow HS^- \rightarrow$ MeS	- anaerob - ubiquitous: marine sediments, biological wastewater treatment plants, … - They can cause corrosion, odour and taste problems in drinking water distribution systems	*Desulfomicrobium spp.* $\rightarrow$ *Desulfovibrio spp.* $\rightarrow$ *Desulfobulbus spp.* *Desulfomonas spp.* …	16S-DSV214 16S-DSV698 both with 16S-63F	Manz et al, 1998 Manz et al, 1998 Marchesi et al, 1998
SPHINGOMONAS SPP.	- aerob, potentially pathogens - widely distributed in nature: soil, sediments, plants, several kinds of water (including drinking water) - well-known slime producers - may corrode copper pipes	*Sphingomonas stygia,* *S. xenophaga,* *S. aromaticivoans,* *S. paucimobilis…*	16S-Sph78F with 16S-Sph422R	Kéri et al, 2002 Kéri et al, 2002
EUBACTERIA (universal)			16S-8V with 16S-1387R	Kéri et al, 2002 Marchesi et al, 1998

4.2. MICROBIAL COMMUNITIES

According to the literature, more than 99 percent of all bacteria in the water live in biofilms, the number of planctonic bacteria is negligible compared to biofilm bacteria attached to the surface. In the course of classic microbiological (cultivation on R2A) and molecular biological (taxon specific PCR) examinations we were not able to differentiate the two main water types because the actual source of bacteria is the distribution system itself. All three investigated bacterial groups (nitrifiers, sulphate-reducers, *Sphingomonas*) were detected by PCR in every sample. Although both types of drinking waters contain only trace amounts of ammonium and nitrite ions it was sufficient for **nitrifying bacteria** to subsist. Occasional customer complaints of ferrous drinking water in Balatonfüred may also be attributed to the presence of **sulphate reducers**. The concentration of organic nutrients is very low in karstic water, by contrast heterotrophic ***Sphingomonas* bacteria** were present. Sphingomonas are well-known slime producers. The slime provides a safe haven for other bacteria and promotes their ability to adhere to pipeline.Biofilm-removal may be performed for example by disinfection or by flushing the distribution system.

Comparing with other samples there were less amounts of biomass than in samples coming from other places near Lake Balaton. It may be linked to systematic flushing program of Balatonfüred. However, our results show that being free from microorganisms has not achieved with this process either. Biofilm microbial community diversity was assessed by restriction analysis of the 16S rDNA (16S-8V/16S-1387R, lenght: 1352 bp) using HinfI and HaeIII enzymes. Composition of the communities at this level of differentiation was similar. Difference was only observed in the intensity of restriction fragments, which may suggest the dominance of different species around the various sampling points. In spite of water quality one of other main factors influence bacterial growth in drinking water distribution systems is the material of pipes. Diameters and materials of pipes are very heterogeneous in Balatonfüred: cast iron, asbestos cement, PVC and PE.

According to water quality and pipe materials circumstances of life of microorganisms are very diverse in distribution net of Balatonfüred.

- The aim of this study was to survey the water-network of Balatonfüred with evaluating of water and biofilm quality and its interaction.
- The two main types of drinking water sources could be separated. According to various water quality (karstic, surface and mixed water) and various construction materials of pipelineslife conditions of microorganisms are very diverse.

TABLE 2. Water quality evaluation results.

PARAMETER	METRIC VALUE	KARSTIC WATER ± standard deviation	MIXED WATER ± standard deviation	SURFACE WATER ± standard deviation
Total organic carbon	mg/l	1,24 ± 0,38	2,64 ± 0,60	6,10 ± 1,10
Chemical oxygen demand	mg O_2/l	0,38 ± 0,13	0,93 ± 0,14	2,50 ± 0,38
m-alkalinity	mmol/l	8,12 ± 0,52	7,28 ± 0,61	5,10 ± 0,72
Hydrogen-carbonate ion	mg/l	496 ± 31,7	444 ± 37,0	311 ± 43,9
Nitrate ion	mg/l	11,9 ± 4,15	4,62 ± 1,27	1,05 ± 0,67
pH	–	7,29 ± 0,15	7,36 ± 0,16	7,75 ± 0,20
Conductivity	mS/cm	704 ± 27,7	732 ± 4,47	765 ± 13,4
Ammonium ion	mg/l	nd	nd	nd
Nitrite ion	mg/l	nd	nd	nd
Turbidity	NTU	0,87 ± 0,69 [0,10–2,79]		
Iron	mg/l	0,16 ± 0,17 [0,02–0,54]		

- However, composition of biofilm samples were similar (according to restriction analysis of the 16S rDNA, at this level of differentation) and all three investigated bacterial groups were detected by PCR. Difference was only observed in the intensity of restriction fragments, which may suggest the dominance of different species around the various sampling points. The Article 5 of 98/83/EC Directive allows Member States to set values for additional parameters not included in Annex I where the protection of human health within its national territory or part of requires it so. In our opinion this kind of examinations may be very useful in evaluation of disinfection efficiency.

References

Amann R.I., Binder B.J., Olson R.J., Chisholm S.W., Devereux R., and Stahl D.A. (1990): Combination of 16S rRNA-targeted oligonucleotide probes with flow cytometry for analyzing mixed microbial populations, *Appl Environ Microbiol.* 56(6): 1919–1925.

Amman R.I., Krumholz L., Stahl D.A. (1990): Fluorescent oligonucleotide probing of whole cells for determinative philogenetic and environmental studies in microbiology, *J. Bact.* 172, 762–770.

Hiorns W.D., Hastings R.C., Head I.M., McCarthy A.J., Saunders R.G., Hovanec T.A., DeLong E.F., (1996): Comparative analysis of nitrifying bacteria associated with freshwater and marine aquaria, *Appl. Environ. Microbiol.* 62(8): 2888–2896.

Kéri K., Gaugetz J. (2002): Applicability of polimerase chain reaction in evaluation of microbial quality of drinking water, Diploma study at Technical University, Budapest.

Manz, W., M. Eisenbrecher, T.R. Neu, and U. Szewzyk. (1998). Abundance and spatial organization of Gram-negative sulfate-reducing bacteria in activated sludge investigated by in situ probing with specific 16S rRNA targeted oligonucleotides. FEMS Microbiol. Ecol. 25:43–61.

Marchesi, J. R., Sato, T., Weightman, A.J., Martin, T.A., Fry, J.C., Hiom, S.J., Dymock, D. & Wade, W.G. (1998). Design and evaluation of useful bacterium-specific PCR primers that amplify genes coding for bacterial 16S rRNA. *Appl Environ Microbiol* 64, 795–799.

SUBJECT INDEX

Accidents industrial 29
Agriculture 201
Agrochemicals registration 193
Air quality 321
Air pollution
    transboundary 321
Alkalinity 204
Aquatic chemistry 173
Analysis
    elemental 149
    isotopic 149
    risk-management 219
Assessment
    environmental 1, 34
    exposure 1, 34
    dose-response 1
    risk 1, 47
Atrazine 173
Attenuation
    chemical 219
    natural 219

Bacteria 271
Bio-indicator samples 149
Biomonitor 163
Bioreactor 271
Biosensor 337
Biota 7
Biotechnology 271
Biotransformation 255
BOD 95
Boundary

Carbamates 173
Carcinogens 98
Chemical
    residues 243
    substances 137
    terror attack 57
Chemicals hazardous 29
Chelation 95
Child population 127
Cognitive abilities 127
Compounds
    organometaltic 96

Complexation 95
Concentrations
    ecosystem 1
Contamination
    land 47, 219, 247, 267
    soil 219
    water

DDT 173
Degradation 247
Diseases 255
Detergents 98

Ecology 137
Ecosystem 1
    marine 7
    seacoast 7
Education programs 281
Education institutions 281
Elemental analysis 149
Effect
    assessment 193
    occurence 1
Emergency response 220
Entity
    biological 1
    chemical 1
    physical 1
Environmental
    analysis 7
    characreristics 1
    damage 34
    emergency 105
    health 1, 417
    concentrations 193
    pollution 29
    protection 281
    samples 149
Enzyme purification 271
European institutions 321
Eutrophication 57, 201
Exposure
    extent 2
    lifetime 4
Extraction 7

Fenitrothion 174
Fetus 137
Food safety 237

Grounwater protection 57

Heavy metals 163
Human health 2
Hazard identification 1

Immobilization 271
Immunity 417
Injury
    human 1
Institutional learning 105
Intelligence Quotient (IQ) 127
Isotopic analysis 149

Laser mass spectrometry 149
Lead 127
Liquid-liquid extraction (LLE) 7
Lindane 173

Malathion 174
Metals
    heavy 271
    toxic 7
Management
    action 1
    assessment 34
process safety 31
    program 219
    risk-based 219
    site 219
Mass spectrometry 149
Mercury 271
Mercury reductase 271
Microwave-assisted extraction (MAE) 7

Nickel 417
Nutrients 201

OCPs 7

PAHs 7, 95
Pesticides 95, 193
    organochlorine 173
    organophosporus 174
PH 201
Phytoplankton 215
Photosynthesis 94

Pisciculture 202
Plant protection products (PPPs) 193
Pirimiphos-methyl 173
Plume 283
Pollutants
    global 94
    inorganic 7
    organic 7
    specific 1
    water 173
Pollution
    air 321
    accidental 163
    modeling 57
    intentional 163
    threats 201
POPs 173
Precipitation 94
Preconcentration 7
Problem identification 2
Public health 281
Pubic awareness 281

Radionuclides 95, 163
Reactions
    oxidation-reduction 94
Release prevention 220
Remediation
    action 220
    phyto 149
    selection 47
    technology 236
Reproductive age 137
Reproductive health 137, 417
Resilience 105
Response mechanisms 105
Risk
    assessment 45, 193, 219
    analysis 3
    characterization 2
    comparison 3
    drivers 223
    factors 137
    management 2, 219
    reduction 223
Risk assessment
    environmental 1
    human health 1
    issues 2
    management 3
Risk factors

Subject index

air 223
   community issues 223
   groundwater 223
   land use 223
   off-site impacts 223
   regulatory factors 223
   soil 223
   surface water 223

Samples
   air 7
   biota 7
   bio-indicator 149
   water 7
   soil 7
   sample pretreatment 7
Sediments 95
Sewage 98
Site
   classification 223
   management 221
   prioritization 223
System boundary 1
System reliability 4
Software
   models 283, 285
   programs 283, 286
   tools 283, 284
Soils
   flow in soils 57
   landscape 247

soil samples 149
solute transport 57
Solid phase extraction (SPE) 7
Solid-phase microextraction (SPME) 7
Stressor 2
Supercritical fluid extraction (SFE) 7

Technology transfer 149
Tier 2
Tiered approach 193
Toxic waste disposal 37
Toxicology 255
Trace elements 98
Training programs 281

VOCs
Volatilization 247, 271
Vulnerability 105

Wastewater 247
Water
   alkalinity 204
   discharge 202
   euphotic 214
   hardness 204
   quality 201
   salinity 201
   samples 158
   synthetic 271

Printed in the United States
59912LVS00002B/5